BIOPROCESS ENGINEERING: SYSTEMS, EQUIPMENT AND FACILITIES

BIOPROCESS ENGINEERING: SYSTEMS, EQUIPMENT AND FACILITIES

Edited by

BJORN K. LYDERSEN

NANCY A. D'ELIA

KIM L. NELSON

A Wiley-Interscience Publication

JOHN WILEY & SONS, INC.

New York / Chichester / Brisbane / Toronto / Singapore

This text is printed on acid-free paper.

Library of Congress Cataloging in Publication Data:

Bioprocess engineering : systems, equipment and facilities / [edited by]
   Bjorn K. Lydersen, Nancy A. D'Elia, Kim L. Nelson.
      p. cm.
   "A Wiley-Interscience publication."
   Includes bibliographical references and index.
   ISBN 0–471–03544–0 (cloth) :
   1. Biochemical engineering.   I. Lydersen, Bjorn K.
II. D'Elia, Nancy A.   III. Nelson, Kim L.
TP248.3.B5873   1994
660'.63--dc20   93-44639

Printed in the United States of America

10 9 8 7 6 5 4 3

PREFACE

The development of genetic engineering and monoclonal antibody technology, which started in the 1970s, has led to the introduction of a large number of new products with application in many different areas. The most highly visible applications have been in the area of human health care, with products such as human insulin, interferon, tissue plasminogen activator, erythropoietin, colony-stimulating factors, and monoclonal-antibody–based products. In addition, significant new products for agriculture, the food industry, fine chemicals, and the environment are also under intense development. Although products analogous to these new biotechnology products had been made previously, the number and diversity of products, and the number of companies involved, has grown tremendously. Experience in the manufacture of these "biotech" products was very limited, and the technology needed was seen by many as a limiting factor in the successful application of biotechnology. Improvements in production technology were required, as well as a rapid increase in the number of people involved in all of the different aspects of production. The field of *bioprocessing* became a very important part of biotechnology.

Bioprocessing encompasses the many steps required to synthesize, isolate, and formulate the products. There is a great deal of diversity in methods, and in the equipment and facilities required. However, several key processes and types of equipment are utilized in the production of most biotech products. The synthesis of biotech products by microbes or by cells from higher organisms is generally carried out in fermentors, although some new products are being synthesized in "bioreactors", or in animals. Alternatively, the synthesis of relatively small molecules can be carried out directly by chemical methods, without the need for biological cells. The isolation of the product from the fermentor broth or animal fluid generally involves initial steps of clarification or concentration of the fraction containing the product; centrifugation, precipitation, and filtration are the most common ways to accomplish this. Subsequent purification and formulation steps are highly dependent on the nature of the product and its application. Chromatographic procedures are used almost without exception to separate pharmaceutical products from the remaining contaminants. For products outside the pharmaceutical area, the final steps in manufacturing vary widely with the type of product.

The field of bioprocessing involves many areas of expertise, and requires the fulfillment of several sometimes conflicting criteria for success. Biological and biochemical expertise are required to develop and conduct the processes for the synthesis and purification of the products. A broad range of engineering expertise is required to design and construct the many aspects of the manufacturing facility. This includes the process equipment, the support systems that, for example, provide pure water, clean steam, and waste handling, and the facilities that house the equipment. Furthermore, an understanding of regulatory requirements from the viewpoints of pharmaceutical standards (for example, Good Manufacturing Practices), environmental concerns, and worker safety is also essential. Success is measured in terms of reliability of the overall process, time from conception to routine operation, cost of production, and satisfaction of regulatory requirements.

The limited past experience with bioprocessing and the need to rapidly develop manufacturing capacity, especially in the new biotech companies, have given rise to a need for information that will allow the many individuals entering this field to develop and operate the production facilities. This book was conceived to assist those people with acquiring or expanding their knowledge of the bioprocessing field.

The book is subdivided into four sections, covering *Process Systems*, *System Components*, *Support Systems*, and *Facility Design*. The first and third of these sections reflect the fact that there are two types of equipment required in the plant: the equipment in which the actual product is synthesized or processed (such as the fermentor, centrifuge, and chromatographic columns) and the equipment that supplies support for the process or the facility (such as the water system, steam generator, waste system, and air conditioning). In the second section, System Components, are described such components as pumps, filters, and valves, which are ubiquitous and not limited to a certain type of equipment. The final section, Facility Design, covers planning and designing the entire facility, including the requirements for containment and validation of the process. There is some overlap in the topics covered, but this is seen as useful because different perspectives and in some cases different points of view are brought out.

Although this book has a wider scope than do most books on bioprocessing, it cannot cover every aspect of bioprocess engineering completely. Where there is ample literature already covering certain specialized fields, such as mammalian cell bioreactors, the authors have been encouraged to refer to existing works, rather than discuss those fields in detail. The emphasis has been on examining the systems and components that are common to most or all biotechnology production processes, instead of going into the particulars of all of the alternatives for each specific step in bioprocessing.

The authors for each of the chapters in this book were requested to provide information that would be of practical use in the design or operation of a bioprocessing facility. The authors have relied heavily on the experience they have gained in the performance of their jobs, and through interactions with their colleagues. In many cases, we found differences of opinion as to the best tactic for solving the problems encountered in bioprocessing. The authors were encouraged to explore the options available, and to give their opinions on what they felt was the best approach. Their differences in opinion demonstrate that the field of bioprocessing is still developing rapidly, which makes it both difficult and challenging. We hope this book will assist the participants in the field to make progress more rapidly.

Bjorn K. Lydersen
Nancy A. D'Elia
Kim L. Nelson
May 1993

CONTRIBUTORS

Dan G. Adams
Quasar Engineering, Inc.
300 Phillips Blvd.
Ewing, NJ 08618

Harry Adey
Life Sciences® International
1818 Market Street
Philadelphia, PA 19103

Robert Baird
Genentech
460 Point San Bruno Blvd.
South San Francisco, CA 94080

John P. Boland
Raytheon Engineers & Constructors
30 South 17th St.
P. O. Box 8223
Philadelphia, PA 19101

Dr. Julian Bonnerjea
Celltech Ltd.
216 Bath Road
Slough SL1 4EN
Berkshire
UNITED KINGDOM

Dr. Marvin Charles
Lehigh University
Chemical Engineering Dept.
111 Research Dr.
Bethlehem, PA 18015

Nancy A. D'Elia
Hybritech Incorporated
P. O. Box 269006
San Diego, CA 92196

Phil De Santis
Raytheon Engineers & Constructors
30 South 17th St.
P. O. Box 8223
Philadelphia, PA 19101

Dennis Dobie
Fluor Daniel
301 Lippincot Centre
P. O. Box 950
Marlton, NJ 08053

Thomas Hartnett
Merck Research Laboratories
P. O. Box 2000
Rahway, NJ 07065-0900

Harry L. Johnson
Raytheon Engineers & Constructors
30 South 17th St.
P. O. Box 8223
Philadelphia, PA 19101

Hyman Katz
Pall Ultrafine Filtration Company
2200 Northern Blvd.
East Hills, NY 11548

Dr. Bjorn K. Lydersen
Hybritech Incorporated
P. O. Box 269006
San Diego, CA 92196

Jeff H. MacDonald
Millipore Corp.
5 Ember Lane
Horsham, PA 19044

Jerry M. Martin
Pall Ultrafine Filtration Company
2200 Northern Blvd.
East Hills, NY 11548

Pio Meyer
Bioengineering AG
Sagenrainstrasse 7
CH 8636 Wald
SWITZERLAND

Dr. Kim L. Nelson
Quasar Engineering, Inc.
300 Phillips Blvd.
Ewing, NJ 08618

Tim Oakley
Celltech Ltd.
216 Bath Road
Slough SL1 4EN
Berkshire
UNITED KINGDOM

Casimir A. Perkowski
Raytheon Engineers & Constructors
30 South 17th St.
P. O. Box 8223
Philadelphia, PA 19101

Maria S. Pollan
Badger Design and Construction, Inc.
8313 Southwest Freeway
Suite 300
Houston, TX 77074

Eric A. Rudolph
Millipore Corp.
5 Ember Lane
Horsham, PA 19044

Paul Schubert
Pall Ultrafine Filtration Company
2200 Northern Blvd.
East Hills, NY 11548

Dr. Bhav P. Sharma
InterSpex Products, Inc.
1155 Chess Drive, Suite 114
Foster City, CA 94404

Rostyslaw Slabicky
Boehringer Ingelheim
Pharmaceuticals, Inc.
900 Ridgebury Rd.
Ridgefield, CT 06877

Dr. Gary L. Smith
Lilly Research Laboratories
A Division of Eli Lilly and Company
Lilly Corporate Center
Indianapolis, IN 46285-3313

Robert Stover
Tri-Clover
25665 Ranch Adobe
Valencia, CA 91355

David A. Stutzman
Raytheon Engineers & Constructors
30 South 17th St.
P. O. Box 8223
Philadelphia, PA 19101

Peter Terras
Celltech Ltd.
216 Bath Road
Slough SL1 4EN
Berkshire
UNITED KINGDOM

P. William Thompson
Rotherwood Engineering
70A Gore Road
Burnham
Bucks SL1 7DJ
UNITED KINGDOM

A. Mark Trotter
Pall Ultrafine Filtration Company
2200 Northern Blvd.
East Hills, NY 11548

Jack Wilson
ABEC, Inc.
6390 Hedgewood Drive
Allentown, PA 18106

CONTENTS

BIOPROCESS ENGINEERING: SYSTEMS, EQUIPMENT AND FACILITIES

PART I

PROCESS SYSTEMS

Fermentor Design

Marvin Charles and Jack Wilson

Contents

In this chapter we consider only agitated vessels for aerated, aseptic fermentations. Many other vessel types with distinct advantages for specific applications have been proposed, and a few of these have been used for limited commercial applications. In most cases, however, their advantages do not outweigh those of the agitated tank (namely, that it is better understood, offers more reliable scale-up, and is easier to adapt to multiple products), which is likely to remain the dominant fermentor type for the foreseeable future. Information about other fermentor types is given in the references [1, 2a, 3a].

We address the basics of sizing and designing fermentor vessels and their agitation and aeration systems. We also discuss popular misconceptions and design pitfalls. We work through a somewhat detailed design example at the end of the chapter.

Finally, this chapter is not a complete design manual. Rather, it is a guide that illustrates how a good design engineer thinks, some of the methods he (or she, understood throughout) employs, and the kinds of information and resources he needs to do his job best. It should benefit not only those beginning to study the basics of fermentor design (students and practicing engineers), but also those interested in buying fermentors and complete systems.

1.1   Design Philosophy

The objective of rational design is to produce the plans necessary to build economical systems that satisfy well-considered performance criteria. Two generalizations worth noting are:

(a)   The purpose of a fermentor is to provide a contained (protected), controlled, homogeneous environment, in which a fermentation can be performed safely and practically to achieve particular objectives.

(b)   Objectives usually are defined rather loosely for lab applications, and tend to become tighter with increasing scale. A high degree of flexibility usually is a major objective in the lab. Reliability, safety, cost, and compliance with regulatory requirements usually are primary factors for production equipment. These transcendent criteria are touchstones for ordering the priorities of more specific performance criteria, such as mixing quality, oxygen transfer, and heat transfer.

The realization of process objectives depends on reliable operation of an entire system, of which the fermentor is only one part; therefore, rational design of the fermentor requires careful consideration of how it is integrated into the process. The designer should, therefore, consider from the outset factors such as plant scheduling, space constraints, relationships between fermentor productivity and throughput rates of downstream equipment, containment and validation requirements, utilities requirements, potential interruptions in normal plant operation, overall labor requirements, and operating versus capital costs. Monitoring/control, sterilization methods, and the number and types of addition vessels also must be considered as early in the design process as possible, particularly with regard to how they affect overall process requirements.

The designer usually must use information obtained at a scale much smaller than the one for which he is designing. Very frequently, he must start while the fermentation still is being developed at lab scale. And so he enters the mystical world of fermentation scale-up. Much has been written about scale-up criteria, but what has not been written usually is more important than what has been written. Published scale-up criteria usually fall into one or more of the following types:

(a)   primarily theoretical, and based on assumptions and gross simplifications that are not valid for most fermentations;

(b)   single-parameter objectives (such as maintenance of oxygen-transfer level);

(c)   experiments in which observations are limited to those required to study single-parameter objectives;

(d)   experiments involving scale-up from very small to small;

(e)   experiments done under very idealized conditions, which are impractical for commercial applications.

Some of the popular scale-up methods (so-called) come in the form of correlations, which either are implied or are inferred to be scale-independent and to have a wide range of applicability. Unfortunately, most are not. Others are cast as rules of thumb, subject to similar implications or inferences. Most of these are based on very limited observations of specific fermentations.

The critical factors missing in all of the methods are integration and the recognition that what is important at one scale may not be important at another. These missing factors can lead to very significant problems, since just about every aspect of a fermentation influences most other aspects. The reality is that sound design requires a great deal of compromise, based on a wide range of real experience doing real fermentations at meaningful scale. Rational scale-up is not a simple textbook exercise.

The various methods and rules alluded to above are presented in many references [4, 5a, 6], and are not repeated here. But you are cautioned to bring to such readings all the critical thinking you can muster, and to seek experienced counsel.

Design is an applied art, not a science. It requires the consideration of a host of interrelated factors, many of which are risk assessments and value judgments; therefore, the design engineer must apply experience and judgment in reaching reasonable compromises without attempting to violate the laws of physics. In general, there is no such thing as the ideal fermentor. The engineer who forgets the basic principle that "good enough is better than perfect", and attempts to optimize a satisfactory design, probably won't see the system built in his lifetime, and probably should be in some other line of work.

The designer must balance theory with the application of quantitative, empirical correlations and various rules of thumb. Lest the novice be led astray, it is important to note right away that many of the so-called quantitative methods leave a great deal to be desired, and their results often must be tempered with a healthy dose of practical experience. There are many reasons for this, but three of the most important are:

- (a) Most of the published correlations (for example, oxygen transfer) are based on work, probably less than systematic, done on very small-scale, unrepresentative equipment at universities.
- (b) Oxygen transfer, heat transfer, and mixing characteristics all depend strongly on physical properties, such as broth rheology, that are seldom included explicitly in published correlations.
- (c) Equipment designers and manufacturers seldom get from their customers any actual performance data, unless, of course, the clients are unhappy with the equipment.

With regard to rules of thumb, the design literature is replete with all manner of "magic numbers". Most of these are based on very specific experience, which often has little or no relationship to the project at hand. All too often, what have been called rules do more to inhibit good design than to assist it. Some examples are:

- (a) *Fermentor Geometry Myths*: Fermentor height:diameter ratio must be 3:1. All impellers must have diameters of between 0.25 and 0.33 of the tank diameter.
- (b) *The One VVM Bag of Wind*: The air flow rate to the fermentor must be equal to one liquid volume per minute (1 VVM), come Hell or high water (which might actually come to pass in large fermentors operated at 1 VVM).
- (c) *The 1500 fpm Tip Speed Run-Around*: The tip speed of a fermentor impeller must not be greater than 1500 fpm (feet per minute), even if the bug is encased in armor plate.
- (d) *The 4 °C Chill Factor*: At the end of a fermentation, the broth must be cooled immediately (if not sooner) to 4 °C. The only way you could succeed at this is if the fermentation temperature is 5–10 °C and if the plant is located at one of the Poles.
- (e) *The High $k_l a$ Gasp*: High $k_l a$ (oxygen mass-transfer coefficient) means a high oxygen-transfer rate, which is THE most important factor in aerobic fermentations; therefore, we must get high $k_l a$, at any cost.
- (f) *The "Shoot Your Wad To Get The Maximum Rate/Cell Mass" Theory*: This theory is founded on the misguided belief that process optimization is achieved by obtaining very high cell mass concentrations at very high growth rates. The fact that this may require the vessel to be built like an artillery shell, with a refrigeration system that might tap most of the output of the Grand Coulee Dam, is beside the point.
- (g) *The Continuous Fermentor, Higher Productivity Boondoggle*: Theory predicts and practice confirms that a continuous fermentor gives higher productivity than a batch or fed batch fermentor. Does this explain why there are so few of them used commercially? Perhaps the theory has overlooked something!

And there are many others, but we've said enough. We'll revisit some of these myths as we proceed.

Finally, design is a work of balanced integration, usually requiring several iterations. Unfortunately, we don't have space here to present all the material in a completely integrated way. We do, however, point out many of the important interactions as we discuss each specific topic. Also, in the next section we give an integrated treatment of oxygen transfer and heat transfer, so as to sensitize the reader to the rather broad implications of the interaction between two key factors.

1.2   The Interactions Between Oxygen and Heat Transfer

Oxygen- and heat-transfer rates are coupled closely in aerobic fermentations. This fact can impose serious constraints not only on fermentor design but also on the process in general. An important and often unappreciated fact is that heat transfer usually is the limiting constraint for highly aerobic fermentations of greater than a few thousand liters, and that this limits the practical levels of cell mass and fermentor productivity. In theory, we could get very large cell masses at rapid growth rates by employing heroic measures to satisfy the oxygen demand; but this would result in very high heat-transfer rate requirements, which would require even more heroic and expensive solutions, if they could be met at all. It is also important to note that providing higher levels of oxygen transfer usually requires very powerful drives, which add mechanical problems and cost as well as additional heat loads.

In most cases, the very high oxygen demands occur during short time periods near the end of a fermentation; therefore, it is important to decide at the outset whether overdesigning to meet peak loads is really worth the additional cost, or whether it would be more economical to simply slow down the fermentation a bit. In most instances, a minor decrease in metabolic rate substantially reduces the heat- and oxygen-transfer loads and costs, without materially affecting overall plant productivity.

The *oxygen-uptake rate* or *oxygen-transfer rate* (OTR) during cell growth is determined by the cell's specific growth rate, μ, the cell concentration, X, and the cell's yield coefficient on oxygen, $Y_{X/O}$:

$$OTR = (\mu/Y_{X/O})X \tag{1.1}$$

There is a similar relation for nongrowing cells that produce products.

The total heat generated during growth, Q_{tot}, is approximately equal to the sum of the metabolic heat generated, Q_{met}, and the heat generated by the agitation required to provide adequate oxygen transfer and mixing, Q_{mech}:

$$Q_{tot} = Q_{met} + Q_{mech} \tag{1.2}$$

In terms of the customary English units and their conversions, this is:

$$Q_{tot} \text{ (Btu/gal·h)} = 1.81 \text{ OTR (mmol/L·h)} + 2540 P_g \tag{1.2a}$$

where P_g is the gassed horsepower input. There are some cases for which heat effects due to gas expansion, evaporation, and chemical reaction (for example, the heat of solution of ammonia gas used for pH control) are important. These should be evaluated as individual cases.

Oxygen-transfer rates requiring heroic measures vary with the nature and the scale of the fermentation. The effects of scale for a simple *E. coli* fermentation are indicated in Table 1.1. For a 1-m^3 vessel, mechanical heroics (such as high pressure and horsepower, oxygen enrichment) become necessary between oxygen-transfer rates (OTRs) of 250 mmol/L·h and 300 mmol/L·h, but heat transfer is not a major problem up to 400 mmol/L·h. For a 10-m^3 vessel, oxygen and heat transfer both start to become major problems between OTRs of 250 mmol/L·h and 300 mmol/L·h. To get much above 300 mmol/L·h, we would have to use subfreezing coolant, internal coils, or both (in addition to mechanical heroics), and these both add considerable cost and operating headaches (see below).

Another alternative, noted previously, is to slow down the fermentation as the level of OTR starts to become prohibitively high. Table 1.2 gives the histories of two fed batch bacterial fermentations run until the cell mass concentrations reached 50 g/L. In Case I, μ is held at 0.6 h^{-1}, and the initial cell mass is 1 g/L. In Case II, the initial μ is 0.6 h^{-1}, but when the OTR reaches 350 mmol/L·h, the growth rate is slowed continuously to hold the OTR constant at 350 mmol/L·h. In Case II, the initial cell mass is 2.0 g/L. The forceful approach (Case I) gives a fermentation time of 7.0 hours. The saner approach of Case II takes only 0.25 hour more. When a turnaround time of 6–8 hours and, if necessary, an induction time of about 4 hours are added to each case, the total cycle times hardly differ. Realistically, this translates to a cycle time of about 18 hours for either case. In most instances, the small percentage improvement accruing from the Case I approach does not justify the increased costs and problems associated with very high levels of heat and oxygen transfer.

"Ah, but you've pulled a fast one by upping the initial cell concentration in Case II!", we hear you protest. True, but the reality is that backing off to 1.0 g/L in Case II would add only another 1.5 hours to Case I time. But, by increasing the initial concentration to 2.0, we've forced the seed culture vessel to share part of the burden. If the working volumes of the production and seed vessels were 1000 L and 100 L, respectively, the seed cell mass concentration would have to be around 20 g/L, which the seed vessel could handle readily, while fitting nicely into a reasonable overall plant schedule.

For larger production vessels, we'd probably have to back down further on the maximum OTR to avoid subfreezing coolant or coils. At some point, however, we'd reach a point at which the productivity loss would become unacceptable, and we'd have to reconsider our approach.

The simplest and most easily controlled way to decrease growth rate is to decrease temperature [7, 8]. The additional heat load for cooling is small in comparison to the metabolic heat load. There are other possibilities and it usually is worth experimenting

Table 1.1 Oxygen- and Heat-Transfer Requirements: Effects of Scale

OTR	Volume[a]	Pressure	% Oxygen[b]	Power	Heat Load	Coolant[c]
mmol/L·h	m³	psig		hp	Btu/h x 10⁶	°F
150	1	15	21	5.0	0.084	40
200	1	25	21	4.9	0.107	40
300	1	35	21	7.1	0.161	40
400	1	35	19	6.9	0.208	40
150	10	15	21	50.2	0.884	40
200	10	25	21	50.0	1.078	40
300	10	35	21	75.7	1.621	22[d]
400	10	35	30	77.0	2.096	5[d]

a. Liquid volume.
b. Mole percent of inlet oxygen.
c. Coolant flow is 35 gal/min for the 1-m³ vessel and 100 gal/min for the 10-m³ vessel.
d. If 40 °F coolant were used instead of 22 °F and 5 °F, 194 ft² and 748 ft² coils would be required, respectively.

Table 1.2 Fermentation Histories: Oxygen and Heat Transfer for Unlimited and Limited OTR

Time	Cell Mass		OTR		μ	
h	g/L		mmol/L·h		h⁻¹	
	I	II	I	II	I	II
0	1.0	2.0	19	38	0.6	0.6
1	1.8	3.6	34	68	0.6	0.6
2	3.3	6.6	62	124	0.6	0.6
3	5.9	11.8	111	222	0.6	0.6
4	10.5	20.0	197	350	0.6	0.56
5	18.3	29.8	344	350	0.6	0.38
6	31.0	39.0	582	350	0.6	0.29
7	50.0	47.8	938	350	0.6	0.23
7.25	–	50.0	–	350	–	0.22

with some before committing to a compromise design. To reiterate, heat- and oxygen-transfer loads should never be considered separately or divorced from their effects on overall plant operation.

1.3   Oxygen Transfer

1.3.1   Theoretical Considerations

The rate at which oxygen can be transferred (*oxygen transfer rate* or *OTR*) from air bubbles to the fermentation broth depends on the driving force for transfer and on a constant of proportionality called the oxygen-transfer coefficient, $K_g a$:

$$OTR = K_g a \, (P_{O_2} - P_{O_2}^*)_{av} \tag{1.3}$$

where OTR is the oxygen-transfer rate, mmol/L·h;

$K_g a$ is the oxygen-transfer coefficient, mmol/L·h·atm;

$(P_{O_2} - P_{O_2}^*)_{av}$ is the pressure average driving force;

P_{O_2} is the partial pressure of oxygen in the gas; and

$P_{O_2}^*$ is the equilibrium partial pressure of oxygen in the broth.

The term $K_g a$ is a composite of an intrinsic mass-transfer coefficient, K_g, and the mass-transfer surface area, a, in units of ft²/ft³. They are combined because existing theories and correlations are not yet capable of handling them separately in practical cases.

Driving forces other than $(P_{O_2} - P_{O_2}^*)_{av}$ have been used, but the units and numerical values of the associated mass-transfer coefficients differ from those of $K_g a$. For example, a definition used frequently is:

$$OTR = k_l a \, (C_{O_2}^* - C_{O_2})_{av} \tag{1.4}$$

where C_{O_2} is the actual dissolved oxygen concentration, mmol/L;

$C_{O_2}^*$ is the dissolved oxygen concentration in equilibrium with the existing P_{O_2} in the gas, mmol/L;

$k_l a$ is the mass-transfer coefficient, L/h; and

$(C_{O_2}^* - C_{O_2})_{av}$ is the concentration average driving force.

$K_g a$ and $k_l a$ can be related through the Henry's law constant, H:

$$k_l a = K_g a \, H \tag{1.5}$$

$$H = P_{O_2}^*/C_{O_2} \tag{1.6}$$

If the broth is mixed perfectly ($P_{O_2}^*$ is uniform throughout the tank), and the gas is in plug flow, then:

$$(P_{O_2} - P_{O_2}^*)_{lm} = [(P_{O_2} - P_{O_2}^*)_{in} - (P_{O_2} - P_{O_2}^*)_{out}]/\ln[(P_{O_2} - P_{O_2}^*)_{in}/(P_{O_2} - P_{O_2}^*)_{out}] \tag{1.7}$$

where $(P_{O_2} - P_{O_2}^*)_{lm}$ is the log mean driving force;

$(P_{O_2} - P_{O_2}^*)_{in}$ is the inlet difference; and

$(P_{O_2} - P_{O_2}{}^*)_{out}$ is the outlet difference.

The log mean usually is adequate for most design applications, despite the fact that the assumptions used to derive it are not met fully in most practical cases.

The OTR also can be expressed in terms of an oxygen material balance (if we assume constant dissolved oxygen concentration):

$$\text{OTR} = F_i\, y_i - F_o\, y_o \tag{1.8}$$

where F_i is the molar flow rate of gas in;
  F_o is the molar flow rate of gas out; and
  y is the mole fraction of oxygen.

In most practical cases, the depletion of oxygen in the gas phase is approximately equal to the increase in its carbon dioxide plus water contents. Hence, F_o and F_i are the same for most practical purposes, and:

$$\text{OTR} = F_i\,(y_i - y_o) \tag{1.9}$$

The $K_g a$ value needed for the OTR calculation can, in theory, be found by applying one or more of the myriad published correlations [5b, 5c, 9, 10a, 11–13]. Most of these have the form:

$$K_g a = a_1\,(P_g/V_1)^{\alpha_1}\,(v_s)^{\beta_1} \tag{1.10}$$

where a_1, α_1, and β_1 are said to be constant;
  P_g/V_1 is the gassed horsepower over liquid volume; and
  v_s is the linear (superficial) velocity of the gas.

Most of these are valid only for Newtonian broths, disk turbine impellers, and specific geometry. Also, for those that are applicable to non-Newtonian broths, the rheological properties seldom are included explicitly, thereby making quite limited the usefulness of such correlations.

Unfortunately, the published correlations must be viewed with some skepticism because most of them were determined for relatively small lab fermentors or under conditions not similar to those found in commercial fermentors. Also, it is an unfortunate fact of life that it isn't possible to maintain geometric, kinematic, and dynamic similarities as the scale of an agitated vessel is changed [14–16]; therefore, the correlations tend to be scale-dependent (a_1, α_1, and β_1 vary with scale). So it is not surprising that different correlations predict different values (sometimes the differences can be up to an order of magnitude) of $K_g a$ or of the power required for a given set of conditions.

To illustrate this point, consider the following case:
Total volume: 20,000 gal (76 m³)
Broth volume: 16,000 gal (61 m³)
Inside diameter: 143 in (3.63 m)
Liquid height: 240.6 in (6.11 m)
Impellers: 2 Rushton turbines

Head space pressure: 2.36 atm abs
Bottom pressure: 2.95 atm abs
O_2 mole fraction in: 0.21
O_2-transfer efficiency, ε: 30%
OTR: 150 mmol/L·h
The molar flow rate of air is obtained as follows:

$$OTR = F\ (y_i - y_o) \tag{1.11}$$

$$y_o = (1 - \varepsilon)\ y_i \tag{1.12}$$

where ε is the overall oxygen-transfer efficiency; therefore,

$$F = OTR/(y_i\ \varepsilon) \tag{1.13}$$

The linear velocity, v_s, of the gas increases from the bottom of the tank to the top (decreasing pressure), but it's sufficient in most practical cases to use the arithmetic average linear velocity to account for this variation:

$$v_{s,\ av} = 22.4\ F/(P_{av}\ A_c) \tag{1.14}$$

where A_c is the tank cross sectional area, and P_{av} is the arithmetic average absolute pressure. For this example, eq 1.14 gives 194.6 cm/min for $v_{s,\ av}$. The log mean pressure driving force (eq 1.7) is 0.45 atm, and the $K_g a$ required is (eq 1.3) 334.3 mmol/L·h·atm. For the preceding values, the COOPER correlation [5b] predicts a gassed horsepower requirement of approximately 15 hp, and the FUKUDA correlation [5c] predicts approximately 34.5 hp. Not only do these differ greatly from one another, but they also differ enormously from known requirements, which range anywhere from 300 hp to as much as 400 hp (and even impractically higher in some cases), depending on the type of broth. The correlations are, at best, only rough guides, and their predictions always should be checked against one's own experience with cases similar to the one under consideration. If the predictions are at considerable odds with your experience, it usually is best to trust your experience, assuming, of course, that you have some. If not, find someone who does.

It should be noted that a few correlations based on commercial, large-scale operation have been published [17], but keep in mind that these are for very specific cases, which may differ considerably from your own.

1.3.2 Increasing OTR, and the Consequences

Equation 1.3 shows that the OTR can be increased by increasing $K_g a$ or $(P_{O_2} - P_{O_2}^*)_{lm}$. $K_g a$ can be increased by increasing P_g/V, v_s, or both. The effect of increasing P_g seems fairly clear, but there is seldom any substantive benefit in going beyond 2.5–3.0 hp/100 gal. The effect of increasing v_s is not so clear (see below).

$(P_{O_2} - P_{O_2}{}^*)_{lm}$ can be increased by increasing the vessel pressure or by oxygen-enriching the inlet air (or a combination of these strategies).

Agitation power and aeration are discussed together because they affect in concert the OTR, mixing patterns, foaming, and gas holdup. Foaming and holdup impose important, practical operating and design problems, which are worth discussing before going on to oxygen transfer *per se*.

1.3.2.1   Foaming

Foam can really ruin your day. The simple reality is that if it can foam, it will. If it does, it can foam over into the air exhaust line and require terminating the fermentation, to say nothing of the ensuing cleaning chore. So we need a way to suppress it.

There are two fundamental types of systems used to control foam: mechanical and chemical. A typical mechanical foam breaker is illustrated in Figure 1.1. It is at heart a high-speed impeller (designed much like the impeller of a centrifugal pump) that breaks the foam by drawing it into the impeller and exposing it to large mechanical forces. The resulting liquid drains back into the vessel. These devices can work well for many types of foams, and should be tried before chemical agents are used. They should not, however, be used for mammalian cell cultures because of the large forces they generate. Also, they must be driven independently: a foam breaker impeller mounted on the same shaft as the agitation impellers won't work.

Chemical antifoams are nothing but surfactants. Many are marketed under a wide variety of names, but all fall into two categories: metabolizable (for example, vegetable oils) and nonmetabolizable (for example, silicones, polypropylene glycols). Many can decrease oxygen transfer, and some (for example, silicones) can have very negative effects on recovery equipment, particularly membrane filters. One should always test for such effects before making a final choice.

Unfortunately, even if antifoam systems are in place and operating, sooner or later there'll be a foam-over. It is therefore wise to design the fermentor air exhaust system appropriately. This entails (a) providing a means to slow down air flow and agitation rates to their minima while still maintaining temperature control; (b) protecting off-gas analyzers; (c) protecting the exhaust air filter and the pressure regulator; and (d) providing a means for simple disassembly and cleaning of the exhaust line. Figure 1.2 illustrates one system that has worked well in practice.

1.3.2.2   Gas Holdup

Gas holdup, h, is defined as:

$$h = (V_{gassed} - V_{ungassed})/V_{ungassed} \qquad (1.15)$$

where V_{gassed} and $V_{ungassed}$ are the respective volumes.

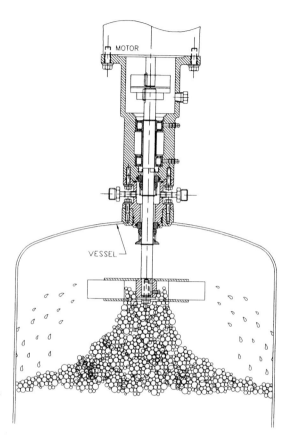

Figure 1.1 Mechanical Foam Breaker. Foam is destroyed by the high speed centrifugal impeller. The shaft should be double mechanical sealed in the same manner as is the agitator shaft.

For the high power levels typical of most aerated fermentors, one can estimate the holdup for Newtonian broths [13, 18] as:

$$h = 1.8 \, (P_{\text{gassed}}/V)^{0.14} \, (v_s)^{0.75} \qquad (1.16)$$

Foaming and holdup require that some fraction of the vessel (the *freeboard*) be left empty. There are no correlations for predicting freeboard, primarily because foaming is so unpredictable. Hence, designers usually employ a rough rule of thumb allowing 20–30% freeboard.

1.3.2.3 Gas Linear Velocity

The effect of increasing gas linear velocity, v_s, is not obvious, because v_s affects not only $K_g a$ but also the driving force (through the material balance). If we assume that

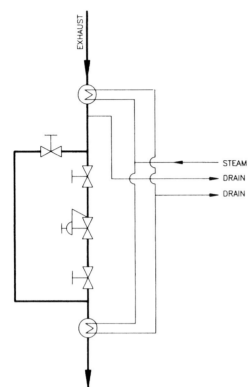

Figure 1.2 Air Exhaust System. Air
exhaust is steam heated to sterilize it
and to prevent condensation in the
back-pressure regulator. A sterile
filter(s) may be added if deemed
necessary.

eq 1.10 is valid, and that the pressure variation across the tank has no substantial
effect, we can show that:

$$\text{OTR} = a_2(v_s A_c P_T/V_1)\,(y_i - y^*\,[1 - \exp\,(-\,\xi\,v_s\,\beta_1^{-1})]) \tag{1.17}$$

where ξ is defined as $V_1\,a_1\,(P/V_1)^{\,\alpha_1}/26.8 \cdot 10^3\,A_c$;
   y^* is the equilibrium gas phase oxygen mole fraction;
   P_T is the absolute pressure (total);
   a_1, α_1, and β_1 are constants from eq 1.10; and
   a_2 is a constant whose value depends on the units used for each of the variables.
   Equation 1.17 predicts that OTR increases with increasing v_s, so long as y_i, y^*, and
ξ are held constant. Note, however, that the equation does not address important
practical constraints that impose limits on the usefulness of increasing v_s:
   (a)   Increasing v_s tends to decrease the gassed horsepower, P_g (see below), which
         requires increasing the impeller speed, impeller diameter, or both, to hold P_g
         constant. There are limits to both.

(b) Foaming and gas holdup increase with v_s. These effects vary with the type of fermentation and the impeller speed, and are difficult to generalize. But we have found that an upper limit of 175–200 cm/min (actual velocity in the tank) usually is safe and reasonable.

(c) Relatively low gas velocities and foam levels can harm delicate organisms, such as mammalian cells, even at low agitation rates [19], and this must be considered in the design of tissue culture reactors. Unfortunately, there is still not enough known to allow any meaningful quantitative generalization, and each case should be considered separately [20–22].

(d) At a fixed actual VVM, v_s increases rapidly with scale. This fact is too often overlooked, and unrealistic values of gas flow, based on lab conditions, are specified. One also must recognize that large vessels operate at a higher pressure than do most lab vessels; hence, the actual VVM at a fixed molar flow rate of gas (measured in standard liters per minute) per unit of liquid volume is smaller in a large vessel than it is in a small one (approximately in the ratio of average pressures at the two scales). Figure 1.3 illustrates these effects.

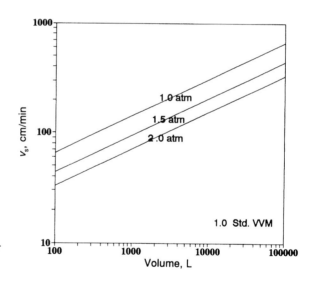

Figure 1.3 Effect of Fermentor Volume on v_s. Values of v_s are based on the gas flow rate at the average tank pressure. The ratio of liquid height to diameter is taken to be constant at 1.5 for all cases illustrated.

1.3.2.4 Gassed Power

Power is delivered to the fluid through two mechanisms: mechanical power from the impeller, and power generated by gas expansion. Which is greater depends primarily on whether the gas flow or the impeller controls flow in the vessel. If the gas flow is relatively high, and the impeller diameter and speed are relatively low, the aerated

impeller has little effect on fluid motion or on gas dispersion, and is said to be
flooded. Under most practical fermentation conditions, the aerated impeller (usually
the lowest impeller) does not flood [18] if:

$$d_i > 1.1 \, [(Q_{gas} \, d_i^{1.5})/N^2]^{0.2} \tag{1.18}$$

where d_i is the impeller diameter, ft;

Q_{gas} is the gas flow rate, standard cubic feet per minute, or scfm; and

N is the impeller speed, revolutions per minute, or rpm.

Equation 1.18 is valid only for Newtonian fluids, standard geometry, and disk
turbines.

The impellers deliver the bulk of the power input under nonflooded conditions, but
gas expansion can account for up to about 15% of the power, depending on the
aeration and agitation conditions; however, the fraction usually decreases with
increasing OTR. The fact that the fraction usually is less than 10% is one of the
reasons that the form of eq 1.6 is satisfactory. Were the fraction larger, the form of
eq 1.6 obviously would not suffice (consider the case of no mechanical power and a
nonzero gas flow).

The power an impeller can deliver decreases with increasing gas flow, primarily
because of the lowered density. Published correlations express this dependence as the
ratio of the gassed horsepower to the horsepower that would be drawn under the same
conditions, but without gassing. The ungassed power, P_o, is defined as [23a, 24a]:

$$P_o = N_p \, N_i \, N^3 \, d_i^5 \, \rho/(550 \, g_c) \tag{1.19}$$

where N_p is the impeller power number (see below);

N_i is the number of impellers;

N is the impeller speed, revolutions per second, or rps;

d_i is the impeller diameter, ft;

ρ is the liquid density, lb/ft^3; and

g_c is the Newton's law conversion factor, ft·lb$_m$/lb$_f$·sec^2.

N_p is a function of impeller geometry, tank geometry, and impeller Reynolds
number (N_{Re} or Re):

$$N_{Re} = N \, d_i^2 \, \rho/\eta \tag{1.20}$$

where η is the broth viscosity.

Classic correlations for N_p as a function of N_{Re} are given in the literature [5d] for
Newtonian fluids.

Before we go on, it is important that you note how strongly the power input
depends on the impeller diameter (fifth power) and the impeller speed (third power).
These very strong dependencies are all too often overlooked, and this oversight is
among the major reasons for motors burning out, gear boxes breaking, and worse: it
is the stuff of which new careers are often made.

Published correlations for gassed horsepower also can be found in the literature [5e, 13]. Note that they are subject to the same caveats as are the published correlations for $K_g a$.

Strictly speaking, such correlations are valid only for the gassed impeller, primarily because they were developed using single, gassed impellers. In most cases involving multiple impellers, only the lower impeller is gassed. The gas flowing by the upper impellers has a different rate per unit area and a different density than the gas flowing by the lower impeller. These differences result from dispersion of the gas and from a decrease in the gas density. Unfortunately, these factors work in opposite directions, which makes it difficult to predict quantitatively their effects on unloading of the upper impellers; hence, there are different schools of thought on how to calculate upper impeller gassed power.

One approach is to use for upper impellers [25]:

$$P_g/P_o = 1/(1 + h) \tag{1.21}$$

Typical values of holdup are about 20%; therefore, eq 1.21 predicts a ratio of about 0.83. Actual values appear to be 20–30% lower, although this is difficult to check. Also, note that eq 1.21 does not take into consideration that gas density, and hence holdup, tend to increase with increasing height.

Another school assumes upper impellers are unloaded to the same extent as is the lower impeller. This obviously underestimates the delivered power, but, given all the uncertainty, we usually opt for this more conservative approach, which usually gives a P_g/P_o between 0.4 and 0.6. The cost increase is seldom significant, and usually is fairly cheap insurance.

Regardless of which method is used, the P_g calculations should focus more strongly on the impeller diameter than on the impeller speed, because P_g is almost a fifth-power function of impeller diameter, and because the impeller diameter is subject to more constraints than is the impeller speed. Also, it is more important that the diameter be correct than for the speed to be correct (in whatever sense), because once the impellers are installed, it is far more difficult to modify them than it is to change the drive speed. Indeed, the drive speed can be used to bring the other values to their desired level. For a given variable speed drive, the speed can be increased up to the centrifugal limit of the motor, or down to the speed at which overheating becomes important. Usually, this provides a range of at least 10–20% on either side of the design speed. The power input depends on the third power of the impeller speed; therefore, a 10–20% speed range translates to a 30–70% range for P_g. We have never encountered a case for which this was not adequate insurance against uncertainties in the calculations.

Finally, there also is the question of the desired power distribution among the impellers in a multiple impeller system. There are several approaches to addressing this question, but in considering their relative merits, one should keep in mind the uncertainties concerning the calculation of gassed power.

One school of thought [23b] suggests that about 40% to 45% of the power should go to the lower impeller, with the remainder being distributed about equally among the rest. Our approach varies depending on the specific case, although it usually results in similar distributions. For oxygen-transfer calculations, however, we do not count as part of the total power the draw of an axial impeller when it is used primarily for providing top-to-bottom mixing, and when the other impellers are all turbines.

1.3.2.5 Pressure

The driving force for oxygen transfer can be increased by increasing the vessel pressure, which increases the oxygen partial pressure. This is a very effective and relatively inexpensive way to increase the OTR. You've already paid for a pressure vessel, and the increased cost of air compression usually is not significant.

Typically, fermentation vessels are rated at about 40 psia to withstand sterilization, and with proper precautions (such as a pressure relief valve in the air supply line before the sterile filter) can be operated safely quite close to 40 psia. This pressure level increases the P_{O_2} by a factor of about 2.7, which increases the overall driving force by a factor between 2.0 and 2.5, depending on the dissolved oxygen concentration and the oxygen-transfer efficiency. Increasing the pressure even more can further improve the OTR, but this leads to very thick-walled vessels, which limits the number of vendors and therefore drives up substantially the cost of the vessels. Also, increasing the vessel wall thickness can decrease substantially the rate of heat transfer. Finally, there are some cases in which increased pressure can cause physiological problems and decrease fermentation rates. Such cases usually involve organisms that are sensitive to higher concentrations of carbon dioxide [2b].

1.3.2.6 Oxygen Enrichment

The oxygen-transfer rate also can be increased by enriching the inlet air with oxygen, which increases P_{O_2}. This technique is used frequently in the lab, but at larger scale it engenders serious problems, associated with explosion hazards and with increased costs, which are seldom justified.

Far be it from us to hinder progress, but in closing this section we feel compelled to remind the reader that before he goes to great lengths to consider ways to increase the OTR, he should consider seriously the concomitant increase in heat load.

1.3.3   Agitation

The gassed horsepower required (at a fixed gas flow) can be obtained using various combinations of impeller speed, size, type, spacing, and number. Each combination has a different effect on the fermentation and on the design. To find a workable combination, one should consider balances among the different factors important in providing agitation.

Obviously, the largest impeller we can use is limited by the vessel internals. However, we must also consider how the impellers will be constructed, and how they will be put into the vessel. This forces consideration of the construction of the top head, which for larger vessels usually becomes a choice between a full-opening head and a manway. Full-opening heads are not usually practical for vessels larger than 500–800 L (diameter about 28–30 in, or 71–76 cm). A manway eliminates the need for an expensive flange, a lot of labor for bolting and unbolting, the inherent difficulty of sealing a large, full-opening head, and about 10% of the vessel cost relative to a full-opening head. The smallest practical manway is about 18" OD (46 cm). The largest is determined by other items on the top head (such as nozzles and drive), but a manway larger than 30" (76 cm) is usually a special item, and can be quite costly.

An impeller having a diameter larger than the manway diameter can be used, but it must be put into the tank in parts, and must be assembled inside by bolting-up or by welding, depending on the compromises one is willing to make with regard to cost and cleanability.

For most microbial fermentations, practical ratios of impeller diameter to tank diameter fall somewhere between 0.35 and 0.45. There are, however, cases that call for ratios outside this range (for example, microbial gums). For mammalian cell cultures, we have found that very large ratios are best.

The fluid exerts several different forces on the impeller, and the torques and bending moments these forces generate increase with increasing impeller diameter. Larger diameter impellers therefore necessitate thicker shafts, heftier bearings and gear boxes, etc., which result in larger costs.

The major factors that limit impeller speed are the natural frequency of the shaft, and standard motor speeds and gear box ratios. The speed should be kept at least 30% below the natural frequency of the shaft to avoid vibration problems. Standard motor and gearbox speeds should be selected to avoid extra cost and long lead times.

The trade-offs in relation to impeller type are primarily among power number, mixing characteristics, and the sizes and types of fluid forces generated by each type of impeller—all of which are interactive. Generally, high-power impellers (*e.g.*, disk turbines) are very good for gas dispersion, but do not produce good bulk mixing, and generate large fluid forces. The opposite is true for low-power impellers (*e.g.*, marine propellors). Also, for a given speed and size, the number of low-power impellers required is greater than the number of high-power impellers. This influences the spacing between impellers, which, in turn, influences both mixing quality and power input.

1.3.3.1 Tip Speed and Shear

Any organism is damaged by fluid mechanical forces, the extent of the damage being dependent on the nature, magnitude, and duration of the force [26–29]. Reliable quantification of these effects is quite difficult, and has not yet been done meaningful-ly. We can, however, generalize to the following extent: single-cell microbes are very resistant, mycelial organisms (particularly in diffuse growth) are less resistant, and mammalian cells are very sensitive—so sensitive that much of the design of tissue culture reactors is constrained significantly by this one factor. But regardless of the type of organism, we have as design guides only rules of thumb and some incomplete theories that have not yet yielded any proven, quantitative correlations. Most popular among these are the tip speed rule and the shear theory.

The *tip speed rule* recommends that the impeller tip speed not exceed 1500 feet per minute (fpm) (which is 762 cm/s). This limit is based on observations made many years ago, which showed that unacceptable damage was sustained by mycelial organisms exposed to tip speeds greater than 1500 fpm. The rule works quite well for the specific fermentations on which it is based, and it is apparent that higher tip speeds lead to larger fluid forces, but no fundamental relationships or meaningful empirical correlations have been developed to relate tip speed to fluid mechanical forces or to the damage they do. Also, the rule puts no constraints on the combinations of impeller diameter and speed that result in the 1500-fpm limit. It has never been shown convincingly that the rule is scale-independent: many organisms can withstand tip speeds well in excess of 1500 fpm, and many others are degraded at speeds well below 1500 fpm. Clearly, the rule is only a very rough guide, and does not apply to most bacterial or yeast fermentations, or to tissue cultures. Interestingly, morphologi-cal changes effected by geneticists also have made the rule less of a concern for fungal fermentations. Finally, in most practical cases, application of the rule does not impose serious constraints on fermentor design, so we can afford to be gracious, obedient disciples.

The *shear theory* is based on the concept that the greater the fluid shear stress, τ, the greater the damage. Shear stress is defined formally as:

$$\tau = \eta \, \sigma \tag{1.22}$$

where η is the fluid viscosity, and σ is the shear rate (velocity gradient).

This approach is misleading in that it implies that the damage is done primarily by true shear stresses (as defined above). In most cases, however, the shear stresses *per se* are relatively small in comparison to the forces associated with turbulence (for example, eddy stresses), the forces that can do the most damage. Also, the effects of time and aeration have not been studied adequately at a basic level, nor have any practical methods been published to evaluate these effects.

Despite all these shortcomings, it is worth considering some quantitative aspects of the two theories, because they are accepted so widely, and are used frequently in engineering design calculations.

The tip speed, v_t, is given by:

$$v_t = 3.14 \, d_i \, N \tag{1.23}$$

This, taken together with the definition of the power number (eq 1.19), imposes fairly rigid constraints on the relationship between the impeller diameter and speed, when the gassed horsepower is specified. For example, for a Rushton turbine with a power number of 6.2 (fully turbulent flow) and a P_g/P_o of 0.4,

$$d_i = 915 \, (P_g/v_t^3)^{0.5} \tag{1.24}$$

and

$$N = 0.0165 \, (v_t^5/P_g)^{0.5} \tag{1.25}$$

The quantitative aspects of the shear theory have their roots in observations [30] that the average shear in the vessel depends only on agitation rate, and is independent of impeller diameter:

$$\sigma = k_1 N \tag{1.26}$$

where k_1 depends on the type of impeller, and is approximately 10 for disk turbines. The maximum shear, which occurs at the impeller tip, does appear to depend on the impeller diameter as well as on the speed; however, no well-defined correlations have been published. Also lacking is a clear picture as to what the "average shear" really means, particularly in the presence of high levels of turbulence. See reference 5e for further discussion.

The defining equations for power, average shear, and tip speed can be cast in forms that relate impeller speed to impeller diameter:

$$N = [32.2 \, P_g \, 550/(N_p \, d_i^5 \, \rho)]^{0.33} \qquad \text{(power)} \tag{1.27}$$

$$N = v_t/(3.14 \, d_i) \qquad \text{(tip speed)} \tag{1.28}$$

$$N = \sigma/10 \qquad \text{(shear)} \tag{1.29}$$

Figure 1.4 is a graphical representation of these equations, and illustrates the constraints they impose. Finally, it is important to note that the nature and magnitude of fluid mechanical forces change with scale. Factors that appear to have little or no effect at a small scale may become very important at a large scale, and vice versa.

1.4   Heat Transfer

1.4.1   Basics

The driving force for heat transfer from one point to another is the temperature gradient, ΔT, between the two points. The rate of heat transfer, Q, is directly proportional to the driving force, and inversely proportional to the resistance, R, to heat flow between the two points. In simplest terms, this can be expressed as:

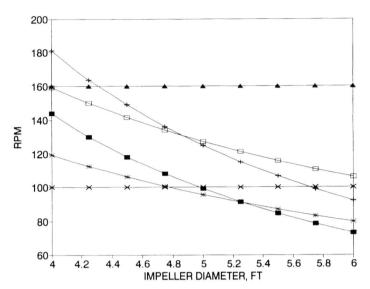

Figure 1.4 Shear, Tip Speed, and Power. Each curve represents the relationship between rpm (revolutions per minute) and impeller diameter for constant shear, tip speed, or power:

shear	✕	0.167/s;
	▲	0.267/s;
tip speed	*	1500 fpm;
	☐	2000 fpm;
horsepower	■	300 hp;
	+	600 hp.

$$Q = \Delta T/R \tag{1.30}$$

The resistance is inversely proportional to the area, A, across which heat is transferred, and inversely proportional to a measure of the ease with which heat is transferred through the material (akin to conductivity in electrical conductors):

$$1/R = UA \tag{1.31}$$

where U is called the *overall heat-transfer coefficient*. It has units of $Btu/h \cdot ft^2 \cdot {}^\circ F$. The value of U depends on the nature of the material across which heat is being transferred, the system geometry, and the nature of any fluid flow involved. Most heat transfer in agitated tanks involves two moving fluids (for example, broth and cooling water) and a solid between them (the tank wall); therefore, three separate resistances are involved. The dependence of U on basic phenomena for this case is quite complex, and the only practical means of evaluating it is to use empirical correlations.

For thin-walled vessels, U for the jacket is given adequately by:

$$1/U = 1/h_i + t/k + 1/h_o + 1/h_{f_i} + 1/h_{f_o} \qquad (1.32)$$

where h_i is the broth side heat-transfer coefficient (inversely proportional to resistance);
   t is the vessel wall thickness;
   k is the metal thermal conductivity;
   h_o is the coolant side heat-transfer coefficient;
   h_{f_i} is the inside fouling heat-transfer coefficient; and
   h_{f_o} is the outside fouling heat-transfer coefficient.

Correlations for h_o can be obtained from vendors. Fouling factors depend on specific operating conditions: every effort should be made to keep them to a minimum (for example, use coolant filters).

Correlations for h_i usually are given in the form:

$$h_i = a \, (d_i^2 \, N \, \rho/\eta)^b \, (c_p \, \eta/k)^c \, (\eta/\eta_w)^d \, f_{geom} \qquad (1.33)$$

where a, b, c, d are constants;
   ρ is the broth density;
   η is the broth viscosity;
   c_p is the broth specific heat;
   k is the broth thermal conductivity;
   η_w is the broth viscosity at the wall temperature; and
   f_{geom} is a function of tank and impeller geometries.

Such correlations usually are accurate to within 20% for unaerated fluids. Unfortunately, no reliable correlations have been published for aerated fermentation broths; hence, one is forced to use the correlations available, coupled with practical experience, unless, of course, experimental values are available for the subject broth (quite rare).

As a rough rule of thumb, it is helpful to know that U for a clean vessel usually has a value between 50 and 200 Btu/h·ft²·°F. The lower values apply to very viscous, non-Newtonian broths (for example, fungi), and the larger values apply to low viscosity Newtonian broths (for example, bacteria and yeasts). See references 3b, 23c, 24b, 31, and 32 for more information concerning heat-transfer coefficients. Note also that these values are for new, clean jackets, and that they decrease as scale builds up. Obviously, the extent of this effect and the speed with which it occurs depends on the care taken to keep the jacket clean.

The form of ΔT in eq 1.30 depends primarily on the system geometry and on the relative directions of the fluid motion. We do not have the space to discuss this in detail (see references 2b, 23b, 25, and 26), but note that for most cases involving heat transfer in agitated-tank fermentors, the appropriate form of ΔT is the *log mean temperature difference* (*LMTD*):

$$LMTD = (T_i - T_o)/\ln \, [(T_f - T_o)/(T_f - T_i)] \qquad (1.34)$$

where T_i is the jacket (or coil) fluid inlet temperature, °F;

T_o is the jacket (or coil) fluid outlet temperature, °F; and

T_f is the fermentation temperature, °F.

Equation 1.34 assumes that the broth is perfectly mixed, and hence can be characterized by a single temperature; however, when the jacket (or coil) temperature is constant (as is the case for condensing steam), the correct form of ΔT is:

$$\Delta T = T_j - T_f \tag{1.35}$$

where T_j is the jacket (or coil) fluid temperature.

We now can illustrate the methods used to calculate values given in Table 1.1. Combining eqs 1.30, 1.31, and 1.34, we get:

$$Q = U A \text{ (LMTD)} \tag{1.36}$$

We next specify the *exit approach temperature*, T_a. This is commonly used jargon for the difference between the vessel fluid temperature and the jacket (or coil) fluid exiting temperature. It is a useful parameter, in that it seldom is outside the range of 5–25 °F, and hence helps one to home in quickly on a practical solution. Substituting the definition of T_a into eq 1.36, we get:

$$A = Q \ln \left[(T_f - T_i)/T_a)\right]/[U (T_f - T_a - T_i)] \tag{1.37}$$

For this example we take typical values of $T_f = 86$ °F, $T_a = 10$ °F, $T_i = 35$ °F, and $U = 150$ Btu/h·ft^2·°F; therefore,

$$A = 10,020 \, Q \tag{1.38}$$

The cooling water flow rate, W_c, now can be calculated from the simple thermodynamic relationship:

$$W_c = Q/[c_p (T_o - T_i)] \tag{1.39}$$

1.4.2   Realities

The calculations above are the easy part. Not so easy is deciding how to cope with large heat loads. Obvious approaches are low coolant temperature, internal coils, or a combination of these. Wouldn't it be grand if life were that simple? Well, it's not!

Internal coils can add about 15–25% to the vessel cost. They also create serious operating problems, including more difficult cleaning, poor mixing, and leaks of nonsterile coolant into the broth. Coolant leaks can be forestalled by covering the welds, as shown in Figure 1.5.

Refrigeration is expensive, and is a significant maintenance problem. For example, a 10×10^6 Btu/h chiller has a purchase cost of about $250,000, but that's not the end

Figure 1.5 Coil Welds.
Coil pipe sections are butt-
welded. The welds are
covered by a tight-fitting
concentric pipe, which is
welded to the pipe
sections.

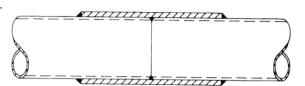

of it. The heat from the chiller must be rejected to something, a cooling tower being typical. A cooling tower for a 10×10^6 Btu/h chiller would have to handle a load of about 20×10^6 Btu/h (accounts for the inefficiencies of the chiller), and would cost at least $100,000. All this does not include the costs of installation, makeup water, and maintenance. And things get worse if a lower coolant temperature is called for. Lower coolant temperatures increase capital and operating costs, and at less than 35 °F (we don't like to go below 40 °F) create the very real possibility of line and valve freezeups, for which you'll pay dearly.

One way to avoid lower coolant temperatures (for a fixed heat-transfer area) is to use higher coolant flow rates. But there are limits, imposed by what are practical levels of pressure drop across the jacket and the coil; the limit is about 15 psi for most applications. For flows that would require a pressure drop higher than 15 psi across a single section jacket, the jacket should be zoned as described in Section 1.7.7.

An alternative often proposed to internal coils and low-temperature coolant is recirculation through an external heat exchanger. This seems like a good idea, and in some cases it is, but most of the time, the problems it creates are worse than the one it solves. The substantially increased risk of contamination is obvious, although frequently underrated. The bigger problem, however, is the pumping itself, a problem that increases rapidly with scale. It is no mean task to find a large pump about which all of the following can be said: it is truly sterilizable and aseptic, mechanically reliable, easily cleaned, capable of handling large quantities of entrained air without cavitating, and "user-friendly" to organisms. There are a few commercially available pumps that satisfy most of these criteria reasonably well (such as the IZ pump developed by Vogelbusch), but their cost is justified only in special cases. Again, we return to the voice of reason: consider slowing down the fermentation.

1.5   Broth Rheology

Broth rheology can have large effects on oxygen and heat transfer, as well as on mixing. There is very little reliable published information and a lot of misconception. A brief introduction to this topic is given in Figure 1.6 and in the following section.

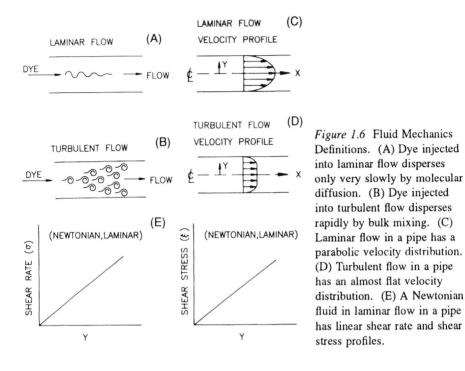

Figure 1.6 Fluid Mechanics Definitions. (A) Dye injected into laminar flow disperses only very slowly by molecular diffusion. (B) Dye injected into turbulent flow disperses rapidly by bulk mixing. (C) Laminar flow in a pipe has a parabolic velocity distribution. (D) Turbulent flow in a pipe has an almost flat velocity distribution. (E) A Newtonian fluid in laminar flow in a pipe has linear shear rate and shear stress profiles.

1.5.1   Definitions

In *laminar flow*, all the fluid appears (on a macroscopic scale) to flow in layers parallel to the direction of the bulk flow. Any motion in other directions is by simple diffusion, and occurs slowly in comparison to the rate of bulk flow. A dye injected (carefully) into a laminar stream appears to flow within a single layer for quite some time, until it disperses significantly through diffusion.

Turbulent fluid flow is very chaotic. There are strong currents in directions other than the direction of the bulk flow, and the fluid appears to be well-mixed at a macroscopic level. A dye injected at any point disperses rapidly throughout the fluid.

Flowing fluids exhibit velocity profiles that arise from the transfer of momentum between adjacent layers of fluid [10b, 33a]. In laminar flow the transfer mechanism is simple molecular diffusion, which results in each layer exerting on adjacent layers forces parallel to the direction of bulk flow. These are called *shear forces*. The magnitude of the shear force divided by the area on which it acts is defined as the *shear stress*. Shear stresses exist also in turbulent flow, but usually are small in comparison to the other fluid forces (for example, eddy stresses) that result from turbulence. It is important that you understand this fact when considering the degradation of cells and of biological molecules by flowing fluids.

Shear rate is defined (for laminar flow) as being equal to the velocity gradient in the direction perpendicular to the direction of flow:

$$\sigma = dv_x/dy \tag{1.40}$$

where x is the direction of flow, and y is the direction perpendicular to x.

Strictly speaking, viscosity is defined for fluids in *simple viscometric flow* [33b] of which laminar flow (under some conditions) is an example. For such flows, viscosity, η, is defined as the shear stress, τ, divided by the shear rate, σ:

$$\eta = \tau/\sigma \tag{1.41}$$

Other types of viscosity (for example, eddy viscosity) have been defined for turbulent flow [34, 35].

Failure to observe these definitions has led to considerable confusion in the literature of biotechnology. Finally, the accurate and meaningful measurement of viscosity *sensu stricto* is not always a simple matter [33c, 36], particularly for fermentation broths [37a, 38, 39].

1.5.2 Rheological Types

The rheologies of various fluids are defined in terms of how their viscosities vary with shear rate and shear stress. There are some fluids that exhibit other behaviors, such as viscoelasticity, but we do not discuss them here because they are not very important in fermentation. The interested reader should consult references 10b and 33a for more information.

The viscosity of a *Newtonian fluid* is independent of shear rate and of shear stress (see Figure 1.7). The viscosity of a Newtonian fluid does, however, depend on temperature, composition, pH, and other physicochemical parameters. Most broths containing single-cell organisms are Newtonian, unless the medium itself is not (for example, media containing significant concentrations of starch, protein, or other macromolecules).

Fluids that are not Newtonian are called *non-Newtonian*. One example (see Figure 1.7) is a *pseudoplastic* fluid. The viscosity of a pseudoplastic fluid decreases with increasing shear rate. There are many published mathematical models for pseudo-plastic behavior [10b, 33a–d, 34–36, 37a, 38–40] but the simplest, the *power-law model*, is usually adequate for practical application to fermentation broths:

$$\eta = K\,\sigma^{(1-m)} \tag{1.42}$$

where K and m are constants.

Broths of mycelial organisms (such as fungi or streptomycetes) and broths containing microbial polysaccharides (for example, xanthan) often exhibit power-law behavior. Examples of viscosity characteristics of pseudoplastic fermentation broths

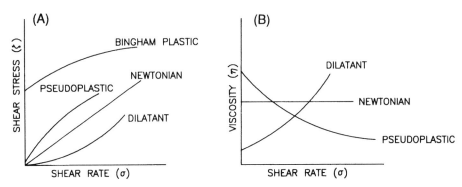

Figure 1.7 Rheology Definitions. (A) The effect of shear rate on shear stress for various types of fluids in simple shear flow. (B) The effect of shear rate on viscosity for various types of fluids in simple shear flow.

are given in Figures 1.8 and 1.9 [37b, 41]. Note that although mycelial and polysaccharide broths both are pseudoplastic, their fluid mechanical behaviors are not the same. This is because in mycelial broths the non-Newtonian behavior is associated with the morphological characteristics of the discontinuous phase: the more diffuse the hyphal growth, the more non-Newtonian the behavior; but in polysaccharide broths, the non-Newtonian behavior arises from the continuous phase.

It is interesting to note that in recent years geneticists have been able to effect changes in the morphologies of several commercially important mycelial organisms (such as *Penicillium*). These changes have decreased the viscosity and extent of non-Newtonian behavior of the broth containing these organisms; the result is that mixing, oxygen transfer, and heat transfer are improved, and the organisms are less sensitive to fluid mechanical forces.

Bingham plastics are fluids to which a minimal, nonzero stress (the *yield stress*, τ_y) must be applied before the fluid begins to move (see Figure 1.7). The viscosity of a flowing Bingham plastic may be Newtonian or non-Newtonian (usually non-Newtonian for fermentation broths). A simple but adequate mathematical model [34] for Bingham plastic fermentation broths is:

$$\tau - \tau_y = k_3 \, \sigma^{(1-n)} \tag{1.43}$$

Bingham plastic behavior has been reported for some mycelial broths [42–44], but there is some doubt [37b] concerning the accuracy of these measurements (as well as many other rheological measurements reported for fermentation broths).

The mixing, heat-transfer, and mass-transfer characteristics of non-Newtonian fluids can differ considerably from those of a Newtonian fluid, the extent of the difference often depending strongly on the type and degree of non-Newtonianism. In principle,

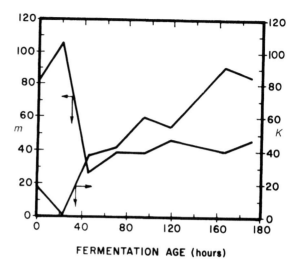

Figure 1.8 Rheology of a Mycelial Broth. The broth was assumed to behave as a pseudoplastic fluid. The constants *m* and *K* are for the viscosity power law (eq 1.42).

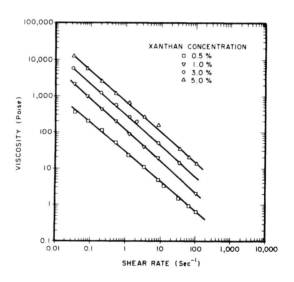

Figure 1.9 Rheology of Xanthan Broth. Xanthan solutions behave as extremely pseudoplastic power-law fluids even at low xanthan concentrations. Xanthan is a very high molecular weight microbial polysaccharide.

therefore, one should use empirical correlations specific to the rheology of the fluid under consideration. Unfortunately, there are very few such correlations available. An example of one fairly reliable correlation for P_o is given in reference 30. It is applicable to a fairly wide range of non-Newtonian fluids, because it contains provisions for considering the specific rheological characteristics of the fluid. An example of a more typical correlation is given in reference 45. It has no means for considering specific rheological characteristics; therefore, it is valid only for the fluid used to develop it, and for other fluids having the same rheological characteristics. For a given case, then, one is left with the problem of having to decide just how non-Newtonian a fluid must be before switching from Newtonian fluid correlations to non-Newtonian fluid correlations. Even when you do decide to use a non-Newtonian correlation, good results are not guaranteed.

1.6   Mixing

Mixing characteristics strongly influence most aerobic fermentations, and fall into two general categories: *bulk mixing* and *micromixing*. Both depend, in different ways, on broth physicochemical characteristics, on impeller type, speed, and size, on vessel geometry, and on gas flow rate. In most aerobic fermentations, the bulk mixing uses a relatively small fraction of the total power input. Most of the power is required to support micromixing and gas dispersion.

Good bulk mixing is required to maintain broth homogeneity, which is necessary for reliable monitoring and control. The *mixing time* is a quantitative measure of the quality of bulk mixing, and is defined as the time it takes for a (presumably) homogeneous broth subjected to a pulse disturbance (for example, addition of acid) to become homogeneous again. Unfortunately, reestablishment of strict homogeneity would take an infinite amount of time. Therefore, practical measurement of mixing time requires a definition that specifies attainment of an arbitrary fraction of the new state of homogeneity, usually 99%.

Several mixing time correlations have been published [46, 47] all of which are, at best, rough guides. In using them one should pay careful attention to the experimental conditions used to develop them.

Micromixing concerns fluid motion at the level of eddies and below, and the nature and extent of fluid turbulence. High levels of oxygen and heat transfer require highly turbulent micromixing, which requires large power inputs; however, these conditions also result in high levels of turbulence-derived fluid forces (such as eddy stresses and cavitation), which are responsible for most of the fluid mechanical degradation of cells and biological molecules.

1.6.1   Impellers and Baffling

Bulk mixing and micromixing both are influenced strongly by impeller type, broth rheology, and tank geometry and internals. Several impeller types are illustrated in references 48–50 and in Figure 1.10. The Rushton turbine is the one used most often for highly aerobic fermentations, because it has among the highest power draws of any of the commercially available impellers, and it is better characterized than the others; hence, its behavior is easier to predict.

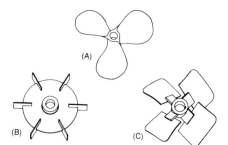

Figure 1.10 Impellers. (A) Marine propellor; may be either left-hand or right-hand pitch. (B) Rushton turbine impeller, 20:5:4 ratio. (C) Lightnin™ A-315 impeller.

However, the Rushton impeller is by no means perfect. One of its major drawbacks is that it provides very little axial flow, resulting in poor overall top-to-bottom mixing. In addition, agitation intensity decreases with distance from the impeller, and this decrease can become very pronounced for viscous, pseudoplastic broths. Furthermore, in multiple impeller systems, partially isolated mixing zones can be established. The extent of the isolation depends on the distance between impellers and on the physical dimensions of the tank and the impellers. It depends also on the fluid characteristics, the air flow rate, and the shaft speed, all of which can change markedly during a fermentation. The result of all this can be high degrees of heterogeneity not only in mixing quality, but also in component concentrations and environmental variables (such as pH), which, in turn, can lead to poor control, decreased productivity, and decreased yields. It also is important to note that it usually is not possible to design a Rushton-only system to cope well with the full range of conditions encountered for fermentations involving broths that become increasingly viscous and non-Newtonian (such as xanthan).

Flow patterns for a typical Rushton impeller combination are illustrated in Figure 1.11 [23d]. The tank in Figure 1.11 is *fully baffled*. This minimizes fluid swirling and vortexing, and is common practice. Full baffling normally requires four baffles, each having a width equal to about 1/10 of the tank diameter.

Figure 1.12 illustrates the flow patterns in an unbaffled vessel [23d]. Note that the vortexing can result in *air-shrouding* of the impeller, which decreases power input, mixing quality, and oxygen transfer. Also, when the power drops to a low enough level, the vortex collapses, at least partially. Forces on the agitator rise rapidly, and the power draw tends to rise sharply. This can result not only in considerable instabilities and in large random forces on all parts of the fermentor and drive, but also in motor burnout and other damage to the drive.

Axial flow impellers (such as the marine propellor) provide good top-to-bottom mixing, but draw relatively little power and therefore do not contribute much to good oxygen transfer. Flow patterns for a marine propellor are illustrated in Figure 1.13 (baffled tank) [23d].

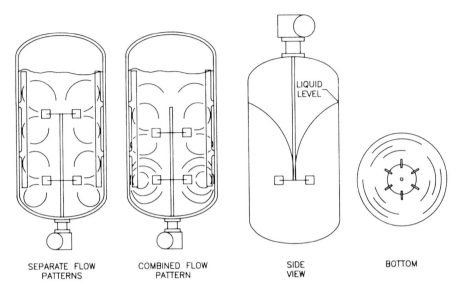

SEPARATE FLOW   COMBINED FLOW   SIDE   BOTTOM
  PATTERNS   PATTERN   VIEW

Figure 1.11 Rushton Turbine Mixing Patterns in a Fully Baffled Vessel. Impeller spacing affects the extent to which independent mixing zones are formed. Closer spacing increases the interaction, but does not completely eliminate zone separation: the overall top-to-bottom mixing still is poor, especially in tall vessels. Closer spacing also influences the power input.

Figure 1.12 Unbaffled Tank Mixing Patterns. The lack of baffles allows uncontrollable liquid swirling and vortex formation, which can be dangerous mechanically. Swirling also decreases power input, mixing quality, and oxygen transfer.

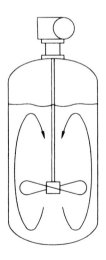

Figure 1.13 Marine Propellor Mixing Pattern in a Fully Baffled Vessel. A marine propellor is an axial flow impeller which provides good top-to-bottom mixing. It is a very low power device which does not provide large oxygen-transfer rates.

One way to improve bulk mixing while maintaining good oxygen transfer is to use Rushton turbines at lower positions in the tank and an axial flow impeller at the top. We have found that marine propellors and A-215s (lighter than a marine propellor) work quite well in this scheme for many applications over a wide range of scales. We expect that the A-315 (hydrofoil design) will work even better because it has a higher power number than the others; hence, it should contribute more to oxygen transfer than the others.

The designer can help to minimize top-to-bottom mixing problems by keeping the ratio of liquid height to tank diameter under 2, and by spacing the impellers properly. If the impellers are spaced too closely, they interfere with each other, thereby decreasing power input, OTR, and mixing quality. If they are spaced too far apart, overall homogeneity suffers. We have found that a spacing of between 1 and 1.5 impeller diameters gives good mixing in most practical cases, and that oxygen transfer is not affected significantly by spacing within this range. Similar observations have been made by others [23e, 51, 52]. These guidelines must be used in concert with power and vessel geometry requirements to determine a workable combination of impeller diameter and speed.

1.6.2   Rheology

The effects of broth rheology on mixing are most pronounced for non-Newtonian broths. This is illustrated in Figure 1.14 for a Rushton turbine used in a pseudoplastic broth. The shear rate drops off rapidly and the viscosity increases rapidly with distance from the turbine tip; therefore, there is good mixing only in the immediate vicinity of the impeller, and air tends to channel around the impeller and to rise up the shaft. Anyone who has ever seen a serious xanthan gum fermentation knows just what we're talking about. One way to overcome this problem is to use very large diameter turbines. While this works quite well it creates high shaft (and, hence, gear box) torques, and it still does not solve the vertical mixing problem.

There are some new impellers that have been designed to give adequate axial flow and very good OTR, particularly for non-Newtonian broths. Indeed, the Prochem Hydrofoil and the Mixco A-315 (also a hydrofoil design) have been shown to do all this, and to require less power than Rushtons for the same OTR [51] at commercial scales.

1.6.3   Mammalian Cells

Agitation and aeration requirements for mammalian cell cultures are very different from those for microbial cultures. OTR requirements are very much lower, but the cells are much more easily damaged by fluid mechanical forces generated by impellers or collapsing gas bubbles. In most cases, the impeller must provide enough mixing

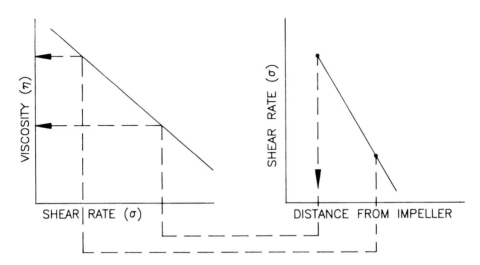

Figure 1.14 Effects of Pseudoplasticity on Mixing. The shear rate decreases rapidly with distance from the impeller. In pseudoplastic fluids this results in rapidly increasing viscosity as the distance from the impeller increases. The result is poor mixing and oxygen transfer in the bulk of the fluid.

to keep cells or microcarriers suspended homogeneously while creating as little fluid force as possible. Oxygen transfer usually is not a significant problem if these conditions are met. It also is undesirable to baffle vessels to be used for mammalian cells.

Many impeller types have been proposed to satisfy these constraints [53, 54]. A few, including marine propellors, have worked well under specific practical conditions (up to several thousand liters), but most simply are not suitable for commercial applications. One type that we have found satisfactory in tissue culture vessels up to 500 L is illustrated in Figure 1.15. This impeller (called the elephant ear) has been shown in commercial applications to scale up reliably and to provide adequate mixing and OTR, with little or no cell damage, at 15–500 L with suspended cells and cells anchored to microcarriers. Swirling and vortexing problems were overcome by mounting the shaft at an angle of 15°. Despite this and other successes, it is fair to say that there is no single best recommendation for the impeller of choice.

Finally, we try to keep the ratio of liquid height to diameter as close to unity as possible. This ratio not only is ideal for good top-to-bottom mixing, but also provides a high ratio of free surface to liquid volume; this helps to make the most of aeration from the free surface, which may be important in tissue culture.

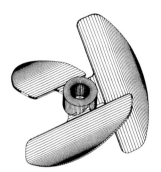

Figure 1.15 Elephant Ear Impeller. This low-speed impeller provides good bulk mixing and oxygen transfer at very low speeds. It distributes the power input over a large fraction of the total volume.

1.7   Fermentor Vessel Design

The reader should find it useful to familiarize himself first with Figure 1.16, which illustrates a typical vessel, and shows some important vessel dimensions and standard jargon.

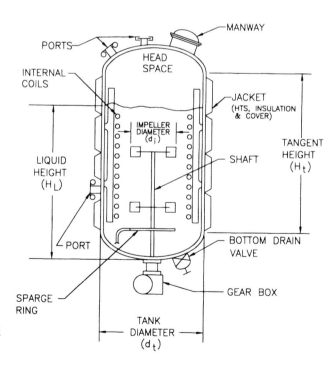

Figure 1.16 A Typical Fermentation Tank.

It also is necessary to recognize some practical matters before getting on to the technical details of vessel design. Some of these are:

(a) *Space requirements*: The vessel dimensions must be chosen to meet plant space limitations. This consideration obviously is most important for vessels that must be placed in existing buildings, but should not be overlooked for new construction. Poor choice of equipment sizes can cause inordinately high building costs. In making size/space allocations, one must keep in mind the space required for mounting pads and other vessel supports, process and utility piping, and space for operators and maintenance. In addition, one must consider space required for getting the vessel into the building, and for installing it. There also are building and safety codes that must be followed.

(b) *Transportation*: Shop-built vessels usually are less expensive and of higher quality than field-built vessels. In most cases, there are no serious transportation problems for vessels having diameters of 12 feet (3.6 m) or less and lengths of 40 feet (12 m) or less. Larger vessels cause a variety of problems (the details of which depend on the sites of fabrication and delivery), including the need for special permits, routings, and handling. With few exceptions, vessels larger than 14–15 feet (4.3–4.6 m) must be built on-site.

(c) *Vendor considerations*: As vessel size and wall thickness increase, the number of qualified vendors decreases. Relatively few vendors can build fermentation tanks having diameters larger than 12 feet (3.6 m). All too often, therefore, one must accept inordinately high prices and compromises on vessel quality. Among other reasons, it is important to keep all this in mind because the vessel price usually is about 40% of the total fermentor price.

(d) *Standard sizes*: One should always try to use standard sizes and materials. To do otherwise results in additional engineering, drafting, and time, all of which add substantively to cost and delivery time. Standard tank diameters are set by standard dish sizes, which usually are available in these increments:

OD, in	increment, in	OD, cm	increment, cm
12–48	2	30.5–122	5.1
48–144	6	122–366	15.2
144–204	12	366–518	30.5

Typical dish specifications can be found in reference 55.

Tangent heights should be rounded to whole inches, and, if at all possible, should be selected to allow the use of standard sheet widths. This may increase the total volume somewhat, but the cost is about the same as for the volume specified originally. The savings in cutting charges usually offset the cost of the additional metal.

(e) *Special heads*: A folklore has developed for tissue culture, which states that a hemispherical bottom gives better mixing and less shear in tissue culture vessels than does a standard dished head. There is really no firm evidence to support this belief, and many tissue culture vessels with standard heads are working quite well in commercial applications. What hemiheads do ensure is that the vessel cost increases easily by 50% or more.

1.7.1   Safety/Codes

The single most important factor in vessel design is that the vessel be safe to operate at design conditions. For safe operation, the vessel must be built and tested in accordance with the ASME Code For Unfired Pressure Vessels [56]. A vessel so designed and inspected carries an ASME stamp. One should never use a vessel that is not stamped. Also, the user should obtain from the vessel manufacturer a copy of the manufacturer's data report, and should keep it on file.

Adherence to the code insures that the vessel has been designed and constructed to withstand all forces generated under the specified operating conditions. All the details of the calculations and specifications should be left to the vessel manufacturer. The interested reader can consult references dealing with pressure vessel design [57].

Finally, if it is absolutely necessary to use high operating pressures in a large vessel, one should consider ways to minimize the metal thickness so as to allow the use of cold rolled sheet rather than plate. This results in better heat transfer, a better interior finish, and a lower price. Two of the ways to achieve this end are: use the jacket construction to strengthen the vessel (half-pipe jackets are best for this purpose), or use belly bands between jacket sections.

1.7.2   Materials

The choice of material(s) should be based on the following:
  (a) *Compatibility with the organism*: This is particularly important for mammalian cells, which are very sensitive to ferrites.
  (b) *Corrosiveness of the fermentation broth*: Acid products, high chloride content, and cleaning compounds can cause considerable corrosion. Clean steam also is extremely corrosive, and usually requires the use of 316 stainless steel (SS). This is particularly important for mammalian cell culture vessels, where use of clean steam is common.
  (c) *Cost and equipment lifetime*: Most vessels designed for aseptic operation are built from SS316, SS316L, SS304, or SS304L [58, 59]. The primary differences among these are concentrations of carbon and alloying materials, such as molybdenum, which affect primarily corrosion resistance and welding characteristics (see below). There are some cases, (such as ethanol fermentations) in which carbon steel is quite satisfactory. Such cases usually do not require aseptic operation. In most cases the specification of "L" adds about 15% to the vessel cost. There is considerable difference of opinion, and often some confusion, concerning the best choice of material for a given application.

Much of this stems from sweeping generalizations, which may have little or no validity in a specific case, and from an inbred desire in most of us to buy the very best. More often than not, the choice is SS316L (after all, it's the most expensive), and more often than not, it's overkill. Also, there are some cases for which it is not the best technical choice (see Section 1.7.3, Welds).

(d) *Materials testing*: R&D people seldom do any materials testing with actual fermentation broth or a growing culture. Indeed, there is precious little such information in the literature. We recommend that a substantial effort be made in this arena. The results could save a lot of money and aggravation.

1.7.3   Welds

The quality of vessel welds is extremely important, not only for code purposes, but also to insure maximum smoothness and cleanability, and to minimize corrosion problems. All vessel welding for aseptic fermentors should be done under inert gas shielding, so as to minimize oxidation and flux residue, and to yield smoother, pit-free welds. Of the acceptable vessel welding methods, we prefer the TIG (Tungsten-Inert-Gas) method because it is not subject to splattering problems as is the MIG method; therefore, it tends to give a better finish. We recommend specification of this method even though it is the vessel manufacturer's responsibility to deliver the finish specified.

With regard to the relationship between welding and vessel material, it is best to use 316L or 304L for multipass welds, to reduce carbide precipitation and consequent pitting near the welds. For single-pass welds, one should use 316 or 304 to minimize ferrites in the metal.

1.7.4   Finish

The smoother a surface is, the easier it is to clean, but, as in all walks of life, smoothness costs. Coarser surfaces are more difficult to clean; therefore, they require more stringent cleaning protocols and greater attention to ensuring that the cleaning is done properly. One way or another you pay: it's simply another of life's trade-offs, and the choice is up to you.

Finishes commonly used are tabulated below:

Grit designation	Microns (μm)
60	6.3
120	3.2
180	1.6
220	0.8
320	0.4
320 EP	< 0.4

Microns is a measure of the surface roughness; the smaller the value, the smoother the surface. A vessel to be electropolished (EP) should be polished to 320 grit.

As with all aspects of the fermentor, one should make a careful evaluation of the finish really needed. There is no doubt that a 320 grit/electropolished surface really shines, and looks like the money. But do you really need it? Also, other than aesthetics, what's the point of electropolishing the exterior of the vessel? Also, keep in mind that the better the finish, the shorter the vendor list.

Finally, one must consider the need for *passivation*. During fabrication and polishing there is some destruction of the surface film that protects stainless steel from corrosion. Welding is the major culprit. A typical passivation process involves cleaning with NaOH and citric acid followed by nitric acid (up to 20% at 120 °F) and a complete water rinse. In addition to restoring the film, this process also removes metal particles, dirt, and welding-generated compounds (such as oxides) [60].

1.7.5   Cleanability

The fermentor must be designed to insure that the *Cleaning-in-Place* (*CIP*) system is not compromised, and that all interior surfaces can be cleaned. This means that careful attention must be paid to the design of all interior surfaces. They all must be designed to prevent the accumulation of solids, to be smooth and free-draining, and to allow easy access for maintenance or repair. There should be no nooks and crannies, and all interior joints should be welded (not bolted) if at all possible. The top of the vessel is one of the places most frequently overlooked. Clearly, a dished head drains more freely than a flat head, but flat heads still are used. Also, the size and placement of the manway affect significantly the positioning of spray balls or hose nozzles, and the ease of access for manual cleaning [61]. Even the best CIP system can't cope with problems such as medium bake-on. And don't forget to check under the rim. A guiding principle is: If it's difficult to clean, it probably won't be cleaned.

1.7.6   Nozzles

Nozzle design must allow for:
(a)   Reliable, aseptic connection to external piping. For sizes up to 4" (10 cm), we have found the best choices to be sanitary clamps (3-A), Ingold nozzles, and vacuum seals (Ultraseal or VCO); however, we have found that ¾" (2-cm) miniclamp connectors tend to leak, and therefore we recommend that they not be used. Above 4" (10 cm), we recommend flanges with flat gaskets, since there is not really anything else available other than welding.
(b)   Free draining. Free draining minimizes the trapping of material in the nozzle, and can usually be achieved by mounting the nozzle at an angle of about 5°–15° off the horizontal (see Figure 1.17).

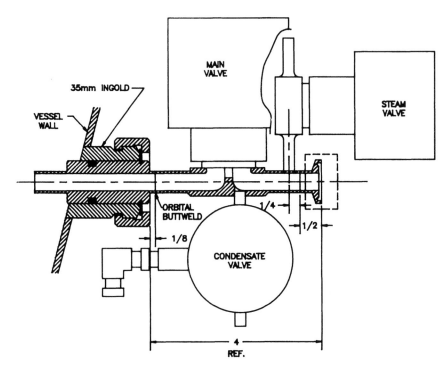

Figure 1.17 Steam Lock and Nozzle. This arrangement allows independent sterilization of the nozzle. A device (feed tank, sampler) may be attached to or detached from the capped end, and the interconnecting piping may be steam sterilized without disturbing the fermentation.

(c)   Simple, reliable cleaning and sterilization procedures. These are achieved most easily if the nozzle cavity is kept as small as possible; therefore, nozzles should protrude only enough to allow convenient making and breaking of connections. Mounting the nozzle at an angle of 5°–15° also helps.

(d)   Additions to be made without the added liquids dribbling down interior surfaces. Figure 1.17 illustrates a satisfactory approach to solving this problem. Note, however, that it does increase slightly the difficulty of cleaning.

1.7.7   Jackets

Various types of jackets are illustrated in reference 50. There are advantages and disadvantages to each type, but the choice usually is not critical, and it usually is best to leave the vessel fabricator some degree of flexibility to allow for his usual shop practices. Some guidelines worth following are:

(a)   It usually pays to use a SS304 jacket cover to extend vessel life, but there is no benefit to going to any higher quality.

(b)   The jacket can be designed to increase vessel strength, as described in Section 1.7.1.

(c)   Jacketing the bottom head usually is not very beneficial for vessels less than 5000 L, because there isn't enough area to make a significant contribution to heat transfer. While jacketing the bottom head adds only about 5% to the vessel cost, this area tends to be less effective than the rest of the jacket. Also, in most cases for which bottom jacketing is considered, an internal coil is necessary. Under these circumstances, it usually is more sensible to make the coil a bit larger, rather than jacketing the bottom.

(d)   Jacketing to the top of the straight side can provide useful cooling area as the air holdup increases, but the jacket should be zoned so that the vessel is not steamed above the unaerated liquid height. This minimizes medium bake-on during sterilization. Note also that jacket zoning usually is necessary to limit pressure drop in the cooling loop.

(e)   We recommend the use of sanitary clamp or flange connections to the jacket inlet/outlet lines. They avoid torquing of the welds between the external pipes and the jacket.

(f)   The jacket should be designed to minimize buildup of solids. Such buildup leads to increased chloride concentrations, which induce stress corrosion. High liquid velocity is recommended, as is filtration of the coolant.

1.7.8   Coils

The operational problems associated with coils were discussed in Section 1.4. If it is absolutely necessary to use them, allow at least 6" (15-cm) spacing between coils, and make certain that the coil is mounted in a way that can withstand the thermal and mechanical stresses generated during fermentations. Joints between pipes should be butt-welded, covered with a welded concentric tube (see Figure 1.5), and leak tested according to the method described by PERKOWSKI [62].

1.7.9   Baffles

It is best to weld baffles to the tank wall. Removable baffles lead to unsealed joints. Baffles should be set off from the wall, so as to minimize solids build up and to simplify cleaning. There should be four baffles, each having a width equal to 1/10 of the vessel diameter. This works at all scales.

1.7.10 Spargers

Spargers are another subject of controversy. In general, we accept the gas distribution arguments favoring ring spargers over single orifice spargers, but beyond that we find very few facts supporting other design criteria (for example, the effects of hole size on oxygen-transfer rate), and have found that the following are adequate for good performance:

(a)   The sparge holes in the ring should be in line with the inner edges of the impeller blades.

(b)   The sparger holes should face downward to minimize medium retention in the sparger.

(c)   Hole diameters should be chosen such that each hole is a critical orifice at maximum gas flow.

(d)   The sparger inlet pipe should be placed so as to allow free draining back into the vessel. Note that this does not eliminate the need for a check valve in the inlet line, distasteful as we find their use.

1.7.11 Manways

The use of manways obviates the need for full-opening heads on larger vessels. Standard manway diameters range from 12" to 28" (30.5 cm to 71 cm), usually in 2" (5.1-cm) increments. A 16" (30.5-cm) manway is the minimum practical size since most folks can't fit easily through anything smaller, but 18" (45.7 cm) is better. The maximum size is determined by the vessel diameter and the other items that must be mounted on the head. The manway collar should be kept as short as possible (3–4", or 8–10 cm) for reasons of cleanliness and cleanability. A typical manway is illustrated in Figure 1.18.

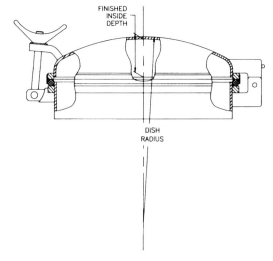

Figure 1.18 Manway. This manway is sealed by an O ring. The head swings open on the pivot shown.

1.7.12 Static Seals

O-ring seals should be used wherever possible. We have found that silicone is the best choice for the head plate, and we recommend ethylene propylene diene monomer (EPDM) for all other static seals. Teflon® has better temperature resistance, but it cold flows and doesn't stretch. Viton® has good temperature resistance and is quite serviceable, but it hardens. A typical O-ring seal is illustrated in Figure 1.17.

1.7.13 Sight Glasses

It is difficult to imagine that much can be revealed to someone looking through a 6" (15-cm) window into several thousand liters or gallons of murky liquid. Perhaps it's a vestige of the good old days in the lab, or a fascination with watching the action in a front-loading washing machine, but there are a lot of folks who like sight glasses. (We don't; could you tell?) If you're one of them, we suggest you use the circular type with sanitary clamp mounting. These are fairly easy to clean and to replace. The long rectangular types are much more expensive, and require special modifications to make them self-draining.

1.7.14 Drain Valve

The best drain valve available is a flush-mounted diaphragm valve. Figure 1.19 illustrates the design of one that has worked very well for many types of fermentations, including mammalian cell cultures. Designs that include a coupling connected directly to the tank should not be used for any aseptic application. There is no way to keep them free of solids buildup or to keep them really clean.

1.7.15 Steam Locks

A good steam lock assembly should satisfy the following criteria:
(a) All tubing, particularly tubing that enters the vessel, should be kept as short as possible.
(b) All steam inlet and trap lines should be piped so as to eliminate dead spots (usually a result of air pockets).
(c) All surfaces that touch sterile media should be steamed directly. Relying on conduction heat transfer to sterilize surfaces is risky.
(d) The assembly should be designed to allow external pipes (such as lines from addition vessels) to be steamed through their entire lengths at any time after they are connected to the assembly.
(e) The assembly should be easily cleaned.

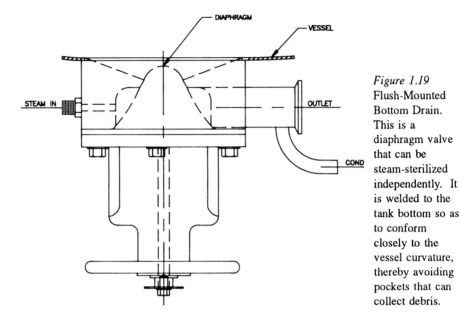

Figure 1.19 Flush-Mounted Bottom Drain. This is a diaphragm valve that can be steam-sterilized independently. It is welded to the tank bottom so as to conform closely to the vessel curvature, thereby avoiding pockets that can collect debris.

(f)   Enough space must be allowed for making and breaking connections easily, and for operation of valves.

(g)   The assembly should be self-draining.

There are several designs that work well, the details varying with manufacturer and specific application. Figure 1.17 illustrates one that we have found to work well for a variety of applications.

1.8   Agitation System Design

1.8.1   Drive Location

The choice between bottom- and top-entering drives is subject to a great deal of controversy. Top-entering drives continue to predominate, but bottom-entry drives have been gaining popularity over the past 15 years. The major arguments of the proponents of top drive are that a top drive poses no threat of a major spill, which might occur through a bottom seal, and that bottom seals are subject to grinding by broth particulates. While it is true that the danger of a leak is greater for a bottom drive than it is for a top drive, the risk of a major spill is so remote as to be an unreasonable consideration. The fact is that if the seal were to fail, the leak rate would be quite small, and corrective maintenance could be applied well before any catastrophic event could occur. With regard to seal grinding: if the seals are designed and maintained properly, this is not a problem. We have installed hundreds of bottom drives over the past 15 years. All of them are still operating, and in no case have we found grinding to be a problem. We also have had no problems with spills resulting from seal failure.

Bottom drives do have some very positive advantages [63]. They have shorter shafts, which means that steady bearings are not necessary in most cases; they have less expensive support structures; they are easier to access and simpler to maintain; and they result in lower overall height. From the point of view of asepsis and of the fermentation *per se*, there is no real difference between top and bottom entry, assuming proper seal design and maintenance. We favor bottom drives solely on the basis of the advantages cited above.

1.8.2   Seals

The best choice is a double mechanical seal [64]. Other types are cheaper, but only in terms of initial cost. We recommend tungsten carbide seats. They are more expensive than the commonly used ceramic and carbon faces, but they give longer seal life and less maintenance. The static shaft seals should be EPDM, except for hydrocarbon fermentations, for which Viton® is recommended.

In the conventional design of double mechanical seals, which derives from the chemical process industries, the seals are mounted back to back [63]. There are several potential problems with this design. The housing pressure for this design must be kept higher than the vessel pressure; therefore, some means is required to maintain the pressure differential under varying conditions. Also, if the top seal leaks, coolant flows into the vessel. Finally, solids can accumulate in the pocket between the vessel flange and the shaft static seal.

The back-to-back design works very well for chemical reactors, but in chemical reactor design asepsis is not a consideration. Figure 1.20 illustrates a design that addresses directly the concern for aseptic operation. Here, the top seal is reversed, and the seal housing is operated at atmospheric pressure. This design eliminates pockets where solids can accumulate, does not allow flow into the vessel, compensates automatically for head pressure variations, and results in lower temperature operation of the seals. Steam flows through the housing during sterilization, and sterile steam condensate flows through it during fermentation to provide lubrication and cooling. The condensate system is illustrated in Figure 1.21. This design is simpler than the one which must be used for back-to-back seals. Obviously, the more complicated design, coupled with the high pressure in the seal housing, increases the likelihood of contaminated coolant entering the fermentor.

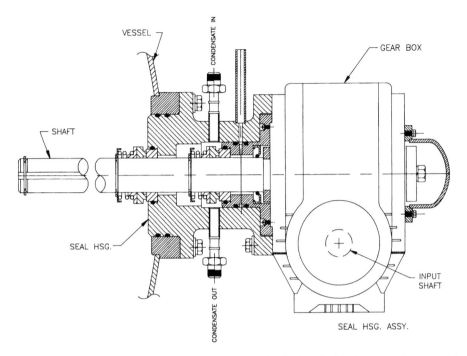

Figure 1.20 Aseptic Mechanical Seal. A double mechanical seal with the top seal reversed is shown. This arrangement eliminates pockets and allows the seal to operate at essentially atmospheric pressure. The arrangement shown is for a bottom-entry agitator, but the same approach may be used for a top entry.

1.8.3   Shaft Vibration

Fermentor vibration can cause serious safety hazards and costly mechanical damage. It is best dealt with by designing it out to the greatest extent practical, and then by performing the best practical construction, testing, and maintenance.

The largest contributor to potentially dangerous vibration is an agitator rotor-bearing system subject to self-induced vibration and bearing instabilities in the design speed range. Minimizing such dynamic instabilities begins with a design that ensures the natural frequency (rotor critical speed) is 50% to 75% higher than the maximum design speed. The stated safety margin usually compensates for secondary effects (such as side loads from hydraulic pumping effects, or decreased damping due to low liquid level), which are not readily predictable, but which tend to reduce the critical speed. In some cases, cost may be used as an argument for a lower safety margin, but the decision to cut costs in that manner should be made with a careful eye to the future.

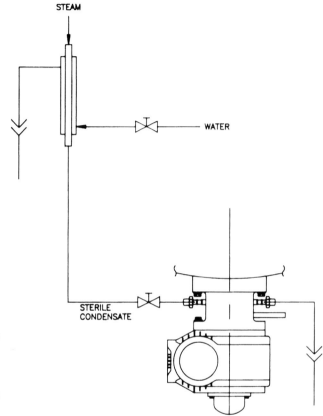

Figure 1.21 Coolant System for Aseptic Seal. This system uses sterile steam condensate to lubricate the seal. It operates at essentially atmospheric pressure.

The details for practical calculation of the critical speed are relatively simple, and depend primarily on whether top or bottom drive is specified, and whether or not a steady bearing is specified. Typical arrangements and nomenclature are illustrated in Figure 1.22. Conventional designs for large fermentors employ top drive and a steady bearing, but both choices are subject to considerable debate in terms of sterility and of mechanical stability.

The rotor critical speed, N_{cr}, can be determined from the shaft stiffness and the total maximum deflection, δ_T, caused by the shaft, impeller, and coupling masses. To a good approximation, the total maximum deflection is equal to the sum of the deflections caused by each mass. If we assume also that the shaft is uniform, that internal shaft couplings have stiffnesses comparable to that of the shaft, and that the couplings are additional masses that can cause shaft deflection, then (see Figures 1.22 and 1.23):

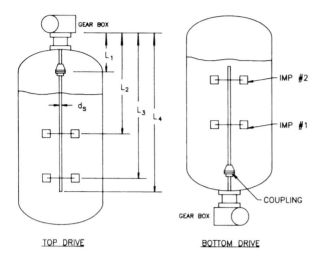

Figure 1.22 Simply Supported Overhung Shafts. Shown are top and bottom entry designs without steady bearings. The dimensions shown are used in eqs 1.44–1.46 for vibration analysis.

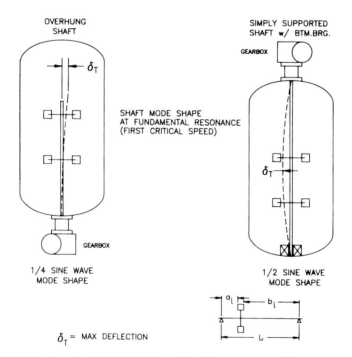

Figure 1.23 Vibration Analysis Definitions. Simply supported (with steady bearing) and overhung (without steady bearing) shafts are shown. Deflections and lengths illustrated are used in eqs 1.44–1.46 for vibration analysis.

$$N_{cr} = (60/2\pi) (g/\delta_T)^{0.5} = 187.7/\delta_T^{0.5} \tag{1.44}$$

where g is the gravitational constant;

δ_T is $(\delta_s + \sum_i \delta_i)$;

δ_s is the deflection of the bare shaft; and

δ_i is the deflection of an impeller or coupling.

Basic beam deflection theory can predict δ_s and δ_i. The following are for the case of a simply supported rotor (the conventional choice for large vessels):

$$\delta_s = (5/384) W_s L/E I_s \tag{1.45}$$

$$\delta_i = W_i a_i^2 b_i^2/(3 E I_i L) \tag{1.46}$$

where W_s is the shaft unit weight, lb/in;

L is the shaft length, in;

E is the modulus of elasticity;

I_s is the shaft moment of inertia, $\{(\pi/64) d_s^4\}$;

d_s is the shaft diameter, in;

W_i is impeller i (or coupling) mass, lb;

a_i and b_i are impeller position variables defined in Figure 1.23; and

I_i is impeller i (or coupling) moment of inertia.

A similar analysis can be done for overhung systems.

Reliable design must be followed by reliable construction, testing, and maintenance, with careful attention to impeller balance, shaft run-out, coupling rigidity, and bearing clearances. Rotor imbalance can cause severe vibration; therefore, all impellers should be balanced statically and dynamically. A bent shaft can behave in a manner similar to that of an imbalanced rotor. For simply supported rotors, where a steady bearing is used (see Figure 1.22), bearing wear results in increased bearing clearance, which can produce significant vibration (subsynchronous). This, in turn, reduces bearing life.

The fermentor should be tested (on the shop floor if at all possible) over its entire speed range at full load, and sources of vibration removed before shipping and permanent installation. After installation, the tests should be repeated to identify and correct sources of vibration that may have arisen during shipping and installation. Finally, things wear out and deform even in a well-designed, well-constructed, well-operated fermentor; therefore, a routine and thorough maintenance program is vital to ensure that a good fermentor doesn't become a quivering, vibrating, leaking piece of junk.

1.8.4   Shaft Orientation

For high-speed agitators used for microbial fermentations, it is best to orient the shaft vertically. For mammalian cell culture, mounting at an angle of about 15° to the vertical is recommended, as described in Section 1.63.

1.9   Design Example

Rational design begins with a design basis that includes all available information having significant influences on the design (such as annual production and annual operating days). The basis is crucial to the success of the design: without a basis, a project is unfocused and unguided.

It is typical that the initial basis has important components missing or cast as best guesses. Distinctions between these conditions must be made clearly. Also, some elements of the initial basis may impose unrealistic constraints, or lead to poor and overly expensive equipment. This is the stuff of which debate and iteration are made, and these are normal parts of the design process.

As stated earlier, the fermentor should be designed giving due consideration to how it fits into the rest of the process. Unfortunately, we cannot take such an approach here, and have to assume that such considerations have been already taken into account.

1.9.1   Design Basis

We assume the following basis:
(a) *Product*: the product is an intracellular compound from a recombinant organism.
(b) *Annual production*: 30,000 kg/yr pure and dry product.
(c) *Fermentation characteristics*:
     Maximum working volume to date: 20 L.
     Maximum cell mass achieved (dry weight): 20 g/L (dw).
     Production cell mass desired: 50 g/L.
     Maximum expression level of product: 0.05 g/g cells (dw).
     Cell yield based on glucose consumed, $Y_{X/S}$: 0.4.
     Cell yield based on oxygen consumed, $Y_{X/O}$: 1.0.
     Sustainable specific growth rate: 0.3 h^{-1} (using present medium at 30 °C and pH 6.5).
     Fermentation medium: salts, technical yeast extract, thiamine, glucose.
     Fermentation mode: all components other than glucose are batch sterilized at the beginning (except thiamine, which is added via sterile filter). Glucose is batch sterilized separately, and is fed so as maintain a fermentor concentration below 5 g/L. At the production scale, the glucose and the medium less glucose are sterilized continuously. They are held in presterilized tanks, and fed according to plant schedule.
     Fermentation temperature: 30 °C.

Fermentation pH: 6.5, held by automatic addition of H_2SO_4 and gaseous ammonia.

Lab fermentation time: 18 h.

Broth rheology: Newtonian; maximum viscosity 2 cp.

Effect of temperature on growth rate: see Figure 1.24.

Effect of temperature on product: none between 22 °C and 32 °C.

Effect of CO_2: none up to total pressure of 4 atm (abs).

Foaming: slight at 1 VVM and 500 rpm (20-L scale).

D.O. (dissolved oxygen): maintained at 30% saturation (relative to atmosphere). Oxygen enrichment required during last two hours. There is no $K_g a$ information available.

Effect of fluid forces: none observed.

Cooling: the broth must be cooled to 4 °C at the end of the fermentation. This should be done within 30 minutes without aeration, in accord with current laboratory practice. There is no information available concerning time/temperature effects on product degradation.

(d)   *Purification efficiency*: 80% overall at laboratory scale.

(e)   *Containment*: BL2 containment will be required.

(f)   *Validation*: the plant must be validatable; it must satisfy CGMPs.

(g)   *Construction*: green field.

(h)   *Operating days*: 330 actual operating days per year; 24 hours per day; 7 days per week.

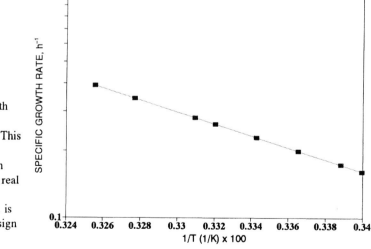

Figure 1.24
Specific Growth
Rate Versus
Temperature. This
is smoothed
specific growth
rate data for a real
recombinant
bacterium, and is
used in the design
calculation.

1.9.2   A Solution

1.9.2.1   Overall Balances and Sizing

Annual production (pure, dry product) = 30,000 kg.
Recovery efficiency = 80%.
Annual fermentor production = 30,000 kg/0.8 = 37,500 kg.
Production of product, g/g cells (dry) = 0.05.
Annual cell production = 37,500 kg/0.05 = 750,000 kg.
Desired cell mass concentration = 50 g/L.
Annual fermentation volume = 750,000 × 1000/50 = 15,000,000 L.
Total cycle time for a reasonable plant schedule = 24 h.
Fermentation time = 16 h.
Turnaround time = 8 h.
Operating days/year = 330.
Cycles/fermentor·year = 330.
Fermentor cycles/day = 1 × number of fermentors.
Fermentation volume/day = 15,000,000/330 = 45,450 L.

The production schedule can be met by one fermentor with a working volume of 45,450 L. This has the lowest fermentor capital cost, but it leads to a large recovery system that is idle a good deal of the time, and it offers no distribution of risk. A better choice is 2 fermentors each with a working volume of 22,725 L. The fermentor capital cost is more than for one large vessel, but the overall and long-term costs are lower. Depending on the recovery process details and other plant operating considerations, there may be good arguments for using 3 (for example, a schedule based on harvests every 8 hours would be very convenient) or perhaps 4 vessels. We settle on 2, however; therefore, the plant schedule is based on a harvest every 12 hours.

If the fermentation is run at 30 °C, the specific growth rate is 0.3 h^{-1}, and the maximum oxygen- and heat-transfer (metabolic only) loads are:

OTR = (0.3) (50) (1000/32) = 469 mmol/L·h.

Q = (0.12) (469) (4) (22,725) = 5.12 × 10^6 Btu/h.

These are unrealistic targets; therefore, we must consider alternatives that allow us to meet the yearly production specified. There are several ways to proceed: one very useful way to begin is to perform preliminary calculations for vessel sizing and for maximum power, oxygen transfer, and heat transfer.

1.9.2.2   Preliminary Sizing Calculations

It is very important that major sizing calculations for long–lead-time items (such as the vessel and gear box) be done as soon as possible. Delivery times for such items can be 6 months or longer.

We begin with the following constraints (see previous discussions):
(a) We will use a standard dish diameter.
(b) Vessel pressure will not exceed 30 psig.
(c) We will not use oxygen enrichment.
(d) Gas velocity will not exceed 180 cm/min (at the temperature and pressure in the vessel).
(e) Horsepower will not exceed 2.5 hp/100 gal.
(f) d_i/d_t will not exceed 0.45. The impeller must fit through largest standard manway that can be put on the vessel. It is desirable that the impeller fit whole through the manway.
(g) The coolant temperature must not be less than 35 °F.
(h) Internal coils will not be used if at all possible.

All the calculations were done using equations presented, along with the following:
(a) Dish depth and volume were taken from standard tables [55].
(b) Turbines were assumed to have twelve blades, and to have a power number of 9.6 [65].
(c) Calculated tangent heights were rounded to whole numbers.
(d) The gassed horsepower requirements were calculated using:

$$P_g = (K_g a/244)^{1.95} V_1 \tag{1.47}$$

(where P_g is in units of hp), which we have found to be reasonably reliable for similar applications under similar conditions. The reader is cautioned that this correlation is applicable over a fairly narrow range, and should not be used when conditions vary significantly from those considered here.
(e) The power was assumed to be distributed equally among the turbine impellers, and the unloading factor was taken to be the same for all the turbines. A marine propellor was used (top) to improve top-to-bottom mixing.
(f) No oxygen-transfer credits were taken for v_s, or for any power input to the axial-flow impeller.
(g) The jacket was assumed to extend over the full tangent height. Note that if this is done in cases in which batch sterilization is performed, the jacket must be zoned to avoid medium bake-on during sterilization, as well as to decrease the coolant pressure drop across the jacket.

Table 1.3 presents the results for five trials at five different sets of assumed operating conditions. The results reveal a sad story: we can't get above 300 mmol/L·h without taking heroic measures. We're not going to be able to satisfy all the constraints listed above. Much as it pains us, we're going to have to make some substantial compromises. We can get a clearer picture of the compromises we'll have to make by calculating projected fermentation histories, including oxygen- and heat-transfer rates.

Table 1.3 Initial Sizing Results for the Design Example

Total volume, m³	32.50	32.50	32.50	32.50	32.50
Liquid volume, m³	22.73	22.73	22.73	22.73	22.73
Tank OD, in	108	108	108	108	108
Adj. tan. ht., in	198	198	198	198	198
Unaer. liquid ht., in	162	162	162	162	162
Unaer. liq. ht./diam.	1.51	1.51	1.51	1.51	1.51
Total ht./diam.	2.20	2.20	2.20	2.20	2.20
Adj. tot. vol., m³	32.52	32.52	32.52	32.52	32.52
Percent fill	70	70	70	70	70
OTR, mmol/L·h	300	300	300	300	300
v_s, cm/min	180	180	180	180	180
Pressure, psig	35	40	15	30	30
O_2 mole frac. in	21	21	37	40	30
O_2-transfer eff., %	33.2	30.3	30.4	19.2	25.7
Req'd. hp/100 gal	4.7	3.7	3.7	1.2	2.4
Tot. gassed hp req'd.	283	223	225	75	147
Total gas flow, std. m³/min	36.5	40.0	22.6	33.0	33.0
Oxygen flow, std. m³/min	0	0	4.6	7.9	3.8
Aerated liq. ht., in	221	212	212	192	204
No. of turbines	2	2	2	2	2
No. of axial impellers	1	1	1	1	1
Shaft speed, rpm	131	122	123	88	108
Turbine diam., in	46	46	46	46	46
Axial impeller diam, in	36	36	36	36	36
Turbine d_i/d_t ratio	0.43	0.43	0.43	0.43	0.43
Turbine tip speed, fpm	1578	1469	1481	1060	1301
Tot. ungassed hp	594	480	491	180	333
Gassed hp (turbine)	283	223	225	75	147
Heat load, Btu/h × 10⁶	3.968	3.812	3.819	3.425	3.624
U, Btu/h·ft²·°F	150	150	150	150	150
Coolant temp., °F	35	35	35	35	35
Coolant flow, gpm	200	200	200	200	200
Avail. jkt. area, ft²	457	446	446	399	425
Coil area req'd., ft²	568	489	493	354	415
Coil pipe diam., in	4	4	4	4	4
Pipe length, in	536	489	493	354	415
Coil spacing, in	4.5	5.5	5.5	8.0	6.5
Coil ht., in	220	211	212	191	198

1.9.2.3   Projected Fermentation Histories

One way to reach the target of 50 g/L in 16 hours, and still have a rational design, is to slow down the fermentation. This can be accomplished by lowering the temperature so as to keep the OTR in check, by changing the medium composition, by changing the pH, or by several other techniques.

The only information we have about such changes is that we can decrease the fermentation temperature to 22 °C without having a significant effect on the product. We'll take this approach because the others require additional research, which is not going to make the investors happy. Note that slowing down the fermentation requires a larger inoculum, which is not a major problem in this case.

The calculations require the use of several material balances and an equation that describes the growth kinetics as a function of temperature. Due to space limitations, we present here only the equations. The derivations are reasonably straightforward, and can be done easily by any genius having 10–20 years of experience in the business. Please note that the following assumptions were used in deriving the equations:

(a)   Time zero is taken as the time at which the tank has been filled to the initial volume less inoculum.

(b)   Inoculation is made at time zero. The volume is now V_0.

(c)   The specific growth rate, μ, obeys a simple Arrhenius relationship:

$$\mu = a_3 \exp(-E_a/R\, T_{abs})\tag{1.48}$$

where a_3 is a constant;
 E_a is the activation energy;
 R is the gas constant; and
 T_{abs} is the absolute temperature.
 The values of a_3 and E_a/R are obtained from Figure 1.24.

(d)   The concentration of sugar in the fermentor remains constant.

The material balance and kinetic equations are, up to the time when the maximum OTR is reached:

$$V = V_0 + (X\,V - X_0\,V_0)/Y_{X/S}\,(S_f - S)\tag{1.49}$$

where S_f is the substrate concentration in the feed, g/L; and
 S is the substrate concentration in the fermentor, g/L.

$$\mu = \mu_m = \text{constant}\tag{1.50}$$

$$X\,V = X_0\,V_0 \exp(\mu_m\, t)\tag{1.51}$$

To simplify the rearrangements, we write:

$$X = (X\,V)/V\tag{1.52}$$

$$\text{OTR} = \mu_m X \tag{1.53}$$

$$T_{abs} = 303 \text{ K (30 °C, constant)} \tag{1.54}$$

The maximum OTR is reached at

$$t = t_m = (1/\mu_m) \ln \frac{Y_{X/S} V_0 (S_f - S) - X_0 V_0}{X_0 V_0 [\mu_m Y_{X/S} (S_f - S)/\text{OTR} - 1]} \tag{1.55}$$

At t_m,

$$V_m X_m = V_0 X_0 \exp (\mu_m t_m) \tag{1.56}$$

and

$$V = V_m = V_0 + (V_m X_m - V_0 X_0)/Y_{X/S} (S_f - S) \tag{1.57}$$

Beyond t_m, the temperature is decreased steadily so as to keep the OTR constant. The following equations apply:

$$XV = V_m Y_{X/S} (S_f - S) [\exp \{Y_{X/O} (\text{OTR}) (t - t_m)/Y_{X/S} (S_f - S)\} - 1] + X_m V_m \tag{1.58}$$

$$V = V_m + (X V - X_m V_m)/Y_{X/S} (S_f - S) \tag{1.59}$$

$$\text{OTR} = 300 \text{ mmol/L·h} \tag{1.60}$$

$$\mu = \text{OTR}/X \tag{1.61}$$

$$T(°F) = 1.8 [6159.9/(19.118 - \ln \mu)] \tag{1.62}$$

$$Q_{tot} = 0.12 (\text{OTR}) V + 2450 P_g + V \rho c_p (T_{j+1} - T_j)/(t_{j+1} - t_j) \tag{1.63}$$

$$W_c = Q_{tot}/c_{pc} (T_f - T_a - T_i) \tag{1.64}$$

where c_p is the broth heat capacity;
   T_{j+1} is the broth temperature at current time (t_{j+1}), °F;
   T_j is the broth temperature at previous time (t_j), °F;
   W_c is the coolant flow rate; and
   c_{pc} is the coolant heat capacity;
   T_f is the temperature of the fermentation, °F;
   T_a is the approach, °F; and
   T_i is the inlet temperature of the coolant, °F.
Note that the heat-transfer equation (1.63) accounts for cooling as well as for metabolic heat and mechanical agitation.
   The results presented in Table 1.4 show that we can stay within the 22 °C limit by taking a maximum OTR of 300 mmol/L·h, but we still can't satisfy all of the original constraints: something or someone has to give. Among the possible alternatives are:

Table 1.4 Projected Fermentation History for the Design Example

A. Overall characteristics

Specific growth rate, μ, h^{-1}	0.30	Max. OTR, mmol/L·h	300
Yield coeff., g/g	0.40	Time OTR max., h	12.7
Feed glucose, g/L	250	Cells at max. OTR, g	528,873
Broth glucose, g/L	5	Vol. at max. OTR, L	16,527
Final volume, L	22,725	Cells at max. OTR, g/L	32.0
Final cell conc., g/L	50	Inoc. cells, g	11,586
Initial cell conc., g/L	1.03	Inoc. volume, L	1,500
Initial volume, L	11,249	Inoc. conc., g/L	7.7
Glucose feed, L	11,476	Nutrient set vol., L	9,749
O_2 yield coeff., g/g	1.0		

B. Detailed History

Time h	Volume L	Cells g/L	OTR mmol/L·h	μ h^{-1}	Temp. °F	Ht. load Btu/h × 10^6
0.0	11,249	1.0	9.7	0.30	86.0	0.24
1.0	11,290	1.4	13.0	0.30	86.0	0.26
2.0	11,346	1.9	17.4	0.30	86.0	0.29
3.0	11,421	2.5	23.4	0.30	86.0	0.32
4.0	11,523	3.3	31.3	0.30	86.0	0.37
5.0	11,660	4.5	41.7	0.30	86.0	0.43
6.0	11,846	5.9	55.5	0.30	86.0	0.51
7.0	12,096	7.8	73.3	0.30	86.0	0.63
8.0	12,434	10.3	96.3	0.30	86.0	0.78
9.0	12,890	13.4	125.4	0.30	86.0	0.99
10.0	13,505	17.2	161.5	0.30	86.0	1.27
11.0	14,336	21.9	205.4	0.30	86.0	1.65
12.0	15,458	27.4	257.2	0.30	86.0	2.17
13.0	16,960	33.7	300.0	0.29	84.6	2.73
14.0	18,705	39.7	300.0	0.25	80.2	3.01
15.0	20,630	45.1	300.0	0.21	76.9	3.32
16.0	22,753	50.1	300.0	0.19	74.2	3.70

(a)   Stick with the 50 g/L and the 16-hour fermentation time. If we do this, we'll have to go back to the vessel calculations to find a combination of specifications that causes as little pain as possible.

(b)   Back off on the 50 g/L and/or the 16-hour fermentation time. Either or both of these leads to a larger fermentation volume and/or larger purification equipment.

(c)   Find another way to slow down the fermentation while retaining the 50 g/L and 16-hour requirements. This requires not only more research but also a larger inoculum.

(d)   Use a larger number of smaller vessels.

At this point a conference with the client is in order. We assume here that this conference results in the following conclusions:

(1)   Alternative (d) is not acceptable to the client because he feels it adds too much complexity at this point.

(2)   Alternative (c) is not acceptable because it will take too much time.

(3)   Alternative (b) is not acceptable because it increases the capital cost too much. There also will be an increase in operating costs.

(4)   Alternative (a) is acceptable with the concession that oxygen enrichment up to 30% is allowed. It is recognized that this increases the operating costs, but the client feels that he will be able to swallow the increase. He will pray for subsequent process improvements that eventually will eliminate the need for oxygen enrichment.

The client also is comfortable with cooling down the fermentation to limit the oxygen- and heat-transfer requirements, but does not want internal coils.

We now go back to the vessel sizing calculations.

1.9.2.4   Sizing Calculations Revisited

Table 1.5 contains the results of the next iteration of sizing calculations. The results show that we can reach the required OTR by using 30% oxygen inlet, a maximum pressure of 30 psig, and a gassed horsepower of 147 (2.44 hp/100 gal). We also can retain our initial vessel sizing, but we'll have to use −5 °F coolant. This choice, while necessary to satisfy the constraints, gives us cause for concern because of the possibility of pipe and valve freezeups, and a higher chiller cost. These concerns will be transmitted to the client, along with a plea to reconsider the other options. But because we've already run on too long, we'll assume that the client has plenty of determination, and sticks to his original decision.

1.9.2.5   Broth Cooldown

The original basis called for the broth to be cooled to 4 °C unaerated within 30 minutes of the end of the fermentation. Table 1.6 presents the results of cooldown calculations based on full use of all of the cooling capacity available. Again, it's bad news. The minimum cooling time is 50 minutes. To decrease this time, and still do

Table 1.5 Second Iteration of Sizing Results for
the Design Example

	Case 1	Case 2
Total volume, m^3	32.50	32.50
Liquid volume, m^3	22.73	22.73
Tank OD, in	108	108
Adj. tan. ht., in	198	198
Unaer. liquid ht., in	162	162
Unaer. liq. ht./diam.	1.51	1.51
Total ht./diam.	2.20	2.20
Adj. tot. vol., m^3	32.52	32.52
Percent fill	70	70
OTR, mmol/L·h	300	300
v_s, cm/min	180	180
Pressure, psig	30	30
O_2 mole frac. in	30	30
O_2 transfer eff., %	25.7	25.7
Req'd. hp/100 gal	2.44	2.44
Tot. gassed hp req'd.	147	147
Total gas flow, std. ft^3/min	33.0	33.0
Oxygen flow, std. ft^3/min	3.8	3.8
Aerated liq. ht., in	204	204
No. of turbines	2	2
No. of axial impellers	1	1
Shaft speed, rpm	108	108
Turbine diam., in	46	46
Axial impeller diam., in	36	36
Turbine d_i/d_t ratio	0.43	0.43
Turbine tip speed, ft/min	1301	1301
Tot. ungassed hp	333	333
Gassed hp (turbine)	147	147
Heat load, Btu/h × 10^6	3.624	3.624
U, Btu/h·ft^2·°F	150	150
Coolant temp., °F	35	−5
Coolant flow, gpm	200	125
Avail. jkt. area, ft^2	425	425
Coil area req'd., ft^2	415	0
Coil pipe diam., in	4	−
Pipe length, in	536	−
Coil spacing, in	4.5	−
Coil ht., in	220	−

Table 1.6 Fermentor
Cooldown in the
Design Example

Time	Temperature
min	°F
0	74.2
5	69.7
10	65.4
15	61.4
20	57.6
30	50.7
40	44.5
50	39.0

all of the cooling in the fermentor, would require a lower coolant temperature, more transfer area, a higher coolant flow rate, or some combination of these. An alternative (although not ideal) is to put a heat exchanger between the fermentor and the next processing step (membrane filtration, in this process). The client is convinced that this is the more rational approach, recognizing that about half of the broth will be at a temperature between 10 °C and 4 °C for about 30 minutes longer than desired (depending on the actual fermentor discharge rate).

1.9.3   Mechanical Design

The mechanical design of the fermentor is based on the operating conditions given as Case 2 in Table 1.5.

1.9.3.1   Vessel

The specifications below are not complete, but enough has been given to provide the reader a very good idea of what is necessary.

(a)   Vessel dimensions: as in Table 1.5, Case 2.
(b)   Vessel pressure: 40 psig/full vacuum.
(c)   Vessel temperature: 300 °F.
(d)   Finish: 320 grit interior; 180 grit exterior.
(e)   Baffles: 4 at 90°; 184" long by 9.6" wide; welded to wall and set off 2"; 320 grit finish.

(f)   Jacket: 198" long, dimple; 3 zones, 50 gpm max per zone; nonchloride-bearing insulation; SS304 clad; 60 psig/full vacuum; −20 °F–300 °F.

(g)   Manway: 22" diameter; O-ring seal; drop-bolt fastened; 320 grit interior finish; 180 grit exterior finish.

(h)   Bottom drain valve: 2" flush-mounted diaphragm valve.

(i)   Static seals: all O rings (EPDM).

(j)   Sight glass/light port: mounted on manway cover.

(k)   Rupture disk: 40 psig; mounted on top head.

(l)   Sparger: ring type, 36" diameter, 40 holes of 0.250" diameter. A 15-psi pressure drop will give full design flow (approximately 38.5 m³/min at 30 psig vessel head pressure).

(m)   Provisions will be made for a CIP system.

(n)   Provisions will be made for 2 aseptic sampling ports.

(o)   Nozzles: acid addition, base addition, antifoam addition, sugar addition, nutrient addition, temperature probes (2), pH probes (2), foam probes (2), sample ports (2), light port, sight port, CIP port, rupture disk, pressure sensor, pressure gauge, drain valve, sparger inlet, and drive flange.

The specification sheet, a vessel drawing, and the agitation system specifications (see below) are sent to the vessel vendor (a preliminary quote already having been obtained). In all likelihood, the vendor returns a drawing and specs that differ slightly from those given above, but these changes probably do not have a significant effect.

1.9.3.2 Agitation System

(a)   Mounting: bottom mounted; vessel vendor to specify details.

(b)   Drive motor: 200 hp, 1780 rpm, 480 volt, 3-phase induction motor; variable-frequency speed controller.

(c)   Gear box: 144 kW, 12.33 speed ratio, 1800 rpm/146 rpm, 14 cm quill.

(d)   Shaft: 4.5" diameter, 174" long; 320 grit finish. With this shaft, the natural frequency is at least 50% greater than the design speed of 108 rpm.

(e)   Impellers: two 46", 12-blade disk turbines and one 36" A-315; Impellers to be mounted approximately 55" apart; all impellers to have 320 grit finish. Impellers must be split to fit through the manway.

(f)   Seals: double mechanical with carbon tungsten on tungsten carbide.

(g)   Steady bearing: none.

1.10   Conclusion

The final design is based on a series of compromises. It does not satisfy all of the original specifications, but it meets the plant production requirements. The important thing to note, however, is that it is based on a rational analysis that considers the fermentor as part of a system, and on sound engineering judgment born of both design and operating experience. The reader should note also that space and scope limitations have prevented us from including explicitly some important details (for example, complete timing of all the process equipment) that were considered in our analysis. Finally, the critical reader may have noticed that some of our final results do conform with some of the magic numbers and myths we lambasted so roundly at the outset. We'll attribute some of that to punishment for (perhaps) a bit too much hubris.

1.11   References

1. WINKLER, M. A.; Applications of the Principles of Fermentation Engineering to Biotechnology; In: *Principles of Biotechnology*; Wiseman, A., Ed.; Surrey University Press, London, U.K., 1984.

2a. CRUEGER, W., CRUEGER, A.; *Biotechnology: A Textbook of Industrial Microbiology*; Science Tech, Inc., Madison, WI, 1984; pp 54–98.

2b. *Ibid.*, p 80.

3a. ATKINSON, B., MAVITUNA, F.; *Biotechnology Engineering and Biotechnology Handbook*; Nature Press, New York, 1983; pp 580–669.

3b. *Ibid.*, pp 838–887.

4. BU'LOCK, J., KRISTIANSEN, B.; *Basic Biotechnology*; Academic Press, New York, 1987; pp 168–172.

5a. WANG, D. I. C., COONEY, C. L., DEMAIN, A. L., DUNILL, P., HUMPHREY, A. E.; *Fermentation and Enzyme Technology*; John Wiley and Sons, New York, 1979; pp 194–211.

5b. *Ibid.*, pp 173–184.

5c. *Ibid.*, pp 184–192.

5d. *Ibid.*, pp 46–56.

5e. *Ibid.*, pp 167–170.

6. AIBA, S., HUMPHREY, A. E., MILLIS, N. F.; *Biochemical Engineering*; 2nd ed., Academic Press, New York, 1973; pp 195–217.

7. BAUER, S., ZIV., E., *Biotechnol. Bioeng. 18*, 81 (1976).

8. BAUER, S., WHITE, M., *Biotechnol. Bioeng. 18*, 839 (1976).

9. BLANCH, H. W.; Aeration; In: *Annual Reports in Fermentation*; Vol. 3, PERLMAN, D., Ed.; Academic Press, New York, 1979.

10a. WELTY, J. D., WICKS, C. E., WILSON, R. E.; *Momentum, Heat and Mass Transfer*; John Wiley and Sons, New York, 1984; pp 669–721.

10b. *Ibid.*, pp 96–110.

11. KAWASE, Y., HALARD, B., MOO YOUNG, M., Liquid Phase Mass Transfer Coefficients in Bioreactors, *Biotechnol. Bioeng. 39*, 1133–1140 (1992).

12. VASCONCELOS, J. M. T., ALVES, S. S., Direct Dynamic k_1a Measurement in Viscous Fermentation Broths: The Residual Gas Holdup Problem, *Chem. Eng. J. 47*, B35–B44 (1991).

13. YASUKAWA, M., ONODERO, M., YAMAGIWA, K., OHKAWA, A., Gas Holdup, Power Consumption and Oxygen Absorption Coefficient in a Stirred Tank Reactor With Foam Control, *Biotechnol. Bioeng. 38*, 629–636 (1991).

14. JOHNSTONE, R. E., THRING, W. M.; *Pilot Plant, Models and Scale-up in Chemical Engineering*; McGraw-Hill, New York, 1957; pp 12–26.

15. HIMMELBLAU, D. M.; Mathematical Modeling; In: *Mathematical Modeling in Scale Up of Chemical Processes*; BISIO, A., KABEL, D. L., Eds.; John Wiley and Sons, New York, 1985.

16. OLDSHUE, J. Y.; Mixing Processes; In: *Mathematical Modeling in Scale Up of Chemical Processes*; BISIO, A., KABEL, D. L., Eds.; John Wiley and Sons, New York, 1985.

17. MURPHY, T. F., WALD, S. A., JACKSON, J. V.; *Oxygen Transfer in a 100,000 Liter Fungal Fermentation*; paper presented at the 188th National ACS Meeting, Philadelphia, August 30, 1984; American Chemical Society, Washington, DC.

18. DICKEY, D. S.; *Turbine Agitated Gas Dispersion*; paper presented at the 72nd AIChE Annual Meeting, San Francisco, 1979; American Institute of Chemical Engineers, Washington, DC.

19. JOBSES, I., MARTENS, D., TRAMPER, J., Lethal Events During Gas Sparging in Animal Cell Culture, *Biotechnol. Bioeng. 37*, 484–490 (1991).

20. HU, W. S., WANG, D. I. C.; Mammalian Cell Culture Technology: A Review From an Engineering Perspective; In: *Mammalian Cell Technology*; THILLY, W. G., Ed.; Butterworths, Stoneham, MA, 1986.

21. GRIFFITHS, J. B.; Overview of Cell Culture Systems and Their Scale-up; In: *Animal Cell Biotechnology*; Vol. 3; SPEAR, R. E., GRIFFITHS, J. B., Eds.; Academic Press, New York, 1988.

22. MACMILLAN, J. D., VELEZ, D., MILLER, L., REUVENY, S.; Monoclonal Antibody Production in Stirred Reactors; In: *Large Scale Cell Technology*; LYDERSEN, B. K., Ed.; Carl Hanser Verlag, New York, 1987.

23a. OLDSHUE, J. Y.; *Fluid Mixing Technology*; McGraw-Hill, New York, 1983; pp 51–53.

23b. *Ibid.*, p 264.

23c. *Ibid.*, pp 278–294.

23d. *Ibid.*, pp 12–22.

23e. *Ibid.*, pp 264–266.

24a. NAGATA, S.; *Mixing: Principles and Applications*; John Wiley, New York, 1975; pp 1–83.

24b. *Ibid.*, pp 85–98.

25. HICKS, R., KIME, D. L., *Chem. Eng. 83* (July 19), 141–150 (1976).

26. TOMA, M. K., RUSHLISHA, M. P., VANAGAS, J. J., ZELTINA, M. O., LEITE, M. P., GALININA, N. I., VIESTURS, U. E., TENGERDY, R. P., Inhibition of Microbial Growth and Metabolism by Excess Turbulence, *Biotechnol. Bioeng. 38*, 552–556 (1991).

27. SMITH, J. J., LILLY, M. D., FOX, R. I., The Effect of Agitation on the Morphology and Penicillin Production of *Penicillium chrysogenum*, *Biotechnol. Bioeng. 35*, 1011–1023 (1990).

28. BELMAR-BEINY, M. T., THOMAS, C. R., Morphology and Cavulanic Acid Production of *Streptomyces clavuligerus*: Effect of Stirrer Speed in Batch Fermentations, *Biotechnol. Bioeng. 37*, 456–462 (1991).

29. NOLLERT, M. U., DIAMOND, S. L., McINTIRE, L. U., Hydrodynamic Shear Stress and Mass Transport Modulation in Endothelial Cell Metabolism, *Biotechnol. Bioeng.* 38, 588–602 (1991).

30. METZNER, A. B., OTTO, R. E., *AIChE J.* 3, 3 (1957).

31. CHAPMAN, F. S., DALLENBACH, H., HOLLAND, F. A., *Trans. Inst. Chem. Eng.* 42, T398 (1964).

32. RADEUZ, I., HUDCOVA, V., KOLOINI, T., Heat Transfer in Aerated and Non-Aerated Mycelial Fermentation Systems in Stirred Tank Reactors, *Chem. Eng. J.* 46, B83–B91 (1991).

33a. MIDDLEMAN, S.; *The Flow of High Polymers*; Interscience, New York, 1968; pp 1–12.

33b. *Ibid.*, pp 8–9.

33c. *Ibid.*, pp 13–83.

33d. *Ibid.*, pp 84–131.

34. BENNETT, C. O., MEYERS, J. E.; *Momentum, Heat and Mass Transfer*; McGraw-Hill, New York, 1974; pp 155–158.

35. FOUST, A. S., WENZEL, L. A., CLUMP, C. C., MAUS, L., ANDERSON, L. B.; *Principles of Unit Operations*; John Wiley, New York, 1980; pp 232–245.

36. VAN WAZER, J. R., LYONS, J. W., LIM, K. Y., COWELL, D. E.; *Viscosity and Flow Measurement*; Interscience, New York, 1963.

37a. CHARLES, M., *Adv. Biochem. Eng.* 8, 1–62 (1978).

37b. *Ibid.*, p 29.

38. METZ, B., KOSSEN, N. W. F., VAN SUIJDAM, J. C., *Adv. Biochem. Eng.* 11, 103–156 (1979).

39. SKELLAND, A. H. P.; *Non-Newtonian Flow and Heat Transfer*; John Wiley, New York, 1967; pp 14–21.

40. KAWASE, Y., KUMAGAI, T., Apparent Viscosity for Non-Newtonian Fermentation Media in Bioreactors, *Bioprocess Eng.* 7(1), 25–28 (1991).

41. DEINDOERFER, F. H., GADEN, E. L., *Appl. Microbiol.* 3, 253 (1955).

42. DEINDOERFER, F. H., WEST, J. M., *Biotechnol. Bioeng.* 2, 165 (1960).

43. RICHARDS, J. W., *Prog. Ind. Microbiol.* 3, 141 (1961).

44. LEONG-POI, L., ALLEN, D. G., Direct Measurement of the Yield Stress of Filamentous Fermentation Broths with the Rotating Vane Technique, *Biotechnol. Bioeng.* 40, 403–412 (1992).

45. TAGUCHI, H., IMANAKA, J., TERAMOTO, S., TAKASATA, M., SATO, M., *J. Ferment. Technol.* 46, 823 (1968).

46. MOO YOUNG, M., TICHAR, K., DULLEN, F. A., *AIChE J.* 18, 178 (1972).

47. BLAKEBROUGH, N., SAMBAMURTHY, K., *Biotechnol. Bioeng.* 8, 25 (1966).

48. OLDSHUE, J. Y.; Agitation; In: *Fermentation and Biochemical Engineering Handbook*; VOGEL, H. C., Ed.; Noyes Pub., Park Ridge, NJ, 1983.

49. OLDSHUE, J. Y., *Chem. Eng. Prog.* 85 (May), 33 (1989).

50. BEHMER, G. J., MOORE, I. P. T., NIENOW, A. W., *Biotechnol. Proc.* 4, 116 (1988).

51. GBEWONYO, K., DiMASI, D., BUCKLAND, B. C.; *The Use of Hydrofoil Impellers to Improve Oxygen Transfer Efficiency in Viscous Mycelial Fermentation Broths*; Paper #21, presented at the Int'l. Conf. on Bioreactors and Fluid Dynamics; BHRA, The Fluid Engineering Centre, Bedford, UK, 1986.

52. BADER, F. G., *Biotechnol. Proc.* 4, 96–106 (1988).

53. YUAN, S., RYU, D. Y., PARK, S. H., Performance Of Mammalian Cell Culture Bioreactor With a New Impeller Design, *Biotechnol. Bioeng. 40*, 260–270 (1992).
54. SEVITZ, J., LAPORTE, T., STINNET, T.; Production of Viral Vaccines in Stirred Bioreactors; In: *Viral Vaccines*; MIZRAHI, A., Ed.; Wiley-Liss, New York, 1990.
55. *TH 5/89*; Tank Head and Components Division, Precision Stainless, Inc., St. Louis, MO, 1989.
56. YOKELL, S., *Chem. Eng. 93* (May 12), 75 (1986).
57. HARVEY, J. F.; *Theory and Design of Modern Pressure Vessels*; Van Nostrand Reinhold, New York, 1974.
58. DILLON, C. P., RAHOI, D. W., TUTHILL, A. H., Stainless Steel for Bioprocessing, Part 2: Classes of Alloys, *BioPharm 5(4)*, 32–35 (1992).
59. DILLON, C. P., RAHOI, D. W., TUTHILL, A. H., Stainless Steel for Bioprocessing, Part 3: Corrosion Phenomena, *BioPharm 5(5)*, 40–44 (1992).
60. BJURSTROM, E., COLEMAN, D., *BioPharm 1* (Nov.), 22 (1988).
61. SEIBERLING, D. A.; Clean-In-Place and Sterilize-In-Place; In: *Aseptic Pharmaceutical Manufacturing*; OLSON, W. P., GROVES, M. J., Eds.; Interpharm Press, Prairie View, IL, 1987.
62. PERKOWSKI, C. A., *Biotechnol. Bioeng. 26*, 857 (1984).
63. WILSON, J. D., ANDREWS, T. E., *Biotechnol. Bioeng. 25*, 1205 (1983).
64. RAMSEY, W. D., ZOLLER, G. C., *Chem. Eng. 83* (Aug. 30), 101 (1976).
65. RUSHTON, J. H., COSTICH, E. W., EVERETT, H. J., *Chem. Eng. Prog. 46*, 395 (1950).

Large-Scale Animal Cell Culture

Gary L. Smith

Contents

2.1   Introduction

Animal cell culture technology has had a long and rich tradition in industry, especially in the production of animal and human vaccines, where such efforts have already reached impressively grand scales to accomodate commercial production [1]. It has more recently become apparent that this technology will also be of value in the pharmaceutical industry, where conventional methods of manufacturing therapeutic proteins through extraction of tissues or fluids will no longer continue to be viable processes for the future. The use of recombinant DNA technology to genetically engineer the common bacterium, *Escherichia coli*, to produce human insulin and growth hormone led to increased optimism and speculation that genetically engineered microorganisms could manufacture all human therapeutic proteins. While the use of microorganisms in this manner continues to be a profitable venture for the production of small, simple proteins, it is currently not possible to use microorganisms as a source for all large, complex proteins such as enzymes, viral antigens, erythropoietin, factor VIII, t-PA or monoclonal antibodies [2].

Microorganisms do not possess the complex biosynthetic machinery to carry out the posttranslational modifications of proteins that are seen in animal cell systems. Bacteria-produced animal proteins are often denatured, and lack the appropriate three-dimensional configuration necessary for biological activity. Nor are these organisms able to secrete human proteins into the culture medium in an active form; they therefore require physical disruption or extraction, followed by complex folding reactions to generate the active molecule. Mammalian proteins of pharmaceutical interest are usually glycosylated, and may have additional modifications, such as phosphorylations or carboxylations, which microorganisms are inherently unable to perform. These modifications may be important to, or required for, the biological activity of the protein. They may also play regulatory roles, or function to control the biological half-life or clearance patterns of the native molecule.

These limitations of genetically engineered microbial systems for the production of pharmaceutically important proteins emphasize the role that animal cell culture technology will play in the emerging future of biotechnology. Continuing advances in recombinant DNA and hybridoma methodologies require that large-scale animal cell culture technology also keep pace through the development of safe, reliable, efficient, and environmentally sound manufacturing practices.

2.2   Animal Cells and Bioreactor Technology

The most striking characteristic of animal cells, which makes their large-scale culture fundamentally different from conventional microbial fermentation, is their fragility. These cells are easily damaged by mechanical stress, and cannot be cultured under conditions of high aeration and agitation that are the hallmarks of turbulent microbial fermentations. Animal cells are considerably larger than their microbial counterparts, and do not possess a rigid, highly resistant cell wall to protect them from the shear-generating environment of a rapidly stirred and aerated tank. Fortunately, animal cells grow slowly in batch systems, do not attain the high cell densities (biomass) seen in microbial systems, have a comparatively low oxygen uptake rate, and therefore do not require the high oxygen inputs characteristic of microbial processes. Thus the inherent fragility of animal cells does not pose any real barrier under moderate agitation and aeration conditions.

Initial considerations of the culture technology of choice for any given animal cell line require examination of many characteristics of the cell line in question. These include growth pattern and kinetics, product stability, sensitivity to mechanical damage, culture medium requirements, and oxygenation demands.

The most basic growth pattern determination is whether the cell line is capable of proliferating in suspension culture, or is anchorage-dependent and requires a solid substrate on which to grow. Providing a solid surface for the growth of anchorage-dependent cells may mean that consideration be given to more sophisticated and nontraditional technologies [3, 4]. These methods will be discussed later in this chapter. For excellent review articles on large-scale process development in mammalian cell culture technology, see those by BLIEM [5], NELSON [6, 7] and ARATHOON and BIRCH [8].

Animal cells that are capable of being propagated in suspension culture have been the most attractive to industry for a variety of reasons, with monoclonal antibody production by hybridoma cells receiving most attention. Such cells have been cultured in large stirred tank reactors for many years. Stirred reactor technology for animal cell culture has benefited from the experience and knowledge gained from the traditional and reliable fermentation industry. Stirred tanks are perceived as being reliable, well understood, and easily scaled up; experience and confidence can facilitate operations in a pilot plant or production setting. Stirred suspension cultures are amenable to various kinds of operations, including batch culture, fed-batch, semicontinuous, or continuous perfusion. Indeed, stirred tanks can support the growth both of suspension cells and of anchorage-dependent cells when used in conjunction with microcarriers. This aspect of operational flexibility has made stirred tank technology the most appealing and dominant animal cell culture practice in industry today.

2.3   Stirred Tank Characteristics

2.3.1 Size and Geometry

The size of the stirred tank for a given suspension cell culture process depends upon previous planning estimates and calculations, which determine the quantity of raw material necessary to meet the demands of production per year. These estimates take into consideration the product titer, yields through the purification process, and losses that may occur during formulation, during final fill operations, or during sampling required for quality control and in process monitoring. This information will provide the total number of liters of cell culture medium required. Based on the mode of operation (batch, semicontinuous batch, or continuous perfusion), the batch size and batches per year can be established.

The largest suspension animal cell cultures in operation today range from 1,000 to 10,000 liters. Reactors of similar design at the laboratory and pilot-plant scale are also necessary, to ensure scalability of critical parameters, provide inoculum for the larger, reactors and supply preclinical and clinical trial material prior to the commercial production phase at the larger scales. Stirred bioreactors for animal cell culture are available or can be designed in virtually any size required to meet the particular need.

Stirred cell culture tanks are almost always cylindrical vessels with a ratio of height to diameter between 1.5:1 and 3:1. Attention should be given to the design of the vessel bottom. Most cell culture vessels have a hemispheric or dished bottom, but other, more creative configurations are also possible. A flat bottom is not considered appropriate, due to the possible presence of stagnant areas that may accumulate solid particulates and require more forceful mixing than is desirable.

2.3.2 Agitation

The fragility of cultured animal cells is of major concern to the engineer assessing the best method to achieve adequate mixing in the stirred bioreactor. The agitators most routinely used are either marine impellers or pitched-blade impellers attached to the end of an agitator shaft, although other impeller designs may prove useful [9].

Baffles are also used to increase turbulence within the reactor. Some practitioners believe that eliminating them is advantageous because it reduces the exposure of the cells to unneeded mechanical damage, but, on the other hand, baffles can prevent the vortex effect observed in stirred systems and the entrainment of bubbles in the vortex, which are believed to be damaging to cells.

Low impeller speeds and agitator power inputs are necessary, especially for cells that are sensitive to mechanical or shear damage, such as some shear-sensitive hybridoma cell lines. Agitator shafts can be inserted in stirred reactors from either the top or the bottom. Neither position has an advantage over the other. Top-mounted agitator shafts have the mechanical seal above the level of the liquid, and are less likely to damage the cells or directly contaminate the culture with debris. Placing a top-driven agitator shaft off center will eliminate vortexing and improve mixing in an unbaffled reactor over that seen in a similar reactor with a center-mounted drive.

Draft tubes have also been incorporated in the stirred tank design to direct flow within the reactor and ensure that the contents are well mixed under low agitation conditions. The draft tube technology is best known for its application in air-lift bioreactors [10]. While the sensitivity of cultured cells to viscometric shear can be easily demonstrated [11, 12], some investigators have grown animal cells under extremely high agitation regimes without notable damage to the cells [13]. GARDNER *et al.* [14] and OH *et al.* [15] observed no effect of agitation at varying speeds, but found that sparging with air had a dramatic detrimental effect under all agitation conditions. Thus, the direct sparging of gases into a stirred reactor may be more critical than agitation to the health of cultured cells [16–19].

2.3.3 Aeration

In standard batch culture conditions, animal cells such as murine hybridomas grow to relatively low cell concentrations (1–5 x 10^6 cells/mL) compared to the high biomass seen in microbial fermentations. In addition, animal cells have dramatically lower metabolic oxygen requirements that those of microbial systems [20]. Nevertheless, oxygen supply in large-scale stirred bioreactors is often the limiting factor in scale-up to larger reactor sizes. This is again due to the fact that animal cells cannot withstand vigorous agitation rates or aeration. Thus, the routine technique of increasing oxygen mass transfer through high agitation rates and liberal sparging is often not possible in animal cell cultures, due to their sensitivity to mechanical damage and shear. These effects need to be minimized by reducing the impeller tip speed and the power input per unit volume as much as is feasible.

From a practical point of view, only a limited number of methods are operationally feasible to deliver oxygen to cells in large-scale stirred reactors: surface aeration, sparging, and membrane diffusion. Surface aeration alone is incapable of supplying enough oxygen to cell cultures of moderate density and size (20–100 liters), but is often used successfully in conjunction with sparging. As the volume of the reactor increases, the ratio of surface area to volume decreases, making surface aeration alone impractical for reactors larger than those of laboratory scale.

Direct sparging of air into the culture medium has proven to be a very effective means to supply oxygen to stirred cell cultures in large tanks. It must be noted, however, that cellular damage due to direct sparging of gas bubbles has been observed by many investigators [16, 18, 19], and precautions to limit the rate of sparging need to be addressed. Some advantage may be gained by using pure oxygen as the sparged gas instead of air.

Foam can also be troublesome in sparged bioreactors, especially if animal serum is used in the culture medium. If cells become entrapped in the foam layer or at the interfacial surfaces of the bubbles, they are likely to be damaged or killed. Sparging and the resultant foam layer may also damage the desired protein product. The use of antifoaming agents or foam-entrapping devices is a possibility, but their practical utility and effectiveness for animal cell culture has not been universally accepted or demonstrated. The nonionic surfactant Pluronic® F68 (a polyol) is used quite commonly to protect cultured mammalian cells from damage caused by the direct sparging of gases into the culture medium [21–27].

Membrane tubing oxygenation systems [28] have also been evaluated in stirred vessels [29] as well as in other bioreactor designs [30, 31]. In this technique to supply oxygen to suspended cells, an appropriate length of silicone tubing is installed in the reactor and pure oxygen is passed through the tubing. Oxygen diffuses through the silicone membrane into the culture medium. This method eliminates the need to sparge gases directly, and alleviates the problems inherent in direct sparging. Another way to accomplish this is to pass the culture through an external silicone tubing loop; oxygen diffuses through the tubing into the recirculating medium.

2.3.4 pH Control

The pH is one of the most important parameters that affects the well-being of cultured cells, and its control needs to be closely attended to [32]. In most traditional cell culture media, it is controlled by a bicarbonate-carbon dioxide buffer system. Bicarbonate ions and carbon dioxide are also used by cells for biosynthesis to some extent [33]. Bicarbonate and carbonic acid are in rapid equilibrium, determined by the pH. Carbonic acid equilibrates with carbon dioxide in both the liquid and the gaseous phases, and it is necessary to maintain an increased concentration of carbon dioxide in the gaseous phase to maintain the pH in the appropriate range. Fortunately, this is usually easier to accomplish in the controlled environment of a stirred cell culture reactor than in small-scale cell culture vessels in an incubator. Sterile carbon dioxide can be sparged directly into the medium or injected into the headspace gases to decrease the pH, or a sterile solution of base, such as sodium bicarbonate or sodium hydroxide, may be added to raise the pH. Automatic control of this parameter is routine in animal cell culture reactors. If stronger buffering capacity in the range of pH 6.8–7.2 is desirable, an organic buffer such as HEPES may be included in the cell culture medium.

2.3.5 Temperature

Temperature needs to be controlled within narrow limits near the normal body temperature of the animal. This is generally 37 °C for mammalian cells; cells from other animals, such as insects or fish, require different temperature set points. Regulation of the temperature should be within 0.5 °C of the set point, but good automated temperature control can be much more precise. Animal cells can withstand much lower temperatures for lengthy periods, but even slightly elevated temperatures, such as 40 °C, for even short periods may cause considerable cell damage and death.

2.3.6 Sterilization

The ability to sterilize large-scale cell culture equipment and auxiliary systems is critical to successful operation. Animal cell culture is characterized by relatively long cycles of operation, which may last weeks to months in the case of semicontinuous or continuous perfusion modes of operation, without the opportunity to resterilize the system. The investment in terms of time and effort required for cell culture scale-up, coupled with the degree of complexity of cell culture operations, and the long-term nature of the cell culture process, make contamination the major risk and source of frustration in large-scale cell culture technology. Many precautionary measures can be taken prior to sterilization, such as preventive maintenance checks of all seals, valves, filters, and O rings. Written procedures should be followed for inspecting the equipment and for performing pressure checks of the equipment after final assembly prior to steam sterilization. Proper sterile design of the equipment, to include the use of steam seals on agitator inlets, double O rings on probe insertions, and steam blocks on inlet and outlet lines, helps ensure sterility.

Resterilizable sample, harvest, and medium feed valve locations are also a prerequisite for the maintenance of long-term sterile operation. Beyond equipment design, the most critical aspect of sterile operation is faithful attention to detail through the use of written standard operating procedures and the adequate training of personnel.

2.4   Support Systems

2.4.1 Water and Steam

Water is clearly the simplest constituent of cell culture medium and associated large-scale operations, but is also one of the most critical. Water can be a source of bacterial endotoxin and other toxic contaminants that could render a cell culture operation ineffective at any scale. Water that is used in the manufacture of cell culture medium, and for rinsing equipment, should be USP Water For Injection (WFI) quality [34]. Water For Injection is produced by reverse osmosis or distillation. A standard water purification system may employ carbon filtration and deionization to pretreat the water prior to distillation. The water produced should be recycled in a hot, closed-loop system, and held at an elevated temperature to discourage microbial growth. The use of a distillation step in the purification process is an advantage over reverse osmosis, since it is more likely to give a high-quality product and may satisfy some regulatory issues. A quality control program should be established to continuously monitor the quality of the purified water through the scheduled implementation of routine assays for microbial contamination and satisfactory conductivity, pH, and chemical checks.

A clean steam generator is necessary to provide a supply of high-quality steam for sterilization purposes, and may serve a dual purpose if it is used in the final purification step for the generation of purified water as well.

2.4.2 Medium Preparation

In addition to the considerations stated above regarding the use of high-quality water, the ingredients of the cell culture medium require careful attention. Medium ingredients such as powdered basal medium, amino acid and vitamin supplements, growth factors, and other additives need to be strictly controlled through the establishment of a quality assurance program, with acceptance criteria for each component [35]. Such specifications may include limits on bioburden, endotoxin, promotion of cell growth and product formation, presence of contaminants, degree of purity, and chemical analysis. These limits may be satisfied by the vendor's certificate of analysis, or through tests on the material performed by the user. This strict control on raw materials used in the preparation of the cell culture medium will make the process more reproducible and reliable. Moreover, some components of the medium, such as serum or animal-source proteins, may be the source of contaminating agents such as viruses. Special precautions may need to be implemented as safeguards to prevent the possibility of introducing such agents into the process. These may include special processing steps by the vendor to circumvent this possibility, such as donor screening, specific virus assays, or virus inactivation steps in the production of the raw material.

Likewise, the preparation of cell culture medium through the addition of powders and other components requires strict adherence to written procedures for medium formulation, and sterilization by filtration. Once a batch of cell culture medium has been prepared, it should be quarantined for a period of time prior to its release for use. Release criteria should include chemical tests such as pH and osmolality [36], and routine assays for critical nutrients such as glucose and glutamine, which is often the major energy source for cultured cells [37]. In addition, the medium should be assessed for its ability to support cell growth in a satisfactory manner.

2.4.3 Inoculum Scale-Up

The laboratory-scale cell culture work necessary to generate a sizable inoculum to successfully seed a large-scale reactor needs to be performed by trained personnel in an appropriate cell culture facility. For suspension cell culture, a volume of 10–20 liters is required to inoculate a medium-sized pilot-plant–scale reactor of 100-liter volume. This can be performed most easily through the use of a small-scale laboratory bench-top reactor as an intermediate step in the scaling up process. The entire contents of the smaller reactor can then be transferred to the pilot plant or production facility to inoculate the next-scale reactor, with the ultimate stepwise generation of inoculum for the production-scale reactor.

Inoculum generation for an animal cell culture process is labor-intensive and time-consuming, and the loss of integrity at any level, leading to contamination, can be a major loss. Therefore, many commercial processes are operated in semicontinuous mode to alleviate the dependence on direct scale-up for every batch. Once the production-scale reactor is initially inoculated, the operator can run it in a self-inoculating mode by harvesting most of the reactor contents but leaving a quantity of the previous batch culture after each harvest to reinoculate the next batch. The initial investment of the inoculum scale-up diminishes with each successive batch harvested. This rationale also applies to continuous perfusion systems. The disadvantage of these modes of operation lies in the requirement for additional equipment in the form of sufficient holding tankage for the poising of fresh sterile medium, as well as harvest vessels, to service the reactors in continuous or semicontinuous operation.

2.5   Downstream Processing

Downstream processing and purification must deal with all of the consequences of the fact that large-scale animal cell culture conditions are optimized primarily to enhance cell growth and elaboration of the desired product. If one is fortunate, these two aspects of the process are not diametrically opposed [38]. However, some aspects of large-scale cell culture, such as the inclusion of antifoams or serum in the medium, can cause trouble downstream. These and similar medium additives can interfere with various purification steps, complicate separation processes, and force the inclusion of additional purification procedures that can have a detrimental effect on yield and therefore on overall cost and efficiency.

Product stability is an overriding factor in any cell culture process. In stirred reactors operated in batch mode, the product elaborated from the cells may have to exist in a potentially degradative environment at 37 °C in the presence of considerable quantities of cell debris for several days prior to harvest. The use of continuous or continuous perfusion animal cell culture reactors offers the possibility of online continuous processing of the product liquor, and therefore of reduced residence time of the product in the broth. However, the practical use of continuous systems remains unproven, and continuous operation is considerably more complex than batch. Again, the conservative approach of using stirred tanks operated in batch mode has gained the widest acceptance in commercial operations, due to the industry-wide familiarity with batch operations, and the regulatory experience in terms of process validation.

2.6   Nontraditional Cell Culture Approaches

Continuous perfusion systems may offer certain advantages over conventional batch cultures utilizing stirred tank technology. Some of the potential advantages include lower costs, higher cell densities, increased productivity, more efficient utilization of expensive medium ingredients, and decreased residence time of the product in the cell culture environment. All perfusion systems are based on techniques that retain the cells inside the reactor while continuously perfusing fresh medium into the reactor and continuously harvesting spent conditioned medium and product at a similar rate. Such perfusion reactors are also compatible with continuous downstream processing technology, but are vulnerable to process upsets, due to their complexity of control and operation. Some of these systems have been developed to pilot plant or production scale, and will be discussed in some detail below. In perfusion reactors, the cell concentration can be raised several fold over that seen in batch culture, and maintained at that level through the addition of perfused fresh medium and the harvesting of cell-free reactor effluent.

2.6.1 Stirred Tank Perfusion Reactors

The stirred tank technology can be adapted to a perfusion format through the incorporation of a cell retention device. The most common mechanism for retaining cells in the reactor has been a spinning filter [39]. This device is a rotating cylindrical screen or mesh mounted on the agitator shaft, but it can also be operated independently of agitation speed for greater control. Many variations of the basic spin filter methodology have been reported in the literature, including the use of ceramic filters, and external spin filter devices coupled to a cell recycle loop [40, 41]. The greatest difficulty encountered with spin filters is controlling the rate at which they become fouled, and therefore inoperable, through accumulation of cells and cell debris. The cell retention efficiency of spin filters changes during the course of long-term operation, and the balance of spin filter efficiency with cell growth and medium feed rates is often difficult to achieve. Spin filters in stirred tanks can also be used to retain anchorage-dependent cells attached to microcarriers [42]. Other methods of cell retention have also been used at small scale in stirred vessels, and include an external tangential flow membrane filter device [43], an external centrifuge [44], or a cell settling zone within the stirred reactor [45] for separation of cells and harvested culture medium.

2.6.2 Ceramic Matrix

The Opticell cell culture system is based on the use of a ceramic matrix [31, 46, 47], which provides a surface for the physical attachment of anchorage-dependent cells or the immobilization of suspension cells by entrapment. The ceramic matrix is a cylindrical core with many channels running the length of the ceramic material. A highly porous and a nonporous ceramic are available for use with the system. The ceramic cores are available in a range of dimensions, to accommodate laboratory through pilot-plant–scale units. Cell culture medium is pumped through the cores from a central medium reservoir in a recirculating loop configuration. Fresh medium is fed into the system, and harvested culture fluid is removed to the medium reservoir. In this nonhomogeneous system, it is not possible to sample the reactor directly to assess the health of the cells sequestered in the ceramic matrix. This is done indirectly through the use of multiple probes to monitor oxygen utilization by the entrapped cells, and pH of the medium within the recirculation loop. Chemical assays to determine nutrient utilization and waste product generation, and enzyme assays, such as lactic dehydrogenase (LDH) to assess cell damage, can be used to further monitor the culture. Oxygenation of the medium is accomplished through the incorporation of inline membrane gas permeators. The Opticell unit is a fully automated system that provides for the integrated control of dissolved oxygen concentration, pH, continuous medium feed and harvest rates, and medium recirculation rate.

2.6.3 Fluidized Bed

The fluidized bed system developed by Verax [30, 48, 49] is a continuous perfusion cell culture process in which the cells are immobilized in a porous three-dimensional matrix of weighted collagen microspheres. These microspheres are suitable for the immobilization of either anchorage-dependent or suspension-adapted cells. The microspheres are placed in the fluidized bed bioreactor and are maintained in a suspended mode by the continuous recirculation of the culture medium through the bioreactor. This design minimizes exposure of the cells to fluid dynamic shear as they remain immobilized in the microspheres in the bioreactor, while culture medium leaves the top of the fluidized bed free of microspheres and is recirculated through a loop to the bottom of the reactor. The recirculating culture medium passes probes that monitor dissolved oxygen concentration, pH, and temperature, and enters a silicone tubing gas exchanger where carbon dioxide is removed and oxygen is supplied to the medium. Fresh culture medium is fed into the recycle loop and an equal volume of harvested conditioned medium is removed continuously. Small samples containing microspheres may be removed from the reactor to assess the physical state of the cells and their density. Verax cell culture systems are available from laboratory scale (150-mL fluidized bed volume) to commercial production scale (24-L fluidized bed volume), capable of producing 300–400 liters of harvested cell culture broth per day.

2.7   Regulatory Issues

Human pharmaceutical products derived from cell culture are regulated by the Center for Biologics Evaluation and Research (CBER) of the FDA. This agency has issued *Points to Consider* documents [50, 51], which advise manufacturers of the general guidelines and information regarding the use of cultured cells to produce licensed products. Adequate characterization and testing of cell lines is necessary to ensure that the cells used to manufacture biological products are free of endogenous viruses, adventitious agents, and mycoplasma contamination [52–55]. The agency may require the implementation of a specific virus inactivation step in the purification process, if the cell line is known to harbor an endogenous virus. Data to demonstrate that the purification procedure will remove viruses may also be necessary. In order to bring a biological to market, a *Product License* and an *Establishment License* must be approved by CBER [56–58]. To gain approval, the production facility and associated systems must be designed and constructed to meet the FDA's requirements for biotechnology facilities [59, 60]. The completed and operating facility must pass inspection by the FDA before any product from that facility is approved for sale.

2.8   Example of a Cell Culture Production Run

Figures 2.1, 2.2, and 2.3 show data generated during a production batch culture of a murine hybridoma cell line grown at the 1000-L scale in a stirred tank. These cells were grown in serum-free culture medium, and cell growth was assessed in the tank by online analysis of the oxygen uptake rate, through the monitoring of exit gas concentrations with mass spectrographic equipment [61]. In addition, daily reactor samples were taken to determine cell density using a hemocytometer count, and cell viability using trypan blue exclusion; chemical assays monitored glucose, glutamine and ammonia concentrations, and antibody titer, using a specific enzyme-linked immunosorbent assay (ELISA).

The reactor was inoculated by the aseptic transfer of the contents of a smaller, 120-L reactor into the larger tank to give an initial cell density of approximately 3×10^5 viable cells per milliliter. The culture was then maintained at 37 °C and pH was controlled at 7.1 by the sparging of CO_2 or the addition of sterile sodium bicarbonate

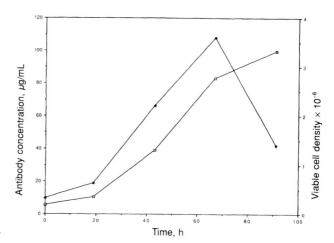

Figure 2.1 Antibody
concentration (open
symbols), μg/mL, and
viable cell density (closed
symbols), cells/mL x 10^{-6},
in a 1000-L batch culture
of murine hybridoma cells.

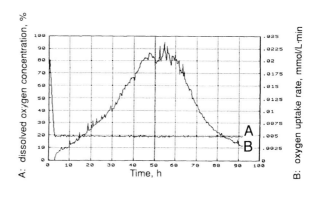

Figure 2.2 Dissolved oxygen
concentration (A), %, and
calculated oxygen uptake rate
(B), mmol/L·min, during a
1000-L batch culture of
murine hybridoma cells.

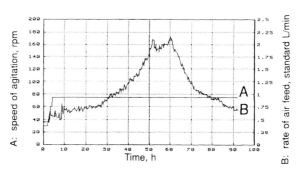

Figure 2.3 Impeller speed
(agitation rate) (A), rpm, and
air sparging rate (B), standard
L/min, during a 1000-L batch
culture of murine hybridoma
cells.

solution.   The dissolved oxygen (C_{O_2}) concentration was controlled at 20% of air
saturation at 4 psi of back pressure by increasing the air sparging rate once C_{O_2} could

not be controlled by oxygen transfer from the head space through the medium surface alone.

Figure 2.1 shows the antibody titer and viable cell density in the culture medium during the course of the batch culture. Figures 2.2 and 2.3 are online computer-generated plots of the dissolved oxygen concentration and oxygen uptake rate, and of the agitation rate and sparged air rate, respectively. The viable cell density in the batch culture increased exponentially for the first 70 hours, until nutrient limitations and waste product accumulation exerted a negative effect on the population growth rate. The death phase ensued quite rapidly thereafter. The oxygen uptake rate and the air sparging rate required to maintain the dissolved oxygen concentration closely paralleled the growth curve of the cells. At the termination of the batch culture, the monoclonal antibody concentration in the culture medium had reached approximately 0.1 mg/mL, or 100 grams of crude antibody in the total culture volume of 1000 liters.

The use of automated process control of the reactor operating conditions facilitates the optimization of cell culture parameters necessary to maximize cell growth and product formation. Stirred tank reactors offer the most straightforward means for the large-scale production of monoclonal antibodies and are easily scaled up, reliable under production plant operating conditions, and amenable to various kinds of operations.

2.9   References

1. RADLETT, P. J., PAY, T. W. F., GARLAND, J. M., *Dev. Biol. Stand. 60*, 163–170 (1985).
2. SPIER, R., *Trends Biotechnol. 6*, 2–6 (1988).
3. MCKILLIP, E. R., GILES, A. S., LEVNER, M. H., HUNG, P. P., HJORTH, R. N., *Bio/Technology 9*, 805–810 (1991).
4. CHIOU, T.-W., MURAKAMI, S., WANG, D. I. C., *Biotechnol. Bioeng. 37*, 755–761 (1991).
5. BLIEM, R., *Pharm. Eng. 8*, 15–19 (1988).
6. NELSON, K. L., *BioPharm. Manuf. 1* (2), 42–46 (1988).
7. NELSON, K. L., *BioPharm. Manuf. 1* (3), 34–50 (1988).
8. ARATHOON, W. R., BIRCH, J. R., *Science 232*, 1390–1395 (1986).
9. JOLICOEUR, M., CHAVARIE, C., CARREAU, P. J., ARCHAMBAULT, J., *Biotechnol. Bioeng. 39*, 511–521 (1992).
10. LE FRANÇOIS, M. L., MARILLER, C. G., MEJANE, J. V.; French Patent 1 102 200 (1955).
11. PETERSON, J. F., MCINTIRE, L. V., PAPOUTSAKIS, E. T., *Biotechnol. Prog. 6*, 114–120 (1990).
12. ABU-REESH, I., KARGI, F., *Enzyme Microb. Technol. 13*, 913–919 (1991).
13. BACKER, M. P., METZGER, L. S., SLABER, P. L., NEVITT, K. L., BODER, G. B.; MBTD Abstract Number 110, ACS National Meeting, 1986; American Chemical Society, Washington, DC.
14. GARDNER, A. R., GAINER, J. L., KIRWAN, D. J., *Biotechnol. Bioeng. 35*, 940–947 (1990).

15. OII, S. K. W., NIENOW, A. W., AL-RUBEAI, M., EMERY, A. N. J., *Bio/Technology* 12, 45–60 (1989).
16. JOBSES, I., MARTENS, D., TRAMPER, J., *Biotechnol. Bioeng.* 37, 484–490 (1991).
17. PAPOUTSAKIS, E. T., *Trends Biotechnol.* 9, 427–437 (1991).
18. BAVARIAN, F., FAN, L. S., CHALMERS, J., *Biotechnol. Prog.* 7, 140–150 (1991).
19. CHALMERS, J., BAVARIAN, F., *Biotechnol. Prog.* 7, 151–158 (1991).
20. GREEN, M., HENLE, G., DEINHARDT, F., *Virology* 5, 206–219 (1958).
21. BENTLEY, P. K., GATES, R. M. C., LOWE, K. C., POMERAI, D. I., WALKER, J. A. L., *Biotechnol. Lett.* 11, 111–114 (1989).
22. MURHAMMER, D. W., GOOCHEE, C. F., *Biotechnol. Prog.* 6, 142–148 (1990).
23. MURHAMMER, D. W., GOOCHEE, C. F., *Biotechnol. Prog.* 6, 391–397 (1990).
24. AL-RUBEAI, M., EMERY, A. N., CHALDER, S., *Appl. Microbiol. Biotechnol.* 37, 44–45 (1992).
25. MICHAELS, J. D., PETERSEN, J. F., MCINTIRE, L. V., PAPOUTSAKIS, E. T., *Biotechnol. Bioeng.* 38, 169–180 (1991).
26. GOLDBLUM, S., BAE, Y.-K, HINK, W. F., CHALMERS, J., *Biotechnol. Prog.* 6, 383–390 (1990).
27. PAPOUTSAKIS, E. T., *Trends Biotechnol.* 9, 316–324 (1991).
28. MILTENBERGER, H. G., HESSBERG, S.; U.S. Patent 4,649,114 (1987).
29. AUNINS, J. G., CROUGHAN, M. S., WANG, D. I. C., GOLDSTEIN, J. M., *Biotechnol. Bioeng. Symp.* 17, 699–723 (1988).
30. DEAN, R. C., KARKARE, S. B., RAY, N. G., RUNSTADLER, P. W., VENKATASUBRAMANIAN, K., *Ann. N.Y. Acad. Sci.* 506, 129–146 (1987).
31. BOGNAR, E. A., PUGH, G. G., LYDERSEN, B. K., *J. Tissue Cult. Methods* 8, 147–154 (1983).
32. EAGLE, H., *J. Cell. Physiol.* 82, 1–8 (1973).
33. GEYER, R. P., CHANG, R. S., *Arch. Biochem. Biophys.* 73, 500–506 (1958).
34. *United States Pharmacopeia XXII/National Formulary XVII*; United States Pharmacopeial Convention, Rockville, MD, 1990.
35. JOHNSON, R. W., *BioPharm* 3 (2), 40–44 (1990).
36. WAYMOUTH, C., *In Vitro* 6, 109–127 (1970).
37. REITZER, L. J., WICE, B. M., KENNELL, D., *J. Biol. Chem.* 254, 2669–2676 (1979).
38. KNIGHT, P., *Bio/Technology* 7, 777–782 (1989).
39. AVGERINOS, G. C., DRAPEAU, D., SOCOLOW, J. S., MAO, J., HSIAA, K., BROEZE, R. J., *Bio/Technology* 8, 54–57 (1990).
40. FEDER, J., TOLBERT, W. R., *Sci. Am.* 248, 36–42 (1983).
41. FEDER, J., TOLBERT, W. R., *Am. Biotechnol. Lab.* 3, 24–26 (1985).
42. GRIFFITHS, J. B., ATKINSON, A., ELECTRICWALA, A., LATER, A., MCENTIRE, I., RILEY, P. A., SUTTON, P. M., *Dev. Biol. Stand.* 55, 31–36 (1984).
43. BRENNAN, A. J., SHEVITZ, J., MACMILLAN, J. D., *Biotechnol. Tech.* 1, 169–174 (1987).
44. HAMAMOTO, K., ISHIMARU, K., TOKASHIKI, M., *J. Ferment. Bioeng.* 67, 90–194 (1989).
45. TAKAZAWA, Y., TOKASHIKI, M., *Appl. Microbiol. Biotechnol.* 32, 280–284 (1989).
46. BERG, G. J., *Dev. Biol. Stand.* 60, 297–303 (1985).
47. PUTNAM, J. E.; In: *Commercial Production of Monoclonal Antibodies*; SEAVER, S. S., Ed.; Marcel Dekker, Inc., New York, 1987; pp 119–138.

48. TUNG, A. S., SAMPLE, J., BROWN, T. A., RAY, N. G., HAYMAN, E. G., RUNDSTADLER, P. W., *BioPharm Manuf. 1* (2), 50–55 (1988).

49. VOURNAKIS, J. N., HAYMAN, E. G.; Verax Technical Bulletin, 1989.

50. *Points to Consider in the Manufacture and Testing of Monoclonal Antibody Products for Human Use*; Center for Biologics Evaluation and Research (CBER), Food and Drug Administration, Rockville, MD, June, 1987.

51. *Points to Consider in the Characterization of Cell Lines Used to Produce Biologicals*; Center for Biologics Evaluation and Research (CBER), Food and Drug Administration, Rockville, MD, November, 1987.

52. PRINCE, D. L., PRINCE, R. N., *J. Ind. Microbiol. 3*, 157–165 (1988).

53. CARTER, M. J., SCOTLAND, R. A., *BioPharm 2* (10), 24–27 (1989).

54. FENNO, J., LUCZAK, J., POILEY, J., RAINER, R., WHITEMAN, M., *BioPharm 1* (8), 36–42 (1988).

55. JONER, E., CHRISTIANSEN, G. P., *BioPharm 1* (8), 50–55 (1988).

56. HILL, D. E., *Pharm. Eng. 8*, 21–23 (1988).

57. CHEW, N. J., *BioPharm Manuf. 1* (7), 21–25 (1988).

58. MILLER, H. I., *Am. Biotechnol. Lab. 10*, 38–41 (1988).

59. HILL, D., BEATRICE, M., *BioPharm 2* (9), 20–26 (1989).

60. HILL, D., BEATRICE, M., *BioPharm 2* (10), 28–32 (1989).

61. BACKER, M. P., METZGER, L. S., SLABER, P. L., NEVITT, K. L., BODER, G. B., *Biotechnol. Bioeng. 32*, 993–1000 (1988).

Cell Separation Systems

Bhav P. Sharma

Contents

3.1   Introduction

In fermentation-based bioprocesses, the isolation and purification of the product is the immediate goal once the fermentation step is concluded. This isolation and purification is often also referred to as *product recovery* or simply *recovery*. The goal of the recovery process is to produce a product of acceptable quality, in compliance with any regulatory and safety requirements, at an acceptable cost. In this chapter, I discuss the unit operations commonly used in product recovery.

The engineer (the term is used here to denote any professional interested in the issues covered in this book) might wish to consider the analogy of a relay race when considering some of the cell separation issues. The final finish line is the production of the desired product at the desired product specifications. The cell separation is but an intermediate lap towards that finish line. The goal in cell separation is to help make the rest of the race as winnable as possible.

3.2   Criteria for Decision

Continuing the above analogy, I suggest that a good beginning for considering the various cell separation alternatives is to understand the final product specification, the finish line. Indeed so, but in addition one must also understand the original medium or broth, which corresponds to the starting line, where the baton must be picked up efficiently. A few of the factors that must be taken into account and their implications are discussed below:

(a) **Cell type (feed type) to be separated.** What type of organism is one working with? Is one going to separate mammalian cells, bacteria, yeast, or fungi? What level of solids does the feed have? Are all the solids unwanted, or do they have a considerable amount of the desired product? Is product entrapped or bound in the solids?

(b) **Intracellular versus extracellular product.** Is the desired product present inside the cells (*intracellular*) or is it in the broth (*extracellular*)? Is the product membrane-bound in a somewhat leaky cell? Is the product made as inclusion bodies inside the cell? Are the cells themselves the product?

(c) **Regulatory considerations.** Are there any hazardous waste or containment requirements for the cell separation process and for the ensuing waste disposal step?

(d) **Product quality and value.** Is one working with a pharmaceutical-type product which is high-value and requires high quality? Or is the product of interest required to be food-grade, or is it a specialty or bulk chemical, and therefore of relatively low value per pound? These factors help dictate the processing cost one can afford in striking a cost/benefit compromise.

(e) **Scale of operation required.** Is the amount of product required a few pounds per month or is it several hundred thousand pounds per month?

(f) **Waste disposal considerations.** Is the waste material to be disposed of into the sewer or is it going to be sent to a landfill? What are the requirements of the appropriate landfill or the treatment plant? Does the Environmental Protection Agency (EPA) require any pretreatments such as inactivation before disposal, or is pretreatment prudent even if not required by the regulations? Is it going to be necessary to document the survival of the cells after the cell separation step?

(g) **Capital versus operating cost constraints.** Sometimes overlooked is the fact that cash flow constraints may require one to minimize the capital investment required. This can be achieved, but may result in higher variable or even higher total operating costs. However, such trade-offs are often acceptable.

(h) **Process development time available.** Process development work is generally time-consuming, because repeated runs with representative broths must be made to define operating performance and its limits. This may occur even when no new technology needs to be developed, but rather, existing technology needs to be customized to fit a given situation. Much of the large-scale equipment has long lead times for delivery and installation. Hence a careful assessing of the time available to make critical process and equipment decisions must be clearly recognized.

A conceptual framework has been outlined as a decision tree in Figure 3.1. The product may be intracellular, extracellular, or the cells themselves. This establishes whether a cell disruption step is required. The major cell separation options are conventional filtration, centrifugation, and cross-flow microfiltration. However, novel, and sometimes potentially useful, approaches for cell separation continue to evolve, and one is well advised to consider the recent literature [1–5].

A key decision for which one's experience is a help is whether any pretreatments of the feed are desirable. These pretreatments may consist of a simple addition of processing aids or chemicals, or may take the form of a more complex operation such as liquid-liquid extraction or affinity separation. The goal of these pretreatments is to make the actual separation of the product from the cell material more successful, whether the separation itself is via filtration, centrifugation, or microfiltration.

A discussion of extraction or affinity separation is beyond the scope of this chapter, but I provide an overview of the other pretreatments.

3.3   Feed Pretreatments

One may encounter several terms that cover the idea of pretreating the fermentation broth before the actual cell separation step. Terms such as *conditioning, thickening,*

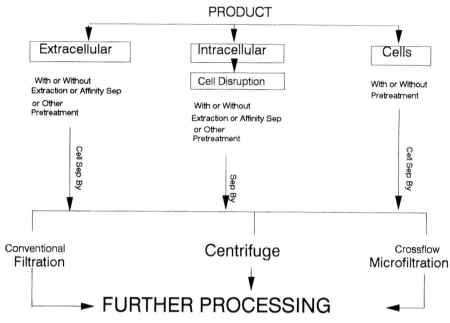

Figure 3.1 A Decision Tree for Cell Separation.

flocculation, and *coagulation* may be used. The general approach of these pretreatments is to use chemical or physical means that have the result of improving the efficiency of the cell separation step.

These pretreatments can have a significant effect on the overall process design. Therefore, the potential of the pretreatments should be exploited as fully as possible before the best option for a given cell separation problem is selected. It is even possible to trade off capital costs for operating costs and vice versa. For example, the use of filter aids adds to the operating costs, but allows the cell separation job to be done with smaller filtration equipment because of the improved throughput. A similar situation exists for centrifugation, where certain flocculents, if used as a pretreatment, add to the operating costs but allow faster throughputs and hence lower capital costs. Note also that pretreatments can and do affect the product quality.

What pretreatments should be considered? The answer depends on the situation, but Table 3.1 provides an idea of the options considered by biotechnologists. Major categories are diatomaceous filter aids, chemicals, and polymeric flocculents (liquid or solid). The cost/benefit of their use has to be evaluated in the context of whether the cell separation is via filtration, centrifugation, or cross-flow filtration.

A wide variety of cationic and anionic polymers is available, and one has to screen them for their utility in a given application. The diatomaceous filter aids come in a

Table 3.1 Some Pretreatment Agents
Useful for Cell Separation

Type	Commercial form
Filter aids	
Diatomaceous earth	Powder
Expanded perlite	Powder
Cellulose	Powder
Aluminum hydrates	Liquid
Lime	Powder
Ferrous sulfate	Granules
Synthetic polymers	Powder or viscous liquid

wide range of particle sizes and hence permeabilities (Figure 3.2). Cellulosic filter aids are also available.

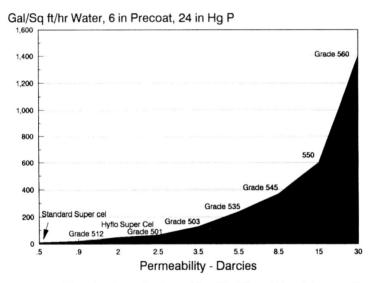

Figure 3.2 Relative Filtration Rates for Some Manville Filter Aids. Other suppliers include Grefco, Eagle-Picher, and Sil-Flo. Filter aids of the type shown are powders derived from diatomaceous earth, expanded perlite, or cellulose. They are available in different particle sizes or grades. The figure shows that by the choice of grade for precoat filtrations, one can control how tight or loose the filter precoat is. Some grades of filter aid lead to a tight precoat but give a relatively low permeability or a low filtration rate; others are more permeable and form a relatively loose precoat.

Most cells in suspension and the diatomaceous earth particles have a negative zeta potential in solution. *Zeta potential* is the electrical potential, usually expressed in millivolts, of the fixed layer of fluid around the cell particle, measured with respect to the bulk liquid, where the electrical potential is neutral. Understanding zeta potential leads to the concept of *coated filter media* [6]. The use of cationic polymers and even alum to coat diatomaceous filter media has been shown to result in a positive zeta potential. This enhances the movement and attachment of the cell particles to the filter media, which leads to better filtration.

How do the pretreatments help? When the cell separation is being done by filtration, the treatments can help by reducing the resistance of the filter medium, reducing the resistance of the biomass layer on the filter medium (usually called the *filter cake*), reducing the feed viscosity, or increasing the particle size.

When the cell separation is being done by centrifugation, the pretreatments can help by reducing the feed viscosity, increasing the density difference between the solid particles and the surrounding liquid, or increasing the diameter of the solid particles to be removed.

From a system design point of view, a few points on the use of the pretreatment aids should be kept in mind. How to add the pretreatment material may present a significant materials-handling challenge for the plant design and also be important for process efficacy. One may need special equipment for receiving, storage, and delivery into the fermentation broth. Direct bulk addition into a tank, or online metering, may have to be employed. One should also investigate whether the method of delivery to the cell separation equipment is satisfactory. In some cases, the shear forces or other flow conditions may serve to negate some of the beneficial effect of the chemical pretreatment.

Factors to consider in the use of pretreatments are: useful pH range of the material for a given application, percentage of the active ingredient in the commercially available material, dose level required for efficacy and the corresponding cost, and any special handling equipment required at the large plant scale.

3.4   CGMP and Regulatory Considerations in Cell Separation

Process development engineers need also to keep in mind the requirements for *Current Good Manufacturing Practice* (*CGMP*) and other regulatory aspects for their particular application. For pharmaceutical products in the United States, the Food and Drug Administration (FDA) is empowered to regulate manufacturing practices. These regulations are published in the *Code of Federal Regulations* (*CFR*) [7] and are generally referred to as CFR, Title 21, Parts 210 and 211. Other parts (Parts 225, 226, 606, 820, 821) are also referred to because they supplement Part 211 for the manufacture of products related to pharmaceuticals. For general CGMP considerations, the reader is referred to publications on the subject [7, 8]. I make a few comments applicable to cell separation here.

The equipment manufacturers generally design the cell separation equipment so as to conform with the CGMP guidelines. The ability to *clean in place* (*CIP*) the cell separation system under consideration is generally a must in process applications. Surfaces that come into contact with process materials should not be reactive, absorptive, or additive. Cooling or lubricating fluids should not contact process materials.

In the United States, guidelines have been established by the National Institutes of Health for containment when recombinant DNA microorganisms are used at a scale larger than 10 liters [9–11]. The guidelines specify *Bio-Safety Levels* (*BL*) for *large-scale* (*LS*) experiments and production. The major aim is to require confinement of the microorganisms in a degree proportional to the potential risk posed to people and the environment.

For recombinant organisms subject to BL1-LS requirements, the main requirements are:

(a)   All operations need to be done in a closed system. This means that all vent streams must pass through a sterile filter.

(b)   The microorganisms should be inactivated by validated procedures before being removed from the closed system. Temperature, pH, or chemical treatments are acceptable.

(c)   Sample collection and additions should be within the closed system.

(d)   Exhaust gases should be vented through sterile filters.

(e)   Before being used for other purposes, the equipment should be sterilized by validated procedures.

(f)   Emergency plans should exist for large spills.

Higher levels (BL2-LS and above) have the above requirements as well as a tighter requirement on leakage, sensing, monitoring, and validation [9–11]. *Validation* means documented demonstration that the approach being used is effective.

3.5   Conventional Filtration

Perhaps the oldest of all the practical options available to the engineer, conventional filtration is often the starting point for separating cells from the fermentation broth. Because this option has been in use for over 100 years, the engineer has a ready-made reference point. One can look at the more modern options and assess whether and why they make a better alternative to conventional filtration.

As a practical definition, conventional filtration is said to occur when cells and other solids are separated from the liquid broth with the aid of pressure or vacuum. The separation is carried out in a filter or a filter press. The typical particle size of the solids removed is in the range of 0.2 μm to several micrometers. There are several variations that are important for solid/liquid separations in biotechnology, and some of these are noted here as a very brief overview.

Clarifying or *polish filtration* is used when the goal is to further clean up a feed stream containing a relatively low proportion of solids. Typical particle size of the solids removed is 0.2–5 μm. *Ultrafiltration* involves concentrating smaller molecules like proteins, which are in the angstrom (0.1 nm) size range. A variation is the use of a spin filter for filtering and retaining cells in perfusion cell cultures [12].

Microfiltration or *cross-flow filtration*, which is an improvement over conventional filtration, can be employed for both cell separation and clarifying filtration. The operation is called *cross-flow* because the feed and concentrate streams flow parallel to the filter surface instead of perpendicular to it, as is the case for conventional filtration.

3.5.1 Rotary Drum Filter

The rotary drum filter is traditionally the most commonly used filter type for large-scale cell separation. Vacuum is usually the driving force for the filtration, and the filter may then be called a *rotary vacuum drum filter*. Pressure is used as the driving force in some installations, and the filter then is called a *rotary pressure drum filter*. Other variations are a *rotary vacuum string filter* and a *rotary vacuum belt filter*. Figures 3.3–3.6 are illustrations of large-scale filters.

A typical rotary vacuum drum filtration system (Figure 3.7) has provision for putting on the drum a layer of filter aid, known as the *precoat*. During the filtration

Figure 3.3 A Large Municipal Waste Treatment Plant Using Ten Coil Filters. Each of the filters shown has an 11.5-ft diameter and a 16-ft face. (Courtesy of Komline–Sanderson Engineering Corp.)

Figure 3.4 Detail of the Discharge Roll, Cloth Medium, Wash Pipes, and Take-Up Roll for a Belt Filter. Flexibelt filter has an 8-ft diameter and a 12-ft face. (Courtesy of Komline–Sanderson.)

Figure 3.5 A Belt Filter with Vapor-Tight Stainless Steel Cover, used for the containment of all process material and vapor. (Courtesy of Komline–Sanderson.)

Figure 3.6 Overview of a Rotary Drum Vacuum Belt Filter System in a Pharmaceutical Plant. (Courtesy of Komline–Sanderson.)

cycle, a knife assembly peels off layers of the precoat in a controlled manner. In addition to the vacuum system, the system piping includes provision for blending the feed with filter aid (called *admix*), controlling the level in the filter trough, discharging the cake, and collecting the filtrate.

3.5.2 Filtration Theory

Filtration theory is seldom used in practice, but is useful for interpreting some practical problems and predicting the effect of suggested changes in operating conditions. The fundamental filtration equation is Poiseuille's equation, which states that the instantaneous rate of filtration per unit filter area is proportional to pressure divided by viscosity and resistance (due to the filter medium and the cake).

The filter cake is said to be *incompressible* if it consists of relatively hard granular particles, and an increase in pressure does not result in deformation of the particles. For incompressible cakes and for many cases where the filter medium resistance can be neglected, the following generalization is derived: the instantaneous rate of filtration per unit filter area is directly proportional to the pressure (or vacuum), and inversely proportional to three factors, the filtrate viscosity, the total amount of filtrate collected, and the average cake resistance.

The filter cake is said to be *compressible* if it consists of relatively soft particles, and an increase in pressure results in deformation of particles. For a compressible cake, the instantaneous rate of filtration per unit area is independent of the operating pressure (or vacuum), but is still inversely proportional to the three factors mentioned above for the incompressible case.

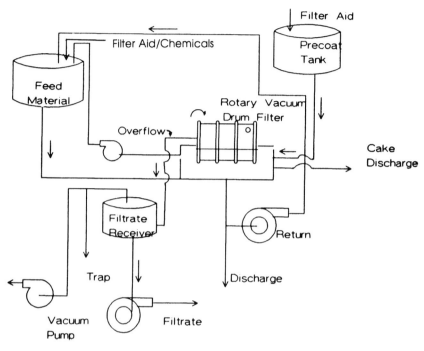

Figure 3.7 Components in a Typical Rotary Vacuum Drum Filter System. A precoat tank is needed to mix the filter aid with water and to lay down a layer of the filter aid on the surface of the rotating drum filter. A vacuum pump and a filtrate receiver system make a uniform layer possible. After the precoat is ready, the feed material is pumped to the filter, where it is filtered through the rotating drum filter with the help of the vacuum system.

The above gives us some insights into the filtration process. The liquid viscosity and the operating pressure or vacuum are obvious operating variables. Achieving a lower viscosity either by dilution or by a change in operating temperature leads to better filtration rates. The theory also tells us that we might benefit by manipulating the cake resistance. By continuously removing a thin slice of the filter cake as the filtration cycle progresses, we can keep the average cake resistance low. That is indeed how a class of filters, called the rotary drum filters, is designed. The filter surface is on a drum that rotates continuously. As the drum rotates, a knife removes a thin slice of the filter cake, renewing the filter surface. This helps keep the filtration rate from decreasing substantially over the filtration cycle.

The effect of particle size on the cake resistance is also significant. In cases where it is practical, one should consider the addition of flocculating polymers, which increase the average particle size, which can then affect both the cake resistance and the cake compressibility.   Other treatments of the feed material should also be considered [13, 14].

3.5.3 Predicting the Size of Filter Required

Projecting the size of the filtration equipment required to do a given cell separation job is a common task for the process development engineer.   Estimating the filter size required for a given job is both an art and a science.   A few practical suggestions are offered here as a starting point.

A preliminary estimate of the filter size required should be done by experimental simulation with as representative a feed material as practical.   The simulation can easily be done with the help of a dip-leaf device in an arrangement like that shown in Figure 3.8.   An alternative is to use the smallest available filter equipment for simulation purposes.   These devices are available from specialized shops or from the manufacturer of the filtration equipment or of the filter aids.   A skid-mounted pilot unit is shown in Figure 3.9.

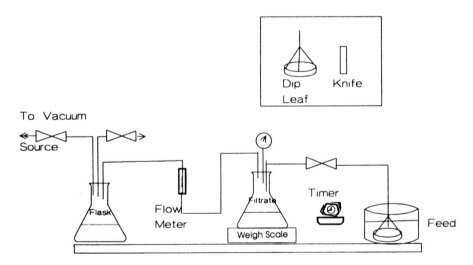

Figure 3.8 Use of a Dip-Leaf Filter for Bench-Top Simulation Tests.   A layer of the filter aid to be tested is laid under vacuum on the surface of the dip leaf.   A typical leaf is 3–4 inches (8–10 cm) in diameter.   The feed mixture to be tested is made ready in a large beaker.   The experimenter simulates filtration conditions by dipping the coated leaf surface (which is kept under vacuum) into the feed material for a measured time, pulling out and drying for a measured time, and then removing a layer of the filter aid from the dip-leaf surface with a knife.   A "dip, dry, cut" cycle can thus be simulated several times.

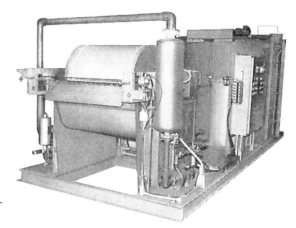

Figure 3.9 A Skid-Mounted, Pilot-Scale Rotary Drum Filter System. (Courtesy of Komline–Sanderson Engineering Corp.)

The experimental simulations should be done with a clear understanding of the large-scale system to be used. One should discuss the application with the manufacturer's representative to help narrow the choices that are realistic in a given situation.

As one of the first steps, data on filtration rates as a function of time (in liters per square meter per hour, or gallons per square foot per hour) should be obtained. This should be done with and without admix (also called *body-feed*), and with and without any other pretreatments being considered, such as flocculation. As representative a feed material as possible should be used. Most filtration equipment and filter aid suppliers have put together very useful summaries on the kind of data one should obtain.

The preliminary data suggested above would allow a judgment of whether the biomass cake is compressible or incompressible, and a rough estimate of the filter size required for the job being considered. Since most cell filtration is done in a semicontinuous manner, it is important to allow for the filter turnaround time, that is, the time required to clean and reset the filter for the next filtration cycle. For many types of filters, the complete turnaround time is a significant fraction of the filtration cycle.

In the next stage of filter size estimation, one should seek to examine the filtration operating variables, such as admix level [15] and type, any other feed pretreatments, and equipment-related variables. In the case of a rotary drum filter, variables such as cycle time or drum speed, drum submergence, cake-forming time, pressure or vacuum level, filter medium, and cell cake washing rate can be evaluated. At this stage, implications of undersizing or oversizing the filter unit should be evaluated and an appropriate safety factor should be selected. The overall process flow for the plant under consideration should be considered, to ensure that the cell separation equipment (after an allowance for the turnaround time) is in acceptable balance with the other unit operations involved.

Commercially available equipment comes in a wide range of sizes. The smallest drum filters start at about 9–10 ft^2 (0.8–0.9 m^2) (diameter about 3 ft or 1 m), although special units as small as 1 ft^2 (about 0.1 m^2) can be obtained for preliminary tests. The largest units are usually about 750 ft^2 (70 m^2) (diameter about 12–14 ft or 3.5–4.3 m, and face length 18–20 ft or 5.5–6.1 m). The size of the small-scale units has a practical implication because it determines the amount of representative feed material required for experiments. One needs to demonstrate that the desired plant throughput rate (after an allowance for the turnaround time for the given filter) can be achieved over a period of time. A variety of feed conditions should be tested to understand the variability of the process.

Depending on the level of confidence in the data, and in the resulting projections of equipment size and operating conditions, additional pilot-scale data should be obtained if time permits.

3.6   Cross-Flow Microfiltration

As shown in the decision tree (Figure 3.1), a valid option for cell separation in almost any circumstance is *cross-flow microfiltration* (often referred to as *microfiltration, MF,* or *tangential flow filtration*). Because of its importance as an emerging technology, and because microfiltration has applications other than cell separation, a separate chapter in this volume has been devoted to it. I offer some comments on the use of this technology for cell separation in this section.

For a number of reasons, microfiltration has the potential to be the most competitive technology for cell separation [16–18]. However, is that the case in your application? One can find out for a given situation only by applying the three rules for success in MF—testing, testing, and testing. The advantages and disadvantages of cross-flow microfiltration are:

(a)   MF systems tend to give a filtrate (also called *permeate*) of excellent quality, a criterion generally harder to maintain with a centrifuge or a precoat filter. This is valuable if the product is in the filtrate.

(b)   The operating costs of an MF system tend to be lower than those for a rotary drum filter or a centrifuge. Of course, life cycle costs should be considered in any cost comparison. One type of cost comparison is suggested at the end of this chapter.

(c)   If the cells themselves are the product, MF could have an edge over the rotary drum filter or the centrifuge. No filter aid is involved, as with a drum filter, and flocculents, which may be needed for centrifugation, can also be avoided. The solids can be washed by diafiltration to get rid of low molecular weight impurities. In addition, very good concentration of the cells can be achieved.

(d) MF has the potential for performing more than just cell separation. Some examples are: simultaneous microfiltration and ultrafiltration (with or without diafiltration), protein separation, affinity chromatography, and sterile filtration. It is ironic that, as a consequence, this asset can sometimes be a liability, because extensive testing is needed to properly define the exact system worthy of a large financial investment based on these ideas.

(e) The MF concept is very attractive when one is working with a cell lysate, as in the case of an intracellular product.

(f) A disadvantage is that sometimes the *passage* (the fraction of the desired product that passes through into the permeate) of the product is not high enough. In those cases, a large dilution is the price for an attractive percent of product recovered.

3.6.1 Parameters for MF System Definition

Collaboration with the manufacturer's representatives is perhaps the most efficient approach to obtaining the kind of data required for a reliable definition of the scaled-up system. In general, the testing work is done with a representative feed and with or without any feed pretreatments. For preliminary work, the following parameters [19] are of immediate interest:

(a) flux (for instance, gallons per square foot per day) as a function of trans-membrane pressure and feed concentration,

(b) percent passage of the product of interest, and

(c) optimal flux as a function of time.

From these kinds of data, a projection of the overall costs can be developed in collaboration with the manufacturer.

A number of very promising systems are on the market for applications related to cell separation. Among these are polypropylene membrane systems (Akzo, Enka), coated sintered steel systems (Carre, Du Pont), ceramic and polypropylene membrane systems (Millipore, Norton), ceramic/acrylonitrile/poly(vinylidine fluoride) membrane systems (LSL), ceramic membrane systems (Alcoa/TWT), cross-flow membrane systems (Filtron), and cellulose acetate and polyolefin membrane systems (Sartorius). Examples of two systems (without implied endorsement) are shown in Figures 3.10 and 3.11.

The real potential of the cross-flow microfiltration technology has yet to be realized. New materials are becoming available, and innovative approaches to the use of such systems are emerging. Sintered metal membranes coated with low-cost materials; long-life ceramic membranes derived from alumina, zirconia, titania, and yttria; rotating membrane systems that reduce fouling [20, 21]; electromicrofiltration [22]; and MF in combination with other approaches such as liquid-liquid extraction, affinity chromatography, ultrafiltration, and diafiltration are all possibilities under

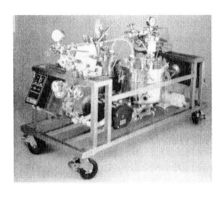

Figure 3.10 A Small Pilot-Scale Cross-Flow Filtration System. Unit shown can be used for simultaneous cell harvesting and protein concentration. Millipore, Akzo, Filtron, Enka, and Membrex are among the suppliers and developers of cross-flow filtration systems. (Courtesy of Sartorius Corp.)

Figure 3.11 A Cross-Flow Filter Unit, with 1000 ft^2 (93 m^2) membrane surface area. (Courtesy of Sartorius Corp.)

exploration. We should see these technologies increasing in commercial use as some of the problems are overcome. These problems have tended to be in the areas of membrane fouling, operating life, product passage, and cost per gallon of product.

3.7   Centrifugation

The concept of using centrifugal force to efficiently separate components of differing densities was put into practice during 1877–1920, first in Sweden by a young inventor, Dr. Carl Gustaf Patrick DE LAVAL, then in Germany, and later in the United States. Over 20 different types of centrifuges are now commercially available, and the centrifugation option is now well established for cell separation.

Some examples where centrifugation is a legitimate tool in the biotechnology industry are:
  (a)   concentration and washing of yeast cells in bakers' yeast production;
  (b)   removal and washing of mycelium in the production of citric acid;
  (c)   removal and washing of bacteria in the production of monosodium glutamate (glutamic acid crystals may also be separated by centrifugation);
  (d)   concentration and recovery of bacteria or yeast in the production of single-cell protein;

(e) removal of cell mass in the production of extracellular enzymes;
(f) removal of cell mass and cell debris in the production of intracellular enzymes;
(g) recovery of penicillin and other antibiotics by countercurrent extraction; and
(h) removal of bacteria in the production of vaccines, such as those for diphtheria and tetanus, and whole-cell vaccines.

3.7.1 Principle of Centrifugation

The fundamental principle that applies to centrifugation is Stokes' law. According to Stokes' law, a particle undergoing sedimentation attains a constant or terminal velocity. This velocity or sedimentation rate (in meters per second) is given by the following expression:

$$v_t = d^2 (\rho_p - \rho_l) g / 18 \mu \tag{3.1}$$

where v_t is the sedimentation rate, m/s;
  d is the particle diameter, m;
  ρ_p is the particle density, kg/m^3;
  ρ_l is the liquid density, kg/m^3;
  g is the gravitational acceleration constant, m/s^2;
  μ is the liquid viscosity, Pa·s.

In the practical application of centrifugation, the Stokes' law equation has no direct use. Its value lies in teaching us the following points for cell separation applications in biotechnology:

(a) The efficiency of separation is proportional to the square of the diameter of the particle being removed. This means that any approaches that lead to cell aggregation (by feed pretreatment) are very worthwhile.
(b) It is important to achieve a density difference between the particle being removed and its surrounding liquid.
(c) Increases in viscosity decrease the separation efficiency. In most cases, a 25–50% dilution of the feed may lead to a 2- to 5-fold reduction in viscosity. This provides another practical tool for the process development engineer.
(d) Centrifugal machines increase their effectiveness by providing a sedimentation rate thousands of times greater than is possible under the normal pull of gravity. An object rotating around an axis experiences a force (G) in accordance with the formula:

$$G = (r / g) (2 \pi n / 60)^2 \tag{3.2}$$

where r is the distance of the particle from the axis of rotation, m; and
  n is the revolutions per minute (rpm).

3.7.2 Evaluating the Centrifugation Option

An evaluation of the centrifugation option usually starts with the kind of questions listed as criteria for decision in Section 3.2. The goal then becomes one of performing enough simulations so that a preliminary process design built around centrifugation can be conceptualized. We want to know the degree of product quality possible with different machines at defined flow rates. These preliminary process data may then be used to do preliminary cost analyses, to gain an understanding of the options that are feasible for a given cell separation problem.

There are at least half a dozen types of centrifuges, available in scores of models from several manufacturers, and there are advantages and disadvantages associated with each option. Fortunately, assistance from an experienced individual can easily narrow the centrifuge options to a manageable number. Among the issues to consider are: feed solids content, biomass particle size, nature or texture of solids, product quality desired, tradeoffs of the product quality with the product yield desired, operating environment, and feed pretreatments.

The values of G for different types of centrifuges are given in Table 3.2. In the next few pages I describe these types and outline the practical steps for evaluating the centrifugation option for cell separation. For large-scale manufacturing purposes, the common bench-top clinical centrifuges and the laboratory ultracentrifuges do not have much relevance; hence, I do not consider them here.

Table 3.2 G Values for Different Centrifuges

Type of centrifuge	Value of G
Common bench-top clinical centrifuges	500–1,500
Basket and filter centrifuges	300–1,500
Decanter centrifuges	1,500–4,500
Disk-stack centrifuges	4,000–13,000
Tube centrifuges	10,000–17,000
Laboratory ultracentrifuges	100,000–higher

The common bench-top clinical centrifuge does have a very useful role in process development, for analytical purposes and for "quick and dirty" guesswork. A practical and easy method for characterizing the feed and the product is to take a test-tube sample of the starting feed material, with or without feed pretreatments, and spin it for a few minutes in the clinical centrifuge. The resulting percentage of volume occupied by the various fractions, the clarity of the supernatant, and the nature of the solids, as well as the solid-liquid boundary, are noted. Such data for different times of spinning can add useful insights to the type of cell separation problem one has on hand. In some situations, centrate or product quality may also be evaluated in a similar manner. An oversimplified and very rough guideline is ventured in Table 3.3 for what process centrifuge may be the best choice based on data from a three-minute spin in a clinical centrifuge at its highest setting. In some cases, an appropriate feed pretreatment can help make a process centrifuge suitable that was unsuitable prior to the treatment.

Table 3.3 Choosing a Large-Scale Centrifuge

% Solids*	Supernatant	Solids type	Solid/liquid boundary	Recommended process centrifuge
10–60	clear	fibrous/sticky	relatively well-defined	decanter
0.5–20	cloudy or clear	fluffy or slimy	boundary not well-defined	disk-stack
< 1	cloudy or clear	not as critical as for the above continuous types		tube
< 5	cloudy	not as critical as for continuous type		chamber-bowl

* To determine the percent solids, centrifuge a sample with a laboratory centrifuge.

3.7.3 The Decanter Centrifuge

The decanter centrifuge is perhaps the most commonly used centrifuge type, due to its versatility and cost. The installed capital cost for a decanter centrifuge is lower than that for a disk-stack–type machine. It is capable of continuous operation, has high reliability in a plant environment, and requires little labor. For these reasons, the decanter centrifuge is available from a greater variety of suppliers than is the disk-stack type. Some of the suppliers are listed at the end of this chapter.

A typical decanter is a horizontal, solid-bowl machine in which a screw-type conveyor helps move the solids to the discharge point. Cell separation in a decanter takes place as follows. The decanter bowl, which is a cylindrical drum, is rotated at 3000–5000 revolutions per minute by an electric motor. Feed liquid is fed into the middle of the machine almost along the axis of rotation. The centrifugal effect causes the solids to settle against the inside wall of the rotating bowl. A screw conveyor, called the *scroll*, rotates in the same direction as the bowl, but at a speed slightly higher than the bowl. The scroll moves the solids up the conical section of the decanter (or *up the beach*) towards the discharge. The liquid phase flows continuously over the outflow weirs at the wider end.

An operating decanter is said to have a *beach zone* and a *pond zone*. Clarification takes place in the pond zone, and dewatering of the solids takes place in the beach zone. Adjustments can be made in the speed with which the bowl and the scroll rotate, and in the overflow weirs. The overflow weir diameter can be adjusted by a regulating ring. Thus two extremes of operating conditions are a "shallow pond, long beach", and "deep pond, short beach". To achieve relatively dry solids one would operate near the shallow pond, long beach condition. To get a clearer effluent, one would operate near the deep pond, short beach condition.

3.7.4 The Disk-Type Centrifuge

A significant advance in centrifugation occurred around 1888–1900 with the invention of the disk-type centrifuge. A German inventor, Clemens VON BECHTOLSHEIM, conceived the idea referred to as the *Alfa patent*. The idea was to place special conical disks inside a rotating bowl, and thus greatly improve the efficiency of separation. The Alfa patent was shortly licensed by the Swedish De Laval company, later called Alfa-Laval.

After the decanter-type machines, the disk-type machines are now the most popular in the biotechnology industry. Although their installed and operating costs are higher than those of other types of centrifuges available, the disk-type machines are often preferred because they give a better quality effluent.

A great variety of machines built around the disk-stack concept are available. Names like *desludger, milk* (or *cream*) *separator, nozzle-bowl* (or *solids-discharging*) *centrifuge, clarifier*, and *solids-ejecting centrifuge* may be encountered. The general principle, however, is similar in all of them. The disks are spaced about 0.5–2 mm apart. Figures 3.12 and 3.13 show two of the variants and how they look from the outside.

In addition to the disk stack, the machines have a solids-holding space where the sediment can collect. The feed to be processed enters along the axis of rotation of the

Figure 3.12 A Production-Scale Disk-Stack Centrifuge, shown without associated piping and instrumentation. The machine shown can have a maximum capacity of 20,000–50,000 L/h, depending on the feed and the degree of clarification required. (Model SB80, courtesy of Westfalia/Centrico Corp.)

Figure 3.13 Two Self-Cleaning Disk-Stack Clarifiers. (Courtesy of Westfalia/Centrico Corp.)

bowl. As the feed comes down the center and flows out to the disk stack, it is distributed into many layers. The centrifugal force pushes the solids to the underside of each disk and then towards the solids-holding space. In the case of nozzle-bowl or solids-discharging machines, the solids are continuously ejected through nozzles. The solids outlet can also be called the concentrate outlet. Different nozzle diameters are available, and the machine may be used with or without recycle of the concentrated solids. Figures 3.14 and 3.15 further illustrate the concept.

In the case of a solids-ejecting or desludger-type centrifuge, the solids are discharged intermittently through ejection ports. The discharge may be once every minute to once every few minutes, and is one of the variables to study during process development. Alternatively, the machine may be equipped with the capability of sensing when the solids-holding space is full and should be opened to discharge the

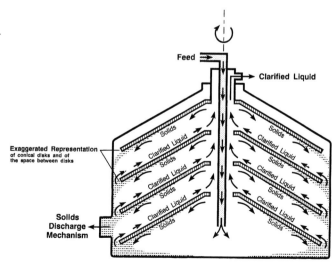

Figure 3.14 A View of a Disk-Stack Centrifuge, showing how the conical disks help in the separation process. As the disk stack rotates, centrifugal action causes heavier particles to build up under the disks, and to move down and outward for discharge. The lighter, clarified fluid moves up along the axis of rotation. (Courtesy of Westfalia/Centrico Corp.)

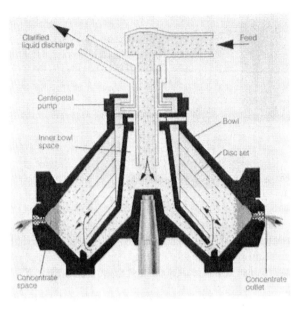

Figure 3.15 Principles of Operation of a Disk-Stack–Type Nozzle Centrifuge. During the separation, the feed enters the rotating bowl and quickly gets distributed into many layers as it enters the region of the conical disks. Heavier solids are forced against the underside of the disks and slide down towards the outlet. The discharge is controlled through nozzles. The clarified effluent is discharged by a centripetal pump. (Courtesy of Westfalia/Centrico Corp.)

solids. The ports open hydraulically. The solids discharge operation has a distinct and loud noise, and a trained operator learns to judge by the sound whether the machine is operating satisfactorily.

As the conical disks and the centrifugal force cause the solids to separate from the liquid and go towards the solids-holding space, the liquid (the *centrate*) moves up towards the liquid discharge. The quality of the centrate is monitored either visually through a sight glass or by more sophisticated means.

One of the questions during process development work is the overall throughput rate for satisfactory operation. This often depends upon whether it is the solids-holding space or the cell separation itself that is the limiting factor in the given situation. For feeds with very high solids (40–60% spun solids), the solids-holding space may be the limiting factor. The solids-holding space depends on the size of the machine. The largest machines have a solids-holding space of about 55 liters, and require a large (70-kW) motor. Too frequent a discharge may mean lower yields, and too infrequent a discharge may mean lower effluent quality.

For the biotechnology industry, an important aspect is adherence to *Good Manufacturing Practice*. One element of that is the ability to *clean in place* (or *CIP*). A number of engineering innovations have been made by the centrifuge manufacturers to allow the inclusion of these machines in CIP processes.

Figures 3.16–3.18 show three installations of industrial-scale centrifuges and their associated equipment.

Figure 3.16 (on left) One of the Largest Disk-Stack Separators Available, as installed in a winery for clarification, and reported to have an effective capacity of 115,000 L/h. The self-cleaning, semicontinuous machine has a sediment-holding space of 57 L. (SA160, courtesy of Westfalia/Centrico Corp.)

Figure 3.17 (upper right) A Nozzle Centrifuge Installation in a Potato Starch Factory in The Netherlands. The machines shown have a maximum bowl speed of 4500 rpm and weigh about 3000 kg each. Maximum capacity ranges from 50 tons per hour to 300 tons per day, depending on the job to be done. (DA100 machines, courtesy of Westfalia/Centrico Corp.)

Figure 3.18 (lower right) An Installation of Disk-Stack Separators for the Pharmaceutical Industry. The centrifuge discharges the solids in an enclosed system, most visible in the picture on the front left. Thus, the solids can be recovered without exposure to ambient air. (SA160 clarifiers on the right, SB80 on the left, courtesy of Westfalia/Centrico Corp.)

3.7.5 Other Industrial-Scale Centrifuges

The imperforate-bowl centrifuge, of which the decanter centrifuge is a specialized example, has two other variations that should be noted. These are the tube-type centrifuge and the chamber-type centrifuge.

The tube-type centrifuge usually consists of a slender tube that is capable of delivering high centrifugal forces. The tubular bowl usually has a diameter in the range of 75–150 mm, and a ratio of height to diameter of 5–7. The G force delivered, of 10,000–20,000, is much higher than those of the decanter and disk-stack–type machines. Tubular centrifuges are easy to operate and have very low capital costs. However, they have a limited utility in large-scale work because of the relatively low solids-holding capacity and low effective separation capacity. In their simplest form, tubular centrifuges are operated in a semibatch mode. The bowl must be manually cleaned after 2–10 liters of solids are collected; hence, their use is very limited. The decanter centrifuge represents a conceptual compromise by providing a continuous conveyor discharge, but at a lower G force.

The chamber-bowl centrifuge is more common than the tubular centrifuge in process-scale applications. The bowl diameter is wider (in the range of 125–525 mm), and the ratio of height to diameter is about unity. The speed of rotation is less than that of the tubular design, due to tensile strength limitations. The G force delivered is about 6000–11,000. In one variation, the bowl has a series of baffles or chamber inserts. The baffles allow the machine to function as several concentric chambers in series. The chamber-bowl–design machines also have limited application because the chamber must be manually cleaned every time it fills with solids. Design improvements have been made in recent years that allow the unloading of solids almost automatically. In these designs, a knife or a plow removes and unloads the solids once the bowl is full. Maximum bowl capacity is about 60–65 liters.

Another variation of the chamber-bowl is the basket centrifuge. The basket or the chamber may have holes in the walls of the bowl, and is called a *perforate-bowl* machine. An *imperforate-bowl centrifuge* then is a machine with solid walls. The largest perforate-basket–type centrifuge in commercial use may be a 1500-mm-diameter and 1000-mm-deep machine made for Dow Chemical by Ametek, Inc. [23], for a dewatering application. Because of the very large size, the machine is designed to deliver a G force of only 770. The chamber-bowl machines find use in the dewatering of sludge in waste treatment applications, and in the chemical and pharmaceutical industry.

Separation of red and white blood cells from blood plasma is another specialized case of cell separation. Centrifugation is extensively used. The best platelet recoveries are achieved at G forces of 3000–4000. Special continuous- and discontinuous-flow centrifuges have been designed. For further information, references on blood fractionation should be consulted [24–26].

3.7.6 Pilot-Scale Centrifugation

Pilot-scale experimentation with several hundred to one thousand liters is recommended to better define the results possible with a centrifuge. The experimenter should seek the following information:

(a) graphical data on the liquid and solid effluent quality as a function of feed flow rate;

(b) graphical data on the liquid and solid effluent quality as a function of feed flow rate for a variety of practical pretreatments;

(c) data on the liquid and solid effluent quality for the likely variability in the feed, or on how sensitive the centrifuge performance is to the feed quality;

(d) data on the most probable operating conditions of the centrifuge.

Depending on the machine type, one would need to experiment with the machine settings, for example, bowl and scroll speeds, pond and beach length, frequency of solids discharge (and whether partial or full discharge), and the settings of regulating rings. A preliminary definition of operating conditions can be achieved fairly rapidly. However, it does take considerable effort and experience to optimize the operation of a particular machine.

3.8 Intracellular Product: Cell Disruption

In some processes, the cell mass itself is the product, as in production of bakers' yeast and single-cell protein. In such cases, the process development concerns after cell separation would move on to issues like product formulations and product quality, which are beyond the scope of this chapter. But what if the product of interest is either inside the cells or is cell-membrane–associated, as is the case with many intracellular enzymes? For that reason, I now discuss cell disruption to release the product in some detail.

The goal of a cell disruption treatment is to release the product at an acceptable yield and an acceptable cost for further separation. The separation of the product from the cell debris then is a specialized case of cell separation [27]. In this section I introduce the options for cell disruption, and then discuss cell debris removal.

At the laboratory scale, the number of options for cell disruption is perhaps limited only by the imagination. Depending on the cell type (bacteria, yeast, mycelium, or mammalian cells), any of a number of methods may allow the release of the protein of interest from the cells. Some of the approaches that have been used are enzymatic degradation of cell walls, osmotic pressure change, temperature change, sonication, and freezing and thawing to promote cell lysis. Any of these methods, alone or in combination, may yield the desired results. These may also be used in conjunction with physical disruption of cells, which I introduce in some detail here. Physical cell disruption with process equipment is appropriate mostly for yeast and bacteria. It is usually desirable to carry out cell disruption operations at low temperatures (such as 4 °C).

Equipment has been available for some time for the mechanical disruption of cells [28]. Reasonable yields and scale-up are possible with assistance from the manufacturer's representatives. Most of the equipment concepts were developed for other industries, but have been adapted to biotechnology. Equipment developed for emulsification has been found to be effective, but temperature control is a key issue. A colloid mill or a Manton–Gaulin homogenizer can be used for cell disruption by high shear.

Other equipment types with an advantage in low-temperature operation are from Microfluidics Corporation (developed by Arthur D. Little) and from Willy A. Bachofen AG Maschinenfabrik of Switzerland. Examples of these machines are shown in Figures 3.19 and 3.20. The machines have been used with bakers' yeast, filamentous fungi or mycelia, algae, and bacteria.

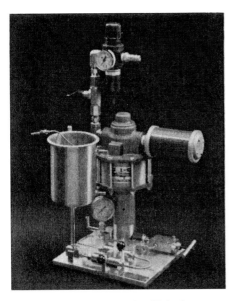   

Figure 3.19 Cell Disruption Unit, from Microfluidics Corp.

Figure 3.20 Model KDL for Cell Disruption from WAB Maschinenfabrik, Switzerland. Grinding chamber size is 0.3–0.6 L; dimensions are 47 cm x 52 cm x 70 cm; weight is 82 kg. (Courtesy of Glen Mills, Maywood, NJ.)

The Microfluidics unit disrupts cells by spraying liquid jets of the feed stream into each other at high pressure.

The working principle in the case of the machines from WAB Maschinenfabrik (trade name Dyno®-Mill) is illustrated in Figure 3.21. The horizontal grinding chamber contains glass beads in addition to the agitator disks. The feed, containing whole cells, is pumped through the chamber. The cells are impacted by the beads,

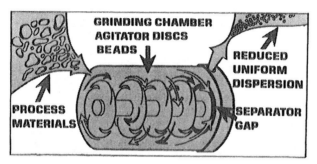

Figure 3.21 Principle of Operation of the Cell Disruption Equipment from WAB Maschinenfabrik, Switzerland. In the grinding chamber, beads and process solids are moved around by the agitator disks, impacting millions of times. A separator retains the beads while permitting the product to exit. (Courtesy of Glen Mills, Maywood, NJ.)

perhaps millions of times. A separator, whose slit is adjustable, retains only the beads and allows the cell debris suspension to escape. The suspension can then be processed further for the recovery of the desired product.

There are a number of advantages with the Dyno®-Mill approach. Chambers ranging from 0.6 liters to 275 liters are available, making process-scale use feasible. The commonly used 15-liter machine (Figure 3.22) can process 180–200 liters per hour of a 14–18% yeast slurry. It is possible to avoid air entrapment, to control the temperature of cell disruption to as low as 4 °C, and to use relatively high-viscosity cell suspensions. The degree of cell rupture can be controlled by varying the cell concentration, bead size and type, shaft speed, temperature control, and pumping rate. Sterilization of the chamber and full containment of aerosols are possible, important considerations for many situations in the biotechnology industry. A larger, 215-liter-chamber machine is shown in Figure 3.23.

Disruption of mammalian cells is a special case. Mammalian cells are easy to break open, but the disruption method must preserve the integrity of the subcellular components and allow effective separation in follow-up steps. Several of the methods are based on shearing effects as a fluid passes through an orifice. Cell disruption by sonication and explosive decompression are also employed [29, 30].

The separation of product from cell debris after cell disruption is usually done by centrifugation. The comments made earlier in this chapter on centrifugation are applicable. However, the main challenge for the process development engineer is to selectively remove the component of interest from the cell debris, so other separation techniques in conjunction with centrifugation should be evaluated. Techniques such as precipitation, affinity ligands, and extraction, in combination with centrifugation, could lead to a cost-effective process.

Figure 3.22 Model KD-15 for Cell Disruption from WAB Maschinenfabrik, Switzerland. Grinding chamber size is 15 L; dimensions are 1.5 m x 1 m x 0.7 m; weight is 550 kg. (Courtesy of Glen Mills, Maywood, NJ.)

Figure 3.23 A Large-Scale Machine for Cell Disruption from WAB Maschinenfabrik, Switzerland. Grinding chamber size is 215 L; dimensions are 1.9 m x 2.2 m x 1 m; weight is 2,150 kg. (Courtesy of Glen Mills, Maywood, NJ.)

3.9   Capital and Operating Costs

Cost information is generally hard to come by. One reason is that costs tend to be very situation-specific, and any attempt at generalization of cell separation costs is likely to lead to trouble. Depending upon the circumstances, cell separation by option A may be better or worse than its alternative option, B. The total operating costs must be worked out for a given situation, if possible by an experienced professional. Obviously, competition also accounts for the lack of good published cost information. Some numbers are provided here for perspective. Any decisions, however, should be made only after consultation with the equipment suppliers and appropriate analyses.

As in any other cell separation system, the costs for a microfiltration system depend very specifically on the job at hand. Such costs should be developed in collaboration with the manufacturer's representative, and should take into account the costs of cleaning, membrane replacement, maintenance, energy (pumping), feed pretreatment, and postprocessing steps. Membranes themselves cost in the range of $25–$300 per square foot. The installed systems with piping, valves, and pumps cost 4- to 6-fold that, and are thus in the range of $150–$1000 per square foot.

Rotary vacuum filters in the range of 1000–1500 ft² tend to have an installed cost of about $800,000–$1,400,000. Large-scale disk-stack–type centrifuges or desludging clarifiers tend to have an installed cost of about $300,000–$600,000 per machine. These costs include the normally required accessories and options, and the cost of installation. Operating costs should be developed for the specific situation at hand.

Delivered costs of cell disruption equipment from WAB Maschinenfabrik are in the range of $18,000–$20,000 for their KDL-pilot unit (1.4-liter grinding chamber), about $45,000 for the KD-15 (15-liter chamber), and about $120,000–$150,000 for the 200-liter machine. Installed costs may be 1.5 to 2 times these amounts, to account for required accessory equipment and utilities. Again, the supplier should be consulted to obtain a more accurate estimate. A simple model for cost comparisons between cell separation options can be constructed on a personal computer spreadsheet along the lines suggested in Table 3.4.

Table 3.4 Cell Separation Economics[a]

COST	Option A	Option B	Option C
Capital costs			
Installed cost[b]	850	1,110	850
Annual depreciation[c]	85	110	85
Operating costs			
Consumables	425	75	375
Waste disposal	110	70	65
Labor	40	30	80
Utilities and maintenance	60	80	60
Other overhead	20	15	25
TOTAL cost per year	740	380	690
AMOUNT OF EQUIVALENT PRODUCT MADE PER YEAR			
Million units	1.90	1.70	1.95
COST PER MILLION UNITS PRODUCT	389	224	354

(a) A possible spreadsheet for comparing alternatives for cell separation. For example, the use of conventional filtration, microfiltration, or centrifugation may constitute three different options; or the options may involve the same separation technique, but may differ in the degree of automation or the type of consumables required. Costs are in arbitrary units, for comparison purposes.

(b) Installed costs: the expected cost of the equipment, accessories, instrumentation, and the cost of their installation at the site of interest.

(c) Ten-year straight-line depreciation is assumed.

Additional levels of sophistication can be brought in by building underlying modules for the key entries. As shown in Table 3.4, consumables, waste treatment, or even utilities may tip the overall balance in favor of units with higher capital cost. Therefore, the installed cost of the equipment is only one of the important factors to consider in the overall operating costs. However, other considerations, such as confidence in the reliability of the process, consistency of the product quality, and regulatory factors, may lead one to select an option that has higher unit costs.

3.10   Acknowledgments

I would like to thank Komline–Sanderson Engineering Corporation, Sartorius Corporation, Alcoa/IWT Company, Sharples Div. of Alfa-Laval Separation, Inc., Westfalia/Centrico Inc., Biotechnology Development/Microfluidics Corporation, and WAB Maschinenfabrik/Glen Mills, Inc., for allowing the use of their illustrations and photographs. I would also like to note that the use of their photographs is for illustration only and does not imply an endorsement. The reader should investigate the options offered by all suppliers.

I am also indebted to my former colleagues at Genencor (T. BECKER, J. DEVENEY, I. HAAIJER, K. HAYENGA, J. KAN, B. LAWLIS, J. LORCH, and T. REVAK) for collaborations which have been the chief source of my understanding of cell separation issues.

3.11   List of Suppliers of Cell Separation Equipment

FILTER AIDS AND PRETREATMENT CHEMICALS

Allied Colloids, Suffolk, VA, 804-934-5700 (polymers/flocculents).
Dow Chemical, Midland, MI, 517-636-1000 (polymers as cell separation aids).
Grefco, Inc. (DICALITE), Torrance, CA, 213-517-0700 (filter aids).
Hercules, Inc., Wilmington, DE, 302-594-5000 (polymers/flocculents).
Manville Corp., Denver, CO, 303-978-2004 (filter aids and other services).
Nalco Chemical Co., Naperville, IL, 312-961-9500 (polymers/flocculents).

CONVENTIONAL FILTRATION EQUIPMENT

Bird Machine Company, Inc., South Walpole, MA, 508-668-0400.
Dorr–Oliver, Stamford, CT, 800-243-8160 (belt filters and other equipment).
EIMCO Process Equipment Co., Salt Lake City, UT, 801-526-2000 (filter presses, belt filters, and other items).
Gelman Sciences, Ann Arbor, MI, 800-521-1520.

Komline–Sanderson, Peapack, NJ, 201-234-1000 (rotary drum filters, belt filters, and other equipment).
Larox, Inc., Columbia, MD, 301-381-3314 (automatic pressure filters).
Perrin, Ajax, Ontario, Canada, 416-683-9400 (filter presses).
R & B Filtration Systems Co., Marietta, GA, (filter presses).
Serfilco, Ltd., Glenview, IL, 312-998-9300 .
Technical Fabricators Inc., Piscataway, NJ, 201-469-7373 (filter media).

CROSS-FLOW FILTRATION EQUIPMENT

Enka America, Inc. (AKZO), Asheville, NC, 704-258-5010.
Filtron Technology Corp., Northborough, MA, 508-393-1800.
Illinois Water Treatment Company Life Sciences Group (ALCOA), Rockford, IL, 815-877-3041.
LSL Biolafitte, Princeton, NJ, 609-452-7660.
Membrex, Inc., Garfield, NJ, 201-546-5242.
Millipore Corp., Bedford, MA, 617-275-9200.
Sartorius Corp., Bohemia, Long Island, NY, 516-563-5120.

CENTRIFUGES

Alfa-Laval, Fort Lee, NJ, 201-592-7800.
Bird Machine Co., South Walpole, MA, 508-668-0400.
Centrico (Westfalia), Northvale, NJ, 201-767-3900.
Dorr–Oliver, Stamford, CT, 203-358-3200.
Krebs Engineers, Menlo Park, CA, 415-325-0751 (cyclones).
Sharples, Inc., Warminster, PA, 215-443-4220 (continuous solid-bowl decanter and chamber-bowl centrifuges).
Tema Systems, Inc., Cincinnati, OH, 513-489-7811.

CELL DISRUPTION EQUIPMENT

Microfluidics Corp., Newton, MA, 617-965-7255.
Willy A. Bachofen AG Maschinenfabrik, Basel, Switzerland; (United States Distributor: *Glen Mills, Inc.*, Maywood, NJ, 201-845-4665).

DIRECTORIES

The following publications include more detailed listings of suppliers:
1. *Thomas Register of American Manufacturers*, Thomas Publishing Co., New York, NY, 212-695-0500.
2. *Chemical Engineering Catalog*, Penton/IPC Reinhold, Stamford, CT, 203-348-7531.
3. *Chemical Processing Guide & Directory*, Putnam, Chicago, IL, 312-644-2020.
4. *Chemical Engineering Buyers' Guide*, McGraw Hill, New York, NY, 212-512-2000.

3.12 References

1. MICHAELSEN, T.; Hapten/Anti-Hapten Affinity Linking in Cell Separation, World Patent WO 9 101 368 (Feb. 7, 1991).
2. THILLY, W. G., TYO, M. A.; Nested Cone Separator, World Patent WO 9 106 627 (May 16, 1991).
3. BLOOMBERG, C. A.; Cell Separation Invention, World Patent WO 9 104 318 (April 4, 1991).
4. OKARMA, T. B.; Device and Process for Cell Capture and Recovery, European Patent EP 405 972 (Jan. 2, 1991).
5. VARY, C. P. H.; Isolating Cells by Elution From Cationically Derivatized Cellulosic Material, World Patent WO 9 102 816 (Mar. 7, 1991).
6. BAUMANN, E. R., Applying "Old" Filtration Technology Effectively, *Chem. Process.* 52 (2), 30–34 (1989).
7. Current Good Manufacturing Practice in Manufacturing, Processing, Packing or Holding of Drugs, *Fed. Regist.* 56 (29), 5671–5674 (Feb. 12, 1991).
8. HARRISON, F. G., cGMPs in Plant Design and Operation, *BioPharm* 2 (3), 24–28 (1989).
9. NIH Guidelines for Research Involving Recombinant DNA Molecules, *Fed. Regist.* 51 (123), 23367–23393 (June 26, 1986).
10. MAIGETTER, R. Z., BAILEY, F. J., MILLER, B., Safe Handling of Microorganisms in Small and Large Scale BL-3 Fermentation Facilities, *BioPharm* 3 (2), 22–28 (1990).
11. SEKHAR, C., Guidelines for Large Scale rDNA Fermentations, *Chem. Eng.* 92 (9), 57–59 (1985).
12. THALER, T., Cell Culture Perfusion Systems, *Genet. Eng. Biotechnol.* 12, 20–22 (1992).
13. TILLER, F. M., WILENSKY, J., FARRELL, P. J., Pretreatment of Slurries, *Chem. Eng.* 81 (9), 123–126 (1974).
14. THOMAS, C. M., Facts About Filtration Pretreatment, *Am. Inst. Chem. Eng. Symp. Ser.* 73 (171), 18–25 (1977).
15. COOK, L. N., Laboratory Approach Optimizes Filter-aid Addition, *Chem. Eng.* 91 (14), 45–50 (1984).
16. MURKES, J., CARLSSON, C.; *Cross-Flow Filtration: Theory and Practice*; John Wiley & Sons, New York, 1989.
17. ANON., Filtration Special Report, *Bio/Technology* 5, 915–919 (1987).
18. MICHAELS, S. L., Crossflow Microfilters: The Ins and Outs, *Chem. Eng.* 96 (1), 84–91 (1989).
19. HAARSTRICK, A., RAU, U., WAGNER, F., Cross-flow Filtration as Method of Separating Fungal Cells and Purifying the Polysaccharide Produced, *Bioprocess Eng.* 6 (4), 179–186 (1991).
20. KRONER, K. H., NISSINEN, V., ZIEGLER, H., Improved Dynamic Filtration of Microbial Suspensions, *Bio/Technology* 5, 921–926 (1987).
21. PARKINSON, G., Novel Separator Makes Its Debut, *Chem. Eng.* 96 (1), 44–48 (1989).
22. KNIGHT, P., Downstream Processing, *Bio/Technology* 7, 777–782 (1989).
23. ANON., Plant Notebook: A Larger Centrifuge Separates More Materials in Less Time, *Chem. Eng.* 96 (1), 149 (1989).

24. STRYKER, M. H., WALDMAN, A. A.; Blood Fractionation; In: *Kirk-Othmer Encyclopedia of Chemical Technology*, 3rd ed.; John Wiley & Sons, New York, 1978; Vol. 4, pp 25–61.
25. KUHN, R. A., COSSETTE, I., FRIEDMAN, L. I., Optimum Centrifugation Conditions for the Preparation of Platelet and Plasma Products, *Transfusion 16*, 162 (1976).
26. BAUER, J., STUENKEL, K. G. E., Isolation of Human B-cell Subpopulations for Pharmacological Studies, *Biotechnol. Prog. 7* (5), 391–396 (1991).
27. ANON., Clarifier Isolates Enzyme By Removing Cell Debris, *Chem. Process. 45* (May), 94–95 (1982).
28. MARFFY, F., KULA, M.-R., Enzyme Yields from Cells of Brewers' Yeast Disrupted by Treatment in a Horizontal Disintegrator, *Biotechnol. Bioeng. 16* (5), 623–634 (1974).
29. ZWERNER, R. J., WISE, K. S., ACTON, R. T.; Harvesting the Products of Cell Growth; In: *Methods in Enzymology*, Vol. LVIII; JAKOBY, W. B., PASTAN, I. H., Eds.; Academic Press, New York, 1979; p 223.
30. CRISTOFALO, V. J., KABAKJIAN, J.; Processing Cells for Enzyme Assays; In: *Tissue Culture*, KRUSE, P. F., PATTERSON, M. K., Eds.; Academic Press, New York, 1973; p 206.

Tangential Flow Filtration Systems for Clarification and Concentration

Eric A. Rudolph and Jeff H. MacDonald

Contents

4.1   Introduction

Our objective in this chapter is to provide engineers and scientists involved in the design of pilot-plant and production facilities with an introduction to tangential flow filtration (TFF) system design and requirements. We attempt to create a basic understanding of TFF so that this technology can be integrated into an overall process scheme. Fundamentals of design are provided in order to better explain the general design logic. From this design basis, processing alternatives are more easily addressed.

4.1.1   Theory

Tangential flow (or *cross-flow*) *filtration* is a process that uses membranes to separate components in a liquid solution or suspension on the basis of their size. Microporous or ultrafiltration membranes are generally used in the downstream processing of recombinant products. *Microporous membranes* have pore diameters in the range from 0.05 to 5 μm, and are used for separations requiring the retention of suspended solids 0.2–10 μm in diameter. *Ultrafiltration membranes* have pores from 0.005 to 0.05 μm, and are used to retain macromolecules whose molecular weights range from 1,000 to 1,000,000 daltons. These membranes are available in various materials of construction and geometries, as seen in Table 4.1.

As in traditional filtration, components smaller than the membrane pores pass through the membrane when pressure is applied. The pressure required varies with the nature of the material being separated, but, in general, microporous membranes require pressures of less than 1 psig up to 30 psig (< 7 to 200 kPa), and ultrafiltration membranes require pressures of 10 to 100 psig (70–700 kPa). For pure solvent streams the filtrate rate increases linearly with applied pressure, but in biological applications the solution being filtered contributes an additional hydraulic resistance to that of the membrane because of the phenomena of gel polarization and membrane fouling. The filtration is described by the relationship:

$$J = \mathrm{TMP}/(R_m + R_g + R_f) \tag{4.1}$$

where J is the membrane flux or rate of filtrate passage through the membrane per unit area (L/h·m^2);

TMP is the transmembrane pressure;

R_m is the resistance to flow through the membrane;

R_g is the resistance to flow through the gel polarization layer; and

R_f is the resistance to flow due to membrane fouling.

Figure 4.1 illustrates tangential flow filtration in which the key parameters of tangential flow, q_f, and transmembrane pressure, TMP, can be described by the feed pressure, P_f, retentate pressure, P_r, and permeate pressure, P_p, as follows:

Table 4.1 Partial List of TFF Media—Materials and Geometries

Material	Pleated sheet	Tubular	Spiral wound	Tubular multichannel	Hollow fiber	Flat sheet
1. Polymers						
Cellulosics	XXX		XXX		XXX	XXX
Polysulfone		XXX	XXX		XXX	XXX
PVDF		XXX	XXX			XXX
Acrylic					XXX	
Polypropylene					XXX	XXX
Nylon	XXX				XXX	XXX
2. Ceramics						
Alumina				XXX		XXX
Zirconia/alumina				XXX		
Zirconia/sintered metal		XXX				
Zirconia/carbon		XXX				
Silica		XXX				
Silicon carbide				XXX		
3. Sintered metal						
Type 316 stainless steel		XXX				
Other alloys		XXX				

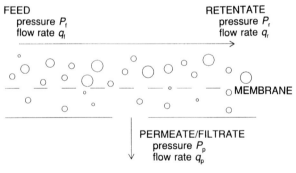

FEED
pressure P_f
flow rate q_f

RETENTATE
pressure P_r
flow rate q_r

MEMBRANE

PERMEATE/FILTRATE
pressure P_p
flow rate q_p

Figure 4.1 Pressure Relationships in Tangential Flow Filtration. The key parameters of tangential flow filtration are the tangential flow rate, q, and transmembrane pressure, TMP, defined as $\{(P_F + P_R)/2 - P_P\}$.

$$q_f \sim \Delta P; \qquad \Delta P = (P_f - P_r) \tag{4.2}$$

$$\text{TMP} = (P_f + P_r)/2 - P_p; \qquad P_p = 0 \text{ (typically)} \tag{4.3}$$

Tangential flow filtration differs from traditional through-flow filtration in that the flow tangential to the membrane surface acts to sweep the retained components from the membrane surface and back into the bulk of the solution, thus avoiding the plugging effects inherent in through-flow filtration.   Further information on the mechanisms and theory of tangential flow filtration and the effects of gel polarization (R_g) and fouling (R_f) can be found in the literature [1–6].

Gel polarization is the concentration of the retained species at the membrane surface that acts as a resistance to flux. One can control the degree of gel polarization by increasing the fluid shear rate along the surface of the membrane, which, in practical terms, is governed by the flow rate, q_f, for a specific module design.   To determine the fluid shear in a TFF device, divide the velocity within the TFF device by its effective channel height.   For modules that operate in the laminar flow regime (flat sheet, spirals, and hollow fibers), the flux increases as the one-third power of the tangential flow or shear:

$$J = K_1 C + K_2 q_f^{1/3} \tag{4.4}$$

where K_1 and K_2 are constants for a given system, C is the solute concentration, and q_f is the tangential flow rate.

In devices that operate in turbulent flow regimes [7], the flux increases proportionally to the tangential flow or shear:

$$J = K_1 C + K_2 q_f \tag{4.5}$$

Fouling of the membrane can be a result either of adsorption of a component (such as antifoam or protein) to the filter, or of the solidifying of the polarization layer [8]. Membrane fouling can be reduced by limiting process time, minimizing or avoiding the fouling agents (by switching or lowering amount of antifoam), or changing to a more suitable membrane chemistry.   In the case of biological processing, the use of hydrophilic membranes is generally preferred.

4.1.2   TFF Applications

TFF applications can be divided into three basic unit operations: concentration, clarification, and diafiltration. *Concentration* applications are those in which the desired product is retained by the membrane, and the solvent passes through the membrane, resulting in a concentrated product.   Concentration applications in biotechnology include the harvesting of yeasts, bacterial cells, animal cells, cell lysates, and protein precipitates, and the concentrating of soluble macromolecules. The keys to successful application of TFF are (a) near-quantitative recovery of the concentrated product from the system, and (b) complete retention of the desired species.   The percent retention, R, of a particular component, C, is expressed as:

$$R = (1 - C_p/C_r)100 \tag{4.6}$$

where C_p and C_r denote the concentrations of the particular species in the permeate and retentate, respectively.

Clarification applications involve the separation of macromolecules (proteins, polysaccharides, or viral particles) from cells, cell lysates, or protein precipitates. These separations require the proteinaceous material to pass through the membrane while the insoluble material is retained, and are affected by the potential formation of a gel layer at the membrane surface, with resultant fouling. The gel layer and fouling are primarily dependent on the membrane chemistry (such as hydrophilic versus hydrophobic), and can be strongly influenced by the process flux or TMP. All of the factors mentioned above are critical to maximizing the percent transmission, T, of the macromolecule, C, through the membrane:

$$T = 100 - R \tag{4.7}$$

Diafiltration applications are extensions of concentration and clarification applications. Diafiltration involves adding fresh solvent or buffer solution to the system, to wash solute through the membrane. This technique can be used to enhance yields in clarification applications, exchange buffers before or after chromatography steps, and wash contaminants away from the retained species. The yield of a diafiltration step is maximized when membrane retention of the product is maximized and the addition of wash solution is matched to the permeate flow rate, in order to provide the most efficient exchange of solution (infinite dilution versus stepwise). The extent of washing is generally referred to in *number of wash volumes*, N, which is the ratio of the amount wash solution added, V_w, to the system operating volume, V_o:

$$N = V_w / V_o \tag{4.8}$$

The final concentration of any species, C_f, after N wash volumes can then be determined by:

$$C_f = C_i^{-N(1-R)} \tag{4.9}$$

where C_i is the initial concentration of that species.

4.1.3   Module Design

Module design, defined as the membrane medium and support structure, is critical to the performance of the TFF system. Factors to consider in module design include flux, solids handling capabilities, optimizing cost, minimizing pumping, and chemical and physical compatibility, as well as scalability. Some combination of these parameters goes into each particular module design. Details of module design can be found in the literature [3, 9, 10].

The TFF membrane, whether it is an ultrafilter or microfilter, is supported by a TFF membrane module. There are two classes of membrane modules, open-channel and turbulence-promoted. *Open-channel devices* are available with unsupported hollow fibers, supported polymer tubes, or rigid sintered ceramic or metal tubes, as well as in the plate-and-frame (flat sheet) configuration. *Turbulence-promoted modules* are available in either cassette (flat sheet) or spiral configurations. Figure 4.2 illustrates the different module geometries.

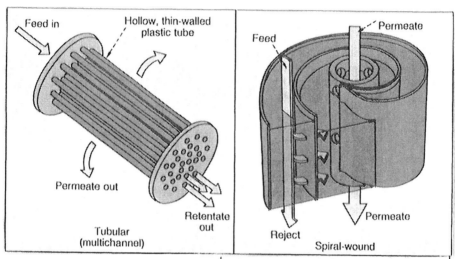

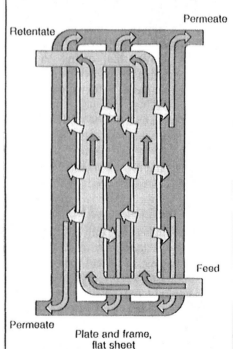

Figure 4.2 Tangential Flow Module Designs. All tangential flow filtration membranes must be supported by membrane modules, which are either open-channel or turbulence-promoted devices. Three typical membrane module geometries are tubular, spiral-wound, and plate-and-frame configurations.

Open-channel devices are generally designed to process fluids containing high concentrations of suspended solids or large particles (usually less than 40% wet weight). However, reasonable filtration rates in open-channel devices require a high degree of recirculation, to create the necessary shear at the membrane surface.

Although highly concentrated suspensions may plug narrower flow channels in turbulence-promoted TFF devices, turbulence-promoted channels are important to minimizing the pumping requirements. Turbulence-promoted devices are generally used for solutions containing dissolved solids but low concentrations (< 1% by weight) of suspended solids. In general, turbulence-promoted devices require one-third to one-fifth the pumping rate of their open-channel counterparts.

Open-channel designs are preferred in applications where cells, cellular debris, and precipitated products are to be retained. If suspended solids are not present, turbulence-promoted or narrow-diameter hollow fibers are preferred, since they preserve the value of the product by reducing the holdup volume losses associated with higher pumping rates and associated pipe sizes. These reduced pumping rates and line sizes may also provide secondary savings in system cost and space conservation.

4.1.4   System Design

The selection of the appropriate system design should be based on the overall process scheme. If the TFF system is to be used in a continuous high-volume operation, a multistage continuous TFF system is best. However, if the TFF system is to be used in a process that is carried out in batches, a batch TFF system is appropriate.

A batch TFF system is shown in Figure 4.3. This most basic system requires a feed tank, pump, pressure gauges, and filter. The tank serves as a reservoir for the solution being processed; as the material is filtered, the volume in the tank is reduced. The pump provides both the pressure to push the fluid through the membrane and the fluid velocity to keep the retained components from settling on and plugging the membrane. The pressure gauges are the most useful instruments for monitoring the recirculation rate, as described in eq 4.2 and 4.3. With batch systems, the process flux is initially high and decreases as the retained species are concentrated. Batch systems provide the most efficient separation, while requiring the minimum membrane area. Batch systems are also simpler and therefore less expensive.

Continuous TFF systems can be either single- or multistage. An example of a single-stage system is presented in Figure 4.4. In the single-stage continuous process design, the retentate is bled from the system at a constant concentration. In situations where the flux falls off dramatically with increases in concentration, this single-stage design may not be appropriate. In these cases, multistage design, where successive stages operate at increasing concentrations, is likely to be more effective. This system is illustrated in Figure 4.5. As additional stages are added, these multistage systems approach the efficiency of a batch system. However, with each stage additional pumping and controls are required, which severely affect the capital cost of the system.

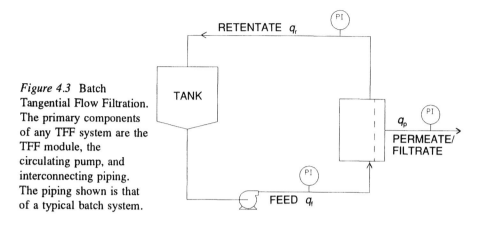

Figure 4.3 Batch Tangential Flow Filtration. The primary components of any TFF system are the TFF module, the circulating pump, and interconnecting piping. The piping shown is that of a typical batch system.

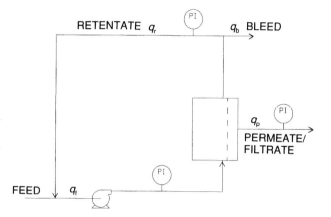

Figure 4.4 Single-Stage Continuous TFF. In the single-stage continuous process design the retentate is bled from the system at a constant concentration. Typical single-stage piping is shown.

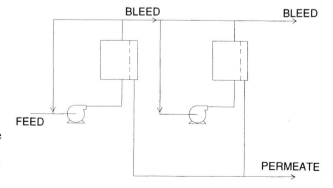

Figure 4.5 Multistage Continuous TFF. In a multistage continuous process system, the retentate from each stage is the feed for the next stage. Typical two-stage piping is illustrated.

4.2   TFF in Biotechnology

Three expression systems are most frequently used in biotechnology: inclusion bodies in *E. coli*, products expressed from yeast fermentation, and products expressed from mammalian cell cultures. There are obviously others, but for the purposes of maintaining a degree of generality, the applications discussion covers mammalian cell culture and bacterial fermentation products. Figures 4.6 and 4.7 show the areas in which TFF can be applied to mammalian cell culture and bacterial fermentation systems for product purification.

4.2.1   Mammalian Cell Culture

Because of the relatively low amount of solids (generally 1×10^7 cells/mL and < 1 mg/mL protein) in mammalian cell culture, the initial harvest of cells and subsequent protein concentration steps are relatively simple. Specific process details and results in the processing of mammalian cell cultures and their related protein purification steps are well documented in current literature [11–15]. Figure 4.6 shows the areas in which TFF can be applied, and Table 4.2 provides process data as well as membrane performance values.

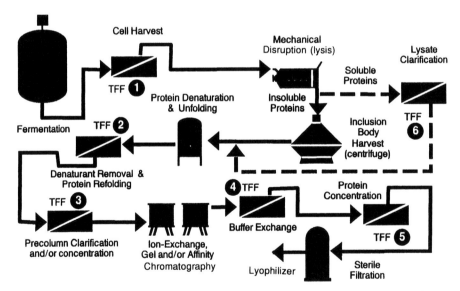

Figure 4.6 Tangential Flow Filtration in Protein Processing from Mammalian Cell Culture. The areas (1–6) in which TFF can be applied in the processing of proteins derived from mammalian cell cultures are shown.

Table 4.2 Mammalian Cell Culture Applications: Processing Conditions

Application	Membrane cutoff, μm or MW	Temp., °C	Maximum solids	TMP, psi	Flux, L/h·m²
Harvesting	0.2–1.0	4	10^8 cells/mL	< 1	30–60
		30			60–120
Clarification	0.2–1.0	4	10 g/L	1–2	50–100
Concentration/ diafiltration	10,000– 1,000,000	4	20%	20–40	10–100

During the initial harvest or the clarification of the extracellular protein, the cells should be subjected to only the mildest of forces. Gentle process conditions (less than 1–2 psi TMP and shear rates below 2000 sec⁻¹) are recommended, in order to keep the cells from lysing. If cell lysis can be avoided during harvesting, the yield does not suffer; and, in the case of clarification, avoiding cell lysis minimizes possible complications to downstream purification steps. In this application, hydrophilic membranes are preferred because they minimize loss of the dilute product; micro-porous membranes are generally used, in order to exploit the difference in size between the cell and the medium constituents; and the module design can be either open-channel or turbulence-promoted, with open-channel devices preferred because the pressure drop on the feed side of the device is lower.

In the initial product enrichment step, the volume of the culture medium can be reduced 10- to 100-fold. Process temperatures for this step are typically 4–8 °C. The degree of concentration obtainable is generally limited either by the amount of serum in the medium, or by the volumetric requirements of the process system.

In subsequent protein clarification steps, the process volume is much smaller and the product values per unit volume are high. At this stage, the use of any prefiltration step prior to column chromatography should be carefully weighed, as a function, on the one hand, of product loss associated with an additional step, and, on the other hand, of the incremental decrease in column life if the product is not prefiltered. TFF can minimize the product losses of prefiltration prior to chromatography by operating under conditions that avoid gel polarization, thus maximizing product transmission through the membrane. Furthermore, the TFF step can incorporate a diafiltration procedure and thus avoid holdup losses.

The concentration and diafiltration (buffer exchange) of the clarified proteins after a chromatography step also involve the handling of small volumes of expensive product. The membranes used in these applications need to be highly retentive of the product, and must allow for the exchange of appropriate buffer solutions. The retention characteristics of the membrane are critical, due to the potential loss of product with each volume of buffer exchanged. For example, if a membrane is 98%

retentive of the product, a five-volume wash results in a 10% loss of product to the permeate. The membrane cutoff employed is generally one-third to one-fifth of the molecular weight cutoff (MWCO) of the target protein. The holdup volume of these systems should be minimized to avoid product losses. Systems that have lower holdup volume losses generally have lower pumping requirements.

4.2.2   Bacterial Fermentation Systems

For processes based on bacterial products, TFF can be used to concentrate the cells prior to disruption, and to clarify the protein solution by removing the insoluble cell debris and inclusion bodies. After the proteins of interest are resolubilized, the applications can be generalized to those of protein concentration and diafiltration, described in the previous paragraphs. The specific process details and typical TFF results for the processing of bacterial fermentations (and subsequent protein purification steps) are well documented in current literature [16–21].

Figure 4.7 shows areas in which TFF can be applied in bacterial processes, and Table 4.3 provides process data, as well as membrane performance values.

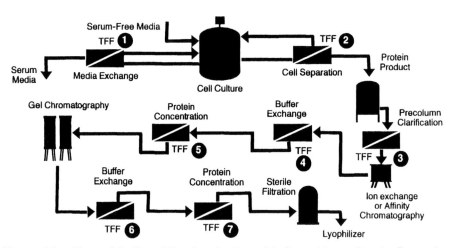

Figure 4.7   Tangential Flow Filtration in Bacterial Recombinant Protein Processing. Opportunities (1–7) are illustrated for the use of TFF technology in the processing of recombinant proteins derived from bacteria.

In the initial harvest of bacterial cells, a high degree of retention is desired, and either microfilters (< 0.2 μm) or ultrafilters (MWCO > 100,000) can be employed. The use of microfilters is recommended if processing times are short (less than 2

Table 4.3 Processing Conditions for Proteins from Bacterial Processes

Application	Membrane cutoff, μm or MW	Temp., °C	Maximum solids (wet)	TMP, psi	Flux, L/h·m²
Harvesting	100k–0.2	4–8	10–60%	10–50	20–80
Lysate clarification	0.1–0.2	4	1–10%	1–2	10–50
Clarification	0.2–1.0	4	10 g/L	1–2	50–100
Protein concentration diafiltration	10,000–1,000,000	4	20%	20–40	10–100

Data presented are derived from personal experience as well as from literature sources. The figures shown are typical of a broad range of applications, but the literature should be consulted for more detailed data in the area of the reader's interest.

hours) or when certain requirements, such as compatibility with live steam, are present. Ultrafiltration membranes can provide a flux that is the same as, or even greater than, that of microporous membranes, and have the advantage of being more easily regenerated [8]. Since bacteria are robust, high-shear and -pressure conditions can be employed to maximize filtration rate. However, the broths may contain antifoam chemicals. If the antifoam present is polyethylene glycol or polypropylene glycol, the medium exhibits an inverted cloud point; in other words, the antifoam comes out of solution and impedes filtration performance. The temperature should be maintained below the cloud point to keep the antifoam in solution and not allow it to deposit on the membrane. If a silicone-based antifoam is present, it tends to foul hydrophobic membranes, reducing the filterability of the broth.

The concentration of the bacterial cells causes the suspension to thicken and its apparent viscosity to increase. Bacterial suspensions are generally concentrated to no more than 40% wet weight, or until the TFF device reaches its maximum operating pressure.

The technique of clarifying cell lysates is similar to that of harvesting mammalian cells. The objective in the processing of cell lysates is to maximize the transmission of large macromolecules through the membrane and the gel layer, which requires matching the product flux with the solvent flux through the membrane and its associated gel layer. Achieving equivalent flux rates requires the minimizing of membrane fouling and the throttling of solvent flux, which in practical terms is done by operating at a low TMP. It is often advisable to incorporate a diafiltration step with the lysate clarification, to ensure quantitative yields. Diafiltration can lead to considerable volume expansion, but this may not be of great concern, since the following step is generally a concentration step with ultrafiltration. The actual increase in processing time due to this subsequent concentration step may be negligible, since the more dilute solution can be processed faster. Furthermore, the two steps could be integrated to reduce the processing time.

4.2.3   Shear

The effect of shear on the denaturation of proteins has been examined in several recent papers [22–25]. The clear result is that shear does not denature the proteins tested, under the conditions used in these studies. Denaturation was found to occur at gas-liquid interfaces, at elevated temperatures, and at high pH. Table 4.4 presents a summary of these results.

Table 4.4 Summary of Shear Effects on Proteins [22–25]

Protein	Shear rate, s^{-1}	Time, h	Temp., °C	pH	Activity loss
Alcohol dehydrogenase	9,000	15	30	7	no
Dehydrogenase	9,000	2	40	8.8	yes
Dehydrogenase	24,000	1	20	7	no
Bovine serum albumin	10,000	*	25	7	no

* Less than one minute.

4.3   TFF System Design

4.3.1   Defining the TFF Application

The most fundamental aspect of any equipment design and installation is to define the application in terms of the fluid to be processed, its physical characteristics, the volumes to be processed, the rates desired, and the operational and environmental constraints involved. Beyond the essential processing information, the nature of the product (its stability as a function of time, temperature, pH, or mechanical shear) needs to be understood. Also, an understanding of the nature of the processing steps before and after the TFF step is required, so that the constraints imposed are consistent throughout the process. Finally, the environment in which the equipment will be operated needs to be defined with regard to the degree of containment and cleanliness.

When the application has been defined, as outlined above, a detailed design integrating the application requirements with the TFF system requirements can be undertaken.

4.3.2   System Design

As seen in Figure 4.3, the primary components of the system are the TFF module, the circulating pump, and interconnecting piping. In this section we describe the selection and specification of the TFF system's primary components.

TFF module selection and configuration is the first consideration. What design of TFF membrane module to choose depends on the application. TFF membrane modules have defined membrane surface areas. The required area for a particular application is calculated as the volume of material to be permeated through the membrane, divided by both the desired process time and the average process flux. Where multiple membrane modules are required, they can be plumbed in series or in parallel to meet the need of the application. The configuration of multiple modules should be discussed with the manufacturer; considerations such as draining, service, and manifold cost need to be reviewed. It is important to minimize the pumping requirements of the system; this may lead to a design in which modules are assembled in series, as long as the pressure limitations of the device are not exceeded.

When the membrane module and its configuration have been specified, the desired pressure and flow parameters for the pump will have been determined, and the pump and piping can then be specified. The physical properties of viscosity and density should be known for the product as it exists both before and after the TFF step.

Pipe sizing is based on meeting the pumping requirements of the TFF system. The line size needs to be of sufficient diameter to meet the suction head requirements of the pump(s), allowing for minimum CIP circulation velocities of 5 ft/s [26]. It is good practice to avoid reducing the pipe diameter below the diameter of that leading to the pump suction. This rule also carries through to the specification of the bottom outlet of the process tank. The process piping on the discharge side of the pump, leading to the membrane module and back to the retentate tank, should be of sufficient diameter not to cause appreciable friction losses. In addition, the line sizes and system piping design should minimize system holdup volume. Table 4.5 provides a guide for line size based on water.

Table 4.5 Characteristics of Sanitary Tubing

OD, in	ID, in	Flow rate, water at 5 ft/s, gpm	Internal volume, mL/ft
0.50	0.37	1.7	21
0.75	0.62	4.7	59
1.00	0.86	9.0	113
1.50	1.36	22.5	284
2.00	1.86	42.1	532
2.50	2.36	67.9	857
3.00	2.86	99.8	1259

4.3.3   Pumps

Pump sizing and specification are discussed in detail in Chapter 8. Issues involving seal selection and drives should be reviewed there. However, TFF applications have specific needs that must be met, since the pumping rates, environment, and time can be more demanding than in other applications. The biopharmaceutical application requirements can include:

(a)   *Aseptic operation*: The pump must be designed to be sanitized or sterilized.

(b)   *Containment*: All pumps are designed to contain the fluid they are pumping, with different pump designs offering various levels of assurance. Pumps without seals, such as tubing or diaphragm pumps, may pose operational hazards if appropriate modifications are not made to allow for the possible rupture of a hose or diaphragm. Double-seal mechanical pumps offer a higher level of assurance than do single-seal pumps, making them ideal for processing hazardous fluids.

(c)   *Low fluid shear/high efficiency*: This requirement is important for pumping mammalian cells and certain solid precipitates, and also minimizes heat input.

(d)   *Dynamic fluid characteristics*: Significant changes in the fluid characteristics of density and viscosity can occur in the course of concentration applications involving macromolecules and cells. The pump should be capable of handling these variations.

(e)   *Ease of operation and maintenance.*

The pumps that are installed in TFF systems must provide the required flow rates and pressures, as well as satisfy a combination of the above criteria. The most common pump types utilized are peristaltic, rotary lobe, and centrifugal.

Peristaltic pumps are good for applications with relatively low flow rate and low pressure. They are available with flow rates ranging from 0.5 mL/h to over 100 L/min, and pressure rating up to 30 psig [27]. Some types of tubing for these pumps can be sterilized by autoclaving. The containment of these systems is generally good, but they are subject to human error in the installation of the tubing and operation of the system: operators either may not secure the tubing, or may turn on the pump with a downstream valve closed. In either case the tube will probably come free, spilling the fluid at the pumping rate of the system. Since TFF systems operate at relatively high volume turnover rates, the process batch can easily be lost in a few minutes. Peristaltic pumps generate little to no shear on the process fluid. Because peristaltic pumps are of the positive-displacement type, they can also handle changes in product viscosity. However, the resulting pressure may be too great for tubing pressure limitations, which can cause the tubing to rupture. These pumps are inexpensive, easy to operate, and easy to maintain.

In general, peristaltic pumps are ideal in the final stages of purification, when the process volumes have been reduced to a few liters, and the autoclaving of small tubing assemblies is manageable (if aseptic processing is essential). In TFF systems, these pumps should be closely monitored, and, if possible, pressure switches should be employed to turn off the pump before a high pressure/vacuum condition is reached.

Rotary lobe pumps are appropriate for applications with high flow rate and high pressure. They have a flow rate range of 4 gpm to 300 gpm, and pressure rating exceeding 100 psig [28]. They can be sterilized by steaming, and condensate removal from the pump cavity is simple. These pumps are appropriate for shear-sensitive fluids and, when fitted with appropriate rotors and operated at lower rpm, can pump mammalian cells without reducing the viability. Rotary lobe pumps are of the positive-displacement type, and therefore can handle the changes in product viscosity seen in most biopharmaceutical applications, while maintaining flow rates and containing the fluid at the increased pressure. These pumps are relatively expensive, and require a lot of maintenance. When equipped with reversible, variable-speed drives, rotary lobe pumps are easy to operate. In general, these pumps are ideal for TFF applications involving cells, cell lysates, or viscous protein solutions.

Centrifugal pumps offer high flow rates and moderate pressure heads, with a flow rate range of 4 gpm to 300 gpm and pressure rating up to 100 psig [29]. They can be sterilized by steaming, with condensate removal from the pump cavity by a casing drain. Because they operate at fixed speed, they aerate fluids upon startup; this aeration has been reported to be a primary cause of product denaturation [12]. Since these pumps generally are operated at high rotor speeds (up to 3500 rpm), they are generally not appropriate for mammalian cells or other products whose activity may be affected by mechanical shear. Centrifugal pumps are not of the positive-displacement type; therefore, if they are not sized for the expected process fluid viscosity changes, their pumping performance may deteriorate. These pumps are inexpensive, reliable, and simple to maintain. Since they have a defined discharge pressure, they are inherently fail-safe and do not require overpressure protection. In general, centrifugal pumps are reliable, and if they are started up in a flooded system they should not affect the activity of proteins [11]. These pumps can be used in steps involving fluids whose viscosities do not change significantly, such as initial broth clarification, and initial stages of clarified solution diafiltration or concentration. These pumps should be strongly considered in cases sensitive to high capital costs. If a variable-speed drive is incorporated, these pumps may be better suited for biopharma-ceutical applications.

Other types of pumps that can be considered include diaphragm, progressing-cavity, gear, and sine pumps. These pumps, although successfully used in the industry, are not seen as often as the three types we have discussed.

4.3.4   Tank Design

During the initial concentration steps of TFF processes, the process volume may be reduced to 1–10% of the original volume, with very high recirculation rates. Therefore, tank design and the proper integration of pump, piping, and the TFF modules are critical. When the volume of fluid has shrunk to 10% of the original volume, tank levels can be difficult to monitor, and may not provide sufficient suction head for the pump. This situation can lead to aeration of the fluid, and possibly to denaturation of the product.

To manage these high recirculation rates, and depending on the application, some combination of the following tank design characteristics, should be incorporated into the TFF system design:

(a)   a conical bottom with steep walls;

(b)   an antivortexing device at the tank bottom port; and

(c)   an inversely tapered dip tube to return the fluid retentate to the bottom of the vessel. The tapered shape reduces the fluid velocity as it leaves the dip tube and enters the tank.

Another way of managing the process volume turnover rates is to return a portion of the retentate directly to the pump suction, satisfying the NPSH requirements of the pump. The ratio of the flow rates to the process tank and to the pump suction depends on the application. This design is referred to as a *modified batch system*, and is shown in Figure 4.8. In general, it is wise to return as much fluid as possible to the process tank to avoid overconcentrating the fluid in the recirculation loop.

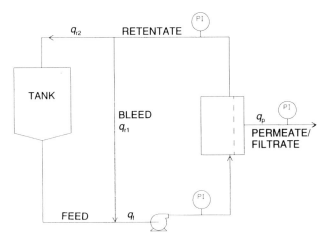

Figure 4.8 Modified Batch TFF. In a modified batch process system, a portion of the retentate is returned directly to the pump suction to manage the process volume turnover rates. A typical system is shown.

4.4   TFF Operation and Its Relationship to Design

A typical operating cycle for a TFF system is laid out in Table 4.6. These process steps may vary with the product/process requirements, and can be grouped into four

Table 4.6 Typical TFF Operating Cycle

Step	Description	Time, hours
1	Flush sanitant	0.25
2	Test integrity	0.25
3	Process fluid	4.00
4	Recover material	0.25
5	Flush system	0.25
6	Clean system	2.00
7	Sanitize system	0.50
	Total	7.5

steps: system preparation, product processing, product recovery, and system regeneration (CIP/SIP).

4.4.1   System Preparation

System preparation involves flushing the storage solution from the system and confirming its removal, either by a chemical test method or by conductivity, and also involves the integrity testing of the filters (performed to confirm that the membrane has not developed any defects prior to introduction of fluid to the system). In the case of operations involving the steaming of membranes, the integrity test should be performed after the steaming cycle, prior to product processing.

Generally, an operation begins with the flushing out of the storage solution. After the flush with reverse osmosis permeate or Water For Injection, the system membranes undergo integrity testing, which involves the application of air pressure at the TFF supplier's recommended pressure, and measurement of the resulting diffusion rate on the downstream side of the filter with an air flow indicator. Diffusion testing is the integrity test method of choice due to the relatively large surface area in the systems [30].

In the validation of aseptic operations, the integrity of the hydrophobic vent filter must also be tested. This can be done if an isolation valve is installed between the vent filter and the process system, and a purge valve between the isolation valve and the vent. The vent filter can then be wetted from the outside, with the alcohol draining through the purge valve. After the vent filter is wetted, a nitrogen line is connected and the vent filter integrity tested. When the vent filter exceeds its bubble point, the TFF filter may be diffusion tested. This arrangement is illustrated in Figure 4.9.

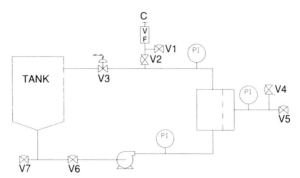

Figure 4.9 Integrity Testing of TFF System with Vent Filtration. The procedures for integrity testing a TFF system with vent filtration are as follows:
- A. Drain system through V5 and V7 (TFF membrane is wetted). Close V2, V3, V5, V6, and V7.
- B. Open V1. Pump wetting agent into vent filter (VF) through connection C and out V1.
- C. Apply air pressure to vent filter at connection C and note bubble point through V1.
- D. Close V1 and open V2. Set air diffusion pressure at connection C and measure diffusion rate through V4.

4.4.2   Product Processing

Process control of a TFF system involves maintaining the desired operating parameters of tangential flow rate, pressure, and fluid temperature.

Tangential flow control depends on the circulating pump selected. With centrifugal pumps, the flow rate is controlled by a control valve (generally diaphragm type) on the discharge of the pump. With positive-displacement pumps, such as peristaltic or rotary lobe pumps, fluid flow is generally controlled by use of a variable-speed drive. The flow can be monitored directly with an appropriate flow indicator, or, if viscosity and specific gravities are well defined, the retentate flow can be monitored by the pressure differential between the feed and retentate pressure gauges. If the system retentate is directed to both the pump suction and the process tank, appropriate throttling valves and flow indicators may be incorporated into the design.

The transmembrane pressure (TMP) of the system is controlled by a constriction that reduces the flow rate on the retentate side of the system. This type of control is typically used for bacterial harvesting, protein concentration, or buffer exchange applications. In some processes, particularly the harvesting of mammalian cells or the clarification of macromolecules from lysates, a finer degree of control is required, to avoid rupture of the mammalian cells or retention of the macromolecules because of excessive gel polarization. To achieve this finer TMP control, the permeate is throttled with either a valve or metering pump.

Temperature control of the fluid in many applications is important. Low temperatures of 4–8 °C are common in order to maintain product activity, which may be compromised due to enzymatic activity, a thermally labile product, or undesirable contaminant propagation. Also, control of the temperature can prevent components (such as antifoam) in the feed solution precipitating, which can decrease process fluxes.

The primary source of heat in the system is the energy of pumping the fluid. To maintain the desired temperature, the heat must be removed, across a heat-exchange area and into a medium (generally glycol). The approximate area of heat exchanger required can be estimated as 1 ft^2 of heat-exchange area per horsepower used by the pump. This rule of thumb assumes that the glycol is supplied at 4 °C below the desired process temperature and at a rate sufficient to keep the rise in glycol temperature to less than 1 °C. If the process operates in a cold room, heat losses to the surrounding environment are adequate for applications not requiring strict temperature control.

The most usual heat-exchange areas are the process tank, an inline shell-and-tube heat exchanger, or a jacketed pipe. In general, jacketed tanks and tubes are limited to the tank size and amount of pipe employed in the TFF system, which is satisfactory for applications requiring less than 1–2 ft^2 of heat-exchange area. Otherwise, shell and multitube heat exchangers provide a greater ratio of surface area to volume and can be incorporated into the TFF system without an appreciable effect on the system size or holdup.

Using a jacketed tank is the most common way of controlling temperature, since the tank jacket can be used for temperature control not only during processing, but also during storage of processed material. However, when assessing the available area, one should note the decreasing heat-exchange area as the volume in the tank drops.

4.4.3   Product Recovery

To comply with sanitary design principles and to ensure maximum yield, all systems in all applications should be easy to drain. All piping should drain easily, from as few ports as possible. The drain port should be the lowest point of the TFF system, and, if physically possible, above the product receiver. The retentate side of the system should be vented to the atmosphere so that it can drain properly. The process pump, or a secondary transfer pump, can be used to move the product to the product receiver. The product can also be blown out of the system using nitrogen or air, but this is not generally recommended because the product might foam in the receiving vessel.

The material not recovered during draining of the system includes the system holdup and the concentrated product adhering to the membrane. Such material can be recovered with a flush of a volume of isotonic buffer equal to the minimum working volume of the system. This flush should be carried out for about five minutes at normal operating cross-flows, with the permeate valve closed. It is important that the system working volume (the minimum volume required for circulation of fluid without causing aeration) be at most half of the final desired volume, so that the final desired product concentration and yield can be attained. After the process is completed, the system is cleaned up and the membrane is regenerated.

4.4.4   Membrane Regeneration

In this section we assume that the cleaning-in-place (CIP) procedures developed and validated for the TFF system are not in conflict with existing in-house methods for CIP. A degree of flexibility, allowing for alternative cleaning agents that are more compatible with the membrane, may be required. An effective and reliable CIP operation, whether it be an existing in-house method or one developed with the TFF system supplier, has to be demonstrated for validation. "A validated cleaning procedure is one that provides a high degree of assurance that a specific cleaning procedure will clean a particular piece of equipment to a predetermined level of cleanliness—a claim that is to be substantiated by specific chemical and microbiologi-cal tests" [31]. A detailed process validation guide, complete with test results demonstrating the ability to clean, sanitize, and depyrogenate process ultrafiltration systems, has been published by Millipore [32]. Millipore's guide is used as the framework for this section, and some parts of this section come directly from that validation guide.

If the filter modules in a TFF system are to be used for multiple process cycles, the maintenance of the system is a prime concern. At the end of the processing cycle, both the membranes and the system must be thoroughly cleaned to ensure reliable, reproducible operations. Several steps must be conducted before and after a process run has been completed. These steps are specific to a particular application, but typically include some of those listed in Table 4.7.

Table 4.7 Possible Steps Involved in TFF System Regeneration

Step	Function
Flushing	Removes the bulk of the process fluid or residual cleaning solution from the system.
Cleaning	Removes the remaining process fluid components from the membrane and system, and restores membrane water permeability.
Sanitizing	Eliminates most of contaminating microorganisms from the system.
Depyrogenation	Eliminates pyrogens prior to processing for injectables.
Water permeability testing	Monitors efficacy of the cleaning procedure.
Sterilization	Kills all living organisms within the system.
Storage	Maintains membrane module integrity and cleanliness.

In certain cases the benefits of reusing the TFF device may not outweigh the potential for improper membrane regeneration and related liabilities. Yet, even in this case, the hardware components of the system and the installed membrane need to be cleaned and rinsed. For most applications the process system is cleaned by the following steps:

(a) System flushing of process material with buffer.
(b) System cleaning with recirculating base.
(c) System flushing of base with reverse osmosis (RO) quality water.
(d) System cleaning with recirculating acid.
(e) System flushing of acid with RO quality water.
(f) Testing efficacy of cleaning by check for water permeability.

The TFF system needs to have the necessary plumbing to carry out the above procedures.

Flushing of the TFF system must remove residual product or cleaning agent from the system. The key operating parameters are flushing velocity and flushing volume. The flushing velocity for the process piping is 5 ft/s, according to the 3-A standards [33]. A flushing volume of less than 2 L per ft^2 of membrane area is adequate [32], but a more conservative 4 L per ft^2 of membrane is preferable. The TFF supplier should be able to specify the most effective flushing conditions [34–36] for the particular membrane module used in the system. The flushing is a one-pass operation, with the flushing water coming from the tank, passing through and across the TFF device, and out to the drain. In flushing, bursts of water should be sent through all ports that are not in the direct flow path of the flushing. The flushing should also pass through piping on the permeate side on its way to waste containment. Water should be warm (40–50 °C), so that flushing cleaning agents stay in solution; buffer should be at ambient temperature, so that the residual proteins are not denatured or precipitated.

Membrane cleaning removes microorganisms and their metabolites, thus maintaining system productivity as well as a clean and sanitary system. Cleaning conditions and the choice of cleaning agents depend on the nature of the foulant and the chemical compatibility characteristics of the membrane. The customer and supplier should jointly select the cleaning agent, as well as the time, temperature, and concentration that are most effective. Table 4.8 is a list of common cleaning agents. The cleaning solutions need to be recirculated throughout the system. Connections should be made from all legs outside the recirculation loop, with returns to the process vessel, so that cleaning solution can pass through all parts of the system. A spray ball should be used to clean the walls and lid of the process vessel.

The effectiveness of the cleaning cycle must be verified by tests of cleaning efficacy. The cleaning efficacy can be evaluated by measurements of the normalized water permeability (NWP) of the TFF module after cleaning; the NWP should be within 10% of its initial value, and can be compared to the NWP measured for the previous use. To measure the NWP, RO water is recirculated at a constant

Table 4.8 Recommended Cleaning Agents for TFF Devices
Cleaning Time of 30–60 Minutes at 50 °C

Process material	Foulant	Cleaning agent	Conc.	pH
Mammalian cell culture	cell debris	NaOH	1 N	14
		or NaOCl,	300 ppm	10–11
		then H_3PO_4	0.1 N	1
Protein for clarification, concentration, or diafiltration	precipitated proteins, adsorbed proteins	NaOCl	300 ppm	10–11
		urea	7 M	8
		NaOH	0.1 N	13
Bacterial cell whole broths	antifoam, adsorbed proteins, cell debris	NaOH	0.1 N	13
		NaOCl	300 ppm	10
		Terg-A-Zyme	0.2%	10
		SDS/Tween® 80	0.1%	5–8
Cell lysates	cell debris, lipopolysac., protein	NaOCl	300 ppm	10
		or NaOH,	0.1 N	13
		then H_3PO_4	0.1 N	1

temperature, a ΔP of 2 psi, and a TMP of 5 psi. (These values are arbitrary, but are chosen because the ΔP of 2 psi and TMP of 5 psi are easy set points on standard 0–100 psig gauges.) Because these pressures are relatively low, the engineer can be sure that the water flux is a function of water permeability and not of permeate piping restrictions. The water rate may be monitored with either a flow indicator or a bucket and stop watch, but a flow meter is recommended to help validate the cleaning cycle. The flow indicator should be of a size that clearly shows whether water permeability has been achieved and the desired fluid velocities have been reached during the cleaning cycle. These specifications may not necessarily coincide with those of the process.

Sanitizing procedures lower the microbial population of the membrane system prior to or after the process run. They do not sterilize a TFF system, but are intended to minimize the chances of microbial contamination during process runs. These procedures must always be performed on membrane systems that have been thoroughly cleaned, since nutrients for microbial growth are no longer present, and cleaned surfaces allow thorough penetration of the sanitizing fluid to areas that might otherwise be inaccessible.

A variety of chemical or thermal mechanisms can do the sanitizing. The selection of a particular sanitizing method is based on its compatibility with the membrane. Recommendations for sanitizing procedures and solutions should be obtained from the TFF supplier [32, 34–37]. Table 4.9 is a list of typical sanitizing agents.

Table 4.9 Recommended Sanitizing Agents and Conditions

Sanitizing agent	Concentration	Temperature, °C	pH	Time, min
Chlorine (as NaOCl)	20–50 ppm (active Cl)	20–50	6–8	15–30
Peracetic acid	100–200 ppm	10–40	3.5	15
NaOH	0.1 N	40–50	13	30
Formalin	1–2%	20–30	5–8	30

Sanitizing procedures have system requirements similar to those of the cleaning cycle, since the fluid has to circulate. An additional vent filter is called for, to keep the system free of contaminants while the system is being drained. The vent should be either on the tank or on the retentate line.

Depyrogenation, defined as the inactivation or elimination of bacterial endotoxins, is typically done prior to the processing of injectable pharmaceutical products. Endotoxins in solution are most commonly inactivated by acidic or alkaline hydrolysis, or by oxidation. Depyrogenation procedures, like cleaning and sanitizing cycles, involve recirculating fluid. Table 4.10 is a list of typical depyrogenation agents. The water used for flushing the depyrogenation agent from the system should be free of pyrogens.

Table 4.10 Recommended Depyrogenation Agents and Conditions

Depyrogenation agent	Concentration	Temperature, °C	pH	Time, min
Chlorine (as NaOCl)	300 ppm (active Cl)	30–50	10–11	30
NaOH	0.1 N	30–50	13	30
HCl	0.1 N	30–50	1	30
H_3PO_4	0.1 N	30–50	1	30

Sterilization of TFF systems is more stringent than sanitizing, and is essential for steps requiring sterile process conditions (when there are no means of terminal sterilization). Steaming in place or autoclaving can be used. In either case, the effectiveness of the sterilization depends on the contact time and temperature of saturated steam, which is required because many spores are resistant to dry heat. The TFF supplier should verify that the membrane system can withstand this treatment and can be properly validated for this use.

Systems that use tubing pumps can be autoclaved. Generally, autoclaving is most practical with a relatively small TFF system (less than 20 ft^2). Due to the relatively harsh environment of an autoclave, instrumentation for the TFF system is generally limited to analog pressure gauges (such as the Anderson Pharmaceutical series).

Steaming in place (SIP) of any size TFF system, in which product contact surfaces are exposed to saturated steam at sterilization temperatures, is the preferred method of sterilization. This allows greater freedom in instrument selection and in sterilization procedures. Standard steaming conditions are 15 psig saturated steam at 121 °C for 15 minutes. Steam should move through all parts of the system. To ensure that proper steam conditions are maintained, condensate and air should be bled from all low points of the system, and the pump casing needs to be purged of condensate. Only if strategically placed thermocouples verify that the complete system is up to temperature, and biological indicators show kill, is an SIP procedure qualified. The steaming of the TFF system is separate from that of the process tank, which is often cleaned and sterilized independently.

After it has reached and sustained the kill temperature, the system is cooled. To prevent the drawing in of contaminants, nitrogen or air is fed into the system through a sterile vent. After cooling is complete, the nitrogen or air line can be removed, with the system left under pressure (5–20 psig).

When TFF systems are used in processes that require both aseptic conditions and containment, all connections made to the system prior to processing and before decontamination should be sterilized. This requires additional valving to allow steam flow between the connections that will be made or broken. Such a *sterile barrier* is necessary at the interface to the process tank (both the suction side of the pump and the retentate return), the filtrate receiver, and retentate drain.

Membranes can be stored in a system that is being shut down for up to 15 days [37]. Bacterial contamination during storage is a concern, so a sanitizing cycle should be performed every 2–3 days. Membrane devices can also be removed from the system, preserved in a suitable long-term preservative solution, and kept in liquid-tight containers for longer periods. Table 4.11 is a list of storage solutions and their effective time periods.

Table 4.11 Recommended Storage Solutions

Storage agent	Concentration	pH	Time period
NaHSO$_3$	1.0%	4–8	2 months
NaCl	5.0%	4–8	1 day
Sodium azide	0.05%	4–8	3 months
Formalin	1–2%	4–8	1 year

4.5   Validation

Validation demonstrates that the process can be used to make a product of expected and defined quality at an expected yield. Validation ensures product safety and quality, and is necessary for approval by the FDA. A TFF system itself is not validated, but rather the process in which it is employed. The validation of the TFF process includes both mechanical and process review. It is initiated with review of the system design and procedures, continued with mechanical and hydraulic testing, and completed with process testing. All testing is done repeatedly, to ensure consistency and to establish a norm and reasonable deviation. In the following section we present an overview of the requirements of TFF system validation.

The following questions related to the mechanical design of the TFF system must be answered:

(a)   Does the system affect product safety and efficacy?
(b)   Does the system add contaminants to the product? and
(c)   Is the equipment chemically compatible with the feed and cleaning solutions?

With respect to product safety, all plastic materials should pass USP Class VI testing as well as USP mouse safety testing. Tests for aqueous extractables and USP oxidizables establish that the system does not add contaminants to the feed. All components, from gasket material to piping, need to be reviewed for chemical compatibility with the cleaning solutions and feed material; chemical compatibility charts for the components of the system should be kept on file and made part of the operating and maintenance manual. The system supplier should make all this information available.

With respect to process design, verification is needed that the membrane and module are intact, and that the system operation is reliable and controllable. Membrane and module integrity testing were discussed earlier in the chapter. These tests are an integral part of validation in that they can detect defective seals, damaged membranes, or improper installation, and are generally performed with every run.

The performance characterization of the system as a part of the TFF operations, including system preparation, product processing and recovery, and membrane regeneration, ensures that the system operation is reliable and controllable, and is the most demanding aspect of the validation procedure.

Validating the system preparation requires that the volume of solution needed to remove the storage solution to acceptably low levels be known, and that a test method for confirmation be in place. The flow rate, pressure, and temperature conditions for flushing must be defined and held constant. The amount of residual material (as total organic carbon, for example) can be monitored as a function of the volume of water flushed through the system, to demonstrate the efficacy of the flushing.

Validating the product processing and recovery operations requires that the system be controllable. This means that the product retention by the membrane must be defined, and, since the separation can be affected by gel polarization and fouling, the processing parameters of pressure, flow rate, temperature, and flux must also be defined. Values of these parameters are generally established over a series of runs; the values used for processing should then be documented in the log of Standard Operating Conditions for the batch, and rigorously adhered to during the biopharmaceutical production process.

The cleaning and sanitizing of the system after processing also need to be validated. The cleaning efficacy in a TFF system is generally acceptable if the system can be returned to the original water permeability (at a given pressure and temperature). After the cleaning agent has been flushed out, the flushing solutions can be tested for residual material. The efficacy of sanitizing can be monitored by viability tests on bacterial cultures before and after sanitizing. Tests on UF systems have demonstrated that contamination of modules with *P. aeruginosa* at 10^6 CFU/mL (colony-forming units per milliliter) can be reduced to less than 1 CFU/mL after 30 minutes at 40 °C in 30 ppm NaOCl [32]. Similar tests of UF systems have demonstrated that pyrogen levels of 10 ng/mL can be reduced to less than 0.1 ng/mL after 30 minutes at 40 °C with 0.1 N NaOH.

In conclusion, the validation of TFF systems is similar to that of other processing operations. Components used in the system should not affect product safety or efficacy by their toxicity, biocompatibility, or extractables; all aspects of the TFF opera-tion must be reliable and reproducible; and test methods must demonstrate efficacy [38].

4.6   Automation

TFF systems can be automated in many ways, to serve a variety of functions. Automated systems operate consistently and with less process variability.

Automation in TFF systems may include control of process conditions with simple single-loop controllers, as well as the actuating of operating cycles, such as CIP and SIP, with simple programmable logic controllers, or TFF systems may be fully automated and microprocessor-driven at all process and regeneration steps.

The basic requirements for manual systems include alarms and safety switches that shut down operations for high pressure, high temperature, and electrical overload, to protect workers, product, and equipment.

Simple TFF process control features include temperature control, level control for continuous diafiltration, and permeation control as a function of TMP for mammalian cell applications and for some applications involving the clarification of macro-molecules from cell or lysate suspensions.

More elaborate automation involves valve sequencing operations for different process and cleaning operations and for data acquisition. Deciding on the level of automation required for a given process can be very subjective. Such matters as the skill level of operators and support people have to be considered, as well as how strongly the desired separation depends on process conditions. A consensus has to be reached among all parties involved in the specification, operation, and installation of the system.

Figure 4.10 shows two 500-ft² spiral UF systems. One system is a simple manual system, the other is fully automated to carry out product prefiltration, product processing, and membrane regeneration.

Figure 4.10 Manual and Automated TFF Systems. (A) This 500-ft² spiral ultrafiltration TFF system, designed with minimal instrumentation, is manually controlled by an operator. (B) This 500-ft² spiral ultrafiltration TFF system is fully automated to perform product prefiltration, product processing, and membrane regeneration.

4.7   System Costs

As can be seen in Figure 4.10, two systems of the same area can differ significantly in the degree of sophistication and associated cost. The factors that influence the cost of a TFF system are: (a) the membrane area, (b) the pumping rate, piping, and type of pump, (c) the degree of sophistication (instrumentation, process controllers, etc.), and (d) the materials of construction required for the system hardware.

The cost of polymeric membrane ranges from $10 to $150 per square foot, and membrane replacement is the principal operating cost of a TFF system. However, the replacement cost as a function of time depends on the service life of various membranes, as well as their performance. It may be wise to budget membrane replacement costs for the first year at the highest figure the process can afford. After you have developed some operating experience, you can more accurately estimate annual replacement costs.

The capital cost of a TFF system depends on the other three items (pumping, sophistication, and materials). The pumping requirements can influence system cost substantially, not only in the type of pump chosen, but also in the cost associated with larger line and valve sizing. The degree of sophistication of the system, and the materials of construction, add incrementally to system cost, but their primary effect is due to their dependence on labor costs; they are relatively insensitive to membrane area.

In general, a manual TFF system with 50 ft^2 of filter area, consisting of pump, filter, manifolds, manifold valves, and instruments for sanitary service, costs from $10,000 to $30,000. Larger systems have an associated economy of scale that is a function of the recirculation rate. Once an initial system cost ($COST_1$) has been established, the relative cost ($COST_2$) for a system of a different size can be estimated by:

$$COST_2(\$) = COST_1(\$) \{(q_{f2}/q_{f1})^{0.6} + A_2/A_1\} \qquad (4.10)$$

where q_f is the relative flow rate required for the area, A, of the filter.

Equation 4.10 applies to manual systems. The cost of instrumentation is relatively insensitive to system size and can be treated as a fixed cost once the degree of sophistication has been determined. Figure 4.11 shows various TFF systems with hollow fiber modules, turbulence-promoted cassettes, and open-channel cassettes.

Figure 4.11 Some TFF Systems. (A) A tubular
Amicon hollow-fiber system used for cell harvesting;
(B) A 100-ft^2 Millipore open-channel plate-and-frame
cassette system used for cell harvesting; (C) A 10-ft^2
Millipore plate-and-frame cassette system used for virus
concentration; (D) A 200-ft^2 Millipore spiral U.F.
system used for protein concentration.   (D)

4.8   Utility Requirements

A TFF system requires electricity, RO water, coolant, and some miscellaneous cleaning chemicals, in quantities that depend on the application and size of the TFF system. Utility requirements are summarized in Table 4.12.

Table 4.12 Utility Requirements of TFF Systems

Item	Purpose	Specifications
Electricity	Powering instruments & pump	As required
RO water	Flushing	4 L per ft^2 membrane per flush
RO water	Cleaning	1 L per ft^2 membrane per cleaning step
Coolant	Removing pump heat	Half process recirculation rate; 8 °F cooler than desired process temperature
Clean filtered steam	Sterilizing	Regulated at 15 psig
Filtered air/nitrogen	Diffusion testing of filters	Capable of being regulated between 5 and 60 psig
Appropriate waste treatment	Disposing of wastewater from flushing and cleaning	As required
Instrument air	Actuating automatic valves	As required

4.9   Installation Tips for TFF Systems

Proper installation makes operation easier. Here are a few suggestions:
(a) Allowing 3 ft on all sides of the TFF skid makes access and maintenance easier.
(b) A TFF system should drain properly and completely. The TFF skid should sit on a level surface, or, if on the production floor, it should have leveling legs to compensate for the grade to drain.
(c) Utilities to the skid (see Table 4.12) should come either from above or one side of the TFF system, so that the area around the system is relatively free of obstructions.
(d) A wastewater drain should be near the unit to carry away the large volume of water used in the cleaning cycle.

4.10   Examples of TFF in Biotechnology

4.10.1   Protein Processing

This example illustrates the 50-fold concentration of a therapeutic protein, from 200 L to 4 L. Due to the low initial concentration (0.1 mg/L), the flux is very high (150 L/h·m² at TMP = 15 psi and ΔP = 10 psi). The protein being concentrated has a molecular weight of 30,000 and is retained (99%) with a 10,000-MWCO polysulfone membrane. The product is valued at $1,000/mg, for a batch value of $20,000.

Process Parameters:
- Concentration: initial = 0.1 mg/L, final = 5.0 mg/L;
- Batch size: initial = 200 L, final = 4 L;
- Process temperature: 10 °C;
- Process time: 2 h;
- Flux: 150 L/h·m²;

Design Calculations:
(1)   Membrane area required to concentrate batch in 2 hours:
   a.   design rate = (initial vol. − final vol.)/process time = 98 L/h;
   b.   membrane area = design rate/flux = 0.65 m²;
      The device comes in 0.46-m² increments, so we use 2 cassettes, or 0.92 m².
   c.   actual rate = flux × area = 138 L/h;
(2)   Pump requirements = 4 L/min per cassette × 2 cassettes = 8 L/min, at 20 psig (0.5–1.0 hp motor).
(3)   According to the heat-exchange rule of thumb, 1 ft² of heat-exchange area is required per motor hp, so the approximate heat-exchange area is 1 ft².

Design Considerations:
(A)   *Pump selection*: A rotary lobe pump or peristaltic pump is the best choice for this application because of low internal pump volume and low flow rate. We prefer the rotary lobe pump because it offers maximum mechanical integrity. We decide against centrifugal pumps because of the flood volume and high-speed operation they entail.
(B)   *System tankage and piping*: For the 8 L/min flow rate to achieve suitable cleaning velocity requires ¾" sanitary tubing (Table 4.5). Larger tubing could be used if the pump has additional capacity to clean the system. A 2-ft diameter tank with 3-ft straight-wall side can hold 200 L. Therefore, running the UF system in batch mode requires a minimum of 10 ft of ¾" line having a flood volume of 0.6 L. The final volume of 4 L (< 2% of the original volume) is difficult to monitor unless we employ an intermediate vessel (< 20 L).

Since 5% of the product (worth $1,000) may be left within the drained system on the wetted surfaces of piping and membrane, we flush it; we must then reduce the final recirculation volume from 4 L to 2 L. Again, we need an intermediate vessel. Figure 4.11C shows a cassette system that can process less than 2 L of material.

(C)   *Temperature control*: The temperature can be maintained with a jacket on the primary feed tank, the intermediate tank, or both.

(D)   *Automation*: Automation improves reproducibility. In this particular example, the physically small size of the system limits us to operating manually; standard valve actuators would overwhelm the small process system. Therefore, to get the needed reproducibility, we must rely on well-tagged instruments and valves, a detailed batch sheet, and standard operating procedures.

(E)   *Utility requirements*: This system requires standard electrical service, cooling supplies, and RO quality water for cleaning. The cleaning protocol should consist of 10 L of 0.1 N NaOH, 40 L of water, testing to assure permeability, 10 L of 0.1 N NaOH for sanitizing, 40 L of water for flushing prior to storage, and 40 L of water before starting to operate. Therefore, this system requires at least 150 L of water, and waste tankage to handle not only this water but also the waste permeated during processing.

4.10.2   *E. coli* Harvesting

This example illustrates the concentration of *E. coli* prior to homogenization. The initial volume is 500 L, containing 3 g/L *E. coli*. The homogenization requires *E. coli* concentration of 60 g/L. The average flux is 60 L/h·m^2 at 8 °C through either a 100k-MWCO UF membrane or a 0.1-μm MF (microfiltration) membrane. The antifoam used has a cloud point of 20 °C. The value of the batch is $50,000.

Process Parameters:
- Concentration: initial = 3 g/L, final = 60 g/L;
- Batch size: initial = 500 L, final = 25 L;
- Process temperature: 8 °C;
- Process time: 3 h;
- Flux: 60 L/h·m^2;

Design Calculations:
  (1)   Membrane area required to concentrate batch in 3 hours:
     a. design rate = (initial vol. − final vol.)/process time = 160 L/h;
     b. membrane area = process rate/flux = 2.64 m^2;
        The device to be used comes in 0.92-m^2 increments, so we use 3 cassettes in series, or 2.78 m^2.
     c. actual rate = flux × area = 166 L/h;
  (2)   Pump requirements = 125 L/min at 60 psig (5-hp motor).
  (3)   Approximate heat-exchange area required = 5 ft^2.

Design Considerations:

(A)   *Pump selection*: A rotary lobe pump or centrifugal pump is the best choice for this application because of the high flow and high pressure capabilities. We prefer the rotary lobe pump because the viscosity of the solution increases as the *E. coli* suspension becomes more concentrated, and rotary lobe pump performance is not as sensitive to changes in viscosity.

(B)   *System tankage and piping*: The 125 L/min flow rate requires 2" sanitary tubing. A 2.5-ft diameter tank with 3.5-ft straight-wall side can hold 500 L. Therefore, running the UF system in batch mode requires a minimum of 14 ft of 2" line having a flood volume of 7.5 L. The final volume is 25 L, but it is difficult manage a recirculation rate of 125 L/min with only 25 L of material, so we specify a modified batch design (Figure 4.8), with $q_r = 125$ L/min, $q_{r2} = 20$ L/min, $q_{r1} = 122$ L/min, and $q_p = 3$ L/min.

Once the material has been concentrated it is not totally free-flowing, and therefore a flush is required. After most of the material has been transferred to a 40-L can, 7.5 L (piping flood volume) can be added to reslurry and recover the remaining *E. coli*. This material (7–8 L) may then be transferred to a 20-L can and used to flush the homogenizer after the initial material has been broken. Figure 4.11A and 4.11B show hollow-fiber and cassette systems that can process 500 L of material.

(C)   *Temperature control*: Temperature control is important here because the process temperature needs to be maintained below the cloud point of the antifoam if we are not to forfeit proper membrane performance and product activity.

(D)   *Automation*: We limit the automation for this application to switches, and alarms for high pressure and temperature. The high-temperature alarm warns the operator of a temperature approaching the cloud point of the antifoam or one that would cause product degradation. The high-pressure switch is essential, since the system is operated with a lobe pump. The volumetric output of the pump is relatively constant, and the pressure rises as the *E. coli* becomes more concentrated; therefore, the high-pressure switch can be used to signal that the batch is finished (if the pump speed is held constant from run to run).

(E)   *Utility requirements*: This system requires standard electrical service, cooling supplies, and RO quality water for cleaning. The cleaning protocol should consist of 30 L of 0.1 N NaOH or 300 ppm NaOCl, 30 L of 0.1N H_3PO_4, 120 L of water, 30 L of 50 ppm NaOCl or 0.1N NaOH (to sanitize), 120 L of water for flushing prior to storage, and 120 L of water for flushing before starting the next operation. Therefore, this system requires at least 600 L of water, and the waste tankage to handle not only this water but also the waste permeated during processing.

4.10.3   Monoclonal Antibody Clarification

This example illustrates the clarification of a monoclonal antibody (MAb) from a mammalian cell culture. The initial volume is 500 L, with 10^7 cells/mL and 1 g/L of protein. The average flux is 100 L/h·m² at 10 °C, through a 0.45-μm hydrophilic MF membrane. The value of the batch is $250,000. A 99% yield of clarified protein is typically required. For this example we specify a 10-fold volume reduction, with a 90% protein yield. To meet the criterion of 99% yield, we plan to use diafiltration to reduce the protein from 1 g/L in the final volume of 50 L to 0.1 g/L. There is no apparent retention of protein by the membrane.

Process Parameters:
- Clarification:

	cells per mL	protein, g/L	protein, g
Initial:	10^7	1	500
Final:	10^8	1	50

- Diafiltration:

	cells per mL	protein, g/L	protein, g
Initial:	10^8	1	50
Final:	10^8	0.1	5

- Batch size: initial: 500 L, final: 50 L;
- Process temperature: 10 °C;
- Process time: 2 h;
- Flux: 100 L/h·m²;

Design Calculations:
(1)   Amount of diafiltration buffer required:
$V_w = -\ln (C_f / C_i) \cdot V_f = -\ln (0.1/1) \cdot 50$ L = 115 L;
(2)   Membrane area required to clarify batch in 4 h:
   a. design rate = $(V_i - V_f + V_w)$/process time = 280 L/h;
   b. membrane area = process rate/flux = 2.8 m²;
       The device to be used comes in 0.92-m² increments, so we use 3 cassettes in series, or 2.78 m².
   c. actual rate = flux × area = 140 L/h;
(3)   Pump requirements = 12 L/min at 1 psig (0.5-hp motor).
(4)   Approximate heat-exchange area required is less than 0.5 ft².

Design Considerations:
   (A)   *Pump selection*: A rotary lobe pump or peristaltic pump is the best choice for this application because these types of pump offer high efficiency and low shear. This process is to be carried out aseptically, and steam sterilization of the system is not required. On the basis of cost we choose the peristaltic pump.
   (B)   *System tankage and piping*: The 12 L/min flow rate requires ¾" process piping. A 2.5-ft diameter tank with 3.5-ft straight-wall side can hold 500 L. Therefore, running the MF system in batch mode requires a minimum of 14 ft of ¾" line having a flood volume of 1 L. The final volume is 50 L, so a recirculation rate of 12 L/min is not difficult to manage. In this case, we use a simple batch design (Figure 4.3).

(C) *Temperature control*: Temperature control with the process tank jackets is adequate.

(D) *Automation*: Transmembrane pressure control requires automation. Since this cell line is presumably fragile, we prevent cell lysis by avoiding excessive pressure. We can control the permeate pressure by throttling the rate of filtration, with a valve or with a second peristaltic pump. At these relatively low flow rates (1 L/min), a peristaltic pump gives better control because it provides a linear flow adjustment instead of the exponential adjustment of a valve. The permeate pump can be dynamically controlled by the transmembrane pressure: if the membrane begins to foul, the increased transmembrane pressure signals the increasing resistance to flow; the controller accordingly reduces the permeate pump speed to avoid gel polarization, reduced product transmission, and further cell lysis.

The second type of automation required in this system is level control in two stages. The first stage of control allows the volume to be reduced from 500 L to 50 L. At that point the level controller meters in fresh buffer to maintain the level during diafiltration.

We also need high-pressure safety switches, to prevent cell lysis or tubing rupture.

(E) *Utility requirements*: This system requires standard electrical service, cooling supplies, and RO quality water. The cleaning protocol is similar to that in the *E. coli* example. Therefore, this system requires at least 600 L of water, and the waste tankage to handle not only this water but also the waste permeated during processing.

4.11 References

1. BLATT, W. F., DAVID, A., MICHAELS, A., NELSON, L.; Soluble Polarization and Cake Formation in Membrane Ultrafiltration: Causes, Consequences and Control Techniques; In: *Membrane Science and Technology*; Plenum, New York, 1970.

2. BELFORT, G.; Membrane Separations; In: *Advanced Biochemical Engineering*; BELFORT, G., BUNGAY, H., Eds.; John Wiley & Sons, New York, 1987.

3. TUTUNJAIN, R. S.; Cell Separations with Hollow Fiber Membranes; In: *Comprehensive Biotechnology*; Moo-Young, M., Ed.; Pergamon Press, Tarrytown, NY; Vol. 2, 1985.

4. CHERYAN, M.; *Ultrafiltration Handbook*; Technomic Publishing, Lancaster, PA, 1986.

5. MURKES, J., Fundamentals of Cross-flow Filtration, *Sep. Purif. Methods 19*, 1–29 (1990).

6. BLAKE, N. J., CUMMING, I. W., STREAT, M., Prediction of Steady State Cross-flow Filtration Using a Force Balance Model, *Membr. Sci. Technol. 68*, 205–216 (1992).

7. COHEN, R. D., Effect of Turbulence on Film Permeability in Cross-flow Membrane Filtration, *Membr. Sci. Technol. 48*, 343–347 (1990).

8. ZAIIKA, J., LEAHY, T.; Practical Aspects of Tangential Flow Filtration in Cell Separations; In: *Purification of Fermentation Products*; LeROITH, D., SHILOACH, J., LEAHY, T. J., Eds.; ACS Symposium Series Vol. 271; American Chemical Society, Washington, DC, 1985; pp 51–69.

9. *An Introduction To Protein Processing by Tangential Flow Filtration*; Process Systems Technical Brief, TB032; Millipore Corp., Bedford, MA, 1988.

10. MICHAELS, S. L., Crossflow Microfilters: The Ins and Outs; *Chem. Eng.* **96** (1), 84–91 (1989).

11. *Harvesting Mammalian Cells with Higher Recoveries and Higher Yields*; SD111; Millipore Corp., Bedford, MA, 1988.

12. RADLETT, P .J., The Concentration of Mammalian Cells in a Tangential Flow Filtration Unit, *J. Appl. Chem. Biotechnol.* **22**, 495–499 (1972).

13. GILMAN, P. E., Harvesting S-49 cells from Culture Media, *J. Biol. Chem.* **254**, 2287–2295 (1979).

14. RICKETTS, R .T., LEBHERZ, W. B., III, KLEIN, F., GUSTAFSON, M. E., FLICKINGER, M. C.; Application, Sterilization, and Decontamination of Ultrafiltration System for Large-Scale Production of Biologicals; In: *Purification of Fermentation Products*; LEROITH, D., SHILOACH, J., LEAHY, T. J., Eds.; ACS Symposium Series Vol. 271; American Chemical Society, Washington, DC, 1985; pp 21–49.

15. BUILDER, S. E., HSU, C. C., LEONARD, L. C., VAN REIS, R., Industrial Scale Harvest of Proteins From Mammalian Cell Culture by Tangential Flow Filtration, *Biotechnol. Bioeng.* **38**, 413–422 (1991).

16. DINARELLO, C. A., Clarification and Concentration of Interleukin-1, *J. Immunol.* **133**, 1332–1338 (1984).

17. PADHYE, D. J., *E. coli* Harvested from Media and Subsequent Concentration and Dialysis of Cytotoxins., *Biochem. Biophys. Res. Commun.* **139**, 424–430 (1986).

18. ANFINSEN, C., Purification of Interferon from Pooled Affinity Fractions, *Proc. Natl. Acad. Sci. U. S. A.* **76**, 5601–5606 (1979).

19. GABLER, R., RYAN, M.; Processing Cell Lysate with Tangential Flow Filtration; In: *Purification of Fermentation Products*; LEROITH, D., SHILOACH, J., LEAHY, T. J., Eds.; ACS Symposium Series Vol. 271; American Chemical Society, Washington, DC, 1985; pp 1–20.

20. BAILEY, F. J., MAIGETTER, R. Z., WARF, R. T., Harvesting Recombinant Microbial Cells Using Cross-flow Filtration, *Enzyme Microb. Technol.* **12**, 647–652 (1990).

21. DEBERNARDEZ, E. R., FELDBERG, R. S., FORMAN, S. M., SWARTZ, R. W., Cross-flow Filtration for the Separation of Inclusion Bodies From Soluble Proteins in Recombinant *E. coli* Cell Lysate, *Ferment. Bioind. Chem.* **48**, 263–279 (1990).

22. VIRCAR, P. D., NARENDRANATHAN, T. J., HOARE, M., DUNHILL, P., Studies of the Effects of Shear on Globular Proteins: Extension to High Shear Fields and to Pumps, *Biotechnol. Bioeng.* **23**, 425–429 (1981).

23. THOMAS, C. R., NIENOW, A., DUNNILL, P. D., Action of Shear On Enzyme Studies with Alcohol Dehydrogenase, *Biotechnol. Bioeng.* **21**, 2263–2278 (1979).

24. CARLISLE, T. G., WOOD, H. G., BENJAMIN, D. C.; *Effects of Fluid Shear on Globular Proteins*; Paper 117–122 presented at ASME Annual Meeting, Charlottesville, VA, 1988; American Society of Mechanical Engineers, Chicago, 1989.

25. LEE, Y., CHOO, C., The Kinetics and Mechanisms of Shear Inactivation of Lipase from *Candida cylindracea, Biotechnol. Bioeng.* **33**, 183 (1989).

26. ADAMS, D. G., AGARWAL, D., Clean-in-Place Design, *BioPharm* **2** (6), 48 (1989).

27. *General Catalog*; Watson-Marlow BTI, Concord, MA, 1988.

28. *Catalog PR-86*; Tri-Clover, Inc., Kenosha, WI, 1986.

29. *Catalog TF-85*; Tri-Clover, Inc., Kenosha, WI, 1985.

30. EMORY, S., Principles of Integrity-Testing Hydrophilic Microporous Membrane Filters, *Pharm. Technol. 13* (10), 36–46 (1989).

31. SADANA, A., Protein Inactivation During Downstream Separation, *BioPharm 2* (5), 20–23 (1989).

32. *Prostak-UF Validation Guide*; Millipore Corp., Bedford, MA, 1989.

33. INTERNATIONAL ASSOCIATION OF MILK, FOOD AND ENVIRONMENTAL SANITARIANS, INC.; *3-A Sanitary Standards*; International Association of Milk, Food and Environmental Sanitarians, Inc., Des Moines, IA; 1947–present.

34. *Maintenance Guide for Tangential Flow Filters and Systems*; SD002; Millipore Corp., Bedford, MA, 1989.

35. *Maintenance Procedures For Pellicon Cassette Filters*; SD001; Millipore Corp., Bedford, MA, 1989.

36. *Maintenance Procedures For Prostak Microporous and Prostak-UF Filters*; SD003; Millipore Corp., Bedford, MA, 1989.

37. *Maintenance Procedures For Spiral Wound Ultrafiltration Cartridges and Systems*; SD004; Millipore Corp., Bedford, MA, 1989.

38. MICHAELS, S. L., Validation of Tangential Flow Filtration Systems, *J. Parenter. Sci. Technol. 45*, 218–223 (1991).

Chromatography Systems

Julian Bonnerjea and Peter Terras

Contents

5.1   Introduction

Advances in recombinant DNA and hybridoma techniques during the 1970s and 1980s have led to the identification of a large number of proteins that have human therapeutic uses. A few such products are already on the market, for example, human insulin, growth hormone, and erythropoietin, with a great many more undergoing development. If clinical trials of these products are successful, then their production needs to be scaled up and routine manufacture begun. This chapter is intended as an introduction to the engineering aspects of scaling up protein purification processes and the design and operation of purification systems. We focus mainly on proteins intended for human clinical use because it is these compounds that must meet the most stringent requirements in terms of product purity and safety.

The purity requirements for recombinant proteins intended for human therapeutic applications are extremely high [1]. Modern analytical methods have the sensitivity to detect impurities present at a concentration of only a few parts per million, and these techniques need to be applied to measure both protein and nonprotein impurities. In addition to foreign impurities, such as growth media components and proteins from the host cell line, the purification process must also remove any aggregated, misprocessed, or chemically altered forms of the product. These may be due to the modification of one or more amino acid residues (for example, by oxidation or deamidation), the cleavage of covalent bonds by protease enzymes, or the incorporation of variable amounts of carbohydrate, if the molecule is a glycoprotein. Such molecules differ only slightly in their chemical properties from the "correct" (or "reference") product and their removal can be a very difficult task. The purification process as a whole also needs to reduce, sometimes by more than 10–15 orders of magnitude, the concentration of DNA and infectious virus particles that could be present in the unpurified material. To achieve this level of purity routinely, reliably, and economically on a large scale is a major technical challenge.

VAN BRUNT has discussed the meaning of the terms *large-* or *process-scale* as applied to the biotechnology industry [2]. In contrast to conventional drugs such as steroids and antibiotics, therapeutic proteins are often highly potent and therefore dosage levels are generally low (for example, micrograms, as opposed to milligrams for small-molecule drugs). Consequently, production of many therapeutic proteins on the kilogram-per-year scale can often be adequate to meet market demand. We use the term *large-scale* to describe processes for the purification of multi-gram amounts (for example, 5–500 g) of purified product per batch.

In addition to having a well-designed and economically viable purification process, the process engineer must also address the aspects of process validation needed to give assurance that the process will consistently produce a safe and efficacious product without batch-to-batch variability. Furthermore, it is important to consider not only the safety of the end user of the product, but also the safety of the process operators, maintenance engineers, etc. If the product being manufactured is a potential health hazard, then containment issues must also be considered [3].

From an engineer's point of view, the key requirements that must be met when scaling up and automating a pharmaceutical manufacturing process are:

(a)   that both the process itself and the equipment used to carry out the process must be completely reproducible and highly robust, that is, insensitive to any small changes that may occur either during purification or further upstream during the cell culture stage;

(b)   that the process, the equipment, and the entire facility must be designed to enable the required standards of safety and hygiene to be maintained during the manufacturing operation; and

(c)   that the process and the equipment must be designed so as to allow adequate data to be collected to demonstrate that requirements (a) and (b) can and will be met.

Chromatography is one of the few techniques that can achieve both the high resolution and the high capacity needed to generate large quantities of highly purified product.   Furthermore, it is relatively easy to increase the scale of operation economically over several orders of magnitude, and to ensure that extraneous substances are not introduced into the process stream.   Other techniques, such as aqueous two-phase partitioning and electrophoretic separation methods, do not have all of these attributes and consequently these techniques are not in widespread use for the purification of therapeutic-grade proteins.

Even high-resolution chromatography is unable to achieve the required purity in a single operation, and therefore several chromatographic steps in succession are usually required. In general, a relatively small number of such steps, each utilizing a different aspect of the product's chemistry (for example, molecular charge, size, or hydrophobicity), can be sufficient to meet the purity required for pharmaceutical applications.

In the following sections we discuss the scale-up of liquid chromatography, the selection of individual items of equipment, and the assembly of entire purification systems for the industrial-scale purification of proteins. The theory of chromatography as applied to protein purification is not reviewed here since it has been covered elsewhere [4].

The engineering aspects of the scale-up of chromatography have been discussed in depth in the scientific literature. JANSON and HEDMAN'S reviews [5, 6] are particularly good examples and are strongly recommended. The specification of equipment and the assembly of purification systems, however, has not been well documented and existing literature on the subject is often written from a manufacturer's, rather than a user's, standpoint.

5.2   Engineering Aspects of Scale-Up

5.2.1 Process Considerations

Over the last few years attention has been focused on the design of large-scale protein purification processes [7, 8]. Although there are no hard-and-fast rules that determine which purification schemes can or cannot be scaled up, certain procedures that perform very well on a laboratory scale may not scale up and automate economically, if at all.

The ideal purification process for economic scale-up involves only a small number of integrated chromatography steps that utilize aqueous buffers and do not require organic solvents. The process should be designed to operate with step changes of buffer composition instead of continuous concentration gradients (to simplify operation) and should not require the collection of numerous fractions across a product elution peak. The buffers used should not support bacterial growth and the chromatographic matrices should be compatible with *clean-in-place* (*CIP*) solutions to allow efficient sanitization and decontamination. The matrices should be able to withstand high flow rates and pressures of several bar in order to reduce purification cycle times. In addition, the purification process as a whole, including buffer makeup, column packing and testing, in-process sampling, etc., should ideally fit comfortably into five eight-hour shifts. Where all or most of these conditions are met, economic scale-up is generally straightforward [9].

Because of its very high specificity and relative insensitivity to operating conditions such as flow rate and product concentration, affinity chromatography (for example, immunopurification, or protein A chromatography) is among the easiest of chromatographic procedures to scale up and automate. Contrary to popular belief, most affinity matrices can be compatible with CIP procedures, provided that sodium hydroxide is replaced with other chemicals for cleaning of the gel. Although affinity gels are more expensive than standard ion exchange gels, careful design of cleaning procedures can allow reuse of the gel over many purification cycles, thereby reducing gel costs. Similarly, the potential disadvantage of leakage of trace amounts of the affinity ligand from the gel into the product can often be overcome by ensuring that subsequent chromatography steps are able to remove any leaked ligand.

One of the goals of process development is to achieve the product purity and yield that is obtainable by affinity chromatography, but using the cheaper and more robust standard adsorption matrices. Even with relatively simple procedures such as ion exchange chromatography it can be possible to design binding, washing, and elution conditions such that a change in buffer composition elutes essentially pure product at high yield. The separation of all the different components present in the unpurified material into well-resolved individual peaks, as in analytical chromatography, is not required. A process-scale purification step should ideally generate only two fractions: pure product, and all other impurities (Figure 5.1). Under these circumstances only one

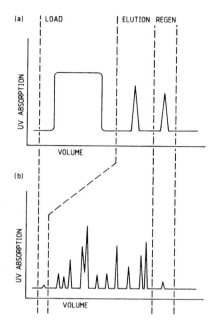

Figure 5.1 Typical UV Absorption Profiles; (a) for process-scale chromatography, and (b) for analytical-scale chromatography, demonstrating the different objectives of the two types of chromatography. In (a) the objective is to separate the compound of interest from all other material; in (b) the objective is to separate all the compounds from one another.

product fraction is collected and it is not necessary to analyze numerous samples prior to proceeding to the next step.

For more difficult purifications, where it is not possible to use standard, low-pressure, soft gels to obtain the required product purity with acceptable yields (for example, where product and impurities have very similar chemical properties), high pressure or high performance liquid chromatography (HPLC) can offer an alternative approach. *High performance liquid chromatography* is generally used to describe chromatography on matrices composed of rigid, small particles (5–25 μm diameter) having a narrow particle size distribution.

Industrial-scale HPLC for the purification of proteins (as opposed to peptides and other small molecules) is still in its infancy. Its main advantage is that it can achieve extremely high resolution, enabling chemically similar molecules to be separated from each other in a single unit operation. In contrast to affinity-based methods, high purities are achieved by utilizing efficiency (for example, up to 50,000 theoretical plates per meter) rather than selectivity [6]. However, HPLC has the disadvantages of requiring expensive chromatography equipment and also expensive matrices that need specialized column packing procedures. Furthermore, when HPLC is used with reversed phase columns, the most common type of HPLC column, organic solvents are required to elute molecules from the matrix. These solvents can irreversibly denature many proteins and also require costly explosion-proof facilities [10] and expensive solvent disposal/regeneration procedures.

In order to achieve good separation between chemically very similar molecules, the buffers and organic solvents used in HPLC operations are almost always applied to the chromatography column in the form of a continuous concentration gradient. The formation of accurate gradients, often with solvents of different densities and viscosities, clearly requires more complex equipment than that used to produce step changes in buffer composition. Hence equipment costs are likely to be considerably higher for HPLC systems, compared to those designed for low-pressure chromatography. Nevertheless, these disadvantages can be offset by the very high resolution and rapid cycle times characteristic of HPLC systems. The short cycle times may also have a secondary advantage in that they may enable the purification to be performed at ambient temperatures. The relatively low stability of protein products (compared to conventional small-molecule drugs) may dictate the use of cold-room facilities for low-pressure chromatography, although the stability of different proteins can vary greatly. Jacketed tanks and columns can be used as an alternative to cold rooms, but they can be cumbersome and difficult to operate in a sanitary manner [11]. If such facilities are not required for HPLC operations, the initial capital costs and the running costs are both reduced. Hence HPLC systems can have many advantages, and there are examples in the literature of the successful use of HPLC for the production of therapeutic proteins [12, 13]. It is possible that intermediate-pressure chromatography, utilizing small, uniform, semirigid particles at pressures of approximately 5–15 bar, will at some future time combine the attributes of HPLC and low-pressure chromatography.

5.2.2 Scale-Up of Adsorption Processes

The overall aim of scaling up a purification procedure is to increase the rate of production in the most economical manner. The column sizes required in the manufacturing operation are generally set by the scale of upstream operations and by any subdivision of one batch of product into successive purification cycles. Cell culture operations determine the overall batch size and also the total number of batches per year. Thus the time interval between batches can be calculated and, ideally, all the downstream processing unit operations, from fermentor harvesting to product formulation and sterile filtration, completed within that time. The process time for air-lift fermentors used in mammalian cell culture, for example, is approximately 10–15 days. Hence the column sizes required can be calculated such that the downstream processing operations are scheduled to have either the same or a faster turnaround time.

The stability of the product at the intermediate stages of the purification process affects any subdivision of one batch into cycles and therefore influences column sizes. Thus, in practice, the first chromatography step in a purification sequence is often performed as quickly as possible in one cycle on a relatively large column, so that any proteolytic enzymes and other deleterious impurities and contaminants are rapidly removed. Later polishing steps such as gel filtration, however, may be performed as a number of successive cycles, provided that sample stability and scheduling requirements allow.

The guidelines that are suggested for scaling up chromatographic column dimensions are different for the two basic types of chromatography: adsorption chromatography and partition chromatography.

In adsorption chromatography (for example, ion exchange, hydrophobic interaction, or affinity chromatography), the product binds to the matrix and is then subsequently eluted by a change in buffer composition (such as pH or salt concentration). With this type of chromatography the bed height is not, in itself, a critical feature in terms of chromatographic performance. Short columns are generally recommended, but this is to maximize throughput and minimize the pressure drop across the bed. However, with partition chromatography (for example, gel filtration), where the buffer composition is kept constant and product and impurities travel at different speeds through the matrix, resolution is critically dependent on bed height and therefore tall columns are recommended for optimum performance.

For ion exchange chromatography, equations have been derived that express the rate of production as a function of operational parameters such as flow rates, column dimensions, etc. [14]. However, scale-up is often an empirical process that involves one or more cycles of increasing the scale of operation (typically 10–100-fold) followed by reoptimization of the chromatographic conditions on the larger scale [15].

The guidelines for scale-up of adsorption chromatography are outlined in Table 5.1, and discussed in detail below:

Table 5.1 Suggested Guidelines for Scale-Up

*   Use exactly the same grade of chemicals and the same gel.
*   Keep the ratio of sample volume to column volume constant.
. *   Maintain the same linear flow rate.
*   Keep the bed height constant, and vary the bed diameter only.
*   For gradient elution, maintain the same ratio of total gradient volume to column volume.

(a)   Use exactly the same gels, the same grade of chemicals, the same quality water and the same environmental conditions (such as temperature) when scaling up. Ideally, even the starting material should be identical and should not vary with scale; for example, the same type and size of fermentors should be used, with the same harvesting and clarification procedures. In practice, however, this is often not possible because process development studies must be done before full-scale manufacture is begun. Therefore a robust process should be developed that is insensitive to small changes in the physical or chemical properties of the feed stream.

(b)   Maintain the same ratio of sample volume to column volume, so as to keep the protein loading per liter of gel bed constant. In practice, process-scale chromatography is generally performed with as high a loading as possible without compromising purity or yield. The acceptable loading range must be determined by process development studies.

(c)   Maintain the same linear flow rate (or velocity of liquid traveling through the gel bed), usually expressed in centimeters per hour.

(d)   Keep the column bed height constant, and increase the volume, and therefore the capacity of the gel bed, by increasing the diameter of the column. Since chromatography columns are available only with certain diameters, it may not always be possible to maintain exactly the same bed height (see Table 5.2).

(e)   If gradient elution is required, maintain the same ratio of total gradient volume to column volume. However, where possible, step changes in solvent composition should be used.

Table 5.2 Scale-Up of Adsorption Chromatography

Bed volume, L	0.1	1.0			10.0		
Bed diameter, cm	3.2	9	10	14	25	30	35
Bed height, cm	12.4	15.7	12.7	6.5	20.4	14.2	10.4
Diameter/height	0.26	0.57	0.78	2.15	1.23	2.12	3.36
Flow rate, L/h	0.8	6.4	7.9	15	49	71	96
Cycle time, h	2.5	3.1	2.5	1.3	4.0	2.8	2.0
Throughput, g/h	0.4	3.2	4.0	7.7	25	36	50

Assumptions: average linear flow rate = 100 cm/h;
            operating gel capacity = 10 g/L;
            one cycle = 20 column volumes.

The shape or *aspect ratio* of the column is an important factor in process-scale chromatography. By increasing the bed diameter and not the bed height when scaling up, it is possible to keep both the cycle time and the pressure drop across the bed constant. However, there is clearly a limit to how wide a column can be before it is necessary to increase the bed height so as to prevent channeling through the column and breakthrough of product prior to elution. The recommended ratio of diameter to height of adsorption chromatography columns for process-scale work is approximately 2.0–4.0. With taller and thinner columns the cycle time is longer and the pressure drop across the bed is higher than necessary. With shorter and fatter columns there may be problems with an uneven flow pattern through the column and breakthrough of product. Therefore column selection is often a trade-off between processing time (that is, throughput), product purity, and yield.

Table 5.2 illustrates the scale-up of adsorption chromatography. The assumptions are that the purification procedure has been optimized using a 3.2-cm diameter column packed with 100 mL of matrix, and that the procedure needs to be scaled up 100-fold in two steps (10-fold each). For the initial scale-up, a bed volume of 1.0 L is required. Columns of approximately 1-L capacity are available from the major manufacturers with diameters of 9, 10, or 14 cm. Selection of the 10-cm column allows the bed height and therefore the cycle time and back pressure to be kept constant. Although the wider 14-cm column would have a shorter cycle time and therefore greater throughput, the purification procedure may need to be reoptimized to ensure acceptable product purity and yield.

For the final scale-up to a 10-L bed volume, 30- or 35-cm diameter columns could be used. Since no intermediate-sized columns are available, it is not possible to maintain the bed height at exactly 12.4 cm. The 30- and 35-cm columns, when packed with 10 L of gel, would have bed heights of 14.2 and 10.5 cm, respectively. Either of these columns could be used. Since throughput is a major consideration with all process-scale operations, the 35-cm column, having a shorter cycle time, would be the preferred option, provided that flow distribution through the column was acceptable. The above example illustrates a scaling up of an adsorption chromatography process by increasing column diameter and maintaining a constant bed height. As a rule of thumb, bed heights of approximately 10–25 cm are used when the full-scale process requires columns of between 10 and 100 L of bed volume.

Table 5.3 illustrates typical column dimensions and running conditions that have been used for process-scale adsorption chromatography applications. Although it is feasible to use chromatography columns with bed volumes in excess of 200 L, such large columns are seldom used for adsorption chromatography. If necessary, a large batch can be subdivided and run sequentially as successive cycles on a smaller column. This reduces the quantity of material at risk should a fault occur, but increases the processing time. It can be seen from Table 5.3 that the time taken to complete one individual cycle is not greatly influenced by the chromatographic bed volume. Note, however, that the time required to pack and then test a column has not been taken into account. Validation of column reuse over several cycles is therefore an important part of process development studies, not only to reduce matrix costs, but also to reduce turnaround time and labor costs.

Modern process-scale gels can withstand high linear flow velocities of several hundred or even a thousand centimeters per hour [4]. However, the kinetics of the adsorption and desorption of the product to the gel may limit the actual flow rate to a much lower figure (such as 100 cm/h), at least for the binding and elution steps. The higher flow rates can be used during the washing, regeneration, and reequilibration steps to decrease the overall cycle time and thereby increase throughput.

Older gels, which were not specifically designed for process-scale use, generally cannot withstand high flow rates and high pressures. The scale-up and operation of large columns packed with these gels can be difficult, and compression of the gel bed

Table 5.3 Typical Column Dimensions and Running
Conditions for Adsorption Chromatography

Bed volume, L	5.0	25	75	150
Bed diameter, cm	25	45	80	100
Bed height, cm	10	16	15	19
Diameter/height	2.5	2.8	5.3	5.3
Flow rate, L/h	49	159	503	785
Cycle time, h	2.0	3.1	3.0	3.8
Throughput, g/h	25	80	250	395

Assumptions: average linear flow rate = 100 cm/h;
operating gel capacity = 10 g/L;
one cycle = 20 column volumes.

can give rise to high back pressure and decreased flow rate. If such gels need to be used, one solution is to employ the recently developed radial-flow columns [16] that can offer easier scale-up of chromatographic separations.

5.2.3 Scale-Up of Partition Chromatography

Gel filtration chromatography as a purification step is inherently a low-capacity technique that is not well suited to industrial-scale use. Conditions that give rise to good resolution separations are generally not compatible with the high throughputs required for process-scale operation. In addition, many matrices produced for gel filtration are mechanically weak and can compress when packed in large columns. However, gel filtration is often used as a final stage in many purification schemes to remove trace amounts of impurities and self-aggregates of the product, and to buffer exchange the sample into the final formulation buffer [17].

In gel filtration, resolution is a function of:
(a)   column length,
(b)   the ratio of the sample volume to the column volume,
(c)   the linear flow velocity, and
(d)   the sample concentration.

On a process scale, individual column lengths of approximately one meter are feasible when using modern gels that have some rigidity. However, with such tall columns, bed diameters may need to be restricted to about 20 cm (due to loss of wall support at greater diameters), and linear flow velocities may need to be limited to avoid compression of the gel bed. For these reasons, and also for convenience and safety during packing and unpacking, very tall columns are not advisable. If a greater column length is required in order to achieve the necessary resolution, or if a wide-diameter gel filtration column is necessary to obtain a high throughput, columns connected in series can be used. These are generally stacked columns with diameter-to-height ratios of approximately 3, but standard taller and thinner gel filtration columns can also be connected in series. In practice, some zone spreading inevitably occurs in the numerous column end pieces and in the piping connecting the individual columns [5]. The effect of this zone spreading on the overall resolution and on the sample concentration needs to be determined experimentally.

The ratio of sample volume to column volume is typically around 5% when gel filtration is used on a process scale for the purification of proteins (as opposed to buffer exchange or desalting). If the impurities that need to be removed consist of either much larger or much smaller molecules than the product, it may be possible to improve the throughput by applying a sample volume that is up to 10% of the column volume. On the other hand, it may be necessary to reduce the sample volume to 2–3% of the column volume for difficult separations.

Linear flow velocities for gel filtration are lower than for adsorption chromatography and lead to much longer cycle times. Flow velocities of 3–30 cm/h (approximately one order of magnitude lower than the flow velocities characteristic of adsorption chromatography) may be necessary to achieve the required resolution, although, as with adsorption chromatography, gels specifically designed for process-scale use should be able to withstand higher flow velocities at least during cleaning and reequilibration.

It is usual to keep the protein concentration of the sample applied to the gel filtration column as high as possible to provide a high throughput. Protein concentrations in the range of 10–50 g/L are typical, but this may vary depending on the buffer system used and the temperature of operation. Very high protein concentrations can lead to uneven flow patterns through the gel due to viscosity effects. Also, when protein concentrations are high in order to maximize the throughput, it is important to ensure that the concentration method used does not produce the very aggregates that gel filtration is designed to remove!

The guidelines for scaling up gel filtration unit operations are broadly similar to those for adsorption chromatography, but careful attention to detail is required since a large number of factors can affect the separation obtained. Process optimization studies need to be done to determine the highest protein concentrations, linear flow velocities, and ratios of sample volume to column volume that can be used without a detrimental effect on the required resolution (with a margin of error) when the process is operated with bed heights of approximately 60–100 cm. These factors are of course interrelated and it may be possible, for example, to increase the flow rate and decrease the ratio of sample volume to column volume. Once these factors have been determined, the column diameter is increased to obtain the required cycle capacity. Where diameters greater than 18–25 cm are needed, conventional stack systems are generally used. The theoretical and practical aspects of the scale-up of partition chromatography have been discussed by JANSON and DUNNILL [18].

5.3   Equipment

5.3.1 General Requirements

There are four fundamental requirements that are common to all items of equipment [15, 19, 20]:

(a)   The materials of construction must be compatible with the proposed buffers, solvents, and cleaning solutions [21]. Gaskets, plastics, and so forth should not leach any plasticizers or other chemicals into the product stream, and stainless steels need to be of an appropriate grade if chaotropic chemicals, high concentrations of salt, or solutions of low pH are used in the cleaning or running buffers [22, 23].

(b)   The equipment must be designed to allow hygienic operation. Internal surfaces should be smooth and highly polished to minimize bacterial adhesion and corrosion. Wherever possible, threaded connections should be avoided, since bacteria and other deposits can accumulate in the threads.

(c)   The equipment must function reproducibly in order to produce a quality product with no batch-to-batch variability [24]. Pumps, for example, should be able to maintain their set flow rate irrespective of back pressure, and monitors should have stable outputs without drift of the signal. Where chromatography gels are reused, product quality and safety should be independent of gel cycle number within the known limits.

(d)   Reliability is of central importance because, with high-value compounds, a single batch of product can have a greater value than the equipment used for its purification. All instrumentation must be serviced and calibrated regularly; all electronic equipment should be immune from electrical noise, and should be able to tolerate high levels of condensation. Critical pieces of equipment such as pumps and UV absorption detectors also need to have extensive alarm functions to safely shut down the system if a failure occurs.

In addition to these four fundamental requirements, each individual component needs to fulfill other specific criteria.

5.3.2 Chromatography Columns

The column itself must be able to withstand the operating pressures without flexing. The end plates need to be carefully designed and constructed to produce uniform laminar flow across the entire width of the gel bed (Figure 5.2). The volume of the flow distribution chamber within the end plate should be as small as possible in order to reduce zone spreading [5]. If the end plate incorporates multiple inlets, then these need to be of a sanitary design and must prevent the formation of air locks. The end

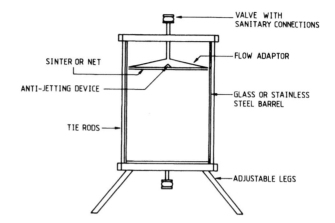

VALVE WITH
SANITARY CONNECTIONS

FLOW ADAPTOR

SINTER OR NET

ANTI-JETTING DEVICE

GLASS OR STAINLESS
STEEL BARREL

TIE RODS

Figure 5.2 Schematic
Representation of a
Typical Process-Scale
Chromatography
Column.

ADJUSTABLE LEGS

connections need to be robust and one should avoid the use of plastic compression-type connections. The seals should contain no potentially contaminating substances, such as oils or hydraulic fluids. Where low pressures are used (up to approximately 3 bar), transparent borosilicate glass columns have many advantages. Obvious faults can be identified, such as the presence of air in the column, which disrupts the liquid flow distribution through the gel, or cracks in the bed due to poor packing techniques. However as column dimensions are increased, transparency becomes less important since only a small proportion of the total gel volume is visible.

For low-pressure chromatography, glass or acrylic columns are frequently specified [25]. Amicon's Moduline™ columns are sturdy, robust, easy to pack, and are good general-purpose variable-bed-height columns. In high-pressure applications, columns are generally constructed of 316L stainless steel. All large columns, and particularly those constructed of stainless steel, can be extremely heavy and may need lifting gear. The advantages and disadvantages of low-pressure chromatography columns available from the major manufacturers are discussed in reference 9.

Although the chromatography gel is not an equipment item, it is one of the most important components of a purification system and must meet a number of criteria. It should be able to withstand high flow rates without compression, and should not swell or shrink with changes in buffer pH, ionic strength, and chemical composition. It should be stable at extremes of pH to enable cleaning, sanitization, and depyrogenation, and must not leach any chemicals into the product. It must be available with a minimum of batch-to-batch variability for as long as the production is likely to continue. The properties of chromatographic gel media required for large-scale protein purification and the importance of supply contracts with gel manufacturers have been discussed by COONEY [26].

5.3.3 Pumps

Pumps should be of the positive-displacement type, such as diaphragm pumps, although with small manual systems (such as in Figure 5.3) peristaltic pumps are sometimes used. The key requirements are:

(a)   the set flow rate should be maintained irrespective of back pressure;
(b)   the handling of the protein solution should be gentle, so as not to shear or denature the protein; and
(c)   the pulsations should be minimized, to avoid disturbance of the chromato-graphic bed.

The major disadvantages of peristaltic pumps are the likelihood of breakage of the flexible tubing used in the pump head, and the difficulty of making sanitary and strong connections between flexible tubing and steel piping. However, where extremely high levels of containment must be achieved, peristaltic pumps are frequently used, since the flexible tubing is the only component in contact with the product. Gear pumps or cog-wheel-type pumps are not recommended for sensitive proteins, since the high shear forces generated in the pump head, possibly in combination with other factors such as air entrainment, can lead to protein aggregation and denaturation [27]. HPLC pumps must be able to deliver low flow rates at high pressures (typically 20–100 bar) and be fully flameproof, due to the frequent use of flammable solvents.

5.3.4 Detectors

Three types of detectors are used to monitor and control the progress of purification procedures: pH, conductivity, and UV absorption detectors. The requirements for all detectors are that they should not cause significant back pressures, should not give rise to dilution or back mixing of the product stream, and should not be a source of contaminants. Due to the high protein concentrations that are frequently encountered in process-scale operation, UV detectors need to have short optical path lengths so as to ensure that the detector is operating in its linear range. Reliability of the UV detector is of critical importance, since its output is frequently used to control the switching of valves and the collection of product. Therefore, extensive alarm functions and monitoring of lamp current should be considered for any system intended for the production of high-value products.

5.3.5 Valves, Piping, and Vessels

Items of auxiliary equipment that need to be considered when specifying a purification system include valves, piping, and tanks. The piping used to connect the individual components should ideally be of stainless steel, although flexible piping is a cheaper alternative. The piping should be seamless, to give gentler fluid handling, and ideally should be able to withstand temperatures up to 150 °C, to allow either steam sterilization or autoclaving. The piping should be polished internally after welding, to eliminate all crevices. The dimensions of the piping should be wide enough not to give rise to a significant pressure drop, but narrow enough to produce slight or moderate turbulent flow to reduce zone spreading [5]. In practice, 6-mm-internal-diameter piping is frequently specified for adsorption chromatography columns up to approximately 10 L in volume, and 10-mm-ID piping for columns up to approximately 100 L (Table 5.4).

Table 5.4 Approximate Flow Rates and Tubing Diameters Used with Adsorption Chromatography Columns

Column volume L	Flow rate L/h	Tubing diameter mm ID
1–10	2–40	6
5–100	25–400	10
50–300	200–1000	25

Valves used to direct the process fluids must completely seal to prevent contamination, and must be easy to clean. A feedback mechanism with an alarm function should be incorporated, to relay information of the actual valve position to the controller. Ball valves should be avoided if possible, as the hollow ball cannot be flushed clean when the valve is closed, because the trapped fluid is not accessible to the cleaning medium. Where the operating pressures allow, weir-type diaphragm valves should be used, as these are relatively easy to clean. However, in high-pressure systems diaphragm valves are not suitable, since the fluid pressure in the system can be higher than the pressure being used to close the valve, resulting in leakage. In these applications the ball valve may be the best alternative, although special attention must be paid to cleaning regimes.

A variety of different vessels are likely to be required in order to operate a chromatography system, such as buffer makeup tanks, supply tanks, and receiver tanks. Where only small volumes of aqueous solutions are used, for example, less than 10–20 L, polypropylene vessels are convenient, compatible with most process streams, and relatively easy to clean and sterilize. For larger volumes, stainless steel tanks are used. Grade 316L austenitic stainless steel with a reflective mirror finish is frequently specified. It is relatively resistant to corrosion, although high concentrations of halide ions, such as those used to clean ion exchange matrices, can lead to pitting corrosion and stress corrosion cracking, particularly when aggravated by low pH or elevated temperatures [23]. Reflective finishes, either electropolished or mechanically polished with 200 grit or finer, are expensive to produce but reduce the risk of contamination, and ease cleaning and sterilization. For liquids that are not compatible with 316L, other alloys or fluoroplastic- or glass-lined tanks are required.

For vessels with volumes up to approximately 100 L, mobile trundle tanks are frequently specified for ease of use. Larger tanks can be difficult (and dangerous) to move around the plant, and therefore fixed tanks are preferred. Ideally all stainless steel tanks should be CIP- and SIP-compatible pressure vessels with spray balls. Depending on the number and type of solutions to be made up, it may be convenient to have separate buffer preparation tanks and buffer supply tanks. Each tank may require agitators, level controllers or indicators, pH/temperature/conductivity probes, sample valves, sight glasses, bursting disks, integral filter housings, and thermostatic jackets.

5.4   Automated Purification Systems

5.4.1 Design: Liquid-Handling Unit

The simplest chromatographic system consists of a pump, a column, a detector, vessels, and interconnecting tubing (Figures 5.3 and 5.4). These components form the core of any purification system, but complete large-scale systems differ in at least two important respects:

(a)   additional components and mechanisms are incorporated to improve product safety and ensure fail-safe operation; and

(b)   automation is used to increase reproducibility and decrease the need for manual intervention.

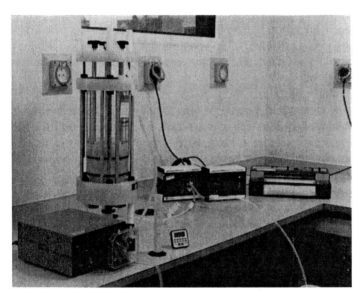

Figure 5.3 A Manual Chromatography System, comprising a pump, chromatography column, UV detector, and chart recorder. (Courtesy of Celltech Ltd.)

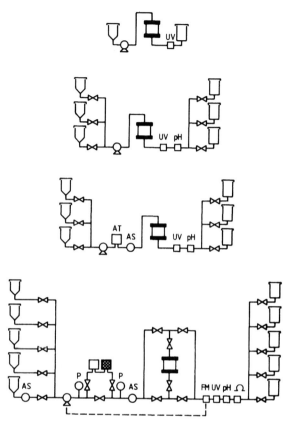

Figure 5.4 Schematic Diagram of Four Chromatography Systems of Increasing Complexity. AS = air sensor, AT = air trap, FM = flow meter, P = pressure sensor, pH = pH monitor, UV = UV monitor, Ω = conductivity meter.

For the manufacture of therapeutic-grade products, complex automated systems are generally used to ensure reliable, fail-safe operation and to collect sufficient data to demonstrate reproducibility.

Automation not only reduces labor costs and increases throughput by allowing 24-h operation, but also enables quality to be built into the product by reducing process variability [24].

The additional components incorporated into production-scale systems (such as those shown in Figures 5.5 and 5.6) are used to ensure that air does not get into the system (air sensors and bubble traps), that pressure does not build up in the system (pressure detectors), and that it operates in as sterile a manner as possible (in-line filters). Further components are then added to control the purification process and to improve the convenience of operation (programmable controller, automatic valves, CIP facilities, etc.).

The detailed design of large-scale purification systems can vary depending on a system's intended use. A multipurpose, multiscale pilot-plant system, for example, is likely to differ significantly from a system dedicated to the routine manufacture of a single product. Furthermore, the details of the purification process itself (for example,

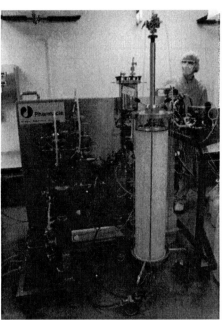

Figure 5.5 An Automated Chromatography System, used with a 6-L ion exchange column for the purification of a therapeutic protein. (Courtesy of Celltech Ltd.)

Figure 5.6 An Automated Chromatography System, used for gel filtration. To reduce dilution of the product before it reaches the column, the product is loaded from a stainless steel vessel adjacent to the column using air pressure, instead of the more conventional arrangement where product is loaded through the pump.

whether reverse elution is required) and consideration of what other tasks the system is to perform (for example, whether it is to be used to pack columns) have a bearing on its design.

Figure 5.4 shows a range of systems of increasing complexity. Process-scale systems generally have a number of buffer tanks and collection tanks connected to the purification system via either two- or three-way valves. The use of automatic solenoid or pneumatic valves, under the control of a microprocessor-based controller, makes it possible to apply various buffers in sequence without manual intervention. An air sensor is frequently used on the sample inlet line so as to allow loading of the entire sample volume onto the column. Only after the air sensor has detected air in the sample inlet line is the next step in the purification cycle initiated. A 0.2-μm sterile filter is often incorporated into the system before the column. If the sample itself is prefiltered off-line, this filter may be bypassed during the loading operation and used only to filter the buffer solutions. A bubble trap located immediately prior to the column serves not only to eliminate any air that may have entered the system, but also serves as a pulse dampener, which may be necessary, depending on the pump design. If it is essential to minimize dilution of the sample (for example, in gel filtration applications) it may be necessary either to bypass the bubble trap during sample loading or to drain the liquid level in the bubble trap to as low a level as possible. However, to minimize the risk of air entering the column this should be avoided where possible.

The piping arrangement around the column depends on whether the abilities to bypass the column and to reverse the direction of flow are required. A column bypass is a feature that is particularly useful during cleaning of the system. The reverse-flow option is frequently used to elute the product in as small a volume as possible [8]. It is also the most efficient method of cleaning and regenerating the gel, since any contaminants and impurities adsorbed near the top of the gel bed or on the sinter can be removed without being forced to travel all the way through the gel.

It is useful to incorporate automatic flow-stopper valves (with manual overrides) immediately adjacent to the column inlet and outlet. If these valves are programmed to close whenever the pump is not operating, syphoning of liquid through the column can be prevented.

When designing the layout of piping, the engineer should minimize or (preferably) completely eliminate dead legs, to ensure that the system can be fully drained. Manual drain valves can be incorporated, although a completely automatic CIP regime is preferable.

The piping layout should also be designed to minimize zone spreading, particularly after the column. Thus the detectors should have low dead volumes, and the piping between the column outlet and the collection tanks should be kept as short as possible. In gel filtration applications, where dilution of the sample must be avoided, both the inlet and outlet sides of the system must be carefully designed to prevent zone spreading. The entire piping layout should be easily accessible; this makes it easy to change filters, see the fluid level in the bubble trap, locate and rectify any leaks, and so forth.

5.4.2 Design: Data-Handling Unit

Although the liquid-handling units produced by most manufacturers often have a similar design, the data-handling units can vary considerably. It is important to consider carefully the requirements of the software and to specify the simplest system that can meet these operational requirements. An overdesigned system with elaborate features and complex safeguards is more likely to malfunction, confuse operators, and be difficult to validate. It is always best to have good operators using a simple system.

The following features can be useful in automated purification systems:

(a)   The data-handling unit should be remote from the wet liquid-handling unit, ideally at room temperature and in a dry environment.

(b)   The software should record the continuous outputs from all the monitors and detectors (UV, conductivity, pH, temperature, pressures at different locations, flow rate, valve positions, liquid levels, etc.).

(c)   The software should be able to integrate the output from the UV monitor to give an estimate of the quantity of UV-absorbing material in each fraction.

(d)   Key parameters (such as UV absorption) should be displayed in real time, to inform the operators of the progress of the run. Further information should be available onscreen, in the event of an investigation being required.

(e)   A real-time mimic of the fluid path showing valve status should be available.

(f)   The system should be able to distinguish between warnings (such as low fluid level in the bubble trap) and alarms (such as incorrect valve position).

(g)   The system should always be fail-safe.

(h)   The software should enable the monitor outputs from a run to be overlaid and compared with a previous run recalled from the system's memory. This is particularly useful for demonstrating batch-to-batch consistency.

(i)   Access to the program should be restricted (for example, by password).

(j)   The facility for manual intervention should be available to supervisors, but all such interventions must be recorded.

(k)   After each run the system should print out a record of all the events and monitor outputs associated with that run for inclusion in the batch records. Estimated volumes and quantities of UV-absorbing material in each fraction, and any manual interventions, alarm overrides, etc., should also be printed out.

(l)   The automated system should be able to communicate with other computer-based systems.

5.4.3 Operation

The operation of automated systems requires more than just the connection of sample/buffer tanks and the initiation of a purification program. A sequence of operations similar to the operation of fermentors [28] must be performed. The analogy is an appropriate one, since many similar operations need to be performed to ensure that the purification system is operated in as sterile a manner as possible. Although complete sterility is not easily achieved, an extremely low level of bioburden is essential to keep pyrogen levels low and ensure that bacterial proteases never come into contact with the product.

Hence, the purification system and the gel must be cleaned before or after each purification run (or both before and after) [29]. Sodium hydroxide solutions are frequently used to clean and store both equipment and gels. Depending on the temperature, the sodium hydroxide concentration, and the contact time, these solutions not only solubilize denatured or precipitated proteins and lipids, but also destroy pyrogens [30] and inactivate most bacteria and viruses. Furthermore, sodium hydroxide is cheap and easy to dispose of [31].

Once the purification system has been cleaned and sanitized, new filters must be installed and their integrity tested, the system primed and checked for leaks, and detectors tested and calibrated. The column needs to be packed and the quality of the packing tested by means of an *HETP* (*height equivalent to a theoretical plate*) test [26] (see Appendix B, Section 5.6.2). If validation studies have been performed to demonstrate that a previously used column can be adequately cleaned and then stored for a period of time, to allow reuse of the column without compromising product safety and efficacy, then repacking of the column with fresh matrix may not be required for each production run. In the absence of such studies, fresh matrix may be needed for each batch.

Column packing for low-pressure chromatography can be more of an art than a science, and for best results the manufacturer's instructions should be followed. The exact procedure depends on the physical properties of the gel/matrix, but in all cases the overall aim is to achieve a uniform packing of the gel throughout the column without the introduction of air into the system. This generally requires the gel to be slurried and poured into the column in one operation (using a column extension piece or packing reservoir if necessary). The slurry should be thoroughly mixed and, ideally, degassed to ensure homogeneity and to remove any air pockets that may be trapped. The gel should then be packed under pressure, either by pumping buffer through the column or by applying gas pressure. Ideally, the pressure should remain constant during the packing operation, and should be 20–50% higher than the maximum pressure encountered during normal operation. Once a stable bed height has been achieved, the upper end plate should be placed on top of the gel surface without introducing air into the column. This operation needs to be performed as soon as possible after the pressure is released so as to prevent the gel reexpanding.

Columns with adjustable end pieces (Figure 5.7) are easier to pack than fixed-bed-height columns, and should be chosen where possible (despite their greater cost) if regular packing and unpacking are required. High-pressure chromatography systems require specialized packing techniques/equipment that are beyond the scope of this chapter.

An HETP test is used to examine the efficiency of the column packing procedure, and is particularly important in gel filtration applications. The HETP test is usually performed by passing a pulse of either UV-absorbing material or high-conductivity solution through the packed column. The HETP value is a measure of the spreading of the pulse as it travels down the column (see Appendix B, Section 5.6.2). As a rule of thumb, the HETP in a well-packed process-scale column should correspond to approximately 2–10 times the average diameter of the particles used as the column

Figure 5.7 A Manual System Used for Column Packing. In this application, a peristaltic pump is used for packing the column, while the automated system is being cleaned and prepared for the chromatography operation.

packing material [32]. The numerical value of the plate height (typically about 0.03 cm for low-pressure chromatography columns) is a function of the nature and volume of the test sample. Therefore, it is necessary to experimentally establish the maximum plate height that gives acceptable purification results and to set an upper limit on the permissible HETP value. Since zone spreading in pipes, valves, etc., affects the HETP value, the plate test should be performed using equipment identical to that used in the actual manufacturing operation. As a final check, an LAL pyrogen assay should be performed on the eluent from the packed, equilibrated column just prior to loading the sample. Further checks may also need to be performed, depending on the design of the system and the purification operation being performed.

5.5   Installation, Commissioning, and Validation

Once purification equipment has been designed and constructed, it needs to be tested to ensure that it operates consistently and reproducibly within the limits required by the process. A validation program involving challenge studies has to be performed, to provide confidence that the equipment operates as expected, and that use of the equipment does not give rise to variations in the product that may affect its safety or effectiveness [33]. In this section we mention briefly some aspects of the validation of purification equipment.

Process validation is a requirement of the *Current Good Manufacturing Practices* (*CGMPs*) for pharmaceuticals. The validation of an entire pharmaceutical manufacturing process requires validation of all the individual components of that process, including analytical procedures, equipment, facilities, etc. [34]. Equipment validation is generally divided into a number of phases through a *life-cycle* approach: a prequalification phase involving the specification and design of equipment, installation qualification (IQ), operational qualification (OQ), process qualification, and ongoing evaluation.

Installation qualification (IQ) has been defined as "documented verification that all key aspects of the installation adhere to manufacturer's recommendations, appropriate codes, and approved design intentions" [33]. Model and serial numbers must be listed and checked to verify that all the equipment in the plant corresponds to the actual equipment specified. Materials of construction (for example, for gaskets, seals and O rings) and pipe material certificates should be checked to ensure that they conform to the appropriate grade of materials. An estimate of the reliable working life of various components, such as pump seals, UV detector bulbs, etc., should be obtained, and an adequate stock of spares established. Calibration and routine maintenance requirements for the equipment have to be determined, and written procedures and schedules drawn up, to ensure that the equipment is adequately and routinely maintained.

Operational qualification (OQ) is "documented verification that the system or subsystem performs as intended throughout all specified operating ranges" [33]. OQ requires protocols to be written before studies are initiated, detailing the scope of the investigation and the tests to be performed. The protocols should specify acceptable ranges for the measured parameters, with clear pass/fail limits. Challenge studies need to be carried out at the limits specified for the system. This *stress testing* should be performed under real or simulated production conditions. The correct functioning of all the component parts of the system must be demonstrated. For example, valves must be shown to change status when instructed to do so, both under manual command from the controller and as part of an automated purification sequence. The fluid path should be shown to change as expected when valve status is changed. Pumps and flow meters need to be tested and their calibration checked with realistic process fluids and under a range of conditions mimicking those likely to be encountered during manufacturing operations. The set flow rate should be increased and decreased to check for hysteresis effects. There must be documented evidence that the flow rate is maintained (within predetermined acceptable limits) irrespective of the back pressures encountered.

Particular attention must be paid to the safety and alarm systems, to verify that they trip when the preset levels are reached and that the system fails safe as intended. All alarms must be shown to cause the expected effects on the system as specified in the functional specification, and not cause any unintended actions to occur during shutdown or while the system recovers to restart its operation. For example, air should be deliberately introduced into the system to ensure that the bubble trap prevents air from reaching the column. Air sensors should also be challenged to verify that any air passed into the sensor is detected and appropriate audio/visual alarms are activated.

Cleaning procedures must be shown to adequately remove the cleaning agents themselves, in addition to impurities, contaminants, and the product itself [35]. The challenge tests need to be repeated a sufficient number of times, not only to assure statistically meaningful results, but also to establish the reproducibility of the process.

In addition to prospective validation, retrospective validation studies involving an analysis of historical batch record data can be useful to augment the validation of those aspects of the process that are difficult to mimic under simulated production conditions. The same concept of validation by challenge studies applies to computer-based systems. The subject of computer system validation is outside the scope of this chapter; some guidance can be obtained from recent publications [36]. Further information on the rationale behind validation studies and their design is available from the literature [33, 34] and from guidelines published by the regulatory authorities [37].

Standard operating procedures (*SOPs*) also need to be written to enable the operators to use the equipment correctly. The instructions must be clear and unambiguous. Where alarm functions are incorporated into the system, the instructions must state what action should be taken and by whom (for example, operator or supervisor) in the event of an alarm being activated or power being interrupted. The engineer may also be required to contribute towards a documented training program for the system operators.

In contrast to conventional small-molecule drugs, biological products are large complex molecules that cannot be fully characterized by physical and chemical tests, even when a wide range of state-of-the-art analytical techniques is used. Furthermore, since many biological products are derived from the culture of continuous or "immortal" mammalian cell lines that have properties similar to cancer cells, the possibility of the purified product transmitting oncogenic DNA or viruses needs to be addressed. Hence, in order to ensure the safety and efficacy of these compounds, it is not adequate just to test the final purified product for purity, integrity, and identity. In addition to final product testing, the entire manufacturing operation for the production of a biological product also needs to be fully characterized. Process validation studies are used to demonstrate the capacity of the purification process to remove both known impurities and potential contaminants, and to demonstrate that the process is consistent and reproducible [38, 39]. Ideally, these studies should be done on the full-scale manufacturing process, and indeed there are some studies that cannot be performed any other way (for example, measurement of bioburden levels at different points in the process). However, there are other studies where a scaled-down version of the manufacturing operation is an acceptable substitute for the full-scale process. For example, in order to determine the capacity of the purification process to remove or inactivate virus particles, an accurately scaled-down version of the process may be challenged with a number of viruses, and clearance factors determined for each purification operation [40, 41]. Such studies, together with a comprehensive characterization of the viral status of the cell line and testing of in-process and final product samples for infectious viruses, viruslike particles, and reverse transcriptase activity, should give reasonable assurance that the purified product has a low risk of containing viruses.

Other validation studies are likely to be required, to demonstrate that DNA, endotoxins, any leached affinity ligand, etc., are adequately cleared by the purification process. A detailed discussion on validation of column chromatography is beyond the scope of this chapter, but a comprehensive account of the subject is in reference 39.

5.6   Appendices

5.6.1   Appendix A: Glossary of Terms

adsorption chromatography—chromatography where one or more components of the sample bind tightly to the matrix. A stepwise or continuous change in solvent composition (for example, salt concentration) is then used to elute the tightly bound component(s).

affinity chromatography—a type of adsorption chromatography where the binding to the matrix is due to a highly specific biological interaction between two molecules (for example, hormone/receptor, antibody/antigen, or enzyme/inhibitor).

channeling—uneven flow in a chromatography column.

contaminants—substances that have been accidentally introduced into the process stream, such as bacteria.

HETP (height equivalent to a theoretical plate)—a term borrowed from distillation theory and used as a measure of column efficiency. More efficient columns have smaller HETP values, that is, they have more plates per unit of column length.

impurities—process-related substances known to be present in the unpurified sample, for example, DNA, host cell proteins, etc.

partition chromatography—a type of chromatography where separation is achieved by the components in a mixture moving through the matrix at different rates relative to the solvent front. In this type of chromatography the solvent composition is kept constant.

resolution—a measure of the separation efficiency of a chromatographic system. The higher the resolution, the better the separation between product and impurities.

zone spreading—the tendency of a pulse or a plug of material to spread out and become more dilute during chromatography. Some zone spreading is inevitable (for example, due to molecular diffusion) but it can be kept to a minimum in a well-designed system.

5.6.2 Appendix B: Calculation of HETP

Well-packed columns are essential for large-scale chromatography. This is particularly true for gel filtration operations. The quality of the packing can be tested by measuring the degree to which a pulse of material, introduced at the column inlet, spreads out while traveling through the gel. The pulse of material can be either a UV-absorbing compound (for example, a dilute solution of benzyl alcohol or tryptophan) or simply a concentrated salt solution. In the latter case, an inline conductivity meter connected to the column outlet is used to measure the spreading of the sample. Where multiple stack columns are used in series, it may be necessary to test individual columns as well as the complete stack. For pharmaceutical applications, it is essential to ensure that the compound used to test the column does not contaminate the final product.

In its simplest form, the height equivalent to a theoretical plate is calculated from three variables: the elution volume of the test sample, the peak width, and the length of the gel bed. The peak width is usually calculated by drawing tangents alongside the rising and falling edges of the peak at half the maximum peak height. The points at which the tangents intersect the baseline define the peak width, as illustrated in Figure 5.8. The plate number and the HETP value are then calculated as shown below.

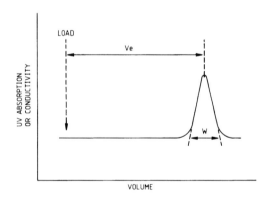

Figure 5.8 Schematic Representation of Elution Volume, V_e, and Peak Width, W.

$$HETP = L/N \qquad (5.1)$$

$$N = 16(V_e/W)^2 \qquad (5.2)$$

where HETP is the height equivalent to a theoretical plate;
   L is the column length;
   N is the number of theoretical plates;
   V_e is the elution volume of test sample; and
   W is the peak width.

For example, if $V_e = 24$ L, $W = 1.9$ L, and $L = 88$ cm, then $N = 16(V_e/W)^2 = 2553$, and HETP $= L/N = 0.034$ cm (or 2900 plates per meter).

5.7 References

1. LUBINIECKI, A. S.; Safety Considerations for Cell Culture–Derived Biologicals; In: *Large Scale Cell Culture Technology*; Lydersen, B. K., Ed.; Hanser Publishers, New York, 1987.
2. VAN BRUNT, J., *Bio/Technology 6*, 479–485 (1988).
3. KEARNS, M. J., *Pharm. Eng. 9*, 17–21 (1989).
4. JANSON, J. C., JONSSON, J. A.; In: *Protein Purification—Principles, High Resolution Methods and Applications*; JANSON, J. C., RYDEN, L., Eds.; VCH Publishers, New York, 1989.
5. JANSON, J. C., HEDMAN, P., *Adv. Biochem. Eng. 25*, 43–99 (1982).
6. JANSON, J. C., HEDMAN, P., *Biotechnol. Prog. 3*, 9–13 (1987).
7. WHEELWRIGHT, S. M., *Bio/Technology 5*, 789–793 (1987).
8. CHASE, H. A.; In: *Discovery and Isolation of Microbial Products*; VERRALL, M. S., Ed.; Ellis Horwood Ltd., Chichester, 1985.
9. VOSER, W., WALLISER, H. P.; In: *Discovery and Isolation of Microbial Products*; VERRALL, M. S., Ed.; Ellis Horwood Ltd., Chichester, 1985.
10. JOYCE, W. G., GAUDET, E. G., SCHRATTER, P., *Bio/Technology 7*, 721 (1989).
11. WATT, J. G.; In: *Frontiers in Bioprocessing*; SUBHAS, K. S., BIER, M., TODD, P., Eds.; CRC Press, Boca Raton, FL, 1990.
12. KROEFF, E. P., OWENS, R. A., CAMPBELL, E. L., JOHNSON, R. D., MARKS, H. I., *J. Chromatogr. 461*, 45–61 (1989).
13. HERSHENSON, S., SHAKED, Z.; In: *Frontiers in Bioprocessing*; SUBHAS, K. S., BIER, M., TODD, P., Eds.; CRC Press, Boca Raton, FL, 1990.
14. SIMPSON, J. M., *Bio/Technology 6*, 1158–1162 (1988).
15. SOFER, G. K., NYSTROM, L. E.; *Process Chromatography—A Practical Guide*; Academic Press, London, 1989.
16. SAXENA, V., DUNN, M., *Bio/Technology 7*, 250–255 (1989).
17. BONNERJEA, J., OH, S., HOARE, M., DUNNILL, P., *Bio/Technology 4*, 954–958 (1986).
18. JANSON, J. C., DUNNILL, P.; In: *Industrial Aspects of Biochemistry*; SPENCER, B., Ed.; Elsevier, North Holland, Amsterdam, 1974.
19. JOHANSSON, H., OSTLING, M., SOFER, G., WAHLSTROM, H., LOW, D., *Adv. Biotechnol. Processes 8*, 127–157 (1988).
20. JOHANSSON, S., JOHANSSON, H., LOW, D.; In: *Modern Approaches to Animal Cell Technology*; SPIER, R. E., GRIFFITHS, J. B., Eds.; Butterworths, Sevenoaks, UK, 1987.
21. COWAN, C. T., THOMAS, C. R., *Process Biochem. 23*, 5–11 (1988).
22. CONNOLLY, B. J., *Br. Corros. J. 5*, 209–216 (1970).
23. DILLON, C. P., RAHOI, D. W., TUTHILL, A. H., *BioPharm 5*, 40–44 (1992).
24. GOPAL, C., *Pharm. Technol. 13* (2), 56–65 (1989).
25. KEHOE, J. A.; In: *Frontiers in Bioprocessing*; SUBHAS, K. S., BIER, M., TODD, P., Eds.; CRC Press, Boca Raton, FL, 1990.

26. COONEY, J. M., *Bio/Technology* 2, 41–55 (1984).
27. VIRKAR, P. D., NARENDRANATHAN, T. J., HOARE, M., DUNNILL, P., *Biotechnol. Bioeng.* 23, 425–429 (1981).
28. BIRCH, J. R., LAMBERT, K., THOMPSON, P. W., KENNEY, A. C., WOOD, L. A.; Antibody Production with Airlift Fermentors; In: *Large Scale Cell Culture Technology*; LYDERSEN, B. K., Ed.; Hanser Publishers, New York, 1987.
29. PHARMACIA, *Downstream* 2, 1–3 (1986).
30. PEARSON, F. C.; In: *Aseptic Pharmaceutical Manufacturing*; OLSON, W. P., GROVES, M. J., Eds.; Interpharm Press, Inc., Prairie View, IL, 1987.
31. CURLING, J. M., COONEY, J. M., *J. Parenter. Sci. Technol.* 36, 59–63 (1982).
32. MCDONALD, P. D., BIDLINGMEYER, B. A.; In: *Preparative Liquid Chromatography*; BIDLINGMEYER, B. A., Ed.; Elsevier, Amsterdam, 1987.
33. CHAPMAN, K. G.; In: *Pharmaceutical Process Validation*; LOFTUS, B. T., NASH, R. A., Eds.; Marcel Dekker, Inc., New York, 1984.
34. KIEFFER, R. G.; In: *Validation of Aseptic Pharmaceutical Processes*; CARLETON, F. J., AGALLOCO, J. P., Eds.; Marcel Dekker, Inc., New York, 1986.
35. HARDER, S. W., *Pharm. Technol.* 8, 29–34 (1984).
36. PMA's Computer System Validation Committee, *Pharm. Technol.* 10 (5), 24–34 (1986).
37. *Guideline on General Principles of Process Validation*; Center for Drugs and Biologics and Center for Devices and Radiological Health, Food and Drug Administration; Rockville, MD, May, 1987.
38. SOFER, G. K., NYSTROM, L. E.; *Process Chromatography—A Guide to Validation*; Academic Press, London, 1991.
39. PDA Biotechnology Task Force on Purification and Scale-up, *J. Parenter. Sci. Technol.* 46, 87–97 (1992).
40. WHITE, E. M., GRUN, J. B., SUN, C. S., SITO, A. F., *BioPharm* 4, 34–38 (1991).
41. BRADY, D., BONNERJEA, J., HILL, C. R.; In: *Separations for Biotechnology*; PYLE, D. L., Ed.; Elsevier Applied Science, London, 1990.

PART II

SYSTEM COMPONENTS

Vessels for Biotechnology: Design and Materials

Pio Meyer

Contents

6.1   Introduction

In all branches of industry, vessels are used for storage and transport of raw material, as reaction vessels in chemical or biotechnological processes, and for many other purposes, such as water storage and liquid waste collection and treatment. In the biotechnology industry, those vessels used for the production and storage of media and product are of most importance.

A great number of different vessel types are available: cylindrical vessels with dished bottom, spherical vessels, conical vessels, cubic tanks, etc. Despite the variety, the cylindrical vessel with a dished bottom and head is predominant in biotechnology. In addition to vessel geometry, several other parameters define the layout and construction of the vessel. These are outlined in Table 6.1.

6.2   Relevant Guidelines and Customer Specifications

Most of the vessels used in biotechnology are designed for sterile operation. Sterility is obtained either by autoclaving or by *in situ* sterilization of the vessel and its appurtenances.

If the vessel is sterilized in an autoclave, the technical construction is much simpler compared to an *in situ* sterilized vessel. The savings in vessel design (pressure and heating systems), however, have to be spent for suitable autoclaves. Autoclaving process vessels is common in certain countries (such as the U.S.A.) for volumes up to approximately 100 L.

In general, the biotechnological vessel is constructed as a pressure vessel to enable *in situ* sterilization at temperatures up to about 130 °C (265 °F) and pressures up to 2.5 bar (37 psig, 255 kPa).

6.2.1   National Guidelines

The technical basis for the construction of pressure vessels is defined by the national guidelines valid at the site where the vessel will be utilized. These guidelines are typically developed by a knowledgeable professional society and then adopted and enforced by governmental agencies. The variety of regulations that apply in different countries is illustrated in Figure 6.1.

In most countries operational pressure together with volume or diameter are used to define the vessel classification. In the U.S.A., for instance, the American Society of Mechanical Engineers (ASME) dictates that any vessel operated above 15 psig (1.03 bar, 100 kPa) and greater than 6" (15 cm) in diameter must be pressure rated. The ASME recommendation is carried further by federal, state, and local regulatory

Table 6.1 Parameters That Define Vessel Design

1. Arrangement:
   (a) Horizontal—on saddles or claws
   (b) Vertical—on legs or supporting frame
2. Installation:
   (a) Fixed—installed at site
   (b) Portable—on wheels or small movable container
3. Pressure Range:
   (a) Vacuum
   (b) Atmospheric
   (c) Pressurizable
4. Material:
   (a) Stainless steel
   (b) Steel
   (c) Light metal
   (d) Exotic metal
   (e) Coated/lined steel vessel
   (f) Plastic/nonmetal
5. Application:
   (a) Liquids
   (b) Gases
   (c) Solids
6. Assembly:
   (a) In workshop—up to 10 m^3
   (b) At site—larger than 10 m^3 from prefabricated parts

agencies, such as the Occupational Safety and Health Administration (OSHA). The ASME has additional guidelines in the ASME Code, Section VIII, that define the way in which the vessel should be designed and fabricated. Vessel manufacturers are required to have ASME welders and designers on site to perform the design and fabrication function. They must maintain a detailed paper trail of the mill spec sheets for the materials used to construct the vessel. Before a vessel can be placed in service, the fabricator must schedule a test and inspection with an ASME inspector. During this site visit, the vessel is hydrostatically tested at 1.5 times its design pressure. The vessel is checked for leaks, condensate, etc. The inspector also reviews the design calculations and material spec sheets. If the requirements are met, the vessel is given its ASME rating.

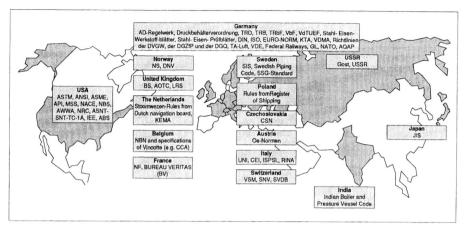

Figure 6.1 Geographical Illustration of National Guidelines.

In Germany, the proper manufacture of a pressure vessel is defined by the *Pressure Vessel Law* (*Druckbehälterverordnung*) in combination with the *Technical Rules for Pressure Vessels* (*Technische Regeln Druckbehälter* or *TRD*) and the *AD Specifications* (*Arbeitsgemeinschaft Druckbehälter* or *AD Merkblätter*). Different testing procedures and acceptance specifications are required, according to the maximum pressure, and the product of maximum pressure and vessel volume (pressure-volume, in bar-liters). Pressure vessels with pressures above 1 bar (15 psig, 100 kPa) and with a pressure-volume smaller than 200 bar-liters are in inspection class 2 (Prüfgruppe 2); vessels with a pressure-volume of 200–1000 bar-liters are in Prüfgruppe 3; and vessels with a pressure-volume above 1000 bar-liters are in Prüfgruppe 4.

Pressure vessels of inspection class 2 require either a manufacturer's certificate for a successful pressure test or final inspection by an expert. Relevant education or experience in pressure vessel manufacture qualify the expert. Pressure vessels of inspection class 3 first have to be examined during the manufacture of the vessel (drawings, materials, design calculations, material specifications, pressure test). After manufacture acceptance, specifications include the examination of the equipment, its final arrangement, and its location. In general, testing is performed by an expert of the *TUEV* (*Technischer Überwachungs-Verein*). Type approvals by TUEV are applicable for the production of a set of pressure vessels. If all vessels are built according to the same specifications, the first examinations (material, drawings, etc.) do not need to be repeated for each subsequent vessel. The pressure vessels of the higher inspection classes require, in addition to the above test procedures, the regular testing and examination of the pressure vessel at the operator's site.

Compliance with the relevant guidelines is straightforward for the manufacturer in his home country, but as soon as vessels are exported to different countries, knowledge of the appropriate guidelines is required. Registration and processing of the required guidelines, and their continuous updating, is labor-intensive for the vessel manufacturer, and these efforts are multiplied with every additional export country covered. For this reason, it seems to be easiest to purchase a vessel in the country where it will be located. Nevertheless, this is not always done due to either financial or, very often, quality considerations. In this case it is of critical importance that the fabricator have the knowledge and means to design and fabricate the vessel according to the regulations of the country where it will be located.

The basic purposes of all national guidelines are safety and quality control, to protect the employees, the population, and the environment. Although the goals of the guidelines, the technical requirements, and the safety of the constructed plants are similar worldwide, the individual guidelines often differ from each other. These differences in national guidelines are due in part to historical differences in the development of industrialization and in part to the degree of legal control (authoritarian rule versus self-responsibility). The units of measurement (English versus metric) may also have played a part.

A few examples of how regulations differ in individual countries are listed below:
- (a)   the Italian guidelines do not accept longitudinal viewing glasses in pressure vessels;
- (b)   in Japan, the isolation jacket of the vessel has to be screwed on; welding is not accepted;
- (c)   in some countries regular examinations of the staff are required for the manufacture of pressure vessels;
- (d)   in the U.S.A., manufacturers must apply for a *U stamp* for vessel construction. Maintenance of the U stamp outside the U.S.A. is rather costly due to the regular examinations and controls required. As a result, a decreasing number of foreign vessel manufacturers possess the U stamp.

6.2.2   Biotechnological Guidelines

In addition to the national guidelines that regulate the construction of pressure vessels for all industries, vessels for biotechnology have other regulations that affect their fabrication. These guidelines are based on the biological process or product (toxic, pathogenic, nontoxic), the field of application (foods, pharmaceuticals, agronomy) and the nature of the process (natural, genetically engineered). Like the national guidelines, these regulations protect the staff, the environment, and the users of the biotechnological products.

Some guidelines, such as those for foods and beverages, were developed from historical data, whereas guidelines for developing technologies are often based on already existing guidelines (such as those of FDA, NIH, etc.) because of the time and knowledge required to develop them. Some countries apply them directly, others adapt them according to their understanding of the technology.

For the vessel manufacturer, these guidelines define material qualities (for example, 316L), and safety requirements (such as sterile seals) [1, 2]. An example of a construction detail influenced by safety regulations is shown in Figure 6.2.

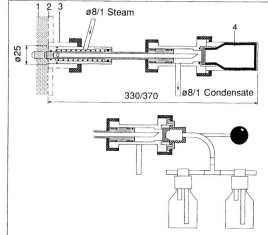

Figure 6.2 Sampling Valve Assembly Designed for Containment and Aerosol-Free Sampling of the Process Fluid.
Key to numbers: (1) vessel contents, (2) vessel wall, (3) side nozzle #42762, and (4) 50–300-ml bottle.

6.2.3   Customer Specifications and Vessel Ratings

Many customers have developed their own specifications, which comply with governmental regulations while defining the arrangement as well as the technical details of vessels constructed for their facilities.

The customer specifications usually incorporate their company fabrication standards with the process requirements to fully define the vessel design and construction. Specific components such as mechanical seals, measurement units, port styles, and mechanical finish are usually defined by company standards. These specifications allow the customer to obtain more consistent equipment, thus allowing for ease of project management and simplified in-house maintenance.

6.2.4   Summary

Governmental regulations specify construction requirements relating to safety and quality, but the vessel design is not complete until the process-related design features have been defined. The vessel rating determines the detailed specifications related to the process and to the biological requirements, such as volume, geometry, stirring, and aeration strategies. Figure 6.3 shows the steps involved between vessel design and vessel acceptance.

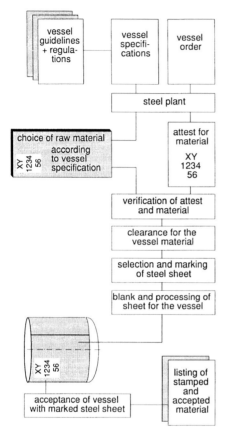

Figure 6.3 Procurement of a Shell Plate with Guaranteed Mark for a Pressure Vessel.

All regulations, specifications, and guidelines are directed to **one** goal: quality control in favor of staff, population, environment, product, producer, and the vessel manufacturer. The challenge is to blend the specifications of government, manufacturer, and customer to ensure that a safe, cost-effective, and functional vessel is fabricated.

6.3   Vessel Material—Stainless Steel

Iron, the predominant element in steel, is very susceptible to corrosion. In stainless steels, the formation of rust is prevented by the addition of up to 12.5% chromium to the iron. The chromium reacts with ambient oxygen and forms a passive layer of chromium oxide on the surface. This layer of chromium oxide protects the steel from corrosion. Stainless steels are defined with a maximum corrosion loss of 0.1mm per year.

The crystal structure of standard chrome steels at room temperature is *ferrite-based* (*alpha* or α *iron*). If ferritic steels are heated to 910 °C, the crystal structure is changed to the nonmagnetic *austenite* (*gamma* or γ *iron*). The change of crystal structure due to high temperature is reversible and reproducible. The stabilization of the austenite structure at room temperature is enhanced by the addition of austenite formers such as nickel, manganese, or carbon. Characteristics of the different types of stainless steel are summarized in Figures 6.4 and 6.5.

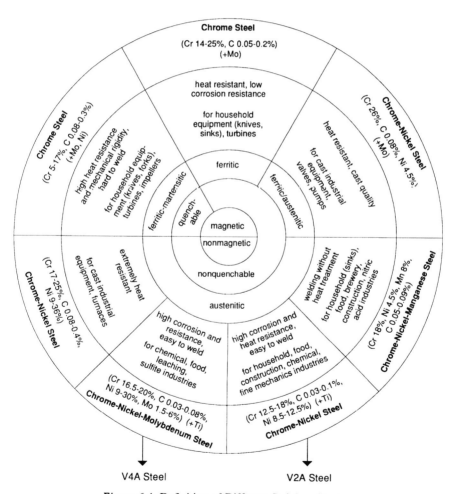

Figure 6.4 Definition of Different Stainless Steels.

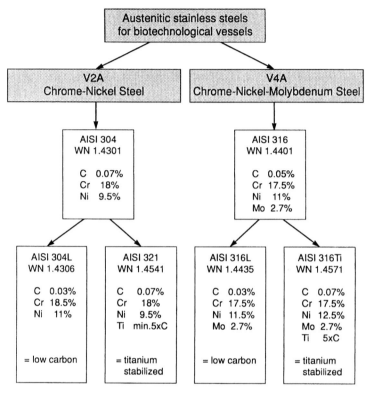

Figure 6.5 Steel Qualities in Biotechnology.

With from some exceptions, austenitic steels (with a maximum of 2.06% carbon) are the material of choice in biotechnology. Compared to the non-austenites, these steels show improved corrosion and heat resistance, and they are nonmagnetic. Other advantages are good tenacity and workability [3].

Quenching or chilling is a standard procedure to harden steels. Since chilling is not applicable for austenitic steel, cold forming or alloying processes are the only possibilities for increasing its strength. However, cold forming of certain alloys (depending on carbon content) creates *martensite* and therewith a low level of magnetism.

The chromium is essential for the resistance to rust, and carbon is required for stabilization of the austenite structure. When steel is heated during welding, the chromium and carbon form chromium carbide. With this reaction, the chromium is removed from the crystal structure and damage along the welding seam occurs. Three different procedures are used to prevent the formation of chromium carbide:

(a)   solution treatment after welding at 1050–1080 °C (1920–1980 °F). This is seldom practiced due to the difficulties in handling;

(b)   lowering of the carbon content in the steel below 0.03% to prevent chromium carbide formation (304L, 316L). The reduction of carbon has to be compensated with other austenite formers;

(c)   stabilization of the alloy with titanium while maintaining the carbon content at a maximum of 0.1% (316Ti). In practice it is easier to add titanium to the melt than to lower the carbon content.

The low-carbonized steels (304L, 316L) are traded worldwide as standard steels. One of the few exceptions is Germany, which produces and uses titanium-stabilized steels (316Ti).

The choice of the appropriate steel quality in biotechnology is usually based on compromises between material costs, availability, and the chemical-physical requirements of the process, unless regulations and guidelines actually define the steel quality.

In general, vessels used in biological processes are manufactured from 316 or 316L steel. The cheaper and less corrosion-resistant steel grades (304/304L) are mostly used in food technology or for certain harvest or storage tanks such as waste tanks. All other parts of the vessel not in contact with the product broth, such as the heating circuit or the heat-transfer jacket, are usually made from cheaper or non-stainless steel materials [1, 4–7].

6.4   Welding of Chrome-Nickel Steel

Welding is used to tightly and irreversibly join metals. The metal is melted in a light arc created by current flashing between the electrode (negative pole) and the welding piece (positive pole). The crystal lattice of the metals may change when alloys are heated or when unsuitable electrodes are used. Small amounts of ferrite are created when the austenitic working piece is not chilled after welding. It is also very important to prevent access of oxygen to the welding puddle.

6.4.1   Welding Techniques

The welding techniques currently in use differ in the nature of the electrode and the method used to prevent access of oxygen to the puddle. Figure 6.6 summarizes these techniques and the acronyms used to describe them in Germany (DIN, Deutsches Institut für Normung) and the U.S.A. (ASME, American Society of Mechanical Engineers).

Shielded metal-arc welding (*SMAW*) utilizes electrodes made from different alloys, according to the steel and weld quality. The electrode is melted in the light arc, with

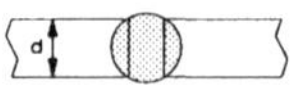

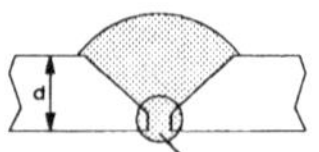

DIN	ASME	Electrode	Puddle Protection	butt weld joint (mostly for pipes)	fillet weld joint (mostly for vessels)
E	SMAW	consumed	sheathed rod	d 1-3mm	d 3-12mm
UP	SAW	consumed	powder	d 2-4mm	d 8-12mm
WIG	GTAW	tungsten	Argon	d 1-3mm	d 4-15mm
MIG	GMAW	consumed	Argon/CO_2	d 2-4mm	d 4-12mm

Figure 6.6 Welding Techniques and Their Application.

the electrode sheath protecting the welding puddle from oxygen. This manual technique requires the smallest expense in apparatus and is applicable for all alloys and steel qualities.

Gas tungsten-arc welding (*GTAW*) utilizes a tungsten electrode, with the welding puddle constantly flushed with argon. The electrode itself is not consumed in the process. If material is required to fill up the welding seam, an additional rod of the appropriate alloy has to be inserted. If the current in this process is pulsed, instead of a steady amperage, it is termed *GTAW pulsing* or *WIA pulsing*. In this fully automated process, the welding seam consists of a succession of partly overlapping fusion points.

Two tungsten electrodes (plus and minus) are used for *plasma-arc welding* (*PAW*). The light arc forms between the two electrodes, and hydrogen protects the weld. This technique is used only for the welding of metal sheets of less than 1-mm thickness.

Gas metal-arc welding (*GMAW*) utilizes an electrode that is automatically drawn from a coil and fused into the arc. The puddle is flushed with either argon (inert gas) or carbon dioxide (active gas).

If the current is pulsed, then the procedure is called *GMAW pulsing*. The melted electrode is dropped into the puddle as the current peaks. This method is also referred to as *controlled drop transfer*.

Submerged-arc welding (*SAW*) is a fully automated welding technique where the electrode is directly fused into the puddle. A powder layer protects the puddle from oxygen, slows down the cooling of the puddle, and protects the environment from splashes or the light flash.

6.4.2   Application in Vessel Construction

The selection of a specific welding technique is based on the nature of the welding material and the variety of welds that have to be applied in a manufacturing process.

The important criteria for the choice of a welding technique are the steel wall thickness and the position and nature of the seam.

In general, *butt welds* are used to connect pipes, and *fillet welds* are used for the thicker steel sheets used for vessels. The root face of the fillet should be located on the inside of the vessel. Thus, the heat-sensitive zones in the vessel are minimized and the danger of fissures is reduced.

Gas tungsten-arc (GTAW) welding is the commonly used method, either manual or automated, for the welding of stainless chrome-nickel steels. The high quality of the welding seam is not at all related to high investment costs. It is the most economic technique for butt welds of 1–5-mm thickness. When thicker steel sheets are welded together, a *fillet weld* is required. In this type of weld, the ends to be joined are sloped at an angle of approximately 20°. The first weld placed is the *root face weld*, at the position where the steel sheets are closest. For high quality, the GTAW method is used. Subsequently, the seam is filled with the GMAW technique, completing the fillet weld. For metal sheets less than 3 mm in thickness, shielded metal-arc welding (SMAW) is applicable for fillet welds, but GTAW welding is superior due to the possible automation in all positions.

GTAW pulsation is used for the fusing of sheets with different thickness, for the fissure-free welding of tubes into plates and for light-gauge sheet steels. Thick sheets are economically fused with the GMAW method, whereas microplasma-arc welding is used for sheets thinner than 1 mm (membranes, bellows).

It is very important to properly select the electrode (GMAG, SMAW) or rod (GTAW) quality, especially when different steel alloys have to be fused. Incorrect alloys change the crystal lattice structure and damage the steel sheets.

6.4.3   Cutting of Chrome-Nickel Steel

The most commonly used cutting method is *plasma-arc cutting*. Cuts that are clean and free of ridges are obtained, especially with a mechanically operated cutting torch. The cutting speeds are relatively high. With a current of 120 amp, 10-mm-thick chrome-nickel steel plates are cut with a velocity of 600 to 1000 mm/min.

6.5   Surface Treatment

A vessel has two different surfaces; the outside surface is exposed to the environment, and the inside surface is in contact with the product. For all stainless steels it is important to treat and clean the surfaces in a way that prevents corrosion at the operating conditions.

In general, both the inside and the outside of the vessel have a defined surface finish that permits the easy and hygienic cleaning of the vessel. Stainless steels are resistant to corrosion due to the microscopically thin, invisible chromium oxide layer.

This layer is formed on clean, metallic, and polished surfaces only. Different methods are used for surface treatment, the final result of the treatments being directly related to the steel quality and the previous working steps [8, 9].

6.5.1   Mechanical Surface Treatment

The surface quality of steels can be greatly improved by mechanical treatment. Parts in contact with product are particularly likely to require specific surface qualities defined in company specifications or guidelines. Required surface finish qualities of the vessel exterior are based more on aesthetic than on functional reasons. However, polished exterior surfaces can allow for easier cleaning. The increasing demands on surface quality are often difficult to obtain, expensive, and extremely time-consuming. It is sometimes questionable whether this expense is made up with improved biological results.

6.5.1.1   Polishing

The quality and smoothness of the mechanically treated surfaces are defined by Ra, Rt, or RMS values, as illustrated in Figure 6.7. In practice, surface definitions such as 240 grit are in use. These definitions are based on the mechanical treatment of the surface with different polish belts of decreasing grain or grit sizes. Standard polish steps or grain sizes are P60, P120, P180, P220 and P320–400. These values are rather inaccurate, as the achieved polish is directly related to the belt quality. New belts have higher abrasion, but at the same time, a 30% decreased polish quality.

The best course is to use the smoothness values rather than polish belt definitions. The smoothness value is easily measured and unambiguous to the manufacturer and the customer.

Surface qualities of Ra approximately 0.6 μm (220 grit) are the minimum generally accepted and approved in biotechnology. The finest polish or abrasive belt (400 grit) reaches smoothness values of Ra 0.28–0.34 μm.

The surface quality obtained with mechanical treatment depends on and is limited by the steel quality. Some examples are:

(a)   the titanium-stabilized chrome-nickel or molybdenum steels, 1.4541 (AISI 321) and 1.4571 (AISI 316Ti), are not suited for high surface qualities;

(b)   the best possible smoothness values are reached with steel grades 1.4306 (AISI 304L) and 1.4406 (AISI 316L);

(c)   for later mirror polishing, a preceding polish to Ra $\approx$ 0.6 μm (P220) is completely satisfactory. Often, finer polishing leads to poorer smoothness values after mirror polishing;

(d)   the steel used for vessels should be cold-rolled sheet material. It is extremely difficult and time-consuming to polish warm-rolled sheets.

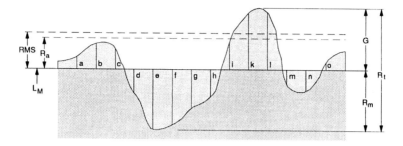

$$R_a = \frac{a+b+c+...o}{z}$$

$$RMS = \frac{a^2+b^2+c^2+...o^2}{z} \approx 1.1\, R_a$$

R_t	Roughness *(Germany)*
R_m	Mean roughness
R_a	Arithmetic mean roughness *(UK)*
RMS	Geometric mean roughness *(USA)*
G	Smoothing depth
a,b,c,.	Distance from profile to medium line L_M
L_M	Medium line
z	Number of distances

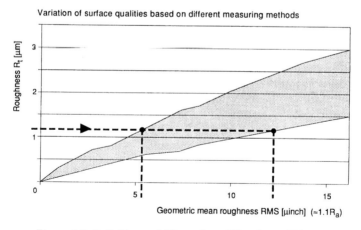

Variation of surface qualities based on different measuring methods

Roughness R_t [µm]

Geometric mean roughness RMS [µinch] $(\approx 1.1 R_a)$

Figure 6.7 Definition and Illustration of Roughness Values.

6.5.1.2   Ball Blasting

Ball blasting is used for special surface finishing and gives various possibilities for optical surface effects with a dull-finished or dull-polished appearance. The ball blasting decreases the risk of tension cracks and corrosion. It is used mostly for the vessel exterior.

6.5.2   Chemical Surface Treatment

6.5.2.1   Cauterization

Impurities and surface imperfections prevent the formation of a perfect passive layer. The purpose of cauterization is to dissolve these flaws with suitable acid mixtures, such as 15–25% v/v HNO_3, and 1–8% v/v HF for steel with a chromium content above 15.5%. The impurities that are cauterized may be oxidation tint, welding slag residues, surface impurities containing unalloyed steel, and fine scales or overlaps generated during the working process.

6.5.2.2   Passivation

The natural passive layer is formed when stainless steel is exposed to air. An artificial passivation with dilute nitric acid is frequently used for corrosion resistance at critical conditions. The oxidizing effect of the nitric acid accelerates the formation of a dense passive layer. Passivation is used either after cauterization or as a final treatment of ground, brushed, or polished steels with special surface structures that would be destroyed when cauterized. New studies on passivation show that alternatives to nitric acid, such as organic acid chelating agents [10], can produce a more long-lasting passive layer.

6.5.3   Electrochemical Surface Treatment

Mirror polishing (*electropolishing*) is an electrochemical treatment to smooth and polish rough and dull surfaces. The vessel or workpiece is immersed in an electrolyte and connected as an anode. The quality of the mirror polishing is only as good as the source material and its mechanical pretreatment. Advantages of mirror polishing are: (a) despite good Ra values, the surface may have roughness in the microscopic range, which is smoothed by the electrolytic process; and (b) corrosion resistance is increased due to exposure to high oxygen concentration, which enables and increases the formation of a dense chromium oxide layer. Disadvantages of electrolytic polishing are: (a) formerly existing imperfections on the material surface are more accentuated after mirror polishing, and (b) it is very difficult to mirror-polish rolled plate steel and impossible to work with titanium-stabilized steels (1.4541—AISI 321; 1.4571—AISI 316Ti). After mirror polishing, the parts have to be thoroughly cleaned to remove remaining electrolyte.

A summary of the surface treatment of vessels is given in Figure 6.8.

6.5.4   Inspection of Surface Treatment

Following the treatments, the quality of the surface and the welds are checked with various methods:

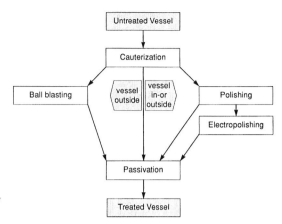

Figure 6.8 Schematic Overview
of Surface Treatments.

(a)   visual examination;
(b)   detection of ferritic contaminations with the ferroxyl test (for cauterized austenitic chrome-nickel steels only);
(c)   palladium testing to check the surface passivation;
(d)   detection of detergent residues;
(e)   detection of chloride or sulfur contamination;
(f)   detection of pickling damage (using a stereo magnifier); and
(g)   electronic surface measuring with a scanning needle (sapphire or diamond) [7, 11].

6.5.5   Storage of the Treated Parts

The surface-treated and dry workpieces have to be stored on a clean and nonabsorbing base support, protected from dust and water spray.   Workshops handling ferritic material are not suited for storage.   Also, locations near railways and tramways (which may produce ferritic brake dust) and pavements in winter (dry sand and salt) should be avoided.   For long storage periods, it is best to cover and wrap up delicate parts, as well as workpieces, in polyethylene foil [12].

6.6   Design Considerations

Compared to the chemical vessel, the biological vessel has to perform one additional task during operation—it has to remain sterile.   This includes the sterilization procedure of the vessel, as well as long-term sterility during process operation.

Assuming that the system is handled correctly, contamination or the inability to sterilize a vessel has two causes: dead spaces or leaks [4, 6, 11–16].

6.6.1   Dead Spaces and Cavities

Installations in the vessel, such as down-tubes, or incorrect constructions at the periphery, create cavities or dead spaces. Organisms or medium may be deposited in the dead spaces, leading to subsequent contamination of the process. In air-filled cavities, the sterilization temperature may not be reached and contaminants may not be killed. The installation of temperature probes that project into the vessel or piping at the critical positions enables the monitoring of the actual temperature to ensure that sterilization occurs. This is illustrated in Figure 6.9.

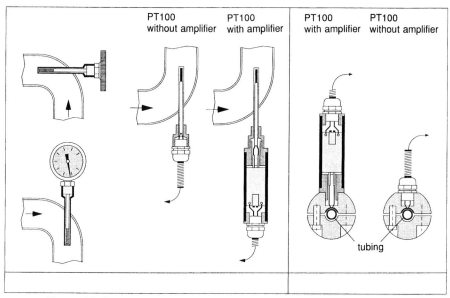

Figure 6.9 Installation of Thermometers and Temperature Probes in Tubes.

6.6.2   Tightness

The vessel is not a hermetically closed system, but consists of different parts: lid, ports for agitation, probes, inlets and outlets, viewing glasses, and items such as foam turbines or rotor filter. The connection of these parts into and out of the vessel has to be maximally tight. Only then are absolute sterility and environmental safety possible.

In practice, absolute tightness in the strictest sense is nonexistent. Tightness is determined with experiments defining the maximum leakage rate under certain circumstances. To quantify the tightness of a closed vessel, the following equation defines the leakage rate:

$$Q = V \Delta P / \Delta t \tag{6.1}$$

where Q is the leakage rate, $m^3 \cdot bar/s$;
ΔP is the measured pressure loss, $= P_{initial} - P_{final}$, bar;
V is the vessel volume, m^3; and
Δt is the time interval, s.

To compute Q in terms of the standard fermentor volume, V_0, we may have to correct to standard pressure:

$$Q = V \Delta P / \Delta t = V_0 (P_{std}/P_{initial}) \Delta P / \Delta t \tag{6.1a}$$

The leakage rate depends on the leak geometry, the composition of the test medium, and the test procedure used. Different leakage test methods are: X-ray, gamma-ray, ultrasound, pressure-drop method, water-pressure test, bubble test (with foaming liquid or in a bell jar), and overpressure leak detection with nitrogen or helium.

For biotechnological vessels, use the following simple procedures to examine the tightness:

(a) pressure test: fill the vessel 2/3 with water, and agitate. Generate an overpressure of about 1.5 bar (150 kPa) with air or nitrogen. Monitor the pressure with a pressure gauge, and observe whether there are any water leaks.

(b) sealing test: fill the vessel with water to cover the mechanical seal, and generate an overpressure of about 1.5 bar (150 kPa). Wait about ten minutes and record the pressure (P_0) and temperature (T_0). Wait approximately fifteen hours and read the pressure (P_1) and temperature (T_1). Determine the number and type of the installed O rings and calculate the total sealing length (total circumference of all the O rings and gaskets) l, in meters, in the vessel. Calculate the pressure difference as follows:

$$\Delta P = P_0 - P_1 \tag{6.2a}$$

when $(T_0 - T_1) < 1 \,°C$;

$$\Delta P = P_0 [(T_1 + 273)/(T_0 + 273)] - P_1 \tag{6.2b}$$

when $(T_0 - T_1 > 1 \,°C)$.

The *specific leakage rate*, L (in grams of air per hour per meter of sealing length), can be determined from the formula:

$$L = kV\Delta P M / R T_0 \Delta t\, l \tag{6.3}$$

where $k = 3.6 \times 10^6$ g·s/kg·h;
V is the vessel volume, m³;
$\Delta P = (P_0 - P_1)$, bar;
M is the molar weight of air (29.3×10^{-3} kg/mol);
R is the gas constant, 8.314 J/mol·K;
Δt is the time, s; and
l is the total sealing length, m.

If the specific leakage rate is less than 0.05 g/h·m, the vessel is considered tight; if it is greater, further effort should be expended to make it tight.

6.7   Components of the Biotechnological Vessel

6.7.1   Rotating Seals

One of the most important and discussed bioreactor components is the construction to seal the rotating shaft. The following sections describe different technical solutions that seal a rotating shaft tightly.

6.7.1.1 Mechanical Seals

The sealing effect of a mechanical seal is generated by a spring and, when the agitator is bottom-driven, by the weight of the medium above it. The sealing effect increases proportionally with the pressure in the vessel, a principle similar to that of a check valve.

In general, mechanical seals are used to seal liquids. If gases are to be sealed, which is the case with top-driven shafts, an inert auxiliary liquid has to be added. Here the check-valve–like sealing effect depends on the leakproof containment of the auxiliary liquid between the secondary seals and the sealing gap. The appropriate choice of sealing materials is based on the medium composition, operating tempera-ture, pressure, and agitation rate. Standard materials are recommended by the seal manufacturers, but guarantees for specific applications cannot be given.

The mechanical seal materials most used in biotechnology are carbon for the rotating ring and ceramics for the seat ring. All mechanical seals should be checked regularly for visible scratches or defects. Difficulties with bottom-driven units can occur when a highly abrasive medium is used. The materials of choice in this case are the more resistant, but also more expensive, silicon carbide or tungsten carbide.

6.7.1.2 Stuffing-Box Packings

The sealing rings in stuffing-box packings are compressed in the axial direction first. With subsequent radial deformation, the installed sealing rings are pressed against the stuffing-box boring and the spindle. Each packing ring has to be deformed and braced within the stuffing box to seal the three classical leak directions. The required compressive stress is transmitted from the outside by an external gland, which enables the tightening of the stuffing box and the elimination of leakage during operation.

Stuffing-box packings are successfully used in large-scale fermentors with volumes of 100 m^3 and more. Due to the difficulty in the manufacture of large-scale mechanical seals (for instance, diameter greater than 220 mm), the stuffing-box packings will not readily be replaced by mechanical seals in these applications.

6.7.1.3 Magnetic Couplings

Permanent magnetic couplings enable noncontact torque transmission to the vessel without any penetration of the vessel. This can be advantageous if containment or maintenance of sterility (for example, in mammalian cell culture) is very important. Magnetic couplings are by their nature protected against overload; the transmission breaks away whenever the load exceeds the design strength of the coupling.

Magnetic couplings seem advantageous because no external bearing is required, but the bearings located in the reaction chamber may cause novel problems. With particulate media, the thin internal cover plates may become abraded and leaks can develop.

6.7.2 Static Bushings

Access to the fermentor contents is required for the supervision of and interaction with the cultivation. Probes and viewing glasses are used for supervision, and inlet and outlet devices enable additions or removals. Sterility is maintained when these installations are tightly sealed with flat gaskets or O rings.

In general, flat gasket rings are used in chemical vessels (except in vacuum and space technology). The sealing material around the opening is squeezed over the complete surface. The sealing effect is created with a constant pressure, compressing up to 20% of the seal thickness. This compression compensates for sealing surface irregularities (approximately 10% of the seal thickness). The construction of the flange is responsible for the constant and sufficient transmission of pressure on the seal.

O rings are the main sealing elements in biotechnology, especially where small port sizes are involved. The O rings are circular in cross section and are made from different sealing materials such as Viton®, ethylenepolypropylene rubber (EP), and silicone. EP is the most widely used and validated material in biotechnology and food technology. EP seals at temperatures from −40 to +160 °C, and is resistant to water, steam, and aging. However, recent studies have shown that EP can become rigid and flake in certain services such as water or steam. The only disadvantage to EP is its instability against mineral oils.

The main advantage of O-ring seals is hygienic sealing without any expenditure of pressure on the sealing area. A disadvantage of O rings is their installation. The O rings are inserted in a defined groove that is designed to enable a slight depression along the total O-ring length. With this required accuracy in construction, the vessel is no longer a welding piece but a precision turned piece.

Sanitary clamp seals are used to connect stainless steel tubing. Two types exist, either sealed with O rings (European) or with flat gaskets (U.S.A.). The sanitary clamp seals with O rings are either screwed or flanged, and the two connected parts are not identical (male and female). In the U.S.A., the sanitary clamp is sealed with a flat gasket combined with an O ring, and is fixed with a *Tri-clamp®*. The two parts of the U.S. sanitary clamp are identical, but the installation of the Tri-clamp® can be difficult. The choice between the different clamp seals is mostly based on available space and company or country standards.

6.7.3   Selected Vessel Components

6.7.3.1   Ports

Ports are required for the installation of probes and inlet and outlet devices. Their size and position have to be defined before the construction of the vessel. To ease and speed up the manufacturing process of the vessel, standardized ports and probes should be built and developed in advance. The predominantly used ports are the 25-mm port (*Ingold port*) for probes, sampling, and addition valves, and the 34-mm port for some addition ports. The bottom port should be designed according to the required harvest valve to give fast harvesting and efficient cleaning. Most of the ports utilized for chemical pressure vessels are too large for use in biotechnology.

6.7.3.2   Viewing Glasses

Viewing glasses (sight glasses) on the lid or in the vessel wall are useful for the observation of the material in a steel vessel. To minimize dead spaces or problems in cleaning, the construction should be flush with the internal wall of the vessel, and have rounded edges to promote drainage. The glass and the flange construction are usually sealed with flat gaskets. Instead of the normal flange construction, viewing glasses are available where the glass is directly melted into the metal frame. This construction is an alternative to the standard flange construction when regulations allow its use. The construction details of the viewing glasses, their frames, size, and geometry are defined and regulated in national pressure-vessel guidelines (for example, DIN 7080, 7081, 28120, 28121).

6.7.3.3 Manholes

As is the case for viewing glasses, the size, position, and details of the manholes are defined in the relevant guidelines. Manholes are needed for cleaning and servicing the inside of the vessel when the vessel cannot be entered through the lid.

The manhole is closed with a metal lid and sealed with an O ring. The manhole must be designed so that dead spaces are not introduced around the perimeter of the cover, and all surfaces are accessible to the cleaning fluids, yet drain readily. Dealing with manholes is made easier by opening devices and central snap closures. Snap closures replace nuts and fix the lid with a central device for easy opening and closing. This type of device is predominant in the autoclave industry, but difficulties with acceptance specifications have retarded its application to pressure vessel construction.

6.7.3.4 Baffles

Baffles are vessel internals that do not require any bushing into or out of the vessel. Their function in an agitated tank is to promote turbulence, and therefore better mixing. Guidelines (DIN 28131) describe the width of the baffles and their distance to the wall of the vessel. The baffles can be permanently welded to the vessel, or they can be installed as removable units if the mixing parameters are expected to change.

6.7.3.5 Heating Jackets

Heating or temperature jackets are used on the one hand for sterilization (heating up to 121 °C) and on the other hand for temperature control during operation. Biological processes are often exothermic reactions, and the heat produced by the biological or mechanical systems has to be led away. In addition, many stored raw materials, in-process solutions, or harvested products require storage at a cold temperature. The layout of the heating system has to be optimized for efficient heating or cooling and accurate temperature control.

The rate of heat transfer is directly correlated to the area of the contact surface and the temperature gradient of the medium. Heating or chilling is performed with fluids such as steam, hot or cold water, or ethylene glycol solutions, which are transported in circuits. The area of contact surface to the vessel and the pressure drop in the system should be optimized for the specific vessel and fluid.

To maintain sterility, the heating circuits are not positioned in the vessel with bushings, but rather on the vessel outside. Three general concepts (illustrated in Figure 6.10) predominate in the layout of external heating system tubing: the double jacket, the full pipe, and the half pipe. The design of the heating circuit should be based on the required heat transfer. The surface area and the heating or cooling medium, as well as the circulation velocity, are calculated according to the heat load [2, 4–6, 13–19].

Type	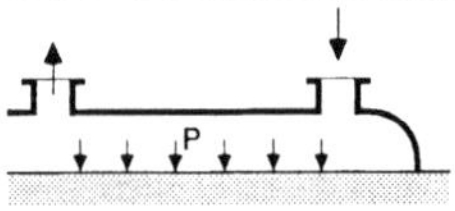 Double Jacket	 Full Pipe	 Half Pipe
advantages	- max. contact surface - min. pressure drop - few welding seams	- totally independent from the vessel, no pressure on the vessel wall	- good contact surface
disadvan-tages	- high pressure on the vessel, vessel wall has to be rein-forced-reduced energy transfer	- high pressure drop - min. contact surface - many welding seams	- high pressure drop - many welding seams

Figure 6.10 Design of Heating Jackets.

6.8   Summary

Vessels intended for the biosynthesis and processing of biotechnology products are usually highly complex structures. They must be carefully designed to meet several sets of requirements. These include: the operational requirements of the process (such as inlet/outlet ports, agitation, ports for monitoring devices, inlets for steam and cleaning solutions, jackets for heating or cooling); the requirements for thorough cleaning and sterilization; the safety requirements for pressure vessels; and the requirement for long-term operation without corrosion or leaching. These requirements can be met only by careful and knowledgeable specification of the vessel at the outset of the project, and by paying close attention to details during the fabrication and testing of the vessel. Due to the complexity of most vessels, the high quality of the materials, and the high level of skill required to fabricate the vessel, the cost of biotechnology vessels is relatively high. To contain costs, and to achieve the quality and characteristics desired, it is essential that vessels for biotechnology be carefully designed at the outset, and carefully managed during construction. An outline of the steps required for the successful procurement of a vessel is shown in Figure 6.11.

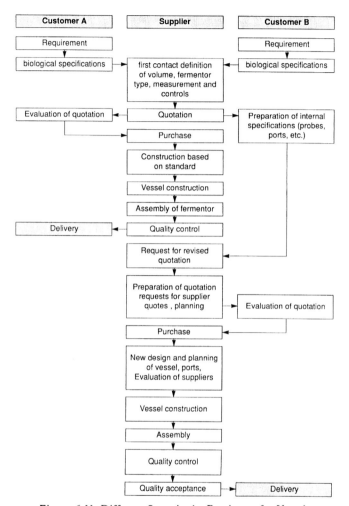

Figure 6.11 Different Steps in the Purchase of a Vessel.

6.9   References

1. DEMAIN, A. L., SOLOMON, N. A.; *Manual of Industrial Microbiology and Biotechnology*; American Society for Microbiology, Washington, DC, 1986.
2. HAMBLETON, P., *et al.*, A High Containment Polymodal Pilot-Plant Fermenter—Design Concepts, *J. Chem. Technol. Biotechnol. 50*, 167–180 (1991).
3. DILLON, C., RAHOI, D., TUTHILL, A., Stainless Steel for Bioprocessing, Part 2: Classes of Alloys, *BioPharm 5* (May), 32–35 (1992).
4. ANON.; *Standardisierungs—und Ausrüstungsempfehlung für Bioreaktoren und periphere Einrichtungen*; DECHEMA, Frankfurt, 1990.
5. CHMIEL, H., Ed.; *Bioprozesstechnik 1*; G. Fischer Verlag, Stuttgart, 1991.
6. EINSELE, A., *et al.*; *Mikrobiologische und Biochemische Verfahrenstechnik*; VCH Verlagsgesellschaft GmbH, Weinheim, 1985.
7. KÜPPERS, W.; *Die Oberfläche von Nichtrostenden Feinblechen—Anlieferung und Weiterverarbeitung*; Thyssen Edelstahlwerke AG, Technische Veröffentlichung 6327/6, Krefeld, Germany, 1976.
8. DILLON, C., RAHOI, D., TUTHILL, A., Stainless Steel for Bioprocessing, Part 1: Materials Selection, *BioPharm 5* (April), 38–42 (1992).
9. DILLON, C., RAHOI, D., TUTHILL, A., Stainless Steel for Bioprocessing. Part 3: Corrosion Phenomena, *BioPharm 5* (June), 40–44 (1992).
10. COLEMAN, D., EVANS, R., Fundamentals of Passivation and Passivity in the Pharmaceutical Industry, *Pharm. Eng. 10* (March/April), 43–49 (1990).
11. GRAF, G., Die Oberflächenbehandlung von Nichtrostenden Chromnickelstählen, *Metall 9* (1987).
12. WALLHÄUSER, K. H., Der Einsatz von Sterilisations und Desinfektionsmethoden im Bereich der Biotechnologie (Teil 1), *Biotech-Forum 4* (2), 90 (1987).
13. BREUKER, E., DREES, U.; *Pilot Fermenters in Industry: Criteria and Development of an Easy-to-Use System*; GIT Verlag, Darmstadt, 1984.
14. JOITT, R., Ed.; *Hygienic Design and Operation of Food Plant*; Ellis Horwood Ltd., Chichester, 1980.
15. KESSLER, H. G.; *Lebensmittel-Verfahrenstechnik, Schwerpunkt Molkereitechnologie*; München-Weihenstephan, Germany, 1976.
16. NARENDRANATHAN, T. J., Designing Fermentation Equipment, *The Chem. Eng. 5*, 23 (1986).
17. MÄRKL, H., *et al.*, A New Dialysis Fermenter for the Production of High Concentrations of Extracellular Enzymes, *J. Ferment. Bioeng. 69*, 244–249 (1990).
18. MAYER, E., Ed.; *Axiale Gleitringdichtungen*; VDI-Verlag GmbH, Düsseldorf, 1977.
19. TITZE, H., Ed.; *Elements des Apparatebaues 2*; Auflage, Springer Verlag, Berlin, 1967.

Piping and Valves for Biotechnology

Harry Adey and Maria S. Pollan

Contents

7.1   Introduction

Development and expansion of biotechnology plants in recent years has created the need for information regarding the design of sanitary, sterile piping systems. Although the design and construction of such piping and valve systems are largely based upon standard design procedures, there are important differences that should be taken into account. In this chapter we present design information particularly relevant to piping and valve systems employed in the biotechnology field, with emphasis on design practices central to maintaining cleanliness and sterility.

The assurance of sterility involves the use of appropriate materials, valves, and fittings, along with special design techniques, such as sterile barriers for fermentors. A working knowledge of these items is essential, but equally important is the ability to determine where to use them. As crucial as the safety of living organisms and processes are, we need to remember that inappropriate use of expensive materials may, through excessive cost, prove fatal to the project itself. Important as materials and techniques are to success, the true foundation for adequately piped systems is in the design stage, where this survey begins.

7.2   Design

There are three phases of design, each with its own appropriate drawings necessary for successful installation:

(a)   the layout phase to produce *Equipment Arrangements*,

(b)   the design phase, which results in *Piping Arrangement Drawings* or *Piping Models*, and

(c)   the detail phase ultimately shown on *Piping Isometric Drawings*.

Computer-aided design and drafting (CADD) is now being used in all stages, including the production of bills of materials, which, when combined with the isometric, usually provide the final installation instructions to the pipe fitter. To date, the thinking functions of design are not performed by computer and these crucial functions still need a systematic approach to ensure that the requirements of the *Current Good Manufacturing Practices* (*CGMPs*) are incorporated into a sound design.

Schedules rarely permit for the ideal situation where all three design phases are complete before construction starts. Frequently, the placing of concrete foundations is under way on some distant site long before the second design phase is over. This necessitates that design be conducted in a manner sufficiently flexible to adapt to the inevitable changes.

7.2.1 Layout Phase

Biotechnology plants have qualities both of fine chemical plants and of pharmaceutical finishing plants; the former are products of conventional piping/equipment layout skills, and the latter involve additional industrial engineering and architectural skills. Therefore, one has to recognize the necessity of involving designers familiar with both types of plants.

Well-designed piping systems require, as a first priority, a plant layout that provides adequate space around equipment to ensure that the piping and associated valving is arranged in an operable and maintainable fashion, with minimal intrusion into operators' space. Such conditions should be a part of the building's design from its conception. The layout that employs a systematic approach produces a workable end result. The end result should reflect due consideration for the construction, operation, and maintenance factors for every piece of equipment involved. This provides a sound foundation for the piping design, which is the best possible compromise between the often conflicting requirements listed in Table 7.1.

Table 7.1 Piping Design Considerations

1. Pipe lengths kept to a minimum.
2. Piping routed so as not to be a hazard or restriction to operators.
3. All valves and instruments located so as to be both easily operable and maintainable.
4. The routing of each pipe made conducive to adequate individual support, and group supports used where possible.
5. All portions of the system designed to be drainable, with few if any pockets, and provided with adequate drain valves.
6. Flexibility for temperature fluctuations incorporated where appropriate.

7.2.2 Concealed Piping

In a *controlled environment*, piping must be designed to meet all the foregoing criteria, but with the added need to be concealed. This need to conceal piping within hollow walls, above ceilings (in interstitial spaces), or in adjacent voids, to improve the cleanliness of the processing rooms themselves, has many space-related concerns. The extra space required to accommodate pipes into suitably arranged rooms affects the floor plan, and the enlargement of interstitial space to facilitate the maintenance of pipes adds to building height. Obviously, building costs are increased, so the extent of concealment required is an important determination to make. The guidelines in Table 7.2 should be helpful in the decision-making process.

Table 7.2 Suggested Pipe Concealment

Area of Plant	Manufacturing	Pilot Plant
Fermentation	none	partial
Cell recovery	none	partial
Purification	partial	maximized
Sterile processing	maximized	total
Sterile finishing	total	total

Where cleaning between batches of different products is involved, it becomes essential to avoid cross contamination, and concealment of piping becomes more desirable, even though the separation of equipment into smaller rooms is often a more practical solution. Where airborne particulates must be minimized for classified areas (Class 10,000 or higher), piping should be concealed as much as possible to facilitate the total cleaning of these process areas.

Alternative methods used to achieve concealment are shown in Figure 7.1. The valves may be hidden within hollow walls, with their stems extended through an escutcheon plate. An adjacent removable plate permits maintenance and inspection. Installation of totally exposed valves within a clean room is not recommended.

7.3   Piping Materials

Information pertaining to piping materials and their specifications is both voluminous and well known. This brief summary of those materials normally considered for biotechnology facilities emphasizes the preeminence of cleanability and sterility concerns. The most commonly used piping materials, in order of usage for biotechnology plants are: (1) stainless steel tubing, (2) stainless steel pipe, (3) thermoplastics (PP, PE, PVDF, etc.), (4) carbon steel, (5) copper tubing, (6) iron, (7) glass, and (8) lined pipe (glass and plastic liners).

In the absence of a code, the guiding document for engineering, design, and installation of *Water For Injection* (*WFI*) and other sterile systems is *CGMP—LVP* (Current Good Manufacturing Practice—Large Volume Parenterals, Part 212.49 of Subpart C, *Water and Other Liquid Handling Systems*), issued by the Food and Drug Administration (FDA) in 1976 as a performance specification [1]. This document specifically mentions stainless steel as the material of construction for WFI and, by inference, sterile services.

Although material selection is always an attempt to find the most suitable material for the flowing media involved, some broad generalizations about each of the materials listed above may help.

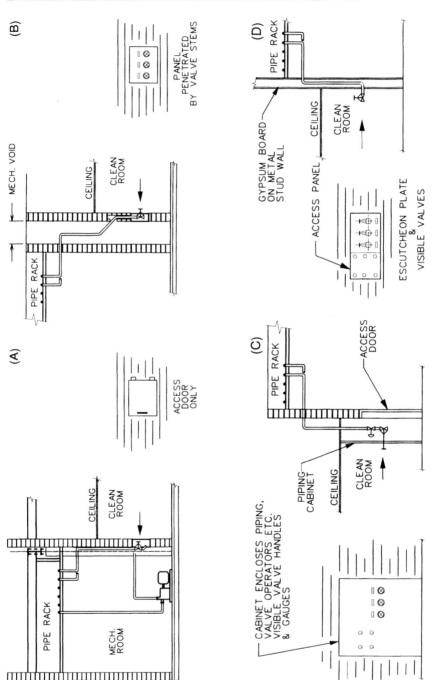

Figure 7.1 Examples of Piping Concealment: (A) mechanical room, (B) mechanical void, (C) piping cabinet, and (D) hollow wall.

(1) Stainless steel tubing (316L), used in various polished and unpolished finishes for sterile and pyrogen-free systems, uses clamped joints with special gaskets instead of conventional flanges.

(2) Stainless steel pipe is most commonly used for process piping where corrosion resistance and cleanliness is of importance. More expensive than carbon steel, it is usually specified in schedules 5S and 10S and frequently in 316L or 304L grades [2].

(3) Thermoplastics (PP, PVDF, polyethylene, etc.) are used for water distribution and drainage systems and for medium-temperature applications, and are gaining ground as an alternative to stainless steel for process piping, but only in certain virgin grades.

(4) Carbon steel, widely used for the distribution of water, steam, and other utility systems in all industries, is a first choice for these services unless considerations of corrosion rule it out.

(5) Copper tubing, used primarily for plumbing, in combination with, or instead of, plastic piping, is limited by price and strength to use in small sizes. It is also used for gases such as oxygen, helium, and argon, as well as for general utilities such as air and glycol solutions.

(6) Iron pipe with good corrosion-resistant qualities (usually ductile iron) is used almost exclusively for underground drainage and sewage systems.

(7) Glass pipe is used for process systems where its high resistance to corrosion outweighs the uncommon installation techniques involved. It is mostly used for small sizes that do not require the added strength of lined pipe.

(8) Lined pipe (glass or plastic in steel) is used where strength must be combined with corrosion resistance. Joints in glass and lined pipe are a cause for some concern in clean systems.

7.3.1 Economics of Selection

What material to use and where to use it can normally be decided on a lowest installed cost basis, with the precautionary emphasis on installed cost rather than material-only cost. Problems with welding or supporting may be overlooked by the designer and ignored by the supplier, only to be a source of consternation to the construction team. Sanitary tubing, although ideal in many ways, has proven vulnerable in these areas. Welders with training and expertise in sanitary tubing fabrication should normally be used if repeatable acceptable welds are to be obtained.

The extent to which to carry the pursuit for cleanliness by the use of expensive materials is a difficult decision to make. The tendency to play safe and pay the price is common, and, to a large extent, it is the inevitable outcome of attempts at CGMP interpretation. Table 7.3, Cleanliness Classifications, should help, if the nature and extent of the risk to product can be ascertained.

Table 7.3 Cleanliness Classifications*

Class	Description	Example	Key Design Features
I	No contamination of any kind can be tolerated	WFI or sterile process systems	Stainless tubing 316L (polished interior); Automatic welding of joints (first preference); Sanitary-style clamped joints where disconnect is necessary (second preference); Diaphragm valves (polished interior); Positive slope to ensure free draining; No high-point vent valves; Valved connections for cleaning in place (CIP) and sterilizing in place (SIP); Pocket-free design (maximum of 6 pipe diameters, preferably 2.5 diameters).
II	Living organisms are at risk or cleanability of lines is important	Fermentation piping	Stainless pipe; Controlled welding (flat and cavity-free); Flanged joints kept to a minimum; Diaphragm valves, flanged or butt-welded; Positive slope to ensure free draining; Valved connections for CIP and SIP; No pockets or difficult-to-clean branch connections; No reduced-size drain connections.
III	Nonprocess piping where some risk to product is recognized	Deionized water supply	Pipe of suitable noncorrosive and nonleaching specification; Valves constructed from noncorrosive materials; Positive slope to ensure free draining; No pockets; Valved connections for CIP and SIP.

* The classes are used for clarity only and have no codal significance, beyond this chapter.

7.3.2 Sanitary Tubing

Pipe fabricated from many different materials has been used traditionally to transport a variety of fluids of widely ranging temperatures and pressures. Unfortunately, pipe must be joined either by welding or by the use of gasketed flanges, and this introduces cleaning problems.   Valves and instruments also contribute to cleaning problems. Conventional piped systems always contain enough cavities and crevices to encourage colonizing by bacteria and depositing of pyrogenic debris if left unsterilized for even short times.

Fortunately, this problem of piping cleanability may be solved by the use of sanitary tubing made from 316L-grade stainless steel with an internal polish. Because joints in this material are automatic fusion welds and clamped ferrule joints, the worst cavity-producing conditions are avoided. The more important features are listed in Table 7.4. This adaptation of a dairy industry material is recommended for use on smaller sized systems (less than 4" or 100 mm diameter) and where severe vibration is not anticipated.

Table 7.4 Sanitary/Sterile Tubing Characteristics

Material	316L stainless steel (seamless or welded). PVDF for cooler systems (not WFI).
Welding	All automatic Gas Tungsten-Arc Welding (GTAW). The machine grips and rotates the tubing to produce flat, cavity-free, repeatable welds with the use of an inert gas purge and without the use of weld filler material.
Polishing	Interior and exterior surfaces, if required.
Passivation	With acid after installation.
Fittings	With parallel extensions for machine clearance and tight ovality tolerances. May be supplied with clamped or welding ends.
Joints	2 Ferrules, 1 clamp, 1 gasket (no nuts, bolts, or threads).
Valves	With polished and/or electropolished interiors and sanitary parts, butt-welded or clamped ends.
Sloping	Positive to ensure free draining.
Pockets	None deeper than 6 diameters.
Supports	Required more frequently than in conventional piped systems.

The use of Kynar® (PVDF) with Tri-clamp® ends for sanitary tubing systems is now becoming more acceptable. For processes sensitive to metal it may be suitable, except where steam sterilization would cause distortion and prevent free draining. No material with the potential to leach toxins into the process stream, either from itself or from jointing materials, should ever be used.

7.4   Polishing

No topic has caused more controversy in clean processing circles than that which surrounds the questions (a) whether to polish, (b) how to polish, and (c) how much polish. Experience has taught that mill finishes for stainless steel may be smoother than mechanical polishes of low grit numbers, but not consistently so. Unpolished tube and the fittings made from it rarely meet the dimensional tolerances required to make acceptable welds, even if surfaces are totally blemish-free. Consequently, a minimum finish of 20–25 microinch Ra (0.51–0.64 μm arithmetic mean roughness) mechanical polish on all interiors is common. To ensure that fittings meet tolerances, some external treatment is advisable.

Tubing and fittings manufactured from tubing may require the following: (a) polishing of internal surface only, (b) polishing of internal and external surface (when tubing is uninsulated but exposed in sterile and clean areas), and (c) polishing of external surface only, for cosmetic appearance of uninsulated lines (least stringent quality).

A smooth surface free of pits, porosity, and surface irregularities is obtained by various methods. *Mechanical polish* is produced by removal of metal using various sizes of abrasive particles in a polishing medium, by one of two different techniques. In the radial process, a rotating tool is passed through the tubing; in the longitudinal process, the tubing is rotated as a polishing belt passes axially through it. However, mechanical polishing "may create a smearing effect on the surface of the metal" [3] and thus hide some irregularities. Internal polish of valves can be obtained by a process that squeezes abrasive particles through them.

The measure of smoothness is sometimes described by polish numbers such as #4 or #7. The numbers provide for a great deal of latitude. For example, depending on the polishing technique, a #4 interior finish may vary considerably, and can be compared to samples made with a polishing agent of 80 grit to 150 grit. Polish numbers have been replaced by grit numbers and by the actual measurement of the surface in microinches or μm, expressed as *root mean square roughness* (*RMS*) or *arithmetic mean roughness* (*Ra*). Surface finish can now be measured by computerized versions of accurate profilometers produced by Hommelwerke GmbH or others. Grit numbers and Ra numbers are not interchangeable.

Using grit or polish numbers for external polish is still acceptable. *Grit numbers* are defined as the number of silicon carbide particles per square inch of abrasive pad, or as the number of scratches per linear inch. In general, the higher the grit number, the finer and more numerous the scratches, and therefore the smoother the finish. The surface finish obtained by polishing with the same grit number polishing agent may vary from vendor to vendor depending on the polishing technique used. Specifications are in Table 7.5.

Until recently, valves used in sterile systems were made from castings, which could have more subsurface defects than forgings or tubing have. Therefore, higher polish requirements were frequently imposed for valves than for tubing. Needless to say, this increased the price of the valves. The introduction of forging, in lieu of casting, improved the quality of surface finish and permitted standardized polish requirements for the system.

Table 7.5 Polish Number Characteristics

Polish #	Grit #	Ra*, inch × 10⁻⁶	Ra*, μm
4	150	30–35	0.76–0.89
7	180	20–25	0.51–0.64
7	240	15–20	0.38–0.51
7	320	9–11	0.23–0.28

* Surface finish expressed as arithmetic mean roughness

ASTM Standard A270-90 still uses grit numbers to describe surface finish, finish R produced by rouge polishing, and finish T produced by electropolishing [4]. However, an ASME committee is investigating materials in bioprocessing, so some updating or additions to existing standards can be expected.

Electropolishing is often used in addition to mechanical polishing. The process removes metal from the surface by anodic dissolution in a suitable electrolyte with an applied current. Surface irregularities are minimized by leveling action. The leveling of projections above 1 μm is called *macropolishing*, and the leveling of small irregularities, greater than 0.005 μm, is called *micropolishing*. Electropolished surfaces have a bright finish, the brightness depending on the degree of micropolish.

Since electropolishing removes metal without smearing of metal or creating folds, it can frequently expose voids at the longitudinal weld not visible after mechanical polish. However, electropolish does not remove these defects; only additional mechanical polishing does. Final analysis of economics versus the safety of product should govern selection and degree of polish required. How much polish, or, more precisely, how smooth the finish should be to make its surface cleanable and sterilizable, may eventually be determined experimentally. A once-popular measure of the degree of polish was to produce a finish that had grooves (or scratches) small enough to be uninhabitable by 0.2-μm-sized bacteria, but biotechnologists have less confidence in this method now. The observable effects of a surface's ability to hold droplets after washing lead to the conclusion that electropolished surfaces drain best, but it is difficult to be certain that the extra cost of droplet-free surfaces is justified. Electropolishing has the inherent benefit of passivating the tubes and fittings at the same time, so it has become a commonly accepted practice. However, since the whole system should be passivated prior to start-up anyway, this benefit should not govern the selection of surface polishing requirements.

7.5   Passivation

It has been common practice for many years to passivate the interior surfaces of sanitary tubing and related vessels with a 35–45% nitric acid solution [5], followed by exposure to air for oxidation. The oxidation process provides a corrosion-preventing protective layer on the stainless steel surface   This practice is now considered somewhat controversial.   The use of nitric acid solutions can be dangerous even without the use of the chromate inhibitors that were once added to prevent *flash attack* of stainless tubing. Nitric acid passivation removes sulfides, but it does not remove aluminum-rich particles "instrumental in the formation of the pits" [6]. In addition, disposal of the nitric acid effluent to municipal waste treatment plants may be restricted by local or even federal legislation, since it can produce hexavalent chromium, a suspected carcinogen. Ammoniated citric acid is a substitute, although it is considered less effective by some. It is easily removable, is water soluble, chelates with metal ions (iron, nickel, copper) formed during the passivation, and prevents formation of insoluble oxides and hydroxides during the cleaning and rinsing operations. Citric acid is obviously safer to use, is less biocidal to the process, and is now used by a number of biotechnology companies. Work now under way by committees (ASME and others) should produce guidelines to help in this area.

The following important instructions apply to the passivation procedure:

(a)   The standard to use is ASTM A-380 (Standard Recommended Practice for Cleaning & Descaling of Stainless Steel Parts, Equipment and Systems).

(b)   To avoid flash attack of surfaces, use not less than 35% or more than 45% HNO_3 solution or 5% ammoniated citric acid (pH 3 to 4). Check the solution with a sample of material before using on the actual system.

(c)   Circulate a detergent solution and rinse thoroughly before using acid.

(d)   Circulate the nitric acid solution for 30 min at 130 °F (55 °C), or citric acid for 6–8 hours at 140–160 °F (60–71 °C), and rinse thoroughly. The final rinse should be with demineralized water.

(e)   Use utmost caution and wear appropriate protective clothing for all procedures involving the use of acid.

7.6   Sizing of Pipes and Tubes

Frequently the words *pipe* and *tube* are used interchangeably, but the distinction is important for line-sizing calculations because their inside diameters are different. Inside diameters vary according to specified *schedule* or wall thickness; whereas *pipes* are labeled by their nominal bore, *sanitary tubing* is labeled by its outside diameter, and has a smaller bore. For example, a 2" stainless steel pipe has an outside diameter of 2.375", and if made of schedule 10S has a resultant bore of 2.157". A 2" sanitary tube has an outside diameter of 2.000" and a 0.065" wall thickness, for a resultant bore of only 1.870". This reduction in available diameter, if overlooked, could be crucial in flow calculations, where diameters raised to the fifth power are used.

Oversizing pipelines increases installation costs, and undersizing increases pressure drop. Excessive pressure drop, paid for in lifetime pumping costs, is avoided by appropriate pipe sizes selected from flow/velocity tables and confirmed by calculations that involve fluid velocities, Reynolds numbers, and friction factors. The procedure illustrated here with an example is first examined for biotechnology implications by a review of the sizing variables involved.

7.6.1 Variables

Very high fluid velocities may shorten the life-span of pipes and possibly risk metal contamination of process streams, which would be totally unacceptable in the biotechnology industry. Unfortunately, although lower velocities avoid erosion problems, they may result in unwanted anchoring of bacteria. This anchoring of bacteria is avoided when the flow through pipes is turbulent rather than laminar, so the need is for velocities high enough for turbulent flow, but low enough to avoid erosion (usually not in excess of 10 ft/s or 3 m/s). Maintenance of velocities of 5 ft/s (1.5 m/s) could lead to excessive pressure drop in smaller tube sizes, so is not necessary if turbulent flow is achieved. The example problem in this section emphasizes this point.

Laminar flow (or *streamline flow*), where fluid particles move in relatively undisturbed layers, in contrast to the irregular swirling motions of particles in turbulent flow, is a condition to be avoided. The more uniform velocity distribution across the pipe diameter found in turbulent flow is desirable for the efficient transfer of heat and the prevention of liquid-solid separation, both important to bioprocessing. Flow is either laminar or turbulent, depending upon pipe diameter, fluid density, fluid viscosity, and the velocity of the fluid through the pipe. These variables are combined in a dimensionless value known as the *Reynolds number* (N_{Re} or *Re*).

For circular pipe, the value of the Reynolds number is calculated using the following formula (if English units are used):

$$Re = dv\rho/\mu = 6.31 \, W/d\mu \qquad (7.1)$$

where d is the internal pipe diameter, in,
   v is the fluid velocity,
   ρ is the fluid density,
   μ is the viscosity, cP, and
   W is the rate of flow, lb/h.

For sizing considerations, flow is considered laminar when the Reynolds number is below 2,000, and turbulent when it is above 4,000. The transition zone between these numbers should be avoided for all forms of flowing media, including those common to bioprocessing.

7.6.2 Pressure Drop

Friction, caused by the interaction of fluid particles in motion through a pipe, causes a pressure drop in the direction of the flow. This pressure drop is calculated by the DARCY formula, which is valid for both laminar and turbulent flow and can be stated as follows (again, if English units are used):

$$\Delta P \;=\; 3.36 \times 10^{-6}\, fLW^{2}\bar{V}/d^{5} \tag{7.2}$$

where ΔP is the pressure drop, lb/in^{2};
   f is the friction factor;
   L is the length, ft;
   W is the rate of flow, lb/h;
   $\bar{V}$ is the specific volume, ft^{3}/lb; and
   d is the inside pipe diameter, in.

Since $\bar{V} = 1/\rho$, and quite often ΔP is expressed in psi/100 ft, the equation is often expressed as:

$$\Delta P \;=\; P/100 \;=\; 3.36 \times 10^{-4}\, fLW^{2}/\rho d^{5} \tag{7.3}$$

Because relatively smooth pipe and tube is widely used, the variable in the DARCY formula that has the most impact on bioprocessing calculations is f, the friction factor. If flow is laminar (Re < 2,000), then

$$f \;=\; 64/\text{Re} \tag{7.4}$$

If flow is turbulent (Re > 4,000), f depends on the relative roughness of pipe, E/d (roughness/diameter), and the Reynolds number (Re). The friction factor is usually determined from graphs, shown in *Flow of Fluids* by Crane [7] or similar handbooks, and varies according to the type of pipe. It is found by using a Reynolds number calculated for an assumed pipe size (see Section 7.6.3).

Table 7.6 provides rules of thumb useful for the calculation of pipe sizes in a variety of conditions. Opinions differ about the content of such rules and there is obviously a need in bioprocessing to avoid the common tendency to err on the safe side by leaning toward larger sizes.

7.6.3 Example Problem

Select a suitable tube size for a required flow rate of 30 gallons per minute of Water For Injection to be circulated at 180 °F.

Table 7.6 Pipe Sizing Rules of Thumb

Type of Line	Pressure Drop, psi/100 ft		Velocity, ft/s
	Average	Maximum	
Pump suction	0.2	0.5	2–4
Gravity flow	depends on available head		
Pipe discharge	2.0	3.0	5–10
Vapors (med. to low press.)	0.1	0.5	50–100
Vacuum	depends on available pressure		
Steam	0.5	1.0	100–115
Water	0.5	2.0	3–10

Special consideration must be given to flow through lines less than 1½" in diameter. Below this size, the pressure drop at a given velocity increases dramatically as line size decreases. For example, in a 2"-diameter line at 5 ft/s (50 gpm), $\Delta P_{100} \cong 2.1$, but in 0.5" diameter line at 5 ft/s (5 gpm), $\Delta P_{100} \cong 10$. Thus, pressure drop becomes more important as a limiting criterion in calculations for smaller sizes.

Step 1: Tabulate known conditions.

Liquid:	WFI
Temperature:	180 °F
Density:	60.6 lb/ft³
Viscosity:	0.35 cP
Flow rate:	30 gpm
Material:	sanitary tubing

Step 2: Assume approximate line sizes and calculate velocities and Reynolds numbers.

$$V = 0.408 \ Q \ /d^2,$$

where Q = flow rate (gpm) and
d = inside tube diameter (in);
and $Re = 6.31 \ W \ /d\mu$,

where W = 30 gpm × 8.34 lb/gal × 60 min/h × 60.6/62.4 = 14,600 lb/h.
Then,

Line size	d (inside)	d^2	V (velocity)	Reynolds no.
1½"	1.37"	1.88	6.51 ft/s	192,000
2"	1.87"	3.5	3.5 ft/s	140,000

Step 3: Determine friction factor, *f*, from standard tables [7] and calculate the pressure drop (*P*/100, psi/100 ft) from eq 7.3:

$$\Delta P = P/100 = 3.36 \times 10^{-4} \, fW^2/\rho d^5$$

For previously selected pipe sizes of 1½" and 2":

Line Size	d	d^5	Re	f	W^2	ρ	P/100
1½"	1.37	4.83	192,000	0.016	2.12×10^8	60.6	3.9
2"	1.87	22.9	140,000	0.017	2.12×10^8	60.6	0.87

The line drop (*P*/100) for 1½" tube is excessive (see Table 7.6 for recommended values for water), so a 2" size is selected even though the velocity through it is below the ideal of 5 ft/s. We have demonstrated that the necessary compromise in line sizing can be justified because the flow is in the turbulent range and the velocity is sufficient to retard bacterial anchoring.

7.7   Connections and Cleanability

Conventional methods used to join sections of pipe have generally evolved on the basis of economics, codes, and convenience. Flanged connections, welded fittings, and threaded couplings (for instrument insertion) have been the predominant methods in conventional processing systems.

The challenge in biotechnology is to make connections that are equally effective, but that do not add to the burdens of line cleaning and sterilization. In fermentation, for example, the process organism is at risk, and both vessels and piping must be designed to eliminate the danger of contamination by other organisms. It is crucial to recognize that any cavity that cannot be washed and sterilized is a risk to be avoided. In cell culture processing, and the pharmaceutical industry in general, stainless steel sanitary tubing has been used extensively. The tubing (described in Table 7.3 as Class I piping), with its automatic fusion welding and clamped joints, provides a smooth interior, in stark contrast to conventional flanged arrangements (see Figure 7.2). Unfortunately, there are areas in large-scale biofacilities where vibration, size, and weight of pipe and machinery make sanitary tubing less acceptable. An example is microbial fermentation using vessels larger than 4,000 L.

Cleanable devices and instruments with O-ring seals or steam barriers are becoming more available, but the best defense against unsterilizable surfaces is a vigilant review of the total system. High purity water, such as WFI, and clean steam are very aggressive media. They attack metal surfaces of vessels as well as piping systems, particularly at elevated temperatures, and produce corrosion in the form of pitting, crevice attack, stress-corrosion cracking, and intergranular corrosion. These media destroy the chromium oxide layer developed during passivation by dissolving

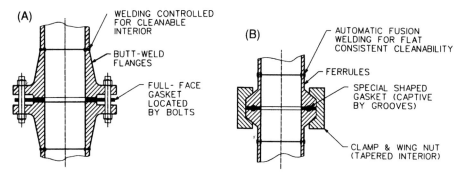

Figure 7.2 Comparison of Connections: (A) Typical flanged connection where manufacturers' tolerances on either gasket, flanges, or bolts may lead to misaligned protrusions; (B) sanitary-style clamped connection where tight tolerances and tapered design eliminate protrusions.

impurities from the metal surface and forming *rouge*, a thin reddish-brown film consisting of ferric oxide with additions of iron, chromium, and nickel [8]. Rouge is easily removed by wiping off. Its occurrence can be minimized by higher surface polish, controlled welding and fabrication, cleaning, and passivation. Water chemistry is not affected by rouging, but periodic cleaning and repassivation of the system may be required.

Although total freedom from cavities is impossible within reasonable cost limitations, some partial solutions are listed in Table 7.7.

7.7.1 Welding

Although welded joints are preferred to flanged ones, the quality of welds is very important. Interiors of vessels are accessible for polishing, grinding, and inspection of welds. Pipes, however, are usually too small for remedial action when radiography or boroscopes reveal flaws. Starting over is often the only remedy. Therefore, it is necessary to insist that welding contractors provide high-quality welds produced to specification. Inspection procedures must ensure repeatability of desirable features such as full penetration and flatness. Weld specifications should reflect a concern for quality with requirements for *Gas Tungsten-Arc Welding* (*GTAW*), thorough inspection procedures, and radiography of 5% to 20% of welds. Welding of sanitary tubing is done almost exclusively by automatic welding machines.

Some contractors may suggest the use of automatic welding of stainless pipe, using the same type of machines used for fusion welding of tubing. The machines are designed to weld pipe, but experience to date indicates that they are most suited for sanitary tubing and that good-quality welds can be made in pipe by other methods.

Table 7.7 Strategies for Obtaining the Best Surface Cleanability

1.	Use the minimum number of flanged connections when butt-welding is impractical (as at vessel connections).
2.	Avoid threaded connections if they have direct contact with the process.
3.	Captive gaskets may be employed in the future. Meanwhile, use full-face gaskets (penetrated by the flange bolts) with bores that match the flange's bore.
4.	Avoid ring-style gaskets (located between bolts).
5.	Use controlled welding and inspection to ensure flat, smooth welds.
6.	Consider pipe bending instead of butt-welded fittings if quantities justify the expenditure of bending machines.
7.	Keep instruments to a minimum, especially those normally inserted into uncleanable couplings.
8.	Use Weldolets® or welding tees instead of couplings to ensure smooth transitions at drain valves and branch connections.
9.	Remember that small deep cavities and crevices are more difficult to clean than large shallow ones.

Some welds cannot be examined by boroscope. Known as *blind welds*, they are either closure welds or in tubing smaller than ½" (13 mm) diameter, and can be examined only by radiography. The number of blind welds should be kept to a minimum.

7.7.2 Joints

The following types of joints have application in biotechnology plants: (a) threaded/socket weld (utility service); (b) flanged (utility service, and on a selective basis in sanitary systems) against equipment or instrument; (c) butt weld (preferred in sterile systems); and (d) Tri-clamp® (sterile systems) for connection to equipment, valves, and instruments. For unavoidable flanged connections, a ferrule with a special adapter flange should be used.

The most frequently used gaskets in sanitary tubing (Tri-clamp®) are Viton® (steam-resistant), and Teflon® envelope with Viton® insert. These gaskets can withstand steam sterilization for short times. Solid Teflon® gaskets are not recommended, since Teflon® is subject to cold flow, which may result in leaking joints. For biotechnology processes using stainless steel pipe, rather than tubing, non-asbestos, 1/16"-thick (1.6-mm), full-face–type gaskets are recommended. However, if flanges are out of tolerance, 1/8"-thick (3.2-mm) gaskets are recommended. These gaskets must be resistant to caustic used in CIP. Even though flanges are raised-face, full-face gaskets are recommended to preclude improper installation and the creation at joints of pockets where bacteria can thrive.

7.8   Piping Applications

The appropriate ingredients of a piped system (materials, sizing, connections, and layout) must be applied in a certain way to ensure that the system produces the desired results. The following selections, taken from a larger group, are those with the most impact on biotechnology and its CGMP requirements.

7.8.1 Pump Suction

Pump suction is always a source of special concern because of the need for pumps to perform efficiently for long, maintenance-free periods. Regardless of the type, pumps work best if suction lines are short and unpocketed, and have a section of straight pipe prior to the inlet connection. To avoid air pockets and cavitation, install reducers of eccentric style with their flat portion on top, even though this may necessitate an additional drain valve. Installed spare pumps when not running provide havens for bacterial growth. Providing complete isolation of the spare pump with valves, in addition to arranging for the pump to run for timed intervals to eliminate the long static periods, is one way to avoid risking this contamination.

7.8.2 The "6*d* Rule"

The "6*d* rule" is one GMP requirement for injectable/pure water systems that is worthy of more general use in bioprocess piping. The rule states that pockets caused by branches in piping headers that are more than 6 diameters deep (6*d*) may cause pools of static liquid to be isolated from the desirable turbulent flow of the main header. In practice, limiting a dead leg to less than 6*d* is advisable. In process utility water systems, for example, it is commonplace to find branches in excess of a hundred diameters, but there chlorination protects against bacterial growth.

The price to pay for eliminating pockets is the added length of headers that must run as close as possible to the point of use before bending back to their place in a pipe rack, as illustrated in Figure 7.3. This also results in an increase in pumping cost, but concerns about water quality override cost concerns. The advantages of this type of installation for biopharmaceutical process streams, especially in the unchlorinated water systems that support them, are gaining recognition.

7.8.3 Sterile Barriers

Sterile barriers, sometimes called *steam blocks* and frequently confused with *block-and-bleed valving*, are used principally to isolate the fermentation process from all

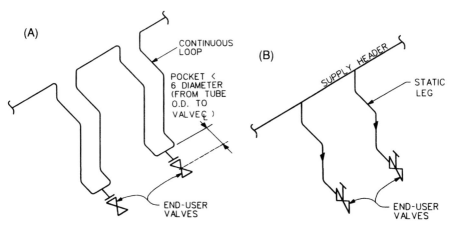

Figure 7.3 Fluid Distribution Piping: (A) continuous loop supply header with restricted pocketing; (B) conventional supply header. The former is preferred to avoid static legs.

other systems connected by piping to the fermentor vessel. The procedure is adopted because of the concern that live organisms can grow through closed valves in certain circumstances, and is designed to guarantee total isolation. This isolation is achieved by feeding steam (approximately 25 psig) into a space between two valves in any line that is considered a risk to the fermentor contents, or when those contents are a risk to health. The presence of steam involves the need for condensate removal by means of a trap, valves, and related piping. Examples of a basic and complex barrier are illustrated in Figure 7.4.

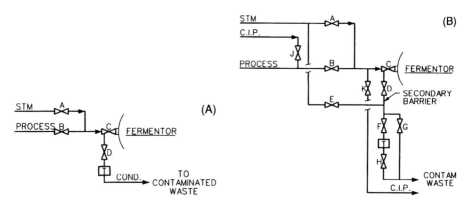

Figure 7.4 Sterile Barriers: (A) simple sterile barrier: valves A and D open when valves B and C are closed; (B) complex sterile barrier: valves A and D are closed and E, F, and H are open, when valves B and C are open; valve G opens for bypass, and valves J and K open for CIP.

The provision of this barrier is complicated by the addition of valves, their actuators, and additional piping required to perform any or all of the following additional functions: (a) to keep the steam trap sterile when the steam valve to the barrier is closed for process transfer, (b) to allow trap to be bypassed, (c) to facilitate cleaning of line with a CIP connection of sufficient size, and (d) to allow off-spec media to be diverted into a contaminated-waste vessel.

The extent to which sterile barriers should be used is the subject of some controversy, and we hope research and experience will ultimately determine their true value and minimize unnecessary usage. Until then, their use should be restricted to where there is obvious risk, or where mandated by safety requirements related to the handling of known pathogens. When the decision is made to use a barrier, it is in the interest of both cleanability and operability to keep it simple.

7.8.4 Sloping of Lines

Sloping of lines is still a source of some confusion due to the belief prevalent in the industry that more is better. Obviously, solids and high-viscosity liquids need slopes that must overcome resistance to flow, but most biotechnology piping is cleaned with dilute caustic or acid solutions and flushed with water. Even vertical pipe remains wet after flushing, so any slope that ensures complete drainage (no residual pools) is sufficient.

An additional concern is that fusion welding of sanitary tubing is difficult where weld surfaces are not perfectly square and the cutting or shaping of weld surfaces may produce unacceptable welds. Slopes of 0.3% to 0.5% are sufficient and effective for post-cleaning and gravity flow of low-viscosity fluids. The degree of slope is not specified in CGMPs, is independent of the direction of the flow, and need only provide a slope toward a low point.

7.8.5 Double Piping

Double piping, as illustrated in Figure 7.5, is an effective method of leak detection and containment assurance where hazardous materials must be conveyed by concealed piping. Hazardous solvents and biowastes en route to kill tanks can employ this method of piping above ceilings, inside walls, or beneath floors. The inner pipe is made from corrosion-resistant material and is totally enclosed by an outer pipe that supports the inner one with rings or pads. This outer pipe is provided with a pocket at its low point to collect and convey leaked material to a device where it can be safely observed and collected. The outer pipe or jacket should have connections for nitrogen purging, so the jacket can be swept, either continuously for biohazards for BL3 level, or temporarily to reduce fire hazards if necessary.

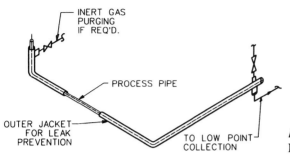

Figure 7.5 Double Piping for Hazardous Service.

7.8.6 Clean-in-Place Connections

Clean-in-place connections are required on most lines involved in bioprocessing, for cleaning solution entry and exit points. They must be located to ensure complete drainability, and protection for process streams in the event of accidental influx of cleaning fluids. The location and method of CIP connections should ensure: (a) that flushing tends to move materials away from pockets with assistance from, and not opposition by, gravity; (b) that drainage of the line is complete through the low-point valve; (c) that permanent connections to nonsterile process lines employ ball valves with tell-tale or similar connections (Figure 7.6A); and (d) that permanent connections to sterile process lines employ double block-and-bleed valving (Figure 7.6B).

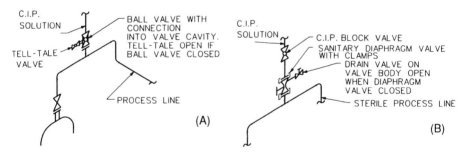

Figure 7.6 Clean-in-Place Entry Connections: (A) for nonsterile process lines, and (B) for sterile process lines.

7.9   Supporting and Insulating Sanitary Tubing

Piping supports commonly used in industry are unsuitable for use with sanitary tubing, but adapted proprietary types are now available. Materials used for hanger components and floor stand supports should be of stainless steel to avoid external corrosion and galvanic action between dissimilar materials. Rubber insert pads (Viton® for higher temperatures and for sterilization) may be used to isolate tubing from the support to reduce vibration. Rigid anchoring should be replaced by floating supports, which avoid distortion of tubing systems due to temperature changes. Support spans vary from 5 ft for ½" tubes (1.6 m for 13-mm tubes) up to 10 ft for 4" tubes (3.1 m for 100-mm tubes). Tubing should be supported on both sides of valves and at all changes in direction. Sanitary tubing as well as stainless steel pipe should be insulated by chloride-free glass fiber conforming to ASTM C547 Class 1, and covered with a PVC solvent-bonded jacket when required by process temperature.

7.10 In-Line Instruments

The main consideration for instruments to be installed in clean systems is to provide types that, together with their connection to pipe or vessel, provide the least possible risk of product contamination.

Instruments should be made from materials considered safe for direct contact between the flowing media and the instrument itself, preferably highly polished 316L stainless steel. Connections to pipe or vessel should avoid cavities or pockets likely to encourage accumulations of solids. Where it is impossible to avoid risk of contamination, a steam bleed may be provided, especially in high-risk locations such as a fermentor vessel.

Pressure gauges and level transmitters should be of the *remote diaphragm seal type* [9], connected to vessels by plate flanges designed to ensure a flush cleanable interior surface, and to pipes by Tri-clamp® connections. Vessel connections should be made to the side or bottom of the vessel, to ensure they are always submerged in product or washing medium.

Temperature sensors should be located in wells welded directly into a pipe or into a vessel flange, as shown in Figure 7.7. Wells located in pipes should be mounted on top of pipe no less than 2" (50 mm) in nominal diameter. It is not advisable to locate wells in elbows because of the difficulty in cleaning, polishing, and sterilizing the weld areas.

A good example of designing for sterility is the in-line pressure-reducing venturi (Badger Meter) shown in Figure 7.8. The conventional restriction orifice plate using a sharp-edged "dirt collector" may now be replaced with a cleanable venturi with Tri-clamp® ends. This is one of a number of improvements recently introduced by an industry that faces especially difficult sterility problems.

Other in-line instruments commonly used are sight-glass assemblies and steam traps [10] available in polished stainless steel construction with Tri-clamp® ends. Recent experience suggests the use of polished sanitary tubing for in-line instruments, rather than conventional stainless steel pipe, which was frequently used in the past. Rust and

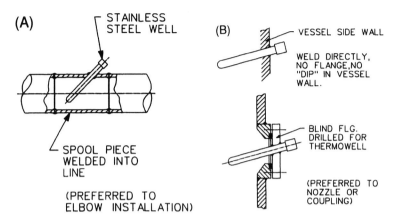

Figure 7.7 Thermowells: (A) in pipe, and (B) in side of vessel.

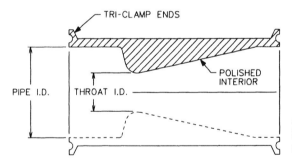

Figure 7.8 Example of In-Line Flow Venturi. (Courtesy of Badger Meter, Inc.)

corrosion in the vicinity of welds exposed to the aggressive environment of clean steam have been observed in instruments made from stainless steel pipe.

7.11   Hoses

Biotechnology plants are large users of hose and flexible connectors. Conventional hoses constructed from rubber, solid Teflon®, or corrugated stainless steel with conventional hose adapters may be used in utility stations and lines not directly involved with sterile processes. Sterilizable sanitary hoses with a variety of end connections are used for flushing process lines, in CIP service, and in place of rigid metal piping in suitable applications.

Flexible, smooth-bore, nontoxic and noncontaminating, reinforced medical-grade silicone tubing can withstand pressure and vacuum application in biotechnology processes, as well as steam sterilization at 25–30 psig (172–207 kPa), corresponding to a maximum temperature of 123 °C. A variety of 316–stainless-steel factory-installed ends crimped onto the sanitary hose are available. These ends minimize crevices that could trap bacteria, and permit quick disconnects to various systems. A disadvantage of flexible hose is its limited life when leakage occurs at the joint between flexible material and end connection due to frequent exposure to steam.

Another type of hose frequently used has a smooth bore, is made from virgin Teflon®, and is reinforced with stainless steel braid. These hoses are available with sanitary fitting ends (such as Tri-clamp®), and with cam-lock, 150-lb ANSI flange and Teflon®-encapsulated fittings.

7.12 Valves

The general principles of valve selection are reviewed here from the standpoint of cleanability. The vast majority of valves installed in all types of plants are used in on/off (isolation) service, to prevent unwanted flow past a given point in a piped system. The remainder are used to control either rate of flow or pressure of the flowing medium, at a particular point in the system. Both the former, frequently referred to as *block valves*, and the latter, called *throttling* or *control valves*, can be installed with activators capable of converting received signals into valve movement.

Utility-system valves are the workhorses of the industry. Gate, ball, plug, and butterfly valves are used in great numbers for isolation service, and globe and butterfly valves are used in throttling service for water, air, steam, or gas systems. With little or no concern for cleanliness, selection is made on the basis of installed cost versus maintenance cost. The one big exception is the concern for valves used in high-purity water systems that feed bioprocess streams.

The body shapes of gate and globe valves render them uncleanable and thus potential havens for bacterial growth in chlorine-free water. The selection of alternative-style valves involves a choice between ball, butterfly, plug, or diaphragm valves. The ball is frequently favored because its valve stem is isolated from the water, it has a quarter-turn operation, and is suitable for sterilization by steam. The diaphragm valve is even more suited to clean operations, but has the disadvantage of a multiturn handwheel operation.

We present in Table 7.8 a short review of the standard valves available on the market and their applications, to help the reader select the appropriate valve for a given service.

Table 7.8 Characteristics of Valves[*]

GATE	**Function**: blocking (isolation).
	Application: utilities; general chemical process; solid wedge type for steam; split wedge type for liquid and noncondensing gas.
	Sizing: ½"–48". Larger sizes frequently replaced by butterfly valves (3" & larger).
	Construction: <u>Bonnet</u>: screwed; union; bolted; <u>Wedge</u>: solid; split; double disk; slide; <u>Stem</u>: rising; nonrising; outside screw & yoke (OS&Y) for steam & high-temperature fluids.
	End Connection: soldered; threaded; socket weld; butt weld; flanged.
	Body Materials: bronze; CI; CS; SS; alloy steels; plastics (PVC).
	Seal/Seat: <u>Metal-to-metal</u>: brass or bronze; CI; stellited CS; EPDM (for plastic valves).
	Remarks: unsuitable for sterile service; heavy in large sizes.
GLOBE	**Function**: throttling; flow control.
	Application: utilities; general service (steam, water, air); needle type for precise regulation of flow.
	Sizing: 1/8"–10".
	Construction: <u>Body type</u>: regular; angle (90°); Y-type; needle; stop-check; <u>Stem</u>: rising; nonrising; OS&Y; <u>Bonnet</u>: screwed; union; bolted; <u>Disk</u>: plug-type; needle-type.
	End Connection: soldered; threaded; socket weld; butt weld; flanged.
	Body Materials: bronze; CI; CS; SS; alloy steels; plastics (PVC, PP).
	Seal/Seat: <u>Metal-to-metal</u> (seats can be integral or renewable) bronze; SS (hardened); Nonmetal- (TFE) to-metal (not recommended for throttling); <u>Disk</u>: solid; composition. All seats subject to erosion.
	Remarks: unsuitable for sterile service except for clean steam. High pressure drop. Tight shutoff requirements.
BUTTERFLY	**Function**: blocking; throttling.
	Application: utilities; general process; selective application in bio-technology.
	Sizing: 2"–144".
	Construction: wafer; lugged; flanged.
	End Connection: between ANSI flanges.
	Body Materials: CI; Duriron®; CS; SS; alloys; plastic-lined; solid plastics (PVC, PVDF).
	Seal/Seat: elastomer-lined seat & disk; metal-to-metal body seal & disk. Sterile-seal type: PTFE lined; SS disk (steam sterilizable); <u>Disk</u>: PP, PVC, PVDF, & TFE; <u>Seats & Seals</u>: Viton®, EPDM, buna N, butyl, neoprene, Hypalon®, natural rubber.
	Remarks: lightweight quarter-turn operation.

Table 7.8 Characteristics of Valves*, *continued*

DIAPHRAGM	**Function**: blocking; throttling. **Application**: general processes; slurries; sanitary valve used in sterile processes & WFI for biotechnology & pharmaceuticals. **Sizing**: ½"–20". (Sanitary style: ¼"–4".) **Construction**: weir; straightway; Dualrange®. **End Connection**: threaded; socket weld; butt weld; flanged; solvent-weld for plastics; Tri-clamp®. **Body Materials**: bronze; CI or DI; CS; SS; alloys (Hastelloy®, Monel®, etc.); Lining material: glass, Tefzel®, PVDF, PVC, CPVC, saran, PP, Hylar®. **Diaphragms**: gum rubber; buna N; butyl (black & white); Hypalon®; EPDM; neoprene; Viton®; TFE; silicone. Backing cushion materials vary but EPDM is typical. **Remarks**: limited by press./temp range of diaphragm. Prevent stem leakage (Clean Air Regulations).
PLUG	**Function**: blocking; directional flow control (multiport). **Application**: general process; sanitary plug valves used in selective application in food industry. **Sizing**: ½"–24". **Construction**: lubricated; nonlubricated (TFE sleeve); lined; eccentric; multiport design. **End Connection**: threaded; socket weld; butt weld; flanged; Tri-clamp®. **Body Materials**: CI; CS; SS; plastic-lined. **Seal/Seat**: metal-to-metal (lubricated); metal-to-elastomer (sleeved); elastomer-to-elastomer; TFE lined. Limited by sleeve material temperature range. **Remarks**: Quarter-turn; good shutoff.
CHECK	**Function**: flow control (prevents backflow); special excess flow. **Application**: utilities, general services. Lift check valves recommended for air compressor systems & reciprocal pumps. Valves with TFE disk, in liquids & gases at low pressure. **Sizing**: ¼"–48" (sizes over 24–36" may be special order). **Construction**: swing check (single disk, double disk); lift check (poppet disk); ball. **End Connection**: soldered; threaded; socket weld; butt weld; flanged; Tri-clamp®. **Body Materials**: bronze; CI; CS; SS; alloys; plastics (PVC, CPVC, PP). **Seal/Seat**: metal-to-metal: bronze, CS, SS; nonmetal- (TFE disk) to-metal. **Remarks**: Tri-clamp® SS sanitary ball checks for limited use in sterile processes (available 1"–3").

Table 7.8 Characteristics of Valves*, *continued*

BALL	**Function:** blocking; directional flow control (3-way). **Application:** utilities; general chemical process; selective application for biotechnology, pharmaceutical and food industry use (stainless steel polished 3-way valve). **Sizing:** ¼"–24". **Construction:** top entry; end entry; floating ball; trunnion ball; 3-way; 3-piece construction. Fire-resistant seats & seals are available, as are special service valves such as chlorine service. **End Connection:** threaded; socket weld; butt weld; flanged; Triclamp®. **Body Materials:** bronze; CS; SS; PVC; PP; PVDF; FRP. **Seal/Seat:** <u>Metal-to-metal</u>: bronze, SS; <u>Soft-Seated</u>: TFE, buna neoprene, etc.; metal ball or all lined; <u>Sanitary</u>: polished with metal ball and TFE or UHMWPE seats. <u>Seals</u>: buna, neoprene, TFE, EPR, Viton®, etc. Applications limited by seat temperature range. **Remarks:** Quarter turn, good shutoff. Economical (small sizes). Low torque. Cavities need special treatment for biotechnology applications (sterile filter or steam).

* **Abbreviations:**
  ANSI: American National Standards Institute
  CI: cast iron
  CPVC: chlorinated poly(vinyl chloride)
  CS: carbon steel
  DI: ductile iron
  EPDM: terpolymer from ethylene-propylene-diene monomer
  EPR: ethylene propylene rubber
  FRP: fiber-reinforced plastic (glass fiber)
  OS&Y: outside screw and yoke
  PP: polypropylene
  PTFE, TFE: polytetrafluoroethylene
  PVC: poly(vinyl chloride)
  PVDF: poly(vinylidene fluoride)
  SS: stainless steel
  UHMWPE: ultrahigh molecular weight polyethylene
  WFI: Water For Injection

7.12.1 Process Valves

What valve to use for clean process service has been reduced to a choice between diaphragm and ball valves in the majority of cases. No valve is cleaner or more easily kept clean than the diaphragm valve, which is the only valve used for WFI systems. The ball valve, on the other hand, has an inaccessible cavity that is difficult to clean, but is nevertheless sometimes chosen because of its quarter-turn operation.

Diaphragm valves are used in a variety of services. The simple construction, variety of available body and diaphragm materials, options for body linings, and positive closure permit their use for a wide range of chemical liquids and slurries. Therefore, it is not surprising that diaphragm valves are found to be best suited for sterile processes. Figure 7.9 illustrates the chief differences between various valves used in sanitary applications. The application of diaphragm valves is limited by the pressure and temperature range of the diaphragm. In addition, the life-span of the diaphragm is limited due to diaphragm flexing and the exposure to the high temperatures of frequent steam sterilization (25 psig {172 kPa} saturated steam for about 30 min).

The materials used for sanitary diaphragm valves are indicated in Table 7.9 and must comply with FDA regulations. The thermoplastic materials used in the valve must be unpigmented, with no extractables. The preference for forged bodies versus castings is justified by: (a) higher internal body finish available, (b) resistance to pitting, and (c) lower ferrite content, and thus higher corrosion resistance of the weld.

Table 7.9 Materials Used for Sanitary Diaphragm Valves

Body:	a.	316L stainless steel forging (preferred);
	b.	316L stainless steel casting. Interior body finish available in 9–35 Ra microinches (previously 150–320 grit, and/or electro-polished);
	c.	plastic: PVDF, PP, Hylar® (limited application such as DI water, USP grade water).
Diaphragm:	a.	TFE (tetrafluoroethylene);
	b.	EPDM (ethylene-propylene diene terpolymer);
	c.	medical grade silicone;
	d.	other materials, such as Viton®, butyl rubber (white and black), buna N (available, but not suitable for sterile service requiring steam sterilization).
Bonnet:	a.	cast iron or ductile iron, coated with epoxy, PVDF, nylon II, or electroless nickel plated;
	b.	stainless steel;
	c.	bronze.
Sanitary Internals:	a.	stainless steel;
	b.	bronze or stainless steel compressor.

The utilization of diaphragm valves in the biotechnology industry is advantageous for the following reasons: (a) they are of pocketless or cavity-free design, (b) they are self-draining, (c) they have built-in drain bosses, (d) they offer in-line maintenance, (e) they have an isolated bonnet that keeps all working components isolated from the

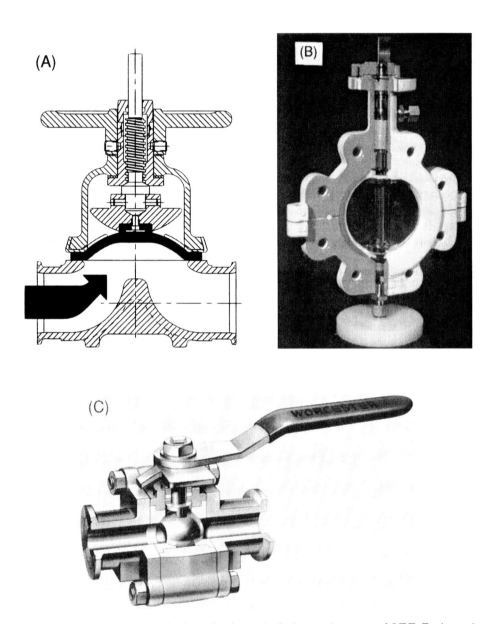

Figure 7.9 Valves Used in Sanitary Service: (A) diaphragm (courtesy of I.T.T. Engineered Valves); (B) butterfly (courtesy of Garlock, Inc.); (C) ball (courtesy of Worcester Controls Corp.).

flowing medium, (f) they offer a choice of end connections (butt weld for OD tubing, and schedule 5S IPS {Iron Pipe Size} with Tri-clamp® ends), (g) they offer a choice of manual or power actuators, and (h) they are available in a range of sizes from ¼" through 4" (6–100 mm), usually sufficient for the average size biotechnology plant.

Since the body and diaphragm are the only surfaces in contact with the process fluid, cleaning and sterilizing are simplified and the top-entry design allows in-line maintenance. The use of butt-welded valves in lieu of valves with Tri-clamp® ends, if disconnection is not required, eliminates unnecessary joints and minimizes downtime of the plant. Positive closure of the valve is achieved by contact of the resilient diaphragm with the metal weir of the valve body. Diaphragms are, of course, the weakest point of the valve, and selection of acceptable material should be based on sterilization temperatures, nature of the process fluid, cost, and experience. TFE and silicone diaphragms are in the same price range; EPDM is much cheaper. TFE can be used in most applications; EPDM is less frequently selected because of its color and tendency to swell when exposed to steam.

All diaphragms have some degree of permeability, depending upon the quality of materials and methods of manufacture. The use of silicone for diaphragms may be limited by its permeability, and by its reduced tensile strength when exposed to steam. Steam exposure also induces elongation, changes in volume [3], and depolymerization, leading eventually to failure.

Historically, fear of diaphragm failure in valves has been frequently used as an argument against their use, but with the recent improvements in diaphragm life expectancy, it is now an unrealistic concern. A requirement that valves in bioprocess service be on a strict routine maintenance program, which includes the regular inspection of diaphragms, should allay all fears about their use, while contributing significantly to system cleanliness.

Diaphragm valves are available with a built-in drain adjacent to their weir. This makes them an obvious first choice for sterile barrier valving, and possible candidates for service in CIP draining. Self-draining is ensured by proper installation of a valve in horizontal pipelines. Each valve has permanent hash marks on its body that indicate the self-draining angle (see Figure 7.10 and Table 7.10). The self-draining angle depends on whether the body is made from a forging or a casting.

The adaptation of ball valves to clean systems has expanded their application in bioprocesses. Sturdy three-piece valves, having castings with a variety of end connections such as tubing, Tri-clamp®, or extended butt weld, with bosses for purge ports on pipe ends, are now available with interior body finish of 6–32 Ra microinches (0.15–0.8 μm), mechanically polished or electropolished. Constructed from 316L stainless steel with Teflon® seats, their main disadvantage is the cavities inherent in the design. Cavities can be eliminated by the use of fillers, but these tend to create crevices that are more difficult to sterilize. The ball cavity can be eliminated by use of a three-way ball in a two-way valve, so material is always discharged downstream.

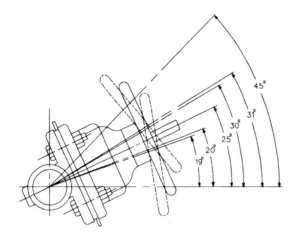

Figure 7.10 Installation Angle for Diaphragm Valves. See Table 7.10 for relationship of size to angle. (Courtesy of I.T.T. Engineered Valves.)

Table 7.10 Self-Draining Angle of Diaphragm Valves

Valve	Cast [1]	Forged [1]	Forged [2]	PVDF [3]
1/4"		30°	25°	30°–45°
3/8"		30°	25°	30°–45°
1/2"	30°	30°	25°	30°–45°
3/4"	30°	30°	30°	30°–45°
1"	31°	30°	25°	30°–45°
1 1/2"	30°	30°	25°	30°–45°
2"	25°	30°	25°	30°–45°
2 1/2"	19°	–	25°	30°–45°
3"	25°	–	25°	30°–45°
4"	20°	–	–	30°–45°

Sources: [1] I.T.T. Engineered Valves; [2] Saunders; [3] SaniTech.

A valve that also has some application in biotechnology plants is the butterfly valve adapted to sterile conditions. It has split body design, is made of 316L stainless steel lined with high-density PTFE liner, and has an electropolished disk. A one-piece stem, with channels for steam sterilization and dual shaft sealing, permits steam sterilization of the shaft. This valve has crevices between shaft and liner, and the wetted surface is very large, but may be of interest for larger lines over 4" (100 mm) in diameter.

Tank-bottom valves until recently were an easier choice because of the lack of suitable options. However, the recent introduction of new types has widened these options somewhat. The adapted ball valve with a flange-type body for welding into the bottom head of the vessel (shown in Figure 7.11) is one of the few suitable valves available that eliminate pockets or nozzles above the valve. The danger of unwelcome growth in the cavities of the valve body can be alleviated by the cavity's steam barrier, designed especially for fermentor harvest valves. Another version of this valve, called a *kettle valve*, is produced by Tri-Clover (and is shown in Figure 7.12). It uses a mushroom-shaped poppet instead of a ball and is a suitable alternative for the cooler processing common to vessels downstream of the fermentor. The adapted ball valve has a quarter-turn operation, but the kettle valve's motion is linear by stem operation. This difference contributes to the added ruggedness of the former over the latter. Nevertheless, the kettle valve has been used successfully in the pharmaceutical industry in WFI service for many years.

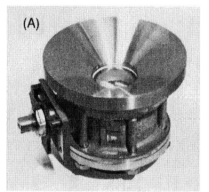

Figure 7.11 Tank-Bottom Valves: (A) ball valve with weld-in flange (courtesy of Worcester Controls Corp.); (B) diaphragm valve with weld-in flange (courtesy of I.T.T. Engineered Valves).

Until recently, truly novel contributions to sterile processing have been unimpressive. At present there are a few innovative designs on the market that adapt diaphragm valves for various applications. Figure 7.13 illustrates some products of I.T.T. Engineered Valves as examples of these efforts. Saunders Valve, Inc., also has available special valve combinations, but at the time of writing, catalogs showing them were not yet available. Both these companies have now introduced diaphragm valves adapted for tank-bottom service that are expected to gain a large portion of this market.

Figure 7.12 Sanitary Diverter Valve, supplied with weld-in flange for tank-bottom service. (Courtesy of Tri-Clover, Inc.)

7.12.2 Control Valves

The need to find valves suitable for throttling service introduces some difficult problems whose solutions lean more towards compromise than to a complete answer. Historically, the quest for the valve with ideal characteristics and wide range often led to the choice of a globe valve, frequently one with a cage-style trim, but such difficult-to-clean valves are unsuitable for clean processing. Most bioprocessing involves moderate-to-low temperature and pressure conditions, so the quest for the most suitable valve again becomes a choice between diaphragm, ball, or butterfly, with the diaphragm a clear favorite for most engineers.

The flow characteristics of a valve are defined as the *inherent capacity* of the valve, measured throughout its travel, with a constant pressure drop across it. Ball valves and butterfly valves (high-performance) tend to have *equal percentage* characteristics, and weir-type diaphragm valves have a mixture of *linear* and *quick-opening* characteristics that contribute to their flow-control advantages (Figure 7.14).

The ratio of maximum to minimum flow over which the valve maintains satisfactory control is defined as *range*, and figures prominently in the assessment of the suitability of a valve for flow-control service. For reasons stated earlier, the diaphragm is the first choice from a cleanliness standpoint, and adaptations such as the Dualrange® valve (I.T.T. Engineered Valves) have given the diaphragm valve fine-throttling characteristics suitable for most applications. These characteristics are compared graphically in Figure 7.14, which demonstrates that the Dualrange® is closer

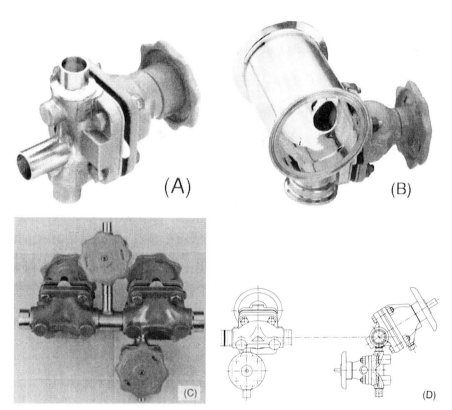

Figure 7.13 Examples of Innovative Diaphragm Valve Adaptation: (A) GMP valve for point-of-use fittings elimination; (B) zero static valve for in-line sampling without dead leg; (C) sterile barrier assembly using minimum space and fittings; (D) sterile access port at shut-off valve. (All courtesy of I.T.T. Engineered Valves.)

to having desirable linear characteristics than ordinary diaphragm valves. This type of valve, which uses an inner compressor to form a contoured opening during low-flow conditions, should perform well in clean processing, especially when fitted with a positioner.

7.12.3 Sampling and Harvesting Valves

The protection of fermentation medium from contamination during sampling, analyzing, or harvesting procedures requires devices that incorporate the following:

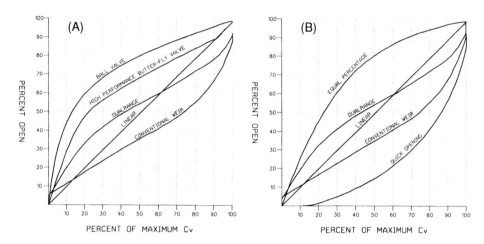

Figure 7.14 Comparison of Flow Characteristics of Valves, using valve coefficient C_v and percentage opening. (A) Ball valve, high-performance butterfly valve, Dualrange® valve, and conventional weir compared to linear characteristic valve. (B) Equal-percentage and quick-opening valve compared to Dualrange® valve, conventional weir, and linear characteristic valve. (Courtesy of I.T.T. Engineered Valves.)

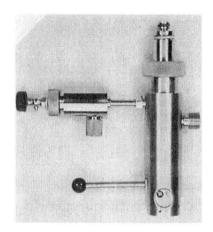

Figure 7.15 Typical Sterilizable Sample or Harvest Device. (Courtesy of L.H. Fermentation.)

cavity-free attachment to the fermentor vessel, provision for steam sterilization of all surfaces (including the probe or sample bottle), exclusion of steam from the fermentor vessel, and safety for the operator and protection for the sample.

Larger sized fermentors normally use a bottom-harvest valve similar to that shown in Figure 7.11A, but with a steam supply to the valve-body cavities. Sampling and low-flow harvesting require a device similar to that shown in Figure 7.15, which is ideal for aseptic fluid transfer. Figure 7.16 shows the Ingold Retractable Probe Device, which is the most widely used device for the measurement of pH, dissolved O_2, and CO_2. The device incorporates a ball valve to isolate the vessel from a lock chamber, where the probe is steam-sterilized prior to its insertion into the fermentor. Retraction is either manual or automatic, and escape of the fermentor medium is prevented by O rings used to seal the probe, and a specially designed vessel adapter.

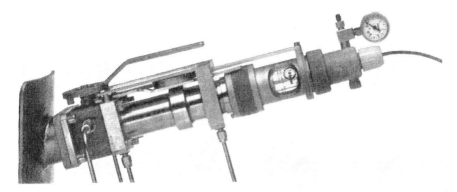

Figure 7.16 Retractable Probe Device for Bioreactors. (Courtesy of Ingold Electrodes, Inc.)

7.12.4 Valve Sizing

Sizing estimates, involving both installed cost and performance considerations, are an attempt to optimize the sizing (or at least prevent the oversizing) of valves. Valve manufacturers use *valve coefficient* charts to provide ratings for their valves; the ratings, in turn, determine valve capacity for given pressure drops. The coefficient (C_v) is measured in gallons of water, at 60 °F for 1 min and 1 psi differential for liquid flow. Gases and steam have equivalent coefficients that take into account their compressibility and critical flow limitations.

For valves smaller than line size in lines smaller than 3" (75 mm), the cost of reducers and additional drain valves is sometimes overlooked in the sizing procedure. Additional fittings in sterile processing are better avoided if possible, and line-size valves are recommended, except for control valves, where oversizing could lead to serious problems. Since C_v factors differ from manufacturer to manufacturer, even for the same type of valve, and computer programs exist to size control valves (Fisher, Masoneilan, Valtek, Worcester, and others), we have not attempted to address calculations in this chapter.

7.13 Acknowledgments

Most of this material is a product of the years of experience with many friends and associates at the Stearns Catalytic Division of United Engineers & Constructors (now Raytheon Engineers & Constructors) and at Life Sciences International, to whom we extend our thanks. We are especially grateful to Harry WILLIFORD (U.E.&C.) for his help with the section on line sizing, Dan ALCAVAGE (U.E.&C.) and Mark GAUZZA (L.S.I.) for their help with the CADD illustrations, and Darlene RAY (U.E.&C.) and Beth Ann HAVEN (L.S.I.) for their patience with the typing.

7.14 References

1. Title 21, *Code of Federal Regulations (CFR-21)*, Part 212.49, Subpart C; *Water and Other Liquid Handling Systems*, CGMP—LVP (Current Good Manufacturing Practice—Large Volume Parenterals); Food and Drug Administration; Superintendent of Documents, Washington, DC, 1976.
2. DILLON, C. P., RAHOI, D. W., TUTHILL, A. H., Stainless Steels for Bioprocessing, Part 1: Materials Selection, *BioPharm 5*, 38–42 (1992).
3. BJURSTROM, E. E., COLEMAN, D., Water for Injection System Design, *BioPharm 1*, 43–44 (1987).
4. *Standard Specification for Seamless and Welded Austenitic Stainless Steel Sanitary Tubing*, ASTM A270-90, 168; American Society for Testing and Materials, Philadelphia, 1992.
5. COLEMAN, D., EVANS, R., Fundamentals of Passivation and Passivity in the Pharmaceutical Industry, *Pharm. Eng. 10*, 43–49 (1990).
6. BLUME, W., Role of Organic Acids in Cleaning Stainless Steel, *ASTM Spec. Tech. Publ. 538*, 48 (1972).
7. *Flow of Fluids*; Technical Paper No. 410; Crane Co., Chicago, IL, 1986.
8. MENON, G., Rouge and Its Removal from High-Purity Water Systems, *BioPharm 3*, 40–43 (1990).
9. VIA, J. W., III, COSTA, D. W., REED, R., Select the Right Sensors for Pharmaceutical Processes, *Chem. Eng. Prog. 87* (10), 38–45 (1991).
10. COLEMAN, D., SMITH, P. J., Effective Steam Trapping for Clean Steam Systems in the Biotechnology and Pharmaceutical Industries, *Pharm. Eng. 12*, 8–14 (1992).

Pumps

Robert Stover

Contents

8.1   Introduction

Fluid transfer is a vital component of almost all industrial biochemical processes. Since the variety of requirements in fluid transfer applications and the variety of pump designs offer such a multitude of possible combinations, the engineer must have a solid and thorough understanding of pump fundamentals. Like other great inventions, most pump designs are deceptively simple, relying on fundamental laws of physics. However, as with most devices that appear simple on the surface, great depths of theory and knowledge underlie pump design and application. Full textbooks cover subjects I deal with here in just a few pages, and some manufacturers are among the best in industry at researching and publishing their data in great volumes. Nevertheless, I have written this chapter for the interested reader to read straight through, each section reinforcing the next, with a hope that a sense of the cohesive whole will result, and that the reader can henceforth make fluid transfer decisions with a new sense of confidence.

8.2   Hydraulic Theory

With the following equations we can accurately determine the resistance to flow imposed by the system. Once the system requirements are established, pump manufacturers can be approached and various secondary concerns such as compatibility with the product and general physical package needs can be addressed. It is easiest to calculate the total resistance one component at a time. This way we can also see the consequences of changes in an individual system component. The variety of possible system configurations is infinite, but, fortunately, back pressure can be applied in only a few ways.

8.2.1   Flow Rate

The flow rate must be specified before any system resistance can be calculated. It's important to be realistic, because of the effects unrealistic estimates can have on the subsequent pressure calculations. Horsepower is directly related to the product of flow rate and pressure, so overestimating can also result in oversized (and overpriced) pumps. The usual units of flow rate are gallons per minute (gpm) or milliliters (mL) per minute. In general, flow rate can be calculated from the following formula:

$$Q = V/t = Av \qquad (8.1)$$

where Q is the flow rate, ft^3/s;
   V is the volume, ft^3;
   t is the time duration, s;

A is the cross-sectional area of pipe, ft^2; and

v is the velocity, ft/s.

One cubic foot is 7.48 gallons or 28.3 liters. The remainder of this chapter uses English units, since most U.S. pump data are in these units.

8.2.2   Head

Head is the general term for pressure, either in a system as resistance to flow, or as a pump's capability to overcome the resistance. It's most easily visualized in terms of a vertical column of liquid, its weight equivalent to this same total system resistance, and is therefore usually expressed in feet of column height. The various individual resistances to flow can all be converted to vertical columns, which when added together give the *total mechanical head, H_m*, also called *TDH*, for *total dynamic head*, expressed in feet. Any pump that can pump against the weight of liquid in this total column can, by definition, pump against the system whose component back pressures were used to calculate it. Head is always given in feet of water, so that pump charts can be universal; corrections are made later for any differences in weight between water and the liquid being pumped.

8.2.3   Bernoulli's Formula

8.2.3.1   Total Dynamic Head

Total dynamic head calculations take the sum of four components, which contain all possible resistances. These resistances are *static head, pressure head, velocity head,* and *friction head*. The total head is most clearly expressed by Bernoulli's formula (modified for pumps):

$$H_m = (Z_2 - Z_1) + (P_2 - P_1) + (v^2/2g) + f(L/d)(v^2/2g) \tag{8.2}$$

where H_m is the total dynamic head or total mechanical head, ft;

   Z_2 is the discharge elevation, ft;

   Z_1 is the suction elevation, ft;

   P_2 is the pressure at discharge, converted to ft;

   P_1 is the pressure at suction, converted to ft;

   v is the velocity in the pipe at maximum reading, ft/s;

   g is the acceleration of gravity, 32.2 ft/s^2;

   f is the friction factor, dimensionless;

   L is the length of run, ft; and

   d is the diameter of pipe, ft.

8.2.3.2 Static Head

Static head, $(Z_2 - Z_1)$ or ΔZ, is the difference in elevation between the pump discharge and suction levels. If the liquid source is a vessel, Z_1 refers to the liquid's surface level; if the pump takes, for example, water from city lines, so no surface datum is available, Z_1 is the center line of the pump. Z_2 refers to the point at which the liquid enters the atmosphere, the surface level of the receiving vessel if the liquid enters below the surface, or where the liquid enters a pipe if it is injected into a line. The point is to calculate the elevation that defines the amount of energy the pump has to overcome. Note that no considerations are made for the maximum height the piping may reach; anything above the final exit point is taken care of by siphon action once flow is begun. The pump must have the capability to pump to the maximum height, but not necessarily at the flow rate demanded. Note also that length of pipe is not considered here; the length is compensated for during the friction loss calculation of the H_m formula.

8.2.3.3 Pressure Head

Pressure head, $(P_2 - P_1)$ or ΔP, represents any differences in pressure between the source and the destination outside the immediate system. P_1 is the pressure in the source vessel, which may be negative pressure (a vacuum vessel), zero pressure (a tank open to atmosphere), or positive pressure (water taken from city lines, for example). P_2 is the pressure at the point of discharge, again potentially any value from vacuum to positive pressure. This difference needs to be converted to feet: 1" Hg (mercury) = 1.133 feet of water; and 1 psi = 2.31 feet of water (not taking into account specific gravity). This component of H_m allows us to determine the energy from outside the immediate system that either hinders or assists pumping.

8.2.3.4 Velocity Head

Velocity head, $v^2/2g$ or H_v, refers to that energy necessary to overcome inertia. It is the portion of the pressure head that is converted to kinetic energy by virtue of the fluid motion. Stated another way, this head is the distance the fluid would have to fall to acquire a particular velocity. Since the denominator, $2g$, is 64.4 ft/s², and fluid velocity in most industrial applications is kept at a maximum 8 ft/s in a discharge line or 6 ft/s in a suction line (to minimize pipe wear), velocity head is rarely important in H_m calculations, and in fact is often not even calculated. The velocity for many flow rates and pipe sizes can be read from Table 8.1.

Table 8.1 Velocity and Friction Loss per Hundred Feet of Pipe as Functions of Flow Rate in Various Sizes of Pipe. (Data courtesy of Sta-Rite Industries, Inc.)

GPM	0.5" Vel.	Loss	0.75" Vel.	Loss	1" Vel.	Loss	1.25" Vel.	Loss	1.5" Vel.	Loss	2" Vel.	Loss
2	2.10	3.47	1.20	0.89								
4	4.23	12.70	2.41	3.29	1.49	1.01	0.86	0.27	0.63	0.12		
6	6.34	26.80	3.61	6.91	2.23	2.14	1.29	0.57	0.94	0.26	0.57	0.09
8	8.45	46.10	4.82	11.80	2.98	3.68	1.72	0.95	1.26	0.45	0.77	0.16
10	10.60	69.10	6.02	17.90	3.72	5.50	2.14	1.44	1.57	0.67	0.96	0.24
12			7.22	24.90	4.46	7.71	2.57	2.02	1.89	0.94	1.15	0.37
15			9.02	37.60	5.60	11.80	3.21	3.05	2.36	1.41	1.50	0.51
18			10.80	50.90	6.69	16.50	3.86	4.28	2.83	1.99	1.72	0.70
20			12.00	63.90	7.44	19.70	4.29	5.21	3.15	2.44	1.91	0.86
25					9.30	30.10	5.36	7.80	3.80	3.43	2.50	1.28
30					11.15	41.80	6.43	10.80	4.72	5.17	2.89	1.80
35					13.02	55.90	7.51	14.70	5.51	6.91	3.35	2.40
40					14.88	71.40	8.58	18.80	6.30	8.83	3.82	3.10
45					16.70		9.65	23.50	7.08	10.90	4.30	3.85
50							10.72	28.20	7.87	13.30	4.78	4.65
55							11.78	33.80	8.66	16.00	5.26	5.55
60							12.87	40.00	9.44	18.60	5.74	6.53
65							13.92	46.70	10.23	21.60	6.21	7.56
70							15.01	53.10	11.02	24.90	6.69	8.64
75							16.06	60.60	11.80	28.20	7.17	9.82
80							17.16	68.20	12.69	32.00	7.65	11.1
85							18.21	77.00	13.38	35.30	8.13	12.5
90							19.30	84.60	14.71	39.50	8.61	13.8
95									14.95	43.70	9.08	15.3
100									15.74	47.90	9.56	16.8
110									17.31	57.30	10.5	20.2
120									18.89	67.20	11.5	23.5
130									20.46	78.00	12.4	27.3
140									22.04	89.30	13.4	31.5
150									23.6		14.3	35.7
160											15.3	40.4
170											16.3	45.1
180											17.2	50.3
190											18.2	55.5
200											19.1	60.6
220											21.0	72.4
240											22.9	85.5
260											24.9	99.2

8.2.3.5 Friction Head

Friction head, H_f, is defined as $f(L/d)(v^2/2g)$, by the DARCY–WEISBACH formula for friction loss in pipe. The friction factor f can be found in various sources for almost all types or conditions of pipe. However, because of the relatively limited types and sizes of pipe popular today, charts of losses already derived from the above definition are readily available (Table 8.1 is an example). To use the tables, determine the total equivalent length of pipe run for each individual size and type of pipe in a system. Each fitting and valve in the run has resistance that can be expressed as an additional equivalent length of straight pipe, so the losses in pressure through either the individual fitting or the listed equivalent length of run is the same (see nomograph, Figure 8.1). Add the actual run and the equivalent of the fittings to obtain what is called the *total equivalent length*. Use this new imaginary length of pipe with a chart such as Table 8.1 that gives losses expressed in feet of head for various flow rates. The greater the flow rate through a particular size of pipe, the more loss per foot of length, and the greater the length, the more total loss incurred. Each portion of the complete system, from the point where the system takes the liquid to its exit from the discharge line, is included in the total system friction head, H_f. When dealing with a system with many branches, remember that pressure in different branches, unlike flow rate, should not be added together; the route providing the most resistance (probably the longest of the group, or the one with the most fittings) is the one that determines H_f. When a system is especially difficult or confusing, it helps to calculate H_f for each recognizably distinct section. Then trace each possible route to find the one with the greatest total losses.

8.2.4   Example Calculation for Total Mechanical Head, H_m

As illustrated in Figure 8.2, a system with the source tank under 5" Hg vacuum is connected through 50 ft of 2" schedule (sch) 40 PVC pipe (including 2 elbows) to the pump. On the pump discharge, 150 ft of 1½" sch 40 PVC pipe with another 4 elbows leads into a line pressurized to 60 psi, 120 ft above the source tank. There is a filter in the discharge line with a maximum pressure drop of 20 psi. Calculate the total head for this system given a desired flow rate of 60 gpm.

We compute each term of Bernoulli's formula (eq 8.2):

- **Static head:**
  $Z_2 - Z_1 = $ **120 ft**.
- **Pressure head:**
  $P_2 = 60$ psi $= 60 \times 2.31 = 139$ ft;
  $P_1 = 5"$ Hg $= 5 \times 1.133 = -6$ ft;
  and therefore
  $(P_2 - P_1) = (139 \text{ ft} - [-6 \text{ ft}]) = (139 \text{ ft} + 6 \text{ ft}) = $ **145 ft**.

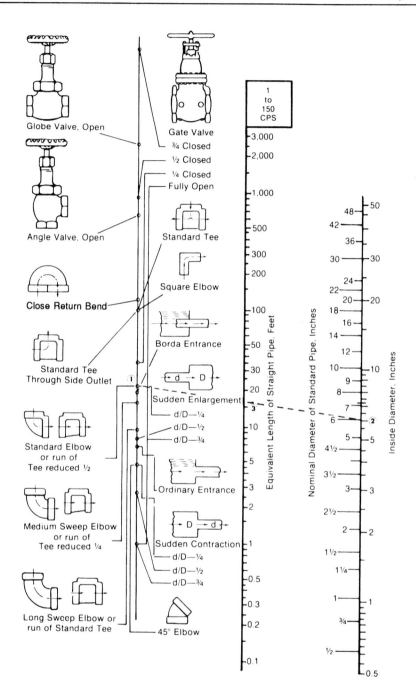

Figure 8.1 (at left) Pipe Fittings and Their Equivalent Lengths of Pipe. In the example, a standard elbow (1) in a 6" pipe (2) is equivalent to 16 ft of straight pipe (3), if the liquid viscosity is between 1 and 150 cP. (Reprinted courtesy of Crane Co.)

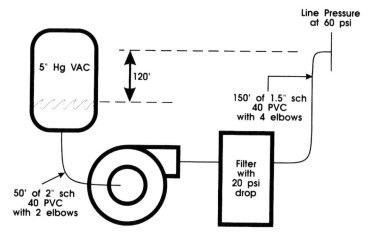

Figure 8.2 System for Example in Section 8.2.4. System with source tank under 5" Hg vacuum and with a filter at pump outlet is used in an example calculation of total mechanical head.

- **Velocity head:**
  $v = 9.44$ ft/s (from Table 8.1), so that $(9.44)^2/64.4 = $ **1 ft**.

- **Friction loss:**

  H_f for suction side (H_{fs}):
     2" elbows @ 5.5 equivalent ft/elbow (from Figure 8.1), so (2 elbows × 5.5 ft/elbow) = 11 ft;
     11 ft + 50 ft (length of run as given in the statement) = 61 total equivalent ft on the suction side.
     Loss for 2" PVC (from Table 8.1) = (6.53 ft loss/100 ft run) or 0.0653 ft loss per foot run, at the required 60 gpm flow rate.
     Therefore, 61 total equivalent feet × 0.0653 = **4 ft** total friction loss on the suction side.

  H_f for discharge side (H_{fd}):
     1½" elbow @ 4 equivalent ft/elbow (from Figure 8.1), so (4 elbows × 4 ft/elbow) = 16 ft.
     16 ft + 150 ft (length of run as given in the statement) = 166 total equivalent ft on the discharge side.
     Loss for 1½" PVC (from Table 8.1) = (18.6 ft loss/100 ft run) or 0.186 ft loss per foot run at the required 60 gpm flow rate.
     Therefore, 166 total equivalent feet × 0.186 = **31 ft** total friction loss on the discharge side.

H_f for filter (H_{ff}):
   Loss through the filter = 20 psi = 20 x 2.31 = **46 ft**.

Total Friction Loss ($H_f = H_{fs} + H_{fd} + H_{ff}$):
= 4 ft (suction) + 31 ft (discharge) + 46 ft (filter) = **81 ft**.

• **Total Mechanical Head (H_m):**
   H_m = 120 ft (static) + 145 ft (pressure) + 1 ft (velocity) + 81 ft (friction)
   = **347 ft**.

A pump that can develop 60 gpm flow rate against a total head of 347 feet should be selected.

8.2.5   Adjustments to Bernoulli's Formula

8.2.5.1   Specific Gravity

Specific gravity, SG, is the ratio of the density of a liquid to the density of water at 60 °F, which is 8.34 lb/gal. Thus, a material that weighs 12.51 lb/gal has a specific gravity of 12.51/8.34 = 1.5; and water at 300 °F, weighing 7.67 lb/gal, has a specific gravity of 7.67/8.34 = 0.92. Specific gravity has a direct effect not only on the horsepower necessary to drive a pump at its chosen speed, but also on individual components of the equation used to arrive at H_m.

A pump impeller with a fixed diameter, rotating at a given rpm, develops a circumferential speed that determines how fast a particle leaves the periphery; any material leaving at that speed goes the same distance. Specific gravity is involved in that the heavier the material, the more horsepower is needed to overcome that weight and to develop the full impeller speed. This greatly simplifies pump selection, as only one pump curve need be selected for all liquids; any changes are made to the motor horsepower driving the pump selected.

The effect of specific gravity on pressure head modifies the previously stated conversion:

$$psi = (feet\ of\ head) \times SG/2.31 \qquad (8.3a)$$

$$feet\ of\ head = 2.31 \times psi/SG \qquad (8.3b)$$

Therefore, a filter with a 20-psi pressure drop, handling a liquid with a specific gravity of 0.8, is equivalent to 20 × 2.31/0.8 = 57.75 ft. This means that a pump capable of delivering 58 ft of head (instead of the 46 ft for water) is required to develop the head necessary to pump against the filter's 20-psi pressure drop. Conversely, a liquid heavier than water requires a shorter column, and therefore less head, to develop the pressure requested.

After all losses expressed in psi are converted to feet, the adjusted H_m is found and a pump selection is made. The specific gravity is then applied to either lower or raise the horsepower necessary to drive the impeller at the chosen speed (see Section 8.2.9).

8.2.5.2 Viscosity

Viscosity is the measure of a material's resistance to flow. There are many units used in particular industries, but in biotechnology, centipoise (cP) is the standard, and Saybolt Seconds Universal (SSU) is well known.

Five types of viscous liquids are usually encountered. *Newtonian* liquids are those unaffected by shear, such as water. *Pseudoplastic* liquids are those whose viscosity decreases with shear, such as molasses and paint. *Bingham plastic* fluids resemble Newtonian fluids in that their viscosity is unaffected by shear; however, they have the additional quality of a threshold of resistance; once the initial resistance is overcome, they flow more freely. An example is catsup. *Dilatant liquids* are those whose viscosity increases with shear, such as taffy. *Thixotropic* fluids also resemble pseudoplastic fluids in becoming thinner at high shear, but additionally they become thinner with time of shearing. An example is mayonnaise.

Of course, the designer must take into consideration the worst conditions that might be experienced due to the nature of the fluid, beyond the normal operating conditions. Special pipe friction loss charts for viscous liquids are given in many handbooks, as are charts giving minimum pipe size for flow of liquids with various cP or SSU ratings. Most importantly, charts exist for adjustments to the values of gpm, H_m, and efficiency for centrifugal pumps handling viscous liquids. As shown in Figure 8.3, viscosity dramatically reduces centrifugal pump performance, especially efficiency. The correction factors are multipliers to be applied to a pump's capacity, constituting losses from standard performance with water (see Section 8.6.2).

With positive-displacement pumps, most adjustments are to the speed the pump can run and still allow the thicker liquid to completely fill the moving cavities. This means a shear-thickening, shear-thinning, or time-dependent material can be pumped with the same pump if the speed can be varied up or down as the viscosity changes. Since various types of positive-displacement pumps differ in their ability to compensate for viscosity, there are no universal charts as for centrifugal pumps, and each manufacturer must specify the maximum rpm under varying conditions, as well as any horsepower adjustments necessary.

8.2.6 Net Positive Suction Head

Net positive suction head (*NPSH*) is that pressure still available at the suction port of a pump after all losses and restrictions to flow have been taken into account.

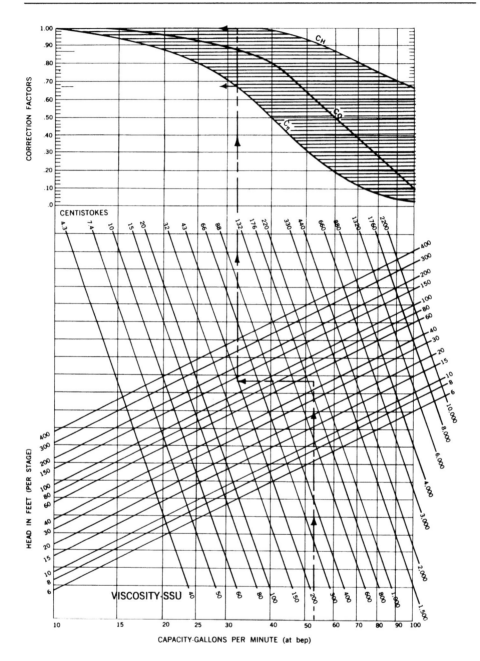

Figure 8.3 (at left) Viscosity Effects on Pump Performance. Viscosities affect centrifugal pump capacity, efficiency, and head. The diagram shows that correction factors are based on a combination of capacity, mechanical head, and viscosity. In the example indicated by the dashed lines, to select a pump for service transporting a fluid with a viscosity of 200 SSU at a desired flow rate of 53 gpm and against 50 ft of head, use correction factors as follows: efficiency correction, C_η, 0.67; capacity correction, C_Q, 0.87; and head correction, C_H, 1.00. (Courtesy of Hydraulic Institute.)

Every pump model has a *net positive suction head required* (*NPSH$_r$*), derived from tests in which the manufacturer reduces pressure at the pump's inlet by increments until the pump performance falls below 97% of its normal value on the performance curve. That pressure still available at the inlet just before this event is the NPSH$_r$ for the flow rate being tested. It is the reduction in total head as the liquid enters the pump. Tests are conducted at multiple flow rates to develop an NPSH$_r$ curve.

Net positive suction head available (*NPSH$_a$*) is the head the system provides to a pump at a particular flow rate. It is the proximity of the liquid to its vapor pressure at the pump suction. This head is determined by a calculation similar to a total head calculation, but on the suction side only. To avoid cavitation, NPSH$_a$ must exceed NPSH$_r$.

$$NPSH_a = H_a \pm \Delta Z - H_{fs} - H_v - H_{vap} \qquad (8.4)$$

where H_a is the pressure available at the applicable elevation, ft (see Figure 8.4);

ΔZ is the elevation difference, ft. As in Bernoulli's equation, positive if the pump is below the source, negative if the pump is above the source;

H_{fs} is the friction loss, ft. Again, as in Bernoulli's equation, but referring exclusively to the suction side of the pump;

H_v is the velocity head = $v^2/2g$, ft; and

H_{vap} is the vapor pressure, ft, from various charts such as Table 8.2, which list the pressure necessary to keep the material in liquid form at the applicable operating temperature.

8.2.7 Example Calculation of Net Positive Suction Head (NPSH$_a$)

In an installation at 6000 feet elevation, water at 180 °F is in a tank where the minimum liquid level is 8 feet above the center line of the pump. Piping from the tank to the pump is 1" schedule 40 PVC, with 2 elbows and a check valve, over a 20-foot run. The required flow rate is 25 gpm. Determine the NPSH$_a$.

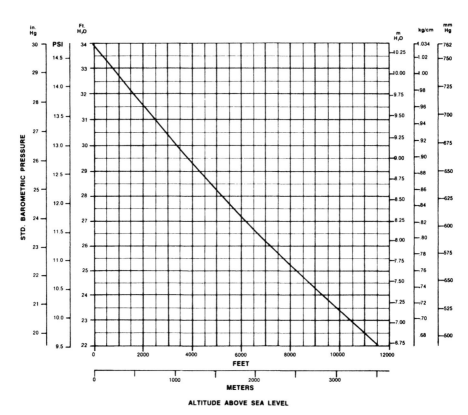

Figure 8.4 Effect of Altitude on Barometric Pressure. The atmospheric pressure (in various units) available at a pump depends on the elevation above sea level, as depicted in this figure. (Courtesy of Hydraulic Institute.)

We compute the terms in eq 8.4:

- H_a = **27.2 ft** (from Figure 8.4 for 6000 ft elevation).
- ΔZ = **+8 ft** (the source is above the pump; therefore, positive for the inlet).
- H_{fs}:

    2 elbows @ 2.5 equivalent ft/elbow (from Figure 8.1);

    2.5 ft/elbow × 2 elbows = 5 ft;

    1 check valve = 7 ft equivalent ft (Figure 8.1);

    5 ft + 7 ft (for elbows and check valve) + 20 ft run (given in the statement)

    = 32 total equivalent ft;

    @ 25 gpm through 1″ schedule 40 PVC = 30.1 ft loss/100 ft run (from Table 8.1)

    = 0.301 ft loss per ft run;

    32 total equivalent ft × 0.301 ft loss/ft run = **9.6 ft.**

Table 8.2 Properties of Water at Temperatures from 32 °F to 705.4 °F. (Courtesy of Hydraulic Institute.)

Temp. F	Temp. C	Specific Volume Cu Ft/Lb	SPECIFIC GRAVITY			Wt In Lb/Cu Ft	Vapor Pressure Psi Abs
			39.2 F Reference	60 F Reference	68 F Reference		
32	0	.01602	1.000	1.001	1.002	62.42	0.088
35	1.7	.01602	1.000	1.001	1.002	62.42	0.100
40	4.4	.01602	1.000	1.001	1.002	62.42	0.1217
50	10.0	.01603	.999	1.001	1.002	62.38	0.1781
60	15.6	.01604	.999	1.000	1.001	62.34	0.2563
70	21.1	.01606	.998	.999	1.000	62.27	0.3631
80	26.7	.01608	.996	.998	.999	62.19	0.5069
90	32.2	.01610	.995	.996	.997	62.11	0.6982
100	37.8	.01613	.993	.994	.995	62.00	0.9492
120	48.9	.01620	.989	.990	.991	61.73	1.692
140	60.0	.01629	.983	.985	.986	61.39	2.889
160	71.1	.01639	.977	.979	.979	61.01	4.741
180	82.2	.01651	.970	.972	.973	60.57	7.510
200	93.3	.01663	.963	.964	.966	60.13	11.526
212	100.0	.01672	.958	.959	.960	59.81	14.696
220	104.4	.01677	.955	.956	.957	59.63	17.186
240	115.6	.01692	.947	.948	.949	59.10	24.97
260	126.7	.01709	.938	.939	.940	58.51	35.43
280	137.8	.01726	.928	.929	.930	58.00	49.20
300	148.9	.01745	.918	.919	.920	57.31	67.01
320	160.0	.01765	.908	.909	.910	56.66	89.66
340	171.0	.01787	.896	.898	.899	55.96	118.01
360	182.2	.01811	.885	.886	.887	55.22	153.04
380	193.3	.01836	.873	.874	.875	54.47	195.77
400	204.4	.01864	.859	.860	.862	53.65	247.31
420	215.6	.01894	.846	.847	.848	52.80	308.83
440	226.7	.01926	.832	.833	.834	51.92	381.59
460	237.8	.0196	.817	.818	.819	51.02	466.9
480	248.9	.0200	.801	.802	.803	50.00	566.1
500	260.0	.0204	.785	.786	.787	49.02	680.8
520	271.1	.0209	.765	.766	.767	47.85	812.4
540	282.2	.0215	.746	.747	.748	46.51	962.5
560	293.3	.0221	.726	.727	.728	45.3	1133.1
580	304.4	.0228	.703	.704	.704	43.9	1325.8
600	315.6	.0236	.678	.679	.680	42.3	1542.9
620	326.7	.0247	.649	.650	.650	40.5	1786.6
640	337.8	.0260	.617	.618	.618	38.5	2059.7
660	348.9	.0278	.577	.577	.578	36.0	2365.4
680	360.0	.0305	.525	.526	.527	32.8	2708.1
700	371.1	.0369	.434	.435	.435	27.1	3093.7
705.4	374.1	.0503	.319	.319	.320	19.9	3206.2

Computed from Keenan & Keyes' Steam Table.

- H_v:

  since $v = 9.3$ ft/s (from Table 8.1), $H_v = (9.3)^2/2g = 86.5/64.4 = $ **1.3 ft**.

- H_{vap}:

  7.5 psi (Table 8.2 for water at 180 °F) = 7.5 × 2.31 = **17.3 ft**, and

- $NPSH_a = 27.2$ ft + 8 ft − 9.6 ft − 1.3 ft − 17.3 ft = **7.0 ft**.

Because it is customary to allow a 2-ft minimum difference between $NPSH_a$ and $NPSH_r$, a pump with an $NPSH_r$ of 5.0 ft or less is necessary to prevent cavitation.

8.2.8   Cavitation

Liquids remain as such only if the atmosphere exerts enough pressure on the liquid surface to maintain the liquid state. This is of particular importance because the movement of liquid in the pump housing causes a low-pressure area to develop on the suction side of the pump. If there is insufficient pressure on the inlet side, the liquid flashes to vapor. The vapor expands and forms small pockets or bubbles, enters the pump with the liquid, and eventually finds itself in a higher pressure area of the housing. At this time the fluid again is under enough external pressure to exist in a liquid state and the bubbles collapse.

These implosions, called *cavitation*, carry tremendous force and are very harmful to pumps. There are many causes of cavitation; one is insufficient $NPSH_a$ or suction *starvation*. Cavitation due to suction starvation has a unique sound. In centrifugal pumps it is a steady noise like a rattling, often compared to gravel in the housing. In positive-displacement pumps it is normally a more percussive or banging sound. Implosions in centrifugal pumps damage impeller vanes, as bubbles collapse and scar the impeller surface. The pitting usually first shows itself on the concave side of the vanes, partway toward the tips. In all pumps the shaft is under stress, the motor probably responds to the stress with a higher amperage demand, and eventually the pump is greatly damaged or destroyed, having all the while performed below its tested capacity.

The same cavitation phenomenon also occurs in most pumps if the inlet flow is valved back (throttled) too greatly. Because of internal velocities or flow patterns not anticipated in the design, low-pressure areas may develop either on the edges of the impeller tips or at the impeller inlet (*eye*), and the liquid can again be under insufficient pressure to stay liquid. Similar erosion takes place at the location of the implosions, but the noise is typically more random and uneven, with louder peaks than with $NPSH_a$ problems. Elbows or turns closer than 6 times the pipe diameter may also cause uneven flow into the impeller eye, therefore producing a low-pressure area and cavitation.

One way to confirm a suspected cavitation problem is to take a vacuum measurement a short distance from the pump inlet. The reading, less H_{vap} (not reflected by the gauge), allows a direct comparison to the $NPSH_r$ stated by the manufacturer (H_a − gauge reading [converted to feet of head] − H_{vap} = $NPSH_a$).

The results of an $NPSH_a$ calculation may not only confirm the problem, but also suggest a direction for possible solutions. If there is cavitation, either the existing pump must be replaced with one needing less $NPSH_r$, or, more commonly, an effort must be made to raise the $NPSH_a$. $NPSH_a$ can be increased by several stratagems: the pump can be lowered, or the source tank raised, to increase ΔZ; a larger suction line can be used, to reduce H_{fs}; elbows and other restrictions can be eliminated, to reduce

H_{fs}; a larger pipe may be used, to reduce velocity and, therefore, H_v; the liquid temperature can be lowered, to reduce H_{vap}; or the pump outlet can be valved back, to deliver less flow, and therefore less NPSH$_r$ (see Section 8.6, Pump Curves and Charts).

In cases where any change is difficult, more marginal corrective actions can be taken: the leading edges of the impeller vanes can be filed, to ease the transition from axial to radial flow, or a small amount of air can be injected into the suction lines to reduce the shock of the implosions (this is very difficult to do successfully because of the narrow margin between absorbing shock and destroying efficiency). This suggestion may surprise some people, because cavitation itself is sometimes mistakenly thought to be air in the suction. Obviously, too much air can interfere with a pump's performance, even vapor lock it, but air does not cause cavitation. Yet another alternative is to live with the condition, but coat the impeller with a resistant material (many elastomers are highly durable under cavitation shock).

In the example in Section 8.2.7, the most direct way to raise NPSH$_a$ is to lower the liquid temperature. If this is not possible, H_{fs} and H_v could be lowered with a larger pipe, a shorter run, or elimination of an elbow. There are also many pumps with NPSH$_r$ under the 5.0 ft available in the system as it stands.

8.2.9 Horsepower

A normal component of a centrifugal pump curve or a positive-displacement pump chart is an indication of the horsepower necessary to drive the pump at the designated speed. Because of differences in specific gravity or viscosity, however, there may be times when it is important to have access to the actual formula for horsepower:

$$\text{HP} = (Q \cdot H_m \cdot \text{SG})/(3960 \times \text{efficiency}) \tag{8.5}$$

where HP is the horsepower, hp;
$\quad Q$ is the flow rate, gpm;
$\quad H_m$ is the head, ft;
$\quad$ SG is the specific gravity; and
$\quad$ efficiency is the unit efficiency expressed as a decimal (for example, 55% efficiency = 0.55).

Note that 3960, the numerical constant, is 33,000 lb/ft·min (the definition of 1 hp) divided by 8.34 lb/gal (the weight of water). Alternatively, if H_m for the total system has been calculated in psi, we can use the formula:

$$\text{HP} = Q \cdot H_{m,\,psi}/(1715 \times \text{efficiency}) \tag{8.6}$$

This is the same formula as eq 8.5, but incorporates eq 8.3 to take into account that $H_{m,\,psi}$ is in psi.

8.3   Centrifugal Pumps

A huge variety of pumps abounds, not only in terms of number of manufacturers (the pump section of the most recent *Thomas Register* contained 193 oversized pages of fine print!), but, more significantly, in terms of the myriad of designs these manufacturers incorporate. Hydraulically, pumps fall neatly into two main groups: *centrifugal* and *positive-displacement*. Each group has advantages and disadvantages, and a thorough understanding of these is more important and more useful than any attempt at a complete knowledge by brand.

8.3.1   General Design

Centrifugal pump design (Figure 8.5) consists of a rotating *impeller*, driven by a shaft extension (or directly by the motor shaft), which captures the liquid and accelerates it. Energy is applied as the liquid flows through the rapidly turning impeller, until the material enters the surrounding casing (*volute*), where the velocity is changed to pressure as the liquid funnels through the discharge port. The only moving parts are the impeller/shaft assembly and the mechanical seal that contains the liquid around the rotating shaft. Clearances are crucial only at the impeller face, or between the outside of the inlet hole (*eye*) of the impeller and the corresponding ring of the casing it fits into. Pressure is determined by the velocity at the impeller's periphery, and flow rate by the depth of the impeller vanes. The advantages and disadvantages of centrifugal pumps are outlined in Table 8.3.

There are many ways to profit from the centrifugal principle, yet configure the pump to more specific applications. In the following sections I describe some of the many styles of centrifugal pumps available.

8.3.2   End-Suction Pumps

The most popular general design is the *end-suction centrifugal pump* (Figure 8.6). The liquid enters the center of the front of the volute through the suction port, normally one pipe size larger than the discharge port (to assist in holding down NPSH$_r$). The direction of the discharge port is perpendicular to the shaft, on a tangent to the volute, to pick up the liquid in its natural path off the impeller. The suction and discharge ports are often cast in the same piece, which allows the volute to stay plumbed while the rest of the pump is pulled out for service.

End-suction centrifugal pumps are available as *close-coupled*, or as *pedestal-mounted* (the pump having its own frame and shaft assembly, bolted to a base, and coupling driven by a motor also bolted to the base). The close-coupled version allows

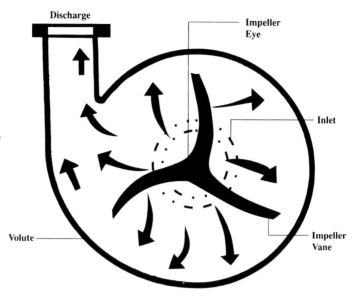

Figure 8.5
Centrifugal Pump
Design. A cross-
sectional view of
a centrifugal
pump volute,
showing the inlet
and discharge in
relation to the
impeller.
(Courtesy of Tri-
Clover, Inc.)

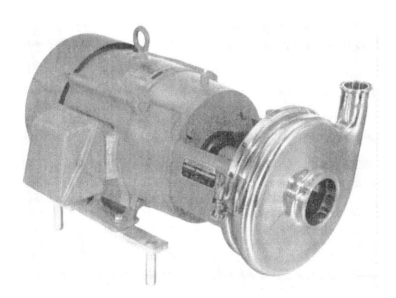

Figure 8.6 End-Suction Centrifugal Pump. This is the most common centrifugal pump design.
(Courtesy of Tri-Clover, Inc.)

Table 8.3 Advantages and Disadvantages of Centrifugal Pumps

Advantages
(a) Low Cost: A centrifugal pump is the least expensive pump, dollar per gallon of flow.
(b) Smooth Flow: A centrifugal pump delivers smooth, nonpulsating flow at uniform pressures, and so is very gentle to equipment downstream. It is, therefore, the first choice in such applications as filtration or reverse osmosis.
(c) Energy Conservation: Since slip is inherent in the design, the horsepower load on the driver decreases as resistance to flow is applied, and the output flow rate drops. As less flow is pumped, less weight has to be moved by the impeller, and less horsepower must be developed by the motor. This characteristic of a centrifugal pump allows it to be matched with a motor that can drive it under full flow and maximum load. Any change in performance can only reduce the horsepower selected. This is called *non-overloading horsepower*.
(d) Easy-to-Control Flow: Because of the significant effects back pressure has on capacity of a centrifugal pump, and the fact that reducing the flow rate also reduces motor load, the flow rate is easily and inexpensively controlled by valves.
(e) Low Probability of Overpressure: A centrifugal pump is characterized by a maximum output pressure, called the *shutoff head*, because of the slip in the impeller. Therefore, a maximum system pressure (pump inlet pressure plus shutoff head) can be established. In fact, in applications where the discharge side may be intermittently closed down, a pressure relief valve may not be needed while the pump continues to run. A small amount of heat goes into the liquid, but this is rarely significant. As explained in (c), the drive horsepower is at its minimum during the zero-flow period.
(f) Wide Range of Flow Rates and Pressures: Many different styles and sizes of impellers exist, so a huge range of flow rates (fractional gpm to hundreds of thousands of gpm) and pressures (barely above 0 to over 1000 psi) can be achieved.
(g) High Flow Rate and Pressure: A centrifugal pump can run at very high speeds (10,000 rpm), so very high flow rates and pressures can come from a relatively small package.
(h) Low Noise: There are no parts rubbing, sliding, or meshing in a centrifugal pump (except the mechanical seal); therefore, it is quiet and requires very little maintenance.
(i) Compact Size: When used in a close-coupled design (the pump mounted directly to the machined face of a motor, the impeller riding directly on the extended shaft of the motor), a centrifugal pump is a very compact package.
(j) Wide Choice of Seals: A great variety of seal designs and materials is available because the centrifugal design is so popular and because where the seal is installed is relatively spacious.
(k) Sanitary Design: The simple, clean design allows for easy adaptation to sanitary specifications.

Table 8.3 Advantages and Disadvantages of Centrifugal Pumps, *continued*

Disadvantages	
(a)	Poor Performance with Viscous Fluids: The impeller slip that permits flexibility makes it impossible for the centrifugal pump to handle viscous liquids efficiently. If the viscosity is much greater than that of water, check individual manufacturers' viscosity charts or general charts (see Section 8.6.2), because the effects of viscosity on pressure and flow rate are complex.
(b)	Low Pressure: Because pressure is limited to that corresponding to the velocity at the vane tips, a centrifugal pump is not by nature a high-pressure device, especially if high pressure is needed at low flow rates. However, impellers can follow each other in a multistage design, each impeller boosting the cumulative total pressure.
(c)	High Shear: Of major importance for biotechnology applications, centrifugal pumps are ineffective at slow speeds, and therefore may not be suitable for low-shear applications or delicate products needing low rpm.
(d)	Abrasive Wear: Because of the mechanical seal, abrasives are a potential problem. Some pump manufacturers have special hardened materials available (silicon carbide or tungsten carbide are the most common) to resist the rapid wear abrasives cause on standard seals. Impeller clearances must be maintained and can wear only at the expense of pressure. Double-seal packages isolate the seals from the liquid being pumped, but the impeller cannot easily be isolated.
(e)	Flow Rate Controlled by Back Pressure: A centrifugal pump is not a constant-flow device, but rather one with a wide range of possible flow rates, determined by the system back pressure. If this back-pressure resistance changes over time and there is no outside flow-control device, the pump does not hold a steady, repeatable flow rate. See Section 8.6.1 for an illustration, and see Section 8.6.4 for a discussion of what determines pump output in practice.

a more compact design, with only the motor bearings providing support. The pedestal-mounted (*long-coupled*) version allows easier replacement of the pump or motor without the need to disturb the other element. Close-coupled is usually preferred because of overall size, simplicity, inherent alignment, and lower price.

8.3.3   Split-Case Horizontal Pumps

Split-case horizontal pumps are used for flow rates greater than those an end-suction pump can deliver because of limitations on the weight that can be supported by the motor shaft alone, or in situations where the initial expense of a high-efficiency split-case design can be justified. They are always pedestal-mounted, with the shaft/impeller assembly supported on both ends. This allows the use of double-sided impellers, which provide better axial balance and extra efficiency, and also means the top half of the casing can be removed for easy servicing. They are available in sizes much larger than end-suction models, and are very rugged and long-lived.

8.3.4   Vertical Sealless Pumps

In *vertical sealless pumps*, the motor's shaft extends vertically through a column down to the impeller. The column is long enough to immerse the impeller yet allow the motor to stay above the liquid level. They can be metal-free (if the shaft of an already all-plastic pump is sleeved with plastic), or can be all metal (if used in a noncorrosive atmosphere). There are no seals needed because the motor is isolated from the liquid by its physical location, and no bushings because the motor/pump shaft is usually one piece from the top of the motor and is rigid enough not to whip during operation. Pumps of this type can run dry indefinitely, can handle liquids with limited loads of solids and abrasives with minimal wear, and can handle very corrosive liquids if built in all PVC, PVDF, or stainless steel.

8.3.5   Magnetically Driven Pumps

The *magnetically driven* design is another way to achieve an all-plastic, sealless pump. The motor shaft drives a large hollow magnet, and, since the end of the pump is bolted to the face of the motor, the magnetized impeller rides in the hollow area of the drive magnet. The impeller is hydraulically separated from the magnet by the plastic body in which it is enclosed. The impeller/magnet is pulled along by the rotating motor magnet, and needs no seal because of the motor's isolation from the material being pumped.

Since the motor shaft and the impeller are not directly connected, magnetically driven pumps can handle only limited specific gravities and pressures without the possibility of *decoupling*. Because of their simplicity they can be very inexpensive in small sizes, but quite expensive when large enough to need special plastics, high-powered magnets, or reinforced casings for rigidity. In some cases, special "soft start" motors are needed to keep the two magnets from decoupling during start-up.

8.3.6   Multistage Pumps

Multistage pumps are no more than standard end-suction centrifugal pumps constructed so that the first impeller to see the liquid has its discharge channeled to the inlet of a second impeller directly behind the first, which discharges into the inlet of a third, and so forth through the multiple stages of that model. Because of a centrifugal pump's ability to boost incoming pressures, very high total pressure can be achieved if the stack of impellers is long enough. The pump cannot, however, provide more flow rate than any individual impeller can pass.

This design is used most often in pumps to be installed in deep wells with narrow casings, and for standard horizontal pumps meant for conditions of high pressure and low flow rate, such as reverse osmosis. These pumps carry all the advantages of any centrifugal pump, except perhaps the ease of repair and variety of materials of construction. They can be supplied close-coupled or pedestal-mounted. They tend to be expensive in large sizes, but competitive with other types of pumps in fractional horsepower or just above fractional. Because a long column is needed to hold and align the stack of impellers, they can take up a lot of space, but they represent one of the few ways to develop conditions of high pressure and low flow rate in a centrifugal pump at standard motor speeds.

8.3.7   Regenerative Turbine Pumps

Regenerative turbine pumps are a modification of the standard centrifugal design, and represent another way to use centrifugal force to generate high pressure. They have a flat impeller with small vanes cut into both sides of its outer edge. Rather than fling the liquid out as a standard centrifugal does, each vane slaps the fluid, which is moved around, jumping from side to side as it moves progressively around the periphery until it is peeled off by a piece called a *splitter* next to the discharge port. The impeller is snugly enclosed by *liners*, which direct the flow around the casing to the outlet.

Because the liquid receives additional energy from successive impulses from the vanes, each boosting the already existing pressure, regenerative turbine pumps can develop higher pressures than regular centrifugal pumps of the same impeller diameter and number of stages. They are very popular as boiler feed pumps, not only because they develop the typical required flow rate and pressure conditions easily, but also because they have inherently low NPSH, and are able to handle liquids with high vapor or gas content, such as very hot water.

Regenerative turbine pumps are not as low-shear as is a standard centrifugal pump of comparable size, they cannot tolerate abrasives at all due to their close tolerances, and they need the same kind of minimum speeds demanded by other centrifugal pumps. They differ hydraulically from centrifugal pumps in that horsepower is directly proportional to the pressure and not to the flow rate (in this respect they are like positive-displacement pumps) but, unlike positive-displacement pumps, they have a maximum shutoff pressure and a maximum horsepower.

8.3.8   Sump Pumps

Sump or *submersible pumps* (Figure 8.7) are usually standard end-suction centrifugal pumps, but they are mounted vertically with the motor sealed for submergence, and have legs supporting the unit off the floor to allow the liquid to enter through the bottom inlet. They often come with built-in level sensors to activate the pump at a preset maximum liquid level, then turn it off when a lower level has been reached.

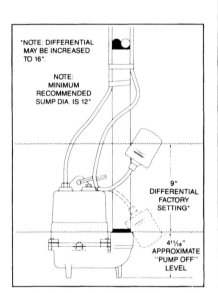 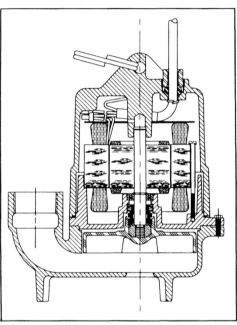

Figure 8.7 Sump Pump. The design is the same as that of a centrifugal pump, except that the pump is mounted vertically, with a sealed motor for submergence. This example has a built-in level sensor to activate the pump. (Courtesy of Sta-Rite Industries, Inc.)

Several designs within this general category exist, including totally immersed versions, column- or pedestal-mounted units with the motor on top of a column, and even those with a column and a flanged volute to allow the pump to be flange-mounted outside the tank. Models are available for handling solids, including vortex types where the impeller is recessed in an area high up in the body, pumping by the vortex or whirlpool they create.

Submerged pumps are subject to motor burnout, if the seal that protects the motor windings from the liquid should fail and the liquid go directly into the motor windings.

In many cases they are available with double seal arrangements, to reduce the probability of this occurrence.

Many grades of pumps are available, from inexpensive intermittent-duty types meant for flooded basement emergency duty, to more expensive continuous-duty versions able to run for long periods under difficult conditions or even to run dry if the level sensor should fail to turn them off. They are most commonly built with impellers that develop high flow rate and low head for the typical sump application, and come in very few materials other than cast iron. Submersible pumps are not suitable for corrosive liquids because the loss of integrity of either the seals or the cable materials could lead to motor failure.

8.4   Positive-Displacement Pumps

8.4.1   General Design

In a *positive-displacement* (*PD*) *pump*, a fixed (positive) amount of liquid is pumped (displaced) every revolution, stroke, or pulse. A piston pump, for example, has a wetted area equal to the available cylinder size, and the flow rate is simply the total volume displaced in one minute. The methods used to move the liquid are divided roughly into categories (see Figure 8.8), but new variations or combinations still arise. Positive-displacement pumps have advantages and disadvantages as a group, independent of their individual designs (Table 8.4.)

8.4.2   Sliding-Vane Pumps

Sliding-vane pumps (Figure 8.9) depend on a piston-like action developed as an eccentrically mounted rotor, with slots holding sliding vanes, is rotated in a housing. The suction port is at the part of the housing where the clearance between vanes and housing is tightest. As the rotor turns away from this area, the individual cells formed by the vanes get larger, create vacuum, and fill with fluid. These newly filled cells continue around the inside of the body until they approach the discharge port. Once again the vanes and the housing are in close proximity. Here the liquid trapped in the cells is squeezed out the discharge port. The vanes progress around to the suction port to create vacuum again.

The vanes provide separation; as they slide in their respective slots they keep contact with the inside of the housing, sealing the cavities and allowing creation of pressure or vacuum. The vanes are the design's biggest advantage, as they provide a continuously adjusting seal despite inevitable wear. However, the vanes can also be the design's biggest disadvantage, since they generate sliding friction, and therefore cause wear on the inside of the body and against the inside of the flat end cover.

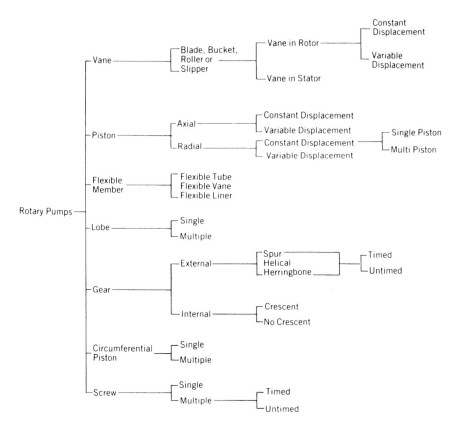

Figure 8.8 Types of Positive-Displacement Pumps. (Courtesy of Hydraulic Institute.)

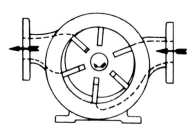

Figure 8.9 Sliding-Vane Pump. An eccentrically mounted rotor creates a piston-like action inside this pump. As the rotor turns away from the suction, the cells created by the vanes get larger and create vacuum, thus transporting fluid from inlet to outlet.

Table 8.4 Advantages and Disadvantages of Positive-Displacement Pumps

Advantages	
(a)	Constant Volume: The flow rate from a PD pump is relatively constant against any head (pressure) the pump is physically able to hold.
(b)	Low Flow Rate, High Pressure: A PD pump delivers low flow rate and high head more economically than a centrifugal pump.
(c)	Good Performance with Viscous Fluids: Viscosity is more easily compensated for with a PD pump. In fact, at times viscous material can seal the close tolerances usually present, making a high viscosity advantageous.
(d)	Ability to Self-prime: A PD pump is self-priming due to the positive nature of the action. As liquid is forced through the housing and out the discharge, a vacuum is formed. When the evacuated cavity moves around to the inlet, the vacuum acts to lift the material. This is explained in more detail in Section 8.7.
(e)	Sanitary Design: Can be of sanitary design.
Disadvantages	
(a)	Overpressure: If flow is restricted sufficiently, a PD pump continues to build pressure until something intervenes. Usually an internal or external pressure relief valve or other pressure-sensing device is required.
(b)	No Maximum Horsepower: Because pressure is directly related to horsepower (in contrast, in centrifugal pumps it is related to flow rate) and because pressure can continue to rise until pump failure, no maximum horsepower can be stated.
(c)	Pulsating Flow: All these pumps have pulses in the discharge flow, related to rpm. Pulsing with gear pumps is minor, because of the large number of cavities opening and closing; pulsing with diaphragm pumps can be major, because there are only one or two pumping chambers moving.
(d)	Low Flow Rate: Speeds are limited because of the physical actions being performed at each revolution; therefore, large flow rates require large pumps.

Sliding-vane pumps are similar to flexible-impeller pumps (discussed in section 8.4.4), but have the advantage of more rigid vanes, and thus, capacity for more pressure, and the disadvantage of more expensive parts. Obviously abrasives accelerate the wear greatly, so sliding-vane pumps are not commonly used with abrasive products. However, the vanes are replaceable and, in some models, the inside of the body can be rebuilt with a slide-in liner. Sliding-vane pumps handle high viscosity well, are efficient self-priming pumps, and are quiet and reliable if given proper liquids to pump. They develop medium pressures and need only one seal or packing box, as they have only one driving shaft. Rugged, they are popular as large transfer pumps because of their ability to keep pumping at constant flow rate over time, compensating for wear by movement of the vanes. Small, compact versions are available for common usages.

8.4.3   Piston Pumps

Piston pumps can develop more pressure than any other pump, and therefore are popular in washing down, injection, and spraying applications. They are also one of several designs for metering pumps. A vacuum develops when the piston, sealed with cups or rings, pulls back within its cylinder housing, opening inlet check valves to the suction line and drawing liquid into the cylinder. Fluid fills the cylinder progressively as the piston continues back, then is pushed out when the stroke moves forward, opening the check valve to the discharge port. Simplex, duplex, and triplex assemblies are common: pulsation is less of a problem as the number of pistons increases. Pulsation dampeners can smooth out the flow considerably.

Because a piston pump has no internal passages for recirculation as pressure builds, and because maximum pressures are limited only by the breaking point of the weakest link in the entire system, pressure relief valves are a necessity. Pressures of over 10,000 psi are possible with the largest pumps, with flow rates of up to 1500 gpm.

It is essential that the liquid pumped by any piston or plunger pump (the difference between the two is the method of sealing the wetted parts) be clean, to prevent wear of parts and scoring of cylinder walls. Reducing the speed of piston travel permits these pumps to be used for high-viscosity fluids to a certain extent, but in general a piston pump is not recommended for viscous liquids or solids handling. They tend to be fairly complicated, especially in the case of multiple pistons. Therefore, they need maintenance, and can be difficult to repair.

8.4.4   Flexible-Impeller Pumps

A *flexible-impeller pump* (Figure 8.10) depends on the same principle as does a sliding-vane pump, except that the sliding vanes are replaced by flexible rubber blades molded onto the impeller hub. The impeller rotates in the center of a housing with a crescent-shaped cam at the top. As the blades strike the cam they are squeezed flat between the cam and the impeller hub, pushing the liquid out the discharge port. As rotation continues, the vanes leave the cam and open up again, creating vacuum to as much as 22" Hg (or 25 ft of water).

Figure 8.10 Flexible-Impeller Pump. This pump depends on the same principle as the sliding-vane pump, except that the vanes are flexible instead of sliding. (Courtesy of Hydraulic Institute.)

Bodies come in standard metals, including stainless steel, and several plastics as well. Impeller materials are limited because the blades must have a memory for shape, but buna, neoprene and Viton®, as a minimum, are available for most models. These pumps are available in sanitary designs, and are conveniently serviced with relatively inexpensive, easily exchanged impellers. Running at slow speeds (although most can take as much as 1800 rpm), they handle viscosity comfortably, abrasives to a limited extent (suffering wear, but able to compensate with the flexibility of the impeller), and small solids. Pressures are limited by impeller elastomers. Above 60 psi, extreme flexing can cause individual blades to snap off. Heat can also be a problem for the types of rubbers used. A temperature of 180 °F is considered a maximum if swelling, and therefore binding, is to be avoided.

Dry running is the biggest problem, with perhaps 30 seconds without liquid as a maximum. However, new impeller materials containing impregnated lubricants may greatly improve the ability of these pumps to withstand dry running.

They are usually small, and therefore portable and very versatile. They are also popular as utility pumps and are suitable as transfer pumps if monitored for dry running. However, they are often limited in practice to intermittent duty, due to their relatively short impeller life.

8.4.5 Lobe Pumps

Lobe pumps (Figure 8.11) act very much like external gear pumps (Section 8.4.6). The impellers rotate counter to each other, creating vacuum at the pump inlet where the lobes separate and create an opening. The liquid drawn into these open spaces is carried around the outside of the impellers, then squeezed out when the lobes come together at the discharge port.

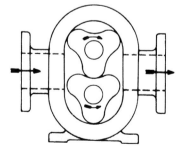

Figure 8.11 Lobe Pump. These pumps are often confused with circumferential-piston pumps. However, here two lobe-shaped impellers rotate in opposition and create vacuum at the pump inlet as they separate. The liquid captured in these open spaces is carried to the discharge.

Like circumferential-piston pumps (Section 8.4.7), lobe pumps are driven by timing gears, with their impellers set to narrowly clear one another: these two types therefore have a longer life than do other PD pumps, for which friction provides much of the motive force. The close tolerances mean they are able to deliver only medium pressure (200–250 psi), and they are unlikely to gall if run dry. They also resemble circumferential-piston pumps in having some pulsation in their flow, although they can be equipped with a larger number of lobes to minimize this problem. They cause very low shear, and are sensitive to abrasives because of the large area between the face of the impellers and the inside of the flat end cover.

Materials of construction are limited, but the rotors can be covered in various elastomers, which allows re-covering of worn impellers rather than complete replacement. A lobe pump needs only two seals because the shafts do not extend beyond the wetted area. Various seal materials and designs are available, including double seal boxes for greater protection from leakage or for protection of the liquid from contamination from the atmosphere. The seal box can contain a compatible liquid to act as a barrier (see Section 8.8.1). This feature, the smooth nature of the internal wetted components, their inherently low shear, and their ability to handle high viscosity makes them particularly well suited for sanitary applications in many fields.

Again like circumferential-piston pumps, they are not inexpensive, but their unique features and long life often easily justify their use in difficult applications.

8.4.6   Gear Pumps

Gear pumps are one of the oldest, and still one of the most popular, types of PD pumps. They are available in two basic designs that share many physically characteristic advantages and disadvantages.

An *external gear pump* (Figure 8.12A) has two gears, with external teeth that mesh. One gear is driven by the motor, and the second one is driven by the first. As the individual teeth separate on the inlet side, a vacuum is created that assists the entry of material into the body, and the liquid is carried around the outside of the gears (they turn in opposite directions). The liquid is squeezed out at the discharge as the teeth mesh once again at their meeting point.

An *internal gear pump* (Figure 8.12B) also uses a gear with external teeth, but mounted a little off center, to mesh with the internal teeth of a larger planetary gear. The larger outer gear is driven by the motor, and the spaces opened and closed due to the eccentric location of the smaller *idler* gear convey the liquid from the suction port to the discharge. A crescent-shaped piece guides the liquid smoothly around the gaps between the gears.

In general, gear pumps are excellent with viscous material because of their efficiency at the slow speeds necessary to move thick liquids. In addition, they deliver comparatively smooth flow with thin material, although like many PD pumps they work more efficiently above a threshold viscosity. They are relatively compact and lightweight, don't need check valves to prevent siphoning or reverse flow, and can pump liquids containing gases or vapors. However, because they have gears meshing

(A)   **(B)**

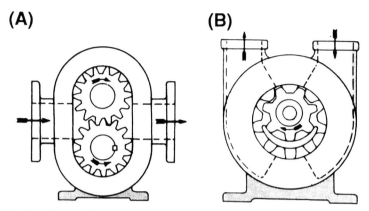

Figure 8.12 Gear Pumps. These are a type of positive-displacement pump. (A) External gear pump, having two gears mounted side by side whose teeth mesh and squeeze liquid from the inlet to the outlet. (B) Internal gear pump, having two gears mounted off center and whose eccentric motion carries liquid from inlet to outlet. (Courtesy of Hydraulic Institute.)

with one another, they do not run well dry. They are often available with built-in pressure relief valves for occasions when flow could be restricted.

The two different types of gear arrangements result in some differences in performance. An external gear pump can run at faster speeds, delivers a higher flow rate from a given model size, and is thus less expensive. Gears come in various styles (spur, helical, herringbone), but the differences affect noise and life of the gears more than general capability. They are available in iron, bronze, stainless steel, Hastelloy®, and some plastics. They can get quite large (in the range of 5000 gpm at as much as 500 psi), and can be very noisy. As in most positive-displacement pumps, because of tight clearances, the alignment of drive shaft to pump shaft is critical when the unit is base-mounted.

Because internal gear pumps must normally run at slower speeds, and because the two gears turn in the same direction, they are quieter, wear less, and are gentler with shear-sensitive liquids. They are, however, physically larger for an equivalent capacity (a maximum of about 1000 gpm at 100 psi) and more expensive. Like external gear pumps, they need clean liquids, but wear more slowly if subjected to abrasives. They are available as standard in cast iron and stainless steel.

8.4.7   Circumferential-Piston Pumps

Circumferential-piston pumps (Figure 8.13) are named for their circular method of providing a piston-like stroke. Sometimes confused with lobe pumps (Section 8.4.5),

Figure 8.13 Circumferential-Piston Pump. Two hemispherical rotors spaced 180° apart turn in opposition to each other and push liquid out the discharge. (Courtesy of Hydraulic Institute.)

these pumps have two (or more) hemispherical rotors spaced 180° apart, which turn in opposition to one another and push the liquid out the discharge port. As each impeller rotates, the trailing edge passes the suction port and creates vacuum, and therefore suction. The rotors are mounted on shafts turned by timing gears in a housing separate from the wetted end of the pump.

This design means the impellers never actually touch one another, which keeps wear to a minimum, and makes this one of the most rugged positive-displacement pumps. The fact that internal parts don't rub one another, however, also limits them to only moderate pressure (150–200 psi). They are excellent with viscous liquids. Although most solids cannot be tolerated because of the close tolerances (in many areas 0.002"), soft compressible solids can usually be handled.

They are available in a limited range of materials, usually iron or stainless steel, in both industrial and sanitary designs. Circumferential-piston pumps have two drive shafts rather than one, because of the timing gears, and therefore need at least two seals or packing arrangements. Since most are built for difficult applications, the shafts are often supported at both ends and need four seals or packing boxes. This fact, plus the general design criteria, mean that circumferential-piston pumps can be expensive. The expense can often be justified by savings in maintenance, downtime, or the less frequent replacement costs.

8.4.8   Progressing-Cavity Pumps

Progressing-cavity pumps (Figure 8.14) utilize a single helical metal rotor rotating inside an elastomeric double helical stator. The cavities formed by this combination progress down the length of the pump as the rotor turns. The action resembles that of a screw pump, with the motion of the rotor inside the stator pushing the liquid ahead toward the discharge. The rotors are available in a few different metals (normally stainless steel), but stators can be found in up to 20 different elastomers.

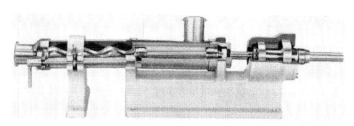

Figure 8.14 Moyno®
Progressing-Cavity
Sanitary Pump. A
single helical metal
rotor turning inside
an elastomeric stator
propels liquid from
inlet to outlet.
(Courtesy of Robbins
& Myers, Inc.)

This design is extremely gentle on the material being pumped, notably on liquids containing suspended solids. If abrasives are present, turning the pump slowly keeps the damage to a minimum, so that despite inevitable erosion, progressing-cavity pumps are sometimes considered acceptable for abrasive duty. Pressure is developed by the tight fit between the rotor and stator. If total slip between these parts with long assemblies is carefully limited, high pressures can be attained. The tight fit means wear on the main components, and means that starting torque has to be considered when a motor/pump package is specified. These pumps can be unwieldy because of their length. Repair parts needed are often so expensive that an industry purveying look-alike but cheaper parts is in existence. These pumps are quiet and lend themselves easily to sanitary design. When gentle handling of sensitive products is essential, progressing-cavity pumps should be included in any comparison.

8.4.9 Peristaltic Pumps

Peristaltic pumps, named after the similar peristaltic action of our intestinal tracts, depend on progressive squeezing of flexible tubing by rollers mounted on a rotating plate. The liquid is pushed towards the discharge as the roller travels its circular path. At the outlet port the roller lifts away from the tubing, and resumes contact at the inlet port. The spring action of the tubing as the roller leaves it creates vacuum to pull in more of the fluid. The major advantage of such a principle is that nothing but the tubing itself is ever in contact with the liquid, and since tubing material selections are varied, a peristaltic pump can handle a great number of fluids with little possibility of contamination. Typical tubing materials are silicone, Tygon®, buna, and Viton®. A candidate material's only limitation is its ability to be easily compressed and then spring back again.

A peristaltic pump can run dry indefinitely, can handle abrasives with little problem, can pass solids within the limits of the tubing's inside diameter, and primes itself very well when tubing is selected in a wall thickness that is sealed completely by the rollers. The liquid end is about as simple a design as possible, needing no seals or check valves, and therefore repair is limited to tubing replacement or repositioning (the fatigued section is moved to the discharge side to prevent its collapse under suction).

The tubing that makes this design unique also provides its disadvantages. Pressure is very limited, usually 20 psi, with a maximum of 40 psi for certain ratios of wall thickness to inside diameter. Although repair is uncomplicated, it is needed often. Flexible tubing is likely to fatigue quickly; fifty hours to several hundred hours of operation can be obtained, depending on the material, the speed of the rollers, and the temperature. Viscous liquids can be handled to a limited extent, but the tubing can balloon as resistance to flow increases. The tubing has a tendency to collapse on the suction side unless the liquid is easily fed to the pump. Flow rates are limited by tubing size and the very slow speeds necessary to give the tubing time to restore itself. Because of their obvious suitability for laboratory duty, small peristaltic pumps are available with many sophisticated controls, and even banks of pump heads operating off the same drive motor.

8.4.10 Diaphragm Pumps

Diaphragm pumps, especially air-driven diaphragm pumps like the one pictured in Figure 8.15, are enjoying a surge of popularity. They come as close to foolproof as anything available, are suitable for a wide range of conditions, and are therefore quite adaptable.

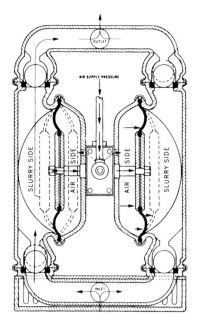

Figure 8.15 Air-Driven Double Diaphragm Pump. A cross-sectional view illustrates how fluid entering the inlet is pulled into one of two cavities when the diaphragm moves forward and backward by air pressure on its nonwetted side. Ball valves prevent backflow as the diaphragm moves. (Courtesy of Wilden Pump Co.)

The general principle of operation depends on a flexible diaphragm, more or less flat, that moves back and forth in a gasketed housing. The inlet and outlet ports contain check valves, either ball-valve- or flapper-type. The inlet valve allows liquid to enter when the diaphragm pulls back and creates vacuum. When the diaphragm moves forward again, the housing develops pressure and the inlet valve closes. The outlet valve operates in opposition, closing when the vacuum pulls it shut, opening when the power stroke pushes it open, and channeling flow out the discharge side. The ball-valve-type check valves are a little more reliable, and usually less expensive, but pass only small solids. The flapper type usually allows any solid that can get through the suction port to pass through to the discharge. Either check valve may not seal completely if particles do not clear the openings, and therefore may allow flow in the wrong direction.

The overriding advantages of the diaphragm pump are the lack of rubbing parts and the loose clearances. They can handle abrasives and slurries, as well as viscous fluids and solids, if properly equipped. Modified diaphragm pumps can develop very high pressure, approaching 1000 psi.

Air-operated double diaphragm pumps are unique in that two separate wetted chambers exist, manifolded by common inlet and outlet ports. The two diaphragms are connected by a shaft inside the air motor, so that when one is on the pressure stroke pushing forward, the other is correspondingly pulling back to create vacuum. Power is developed as the area behind a diaphragm fills with air (or hydraulic fluid in higher pressure versions), the volume of air available determining how quickly the cavity fills and a new power stroke is begun. Air pressure from the compressor determines how much force the diaphragm can develop during the forward movement. Normal installation includes an air lubricator, a filter, and sometimes a dryer to guarantee a supply of clean, dry, lubricated air. A regulator is also necessary to adjust the air feed rate, and therefore the stroke rate and the flow rate.

The advantages of double diaphragm pumps are the more balanced action, their ability to stall if run against too great a head pressure (or starved on the inlet side), their dry running capability, the lighter loads on the diaphragms themselves (the air or hydraulic fluid, rather than a mechanical contrivance, provides the force that causes the diaphragm to move), and the inherent variable flow rate capability. Disadvantages are the need for very clean, dry, usually lubricated compressed air to prevent the air valve assembly from freezing or locking up, the fundamental inefficiency of a typical compressor/pump system, and the higher maintenance and repair such pumps usually demand in practice.

Metering pumps not using the piston design usually use the diaphragm principle: they are driven by electric motors mounted to gear boxes to provide the typical slow speeds needed by any diaphragm pump. Diaphragm pumps make ideal metering pumps, because of their relatively gentle action and ability to deal with viscosity and abrasives, as well as their limited number of components in contact with the fluid (which means they present few difficulties of chemical compatibility).

Diaphragm pumps are commonly available with bodies of several plastic materials, stainless steel, aluminum, or cast iron, and with diaphragms of buna, neoprene, Viton®, or chemically resistant Teflon®. These pumps are frequently used in biotechnology processes as sanitary transfer pumps.

8.4.11 Summary

Table 8.5 summarizes and compares the features of the various types of PD pump. As the chart shows, it is rare that an application dictates a specific pump among the variety of PD pumps available, but it helps narrow down the choices. A certain amount of the decision hinges on recommendations from trusted vendors, due to the huge number of manufacturers within each category. Obviously, what choice is made may differ even when flow rate and pressure conditions are identical, due to differing priorities of capital expense, duty cycle, expected life, and availability of qualified service.

8.5   Comparative Evaluation of Centrifugal and Positive-Displacement Pumps

For the decision between a centrifugal and a PD pump, there are some general statements that can be made, even though most of these can be overruled in specific circumstances. All else being equal, centrifugal pumps are the first choice. They are simpler, quieter, less expensive, and easier to repair; they have fewer wearing parts, do not surge or pulse, and are available from more sources in more different styles and materials of construction. The conditions usually overruling these advantages are:

(a)   Viscosity. A rule of thumb is that 2000 SSU (440 cP for a fluid with specific gravity of 1.0) is a maximum for centrifugal pumps. Consulting a viscosity correction chart for each set of conditions is recommended. A decision is often made based on the efficiency sacrifice, and, therefore, the extra horsepower needed to stay with a centrifugal pump.

(b)   Ratio of pressure to flow rate. A centrifugal pump is better suited for medium to high flow rates in relation to the pressure. For example, a centrifugal impeller is a bad choice to deliver 1 gpm at 1000 psi, because of the way a centrifugal pump builds pressure; any impeller big enough to deliver 1000 psi is too large to be efficient at low flow rates like 1 gpm. In certain circumstances this can be overcome by a multistage pump, but there are limits to the number of impellers economically staged. Where a centrifugal pump seems the clear choice but the ratio of pressure to flow rate is extreme, bypassing the extra flow at the pump discharge is perfectly acceptable, although inefficient.

Table 8.5 Comparison of the Characteristics, Advantages, and Disadvantages of Various Positive-Displacement Pumps Suitable for Sanitary Service. (Courtesy of Tri-Clover, Inc.)

	Price vs. Gallonage	High Pressure Capability	Solids Handling	Viscosity Handling	Abrasives Slurrys	Ease of Repair	Pulsation	Shear	Dry Running Ability	Sanitary Availability	Other Advantages	Other Disadvantages
Circumferential Piston	D	C	C	A	B	D	B	A	A	Y	Most Rugged	Four (4) Seals Needed
Diaphragm (Air Operated)	B	C	A	A	A	B	D	A	A	Y	Versatile Explosion Proof	Inefficient
Flexible Impeller	A	D	B	B	C	A	A	B	D	Y	Simple Design	Limited Impeller Life
External Gear	A	A	D	A	D	A	A	D	D	N	High Speeds O.K.	Noisy
Internal Gear	A	B	D	A	D	B	A	C	C	N	Quieter than External Gear	Metal to Metal Contact
Lobe	D	C	C	A	D	B	B	A	A	Y	No Metal to Metal Contact	Needs Two (2) Seals
Peristalic	B	D	A	C	A	A	B	A	A	Y	Only Tubing Touches Fluid	Limited Tubing Life
Piston	B	A	D	C	D	D	D	C	D	N	Very High Pressures	Complicated High Maintenance
Progressing Cavity	D	B	A	A	B	B	A	A	D	Y	Most Gentle Type	Expensive Wearing Parts
Sliding Vane	C	B	C	B	D	B	A	B	B	N	Self-Adjusting For Wear	Vanes Slide Therefore Wear

A = Best ... D = Worst
Y = Yes, N = No

(c) Shear-sensitive material. Because centrifugal pumps need a fairly high minimum speed to deliver pressure, they may be unsuitable for foaming or fragile fluids. Few numerical values can be cited to help with evaluation; experience is the greatest source of information, but in some cases pump manufacturers, or the supplier of the liquid, may be able to help. For instance, a noncavitating centrifugal pump at a relatively slow speed of 1200 rpm is suitable in many applications and is commonly used because of its other advantages; however, since many of the typical applications of this type don't need a large flow rate, PD pumps should be considered.

(d) Solids and abrasives. Centrifugal pumps can handle solids if an open or recessed impeller is used. They can handle abrasives if special seal materials such as silicon carbide are used and wear on the impeller can be tolerated. However, neither case is ideal for a centrifugal pump. Only when their other qualities, such as high flow rate at low cost per gallon, outweigh the expense, inefficiency, or wear, should they be considered.

(e) Constant flow rate. A centrifugal pump is sensitive to back pressure. If a specific, accurate, or repeatable flow rate is needed despite changes in resistance, and controls such as flow regulators are not feasible, a PD pump is indicated. A positive-displacement pump tends to deliver the flow rate dictated by its speed, regardless of changes in back pressure (see Section 8.6). There is slip, but it is minor compared to that shown by centrifugal pumps.

If a centrifugal pump has not been eliminated from selection by any of the above, many choices are left. There are huge numbers of centrifugal pumps available, some with esoteric specialty applications, but nonetheless excellent pumps. Because centrifugal pumps are relatively simple, and design features are rarely sufficient to force a choice between competitors, factors such as efficiency, availability of parts, warranty periods (typically one year, parts and labor, FOB the factory), and the quality of local representation should be major considerations, in addition to the obvious factors of price and availability.

Although PD pumps are common in biotechnology, they are usually considered only after centrifugal pumps have been eliminated. The choices are more difficult because of the variety of general hydraulic principles to consider, and complicated by the wide range of special features and inherent limitations of various positive-displacement pumps. Only rarely are the requirements for a positive-displacement pump so particular that a specific generic type is an obvious choice, but there are few manufacturers of each so the choice narrows down quickly once a type is selected. As with centrifugal pumps, factors of warranty, parts availability and especially quality of local representation are significant, but general suitability of the design probably takes first precedence.

Despite the variety of centrifugal and PD pumps available, an informed selection should be possible if the engineer has some knowledge of hydraulic theory and of PD and centrifugal principles, and a thorough understanding of the application.

8.6   Pump Curves and Charts

8.6.1   Pump Curves

Once a set of conditions has been established, and a general type of pump selected, a way to translate these parameters into the choice of a specific pump is needed. If the pump application has indicated a centrifugal pump, then a curve is used as a universal means of representing the pump data. There are several components to centrifugal pump curves, and, in some cases, multiple ways to illustrate them.

The *H-Q curve* (so called from the usual symbols for head and flow rate) is the first one inspected, as it directly indicates the ability of the pump to meet the established conditions. The x axis indicates flow rate, Q, and the y axis indicates pressure, H. The manufacturer establishes curves for a variety of pressure–flow rate conditions for each representative pump by direct testing. Measurement of the *shutoff* (no flow) pressure gives a point on the y axis; the maximum flow rate is found if the pump is run until a back pressure is reached below which no further flow is produced. To find the expected flow rate for any pump from the curve, locate the expected back pressure on the y axis, go horizontally until the curve is intersected, then straight down to the flow rate units on the x axis.

A particular flow rate can result from one and only one back pressure, although the reverse is not necessarily true: some curves are domed, the pressure first increasing and then decreasing as the flow rate increases, so pressures near the shutoff point can result in two different flow rates. In Section 8.6.2 I show why a pump rarely seeks between two choices on a curve, but it is physically possible. The shape of the *H-Q* curve can be steep (flow rate changes relatively little with pressure changes) or flat (large fluctuations in flow rate for small pressure changes). The shape is determined by impeller vane shape, number of vanes, and other impeller and casing features. Impellers of a standard radial design can be machined to smaller diameters to provide curves parallel to the first but capable of delivering lower pressures.

Other data present on a typical curve are horsepower, NPSH$_r$, and sometimes efficiency. These are all read in the same general fashion as the *H-Q* curves: from the known operating intercept, move vertically or horizontally to the curve in question, then directly to the scale with the desired units. Horsepower can be shown by steep lines crossing the *H-Q* curve, or on a curve of its own. Efficiency may be shown by a series of lines crossing the *H-Q* curve, or also on a curve of its own. The NPSH$_r$ curve can be at the extreme bottom or top of the grid area, its units in a vertical column on the right or left side. Even if several impeller diameters are shown, there is normally only one NPSH$_r$ curve (impeller diameter affects only discharge conditions, and NPSH is a suction-side condition). In practice, the ideal pump has its *best efficiency point* (*BEP*) near the center, on a part of the curve that has some slope. Potential dangers of operating a pump at either extreme end are cavitation or lack of flow control, but operation there is not unusual in real life if other safeguards are in place.

I include two example curves to illustrate the use of pump curves. The curve in Figure 8.16 has five impeller diameters charted, each indicated on the left; the efficiency is included directly on the curves, and horsepower is shown by dashed lines. This is sometimes called an *oak tree curve*. Values of horsepower between dotted lines can be calculated with the standard horsepower formula (eq 8.5), using the efficiencies indicated. Conditions not met by any indicated impeller can still be handled by a trimmed impeller. For example, if 300 gpm at 30 ft total head were desired, the 6½" impeller could be trimmed to 6¼" to keep the horsepower at a minimum. If the impeller were trimmed, 3 hp would be needed, as opposed to 5 hp (the next available size) if the 6½" impeller were used.

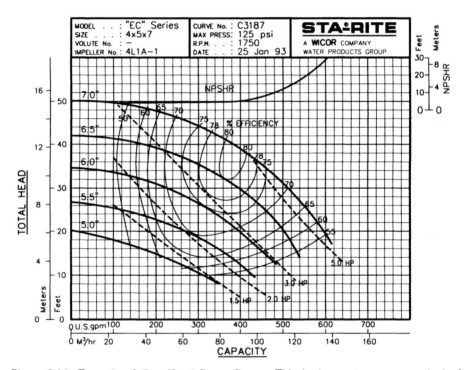

Figure 8.16 Example of Centrifugal Pump Curve. This is the most common method of depicting centrifugal pump data. The curve shows impeller sizes, efficiency, horsepower, and NPSH as functions of head and capacity. (Courtesy of Sta-Rite Industries, Inc.)

The curve shown in Figure 8.17 shows efficiency as a separate curve, in addition to the typical *H-Q* curve itself. Values of horsepower are shown on their own curves as well.

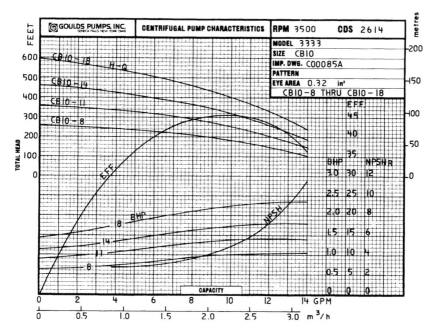

Figure 8.17 Alternative Style of Centrifugal Pump Curve. This diagram depicts horsepower and efficiency on separate curves. (Courtesy of Goulds Pumps, Inc.)

If conditions point to the choice of a positive-displacement pump, charts equivalent to centrifugal pump curves can be consulted. No universal chart has ever been agreed upon, so styles range widely. Since PD pumps rely on rpm to establish flow, the extra variable of speed must be incorporated into the charts. One way of dealing with this is to provide separate charts for a variety of speeds, with the applicable gpm, psi, and hp capabilities; values for speeds between those selected must be interpolated. This is feasible because for PD pumps the only variables in relation to speed are flow rate and horsepower (usually in direct proportion to rpm). Pressure capability is unaffected by speed. Some manufacturers have become quite creative in chart design, incorporating the standard gpm, rpm, psi, and hp all onto a single grid. However, since PD pumps are often used with viscous liquids, several charts still need to be supplied per model if viscosity is to be taken into account.

As with the curves for centrifugal pumps, PD charts visually reinforce the principles discussed earlier. The principles are: (a) horsepower increases directly with pressure; (b) pressure capability stays relatively constant with speed changes; (c) efficiency often increases with viscosity; and (d) the main effect of viscosity is to dictate slower speeds, so that fluid has time to enter the rotating cavities.

A typical gear pump chart is shown in Figure 8.18. The viscosity adjustments are handled by a separate factory-supplied correction factor.

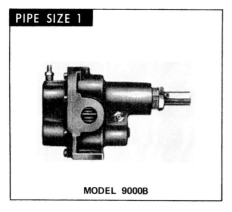

| PIPE SIZE 1 |
| MODEL 9000B |

CAPACITY — Water 60°F

R.P.M.	FT. HD. P.S.I.	0 0	46 20	92 40	138 60	184 80	231 100	290 125*	346 150*
400	G.P.M.	5.00	4.49	3.99	3.48	2.98	2.48	2.05	1.60
	H.P.	.25	.33	.40	.50	.65	.75	1.00	1.25
	MOTOR	1/4	1/3	1/2	1/2	3/4	3/4	1	1 1/2
600	G.P.M.	7.50	7.08	6.65	6.23	5.80	5.38	4.90	4.05
	H.P.	.35	.40	.55	.75	.90	1.10	1.32	1.60
	MOTOR	1/3	1/2	1/2	3/4	1	1	1 1/2	2
800	G.P.M.	10.17	9.82	9.47	9.12	8.77	8.41	7.80	7.30
	H.P.	.40	.60	.70	1.00	1.15	1.40	1.75	2.05
	MOTOR	1/2	3/4	3/4	1	1 1/2	1 1/2	2	2
1000	G.P.M.	13.00	12.65	12.30	11.94	11.58	11.22	10.40	9.90
	H.P.	.50	.70	.90	1.20	1.45	1.72	2.10	2.50
	MOTOR	1/2	3/4	1	1 1/2	1 1/2	2	2	3
1200	G.P.M.	16.00	15.67	15.34	15.00	14.67	14.33	13.60	13.00
	H.P.	.60	.80	1.14	1.45	1.85	2.20	2.70	3.20
	MOTOR	3/4	3/4	1	1 1/2	2	3	3	5
1600	G.P.M.	21.50	21.12	20.74	20.35	19.97	19.58	18.70	18.00
	H.P.	.80	1.07	1.38	1.70	2.03	2.41	2.90	3.40
	MOTOR	1	1	1 1/2	2	2	3	3	5
1725	G.P.M.	23.33	22.93	22.52	22.11	21.71	21.30	20.50	20.10
	H.P.	.90	1.19	1.53	1.92	2.25	2.70	3.15	3.70
	MOTOR	1	1 1/2	1 1/2	2	3	3	3	5

H.P.=ACTUAL HORSEPOWER
MOTOR=CONVENIENT FRACTIONAL SIZES
G.P.M.=GALLONS PER MINUTE
P.S.I.=LBS. PER SQUARE INCH PRESSURE
FT. HD.=EQUIVALENT PRESSURE IN FEET OF WATER
R.P.M.=REVOLUTIONS PER MINUTE

Figure 8.18 Typical Gear Pump Chart. Data for positive-displacement pumps are not described by standardized curves. This diagram displays data for a rotary gear pump, Oberdorfer model 9000B, with bronze bearings. (Courtesy of Oberdorfer Pump Co.)

A more complex set of curves, Figure 8.19, is shown for a circumferential-piston pump, which typically handles very high viscosities. The flow rate is on the y axis, and speed on the x axis. The power is relatively unaffected by viscosity unless the viscosity is very high; also, high flow rates of thin materials are difficult to pump at high back pressures because of slip. Figure 8.19 shows that to achieve 10 gpm at 150 psi back pressure takes 300 rpm if pumping a water-like fluid of 31 SSU (1 cP), but only 90 RPM with a fluid of 5000 SSU (1100 cP for a fluid with specific gravity of 1.0). Horsepower demands are similarly affected. NPSH$_r$ figures are given on a separate chart because most applications with this type of pump are gravity fed and NPSH$_r$ problems are rare.

8.6.2   Viscosity Adjustments

As mentioned previously, the rate at which PD pumps can be operated is affected by the viscosity of the fluid. The performance of centrifugal pumps is also affected by viscosity. Several manufacturers supply their own adjustment data, but the most widely accepted material comes from the Hydraulic Institute [1]. For example, Figure 8.3 gives correction factors based on a combination of desired flow rate, H_m, and viscosity. To use the chart, find the desired flow rate on the x axis, travel vertically

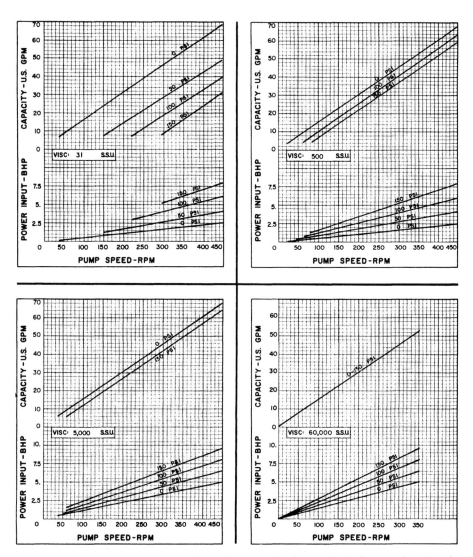

Figure 8.19 Circumferential-Piston Pump Curves. Another method of presenting data for positive-displacement pumps is with a series of performance curves. This is for Tuthill model 65 stainless steel sanitary series process pumps with standard clearance (to 150 °F), based on 15" Hg max. suction vacuum, for liquids of four viscosities. (Courtesy of Tuthill Pump Division, Chicago, IL.)

to the known H_m, horizontally to viscosity, then finally up to intersect the correction factor curves for efficiency (C_η), flow rate (C_Q), and pressure (C_H).

The correction factors represent losses in terms of a multiplier; with their use the engineer can calculate the extra performance necessary to guarantee the required conditions. For example, Figure 8.3 uses a flow rate of 53 gpm at an H_m of 50 ft, and a viscosity of 200 SSU. Following the dashes, we arrive at an efficiency correction of 0.67, a flow rate correction of 0.87, and a head correction factor of 1.00. To achieve the desired conditions, we should select a flow rate of 61 gpm (53/0.87), a head of 50 ft (50/1.00), and horsepower determined by eq 8.5:

$$(53 \times 50)/(3960 \times \{\text{efficiency} \times 0.67\})$$

The chart reinforces my point that centrifugal pumps in fact operate at medium viscosities of 2000 to 10,000 SSU (440–2200 cP), but at a heavy cost in efficiency because the cumulative losses in head and flow rate must be factored in. For this reason, they are not usually recommended for these applications. In the example used with Figure 8.3 above, the cumulative loss of the system due to viscosity is (0.67 x 0.87 x 1.00) or 0.58, approximately doubling the horsepower necessary.

8.6.3   System Curves

Several kinds of problems can be analyzed in detail and various solutions considered most easily if the system description is put into the curve form. Examples of such problems are:

   (a)   a system has grown beyond the existing pump's abilities; options being considered are reducing H_m by changes in piping, speeding up the existing pump, or installing a new pump;

   (b)   a pump is to be installed in an existing system to provide the maximum flow rate possible, staying within the pressure limits and the amperage available;

   (c)   the most economical choice has to be determined in a system needing a higher flow rate; options are to speed up the existing pump or add a second pump either in parallel or in series.

A *system curve* takes the data used in BERNOULLI's formula calculations (Section 8.2.3), for many flow rates, to depict graphically the potential of the whole system. This gives a curve rising to the right from the origin, indicating greater H_m as Q is increased. See the next section for the derivation of the data presented in Figure 8.20.

8.6.3.1 Example Derivation of System Curve

A transfer application requires taking water from a sump to a tower 60 feet above it, through 370 feet of 2½" schedule 40 PVC pipe, including 6 elbows and an inlet strainer with a ΔP of 2 psi. The data to be used are based on flow rates arbitrarily selected, but normally starting with 0 gpm and continuing either to where the curve approaches vertical, or just beyond the flow rates to be considered. In the case of 2½"

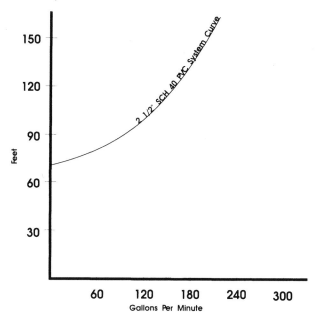

Figure 8.20 System Curve. To produce this graphical depiction of the pumping potential of a whole system, take the data for the Benoulli calculation of head, developed over a range of flows, for the system in question. (Courtesy of Tri-Clover, Inc.)

PVC, flow rate falls off the friction loss chart (Table 8.1) at 400 gpm, so flow rates from zero to 400 gpm will be considered. Data for the system are as follows:

$\Delta Z = 60$ ft for all flow rates;

$\Delta P = 0$ for all flow rates (atmospheric pressure on both ends);

$H_v = v^2/2g$, which varies with flow rate (available from Table 8.1);

H_f varies with flow rate. The strainer component of H is 2 psi, or 5 ft for all flow rates. The equivalent feet are 370-ft run, plus (6 elbows x 5.5 ft/elbow) = 33 ft (from Figure 8.1), for a total equivalent run of 370 ft + 33 ft = 403 ft.

Each component of the Bernoulli equation (eq 8.2) is computed to derive the chart:

Flow rate	ΔZ (ft)		ΔP (ft)		H_v (ft)		H_f (ft)		strainer (ft)		H_m (ft)
0 gpm:	60	+	0	+	0	+	0	+	0	=	60
150 gpm:	60	+	0	+	2	+	48	+	5	=	115
200 gpm:	60	+	0	+	3	+	82	+	5	=	150
300 gpm:	60	+	0	+	6	+	172	+	5	=	243
340 gpm:	60	+	0	+	8	+	218	+	5	=	291
400 gpm:	60	+	0	+	11	+	294	+	5	=	370

A system curve can now be plotted as flow rate versus H_m (see Figure 8.20).

8.6.4   Analysis of System Curves and Pump Curves

One of the most important uses of a system curve is that we may now overlay a pump curve and know precisely what set of conditions that pump can deliver when installed in that system. For example, Figure 8.20 can be laid over a sample pump curve (Berkeley model B2ERHH). The results, shown in Figure 8.21, indicate that the two curves intersect and that the Berkeley model B2ERHH can deliver 150 gpm at an H_m of 115 ft if installed in this system.

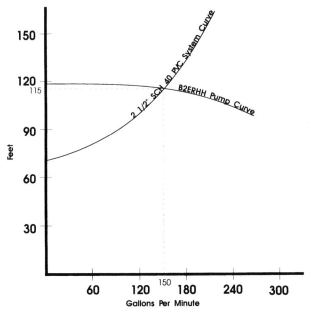

Figure 8.21 Combined System Curve and Pump Curve. To determine the conditions (head and capacity) at which a pump can deliver, overlay the system curve and the pump curve. (Courtesy of Tri-Clover, Inc.)

A new system curve can be drawn for the same system with a change of any component, and comparisons made for performance improvements with the existing pumps. Therefore, comparisons can be made of the cost of refurbishing, on the one hand, versus savings from the extra flow rate the improvements would provide, on the other. As an example, the curve has been redrawn in Figure 8.22 using 4" diameter sch 40 PVC pipe in place of the 2½" pipe. The system allows a flow rate of 275 gpm at a lower H_m of 83 ft with the same pump.

A comparison can now be made of the savings gained by an extra 125 gpm through the system versus the cost of laying the 4" pipe (also keeping in mind the extra

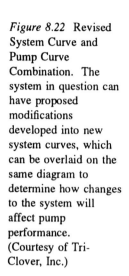

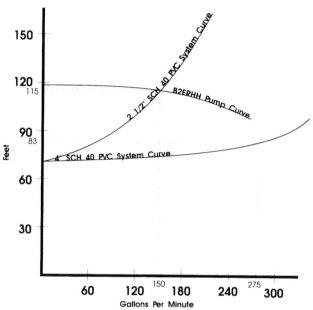

Figure 8.22 Revised System Curve and Pump Curve Combination. The system in question can have proposed modifications developed into new system curves, which can be overlaid on the same diagram to determine how changes to the system will affect pump performance. (Courtesy of Tri-Clover, Inc.)

horsepower demanded by the pump when pumping at the higher flow rate). It is also possible to compare the cost of replacing the existing pump to deliver 275 gpm without repiping (according to the system curve, that results in 222 ft of head with the 2½" pipe), or even to add another pump in series to boost the 275-gpm capacity at 83 ft of the existing pump to the 222-ft demand of the system, that is, adding another 275 gpm at 139 ft (222 ft − 83 ft) of head.

When several pumps connected in series are represented on a curve, pressures at several common flow rates from each individual curve are added, and the new cumulative heads for these flow rates are used to draw a new, higher head curve. When pumps are connected in parallel, the flow rates at several common heads are added, and the cumulative flow rates at these sample heads are used to draw a new, higher flow rate curve representing the two pumps combined. This new cumulative curve can, of course, be laid over the system curve to arrive at a performance for the multiple pumps in the application.

The analysis of system curves and pump curves can get cluttered, with various combinations of system changes, pump modifications, additions, or repiping all overlaid, but this method represents the clearest way to graphically depict the variable nature of a typical hydraulic system.

8.7   Self-Priming Pumps

Many applications require the pump to be mounted above the liquid source. If a pump is capable of operating in such a position it is called *self-priming*; this is not an inherent feature of all, or even most, pumps.

Even though pressure at normal environmental conditions is commonly referred to as 0 psi, and reads so on a standard gauge, there is positive atmospheric pressure at all points on the surface of the earth. A self-priming pump takes advantage of this pressure by creating vacuum on its inlet side and through an air-tight suction pipe down to the source, so that this outside atmospheric pressure now exceeds the pressure inside the pipe. Pressure on the surface of the liquid pushes a column of liquid into the lower pressure suction (inlet) pipe until the pressures are in equilibrium. For example, if the pressure difference between the inside of the pipe and the atmosphere is 5 psi, the column of fluid in the pipe can be $(5 \times 2.31) = 11.5$ ft high in the pipe (assuming the liquid has a specific gravity of 1.0). If the pump can develop enough vacuum so that atmospheric pressure pushes the liquid all the way to the pump inlet, the pump is self-priming for that height. Since the pressure at sea level is 14.7 psi or 34.0 ft of water above a perfect vacuum, a pump able to develop perfect vacuum could lift liquid a maximum of 34.0 feet. However, since no liquid pump comes at all close to developing a perfect vacuum, and in a dynamic situation there are losses due to friction, few pumps prime over 25 ft at sea level.

Positive-displacement pumps in all cases are self-priming within the limits of their internal tolerances. Circumferential-piston pumps, for example, are better self-primers if used with viscous fluids that help close the clearances machined into the impellers, but still prime if used with thin liquids, although to a smaller height. Peristaltic pumps are excellent primers because anything, gas or liquid, inside the tubing can be pushed out the discharge, and the same is true of diaphragm pumps. Gear pumps can prime well but, like flexible-impeller pumps, tend to be damaged if they run dry for too long while building the necessary vacuum. PD pumps with rubbing internals have the extra advantage of what amounts to a built-in check valve: on the next start-up, liquid is still in the pump suction line, so flow is available immediately.

Centrifugal pumps are not by their nature self-priming. The design of their volute and impeller has open spaces, so they can develop pressure only if the fluid to be pumped is incompressible, and therefore they won't pump gases or vapors to create vacuum. However, the need for high-flow pumps that can self-prime has prompted the development of several modifications to centrifugal pumps. If the suction line and pump casing can be filled with liquid all the way down to the source, a centrifugal pump creates vacuum by discharging the liquid in its casing.

The most direct way to modify a centrifugal pump so that it becomes self-priming is to install a check valve or a foot valve (a check valve with a strainer, meant for use on the end of a pipe), and mount this valve below the source liquid level. The suction line must be filled with fluid to prime the pump initially. Thereafter, the check valve holds liquid in the suction line all the way up through the pump. The next time the pump is started it is already primed, and begins pumping immediately. The danger is that a leak in the check valve can allow the level of the column of liquid to eventually drop below the pump housing, but as long as the pump casing itself and the complete suction line are filled, the system is primed and the pump can begin pumping. It is also possible to mount a pump (such as a hand-operated diaphragm pump) above a standard centrifugal pump, and use the auxiliary pump to pull vacuum all the way through the centrifugal pump, fill the suction pipe, and prime the system.

Another modification of a centrifugal pump makes it truly self-priming, again after the initial filling of the casing for the first start-up. A centrifugal volute can be redesigned to face the problem of eliminating air from the suction line and therefore allowing atmospheric pressure to push an intact column of liquid to the pump inlet. When the impeller (mounted in a housing that does not allow the initial priming liquid to drain back or siphon out) starts to turn, air becomes entrained. The air bubbles can be channeled around a baffle so that they enter a separate discharge area, and eventually rise and exit. Without a column of liquid in the suction line, flow cannot begin, so the initial liquid stays in the pump's housing, and continues to pick up air and funnel it to the discharge. As the air is carried through, a vacuum develops, and fluid flow starts when the suction column reaches the pump inlet. This process can take some time—in extreme cases with very high lifts in large-diameter pipes, as much as 20–25 minutes, but in most systems only 1 or 2 minutes. If a check valve or foot valve has been installed (as described above for unmodified centrifugal pumps) and if the pump is as close to the source as possible (to reduce the overall volume of the suction pipe) the start-up time is shorter. If the pump is a true self-priming centrifugal, it can prime itself even if the check valve leaks.

Self-priming centrifugal pumps tend to be a little less efficient and more expensive than standard pumps, because of the extra chamber for the air flow, but they are popular because of their reliability and the many other advantages of centrifugal pumps.

8.8   Materials of Construction and Design Features

The application often defines the materials of construction or other characteristics even if the inherent design of the pump does not. For instance, special body materials may be necessary to provide adequate chemical resistance to the liquid being pumped, particular seals or packing may be required, and regulatory boards or agencies may require certain features such as sanitary design.

A technology-based field like biotechnology, that uses many types of pump to handle a large variety of liquids, has no universally accepted materials of construction. An amazing variety of metals, plastics, and elastomers is available for pump construction, from the most common cast or ductile iron to 316L stainless steel and Hastelloy C®, from the standard PVC or polypropylene to PVDF and acrylics, and from the usual buna or neoprene elastomers to Viton® and Teflon®. Therefore, most pump manufacturers offer a chemical resistance chart, with letter ratings for various chemicals plotted against their standard options. The particular way the elastomers are used (as a flexible impeller, or as a section of peristaltic tubing, for example) may be significant, and manufacturers' standards for castings vary; therefore, specific manufacturers' charts should always be consulted when available. When a particular chemical is not on the list, the manufacturer should be consulted before a final selection is made based on another manufacturer's information. Typically, 316L stainless steel is used for sanitary biotechnology applications.

8.8.1   Seals versus Packing

Whether packing, mechanical seals, or sealless design is most suitable can be a serious question for many applications in biotechnology. Both seals and packing are designed to support the shaft and keep it in proper alignment, while limiting or eliminating leakage around the rotating shaft.

Packing (Figure 8.23) was the earliest effective solution. It consists of a woven material such as cotton, impregnated with a substance such as asbestos or graphite that can tolerate the heat generated. This is formed into individual rings and pressed into a cavity (the *stuffing box* or *packing box*) through which the shaft enters the body. The pressure compressing the packing rings is adjusted so that the shaft is supported but leakage is kept to a minimum (3–5 drops per minute, as a rule of thumb). The leakage can be captured and disposed of, but it can be a nuisance, expensive, or even dangerous if the process liquid is toxic or volatile. Packing nevertheless is still popular if a sudden seal failure must be avoided, or if the lower expense of both the packing material itself and the ease of its maintenance take a high priority.

Mechanical seals (a typical cutaway is shown in Figure 8.24) were developed to provide leakproof sealing, and have become standard for most pumps. The challenges in their original development were the potentially high rotational velocities of the faces, the need for self-adjustment because of inevitable wear, and the multiplicity of materials of construction necessary to handle the variety of liquids to be pumped. The solution, and still the most popular seal design, consists of a stationary component, usually pressed into the pump body, against which a spring-loaded rotating member slides. The spring-loaded component compensates for wear and is held tightly on the shaft by a rubber bellows. The rotating element and the stationary element have highly polished and very flat faces (to within 1 angstrom, 0.1 nm) sliding against one another, so that only normally undetectable amounts of the liquid can get between them. Liquid cannot get around the stationary seat tightly pressed into the body, and cannot get down the shaft because of the elastomer's fit around the shaft.

There are many designs to fit particular seal cavity shapes, but almost all consist of the components listed above. The stationary element, called a *seat*, usually consists of highly polished ceramic, sitting in a buna rubber *cup* to help seal it in the machined

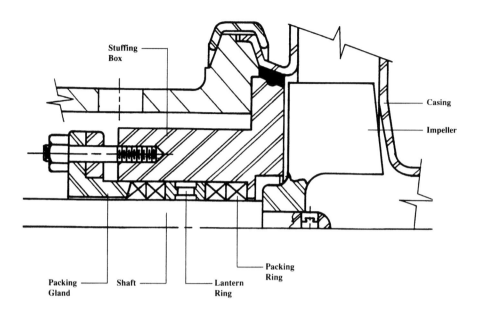

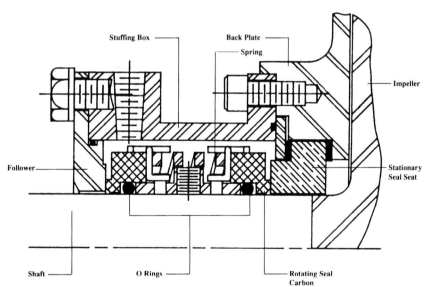

Figure 8.23 (above) Packing.
This is a cross-sectional view of
packing used to support and seal a
rotating pump shaft. (Courtesy of
Tri-Clover, Inc.)

Figure 8.24 (below) Mechanical Seal. This is a
cross-sectional view of the mechanical seals used to
provide leak-proof sealing and support of a rotating
pump shaft. These have become standard on most
pumps. (Courtesy of Tri-Clover, Inc.)

cavity. The spring used to keep the stationary and rotating faces together is typically made of 316 stainless steel. The elastomeric bellows is usually made of buna or neoprene for standard applications, and is often made of Teflon® for biotechnology applications. The *washer* (the other polished face) is usually made of carbon, and the components holding the rotating assembly together are usually bronze or stainless steel.

All of these are available in other materials such as Ni-resist (a stainless cast iron) or silicon carbide for the seat, glass-filled Teflon® or silicon carbide for the washer, and a large number of elastomers for the bellows, including EPR and Viton®.

The advantages of a mechanical seal are that they do not leak or need maintenance. A seal can, however, be a fragile device, subject to thermal shock, physical blows, dry-running damage, premature wear from abrasives, or chemical attack. It carries with it the omnipresent danger of breakdown and sudden leakage, and is not as easily serviced as packing, should failure occur. Adaptations are available that limit the extent of failure, but obviously the unwanted occurrences have to be anticipated.

Thermal shock can affect the seal if a pump has run dry long enough to heat up the two faces, and then liquid is introduced. Since it's usually the ceramic seat that fractures, other materials, such as Ni-resist, which is less subject to fracture, can be used. The advantage of using alternative materials is often compromised by shorter wear life, however. Premature wear from abrasives can be avoided if the wear faces are made of especially hard (and especially expensive) material, such as silicon carbide or tungsten carbide. Chemical attack can be eliminated if a suitable selection is made from the wide variety of available materials.

Double-seal arrangements can resolve problems associated with dry-running damage, eliminate safety hazards resulting from potential leaks of toxic liquids, and reduce high risks surrounding loss of costly liquids through a standard single seal. Two seals are installed in a special deep seal cavity with wear faces outward, rubbing against two stationary seats. The seal face on the impeller side is in contact with the liquid being pumped, but the rest of the assembly is away from the fluid, and is lubricated by a separate source such as sterile condensate so that it can tolerate dry running. If the pressure in the double-seal box is set to just above or just below the wetted-end pressure, any seepage across the faces can be directed.

8.8.2   Sanitary Standards

Avoiding contamination is the essence of sanitary pump design and material considerations. To ensure proper and consistent design, various organizations have promulgated standards (such as the 3-A Standards of the International Association of Milk, Food and Environmental Sanitarians, Inc.) for pumps and equipment used in the food, pharmaceutical and bioprocessing industries [2]. These criteria are intended to guarantee that the pumps can be thoroughly cleaned and to reduce the threat of contamination. The FDA has set forth *Good Manufacturing Practices (GMPs)* [3–6] detailing acceptable design criteria for equipment used in the manufacture and processing of drug products and medical devices.

Several aspects of pump construction are addressed. The particular standards to be applied depend on the specifics of the projected use, but some generalities can be stated. In centrifugal pumps, an open impeller design (an impeller with no shrouds on either side) is preferred, because this reduces the number of places where contamination may lurk. Pump casings are often furnished with a condensate removal drain, so that condensate from steam sterilization can be removed. The product-contacting surface of the pump may be finished to 150 grit (roughness of approximately 35 μin or 0.89 μm RMS), or even as high as 320 grit (approximately 6 μin or 0.15 μm RMS), with or without electropolish, to guarantee a crevice-free surface. Many products manufactured in the biotechnology industry may require the use of pumps with double seals, to prevent unwanted leakage and in some cases to prevent contamination of the product by the atmosphere. If single seals are acceptable, an external seal is usually preferred, so that the seal faces can be inspected visually, and to keep the assembly separate from the product to minimize contamination.

Table 8.5 is a list of the types of PD pumps that are inherently suited for (or can be modified to meet) sanitary design requirements dictated by various regulatory and licensing boards. All requirements should be specified to the pump supplier, so that proper regulations can be researched and the appropriate pump supplied.

8.9 Pump Motors

Most of the specifications for a motor are, of course, dictated by pump needs, since the drive's function is to power the pumping mechanism at the proper speed with the horsepower necessary to achieve that speed. Improvements in motor insulation and design and the widespread availability of AC power, as well as factors of economy, convenience, ease of replacement, and maintenance, have made the electric motor, particularly the AC induction motor, the drive of choice for the vast majority of pump applications. Typically, only when AC power is unavailable or unsuitable are other choices considered. Among the other choices are gasoline engines, diesel engines, hydraulic motors, air motors, and DC motors.

AC motors not only are the most common type of drive but also have the advantages of long life, simple design, excellent availability, and a wide variety of configurations, horsepower, speeds, and enclosures. They are relatively quiet; if minimal considerations are taken into account they are interchangeable; and they are the most efficient type of drive (up to 95% efficient, depending on the horsepower rating and the manufacturer).

As we have seen, the pump dictates the horsepower rating, rpm, and physical configuration of the drive, but the drive must be compatible with an existing electrical system, and there are several factors to consider when the two are matched. Although most plants have both single-phase and three-phase power, there are several differences in the designs of the motors that run on these two different sources of power.

8.9.1   Single-Phase Motors

Single-phase power is the typical two-hot-wire-plus-ground current available in homes. Voltages available in the U.S.A. are 115 and 230, sometimes 208. Amperage depends inversely on voltage, so the higher voltage should be used if possible, since lower amperage allows lighter gauge wiring and other amperage-related equipment. Motors that run on 115 volts are standard only up to 3 hp, but 230-volt motors are standard catalog items up to 15 hp. A single-phase motor becomes more expensive than a three-phase one at horsepowers of approximately 0.5–1 hp.

8.9.2   Three-Phase Motors

Three-phase power is found only in industrial areas, and has three hot lines, plus a ground. Three-phase motors are desirable because of the simplicity of their internal starting devices. However, they have the disadvantage of needing a separate starter box, since three-phase motors don't come with toggles or push buttons as do some single-phase motors. The starter box usually has some sort of overload protection and can be ordered with many automatic controls or remote devices, so the starter package can actually be considered a convenience. Common three-phase voltages are 208, 230, 460, and 575 (in Canada).

Either a single-phase or a three-phase motor can run under the manufacturer's full warranty within a ±10% range from the nameplate voltage. A 460-volt motor is acceptable, therefore, for voltages from 414 to 506 volts, which covers the 440-, 460-, and 480-volt systems.

8.9.3   Frequency

A frequency of 60 hertz is standard in the U.S.A., but 50 hertz is more common elsewhere. The final shaft speed changes in direct proportion to changes in frequency, so losses in pump performance can be anticipated. With AC motors, changing the voltage and frequency proportionally is usually not harmful because the magnetic condition remains unchanged. Therefore, a motor wound to operate on 60-hertz, 460-volt power can probably run successfully on 50-hertz, 380-volt power (a common combination), if the same 1/6 loss is applied to the nameplate horsepower. As with voltage, some flexibility in the frequency, usually ±5%, is acceptable to manufacturers, with the caveat that the combined change in frequency and voltage cannot exceed ±10%.

The electrical characteristics the motor must meet are dictated, or at least limited, by the existing system. There are also mechanical choices just as rigidly confining.

8.9.4   Enclosure

Most manufacturers stock a limited range of types of motor enclosure, including *open drip-proof, totally enclosed,* and *explosion-proof,* with a few specials such as *chemical duty* and *sanitary duty* (smooth surfaces, easily cleaned).

Open drip-proof (ODP) motors have openings front and back, and sometimes in the sides, to allow the fan blades on the rotor to pull in ambient air. This air is channeled by a diffuser over the windings to take up heat, then is pushed out again, normally through the front vents, to keep any seal leakage from entering the motor. Open motors can in theory be exposed to liquids falling at an angle of 15° from the vertical (therefore drip-proof), but most manufacturers prefer protection from atmospheric moisture or dirt because the motor pulls in airborne contaminants or moisture as it cools itself. ODP motors run quieter than enclosed motors (averaging 80 to 90 decibels), run cooler because of superior heat dissipation, draw slightly fewer amps, and, therefore, are always preferred if the area is protected.

Totally enclosed motors, either fan-cooled *(TEFC)* or nonventilated *(TENV),* are not vented to the atmosphere. They may have an external fan to drive cooling air over the frame, or may lack the fan if the manufacturer finds the motor is cooled adequately without it. Both types often have external ribs that act as heat sinks and enlarge the surface area to assist in heat removal. They are neither waterproof nor hermetically sealed, but are commonly installed in outdoor applications with no further protection from the elements. The vulnerable spots are the motor bearings at either end and the opening where the leads exit the stator frame into the outlet box, where water under pressure could be blown into the bearings or the windings. As one would expect, they are noisier than ODP motors (averaging 90 decibels for small motors to 100 decibels for larger ones) and draw slightly higher amperages. Some motor manufacturers are making TEFC the standard for their line of integral horsepower motors since TEFC is usually acceptable to customers with open drip-proof environments.

Explosion-proof enclosure motors come in several UL (Underwriters Laboratories) classifications for various hazardous atmospheres: *Class I, group D* (acetone, acrylonitrile, alcohol, ammonia, benzene, butane, ethylene dichloride, gasoline, hexane, methane, propane, naphtha, natural gas, propylene, styrene, vinyl acetate, vinyl chloride, or xylenes); *Class II, group E* (metal dust); *Class II, group F* (carbon black, coal, or coke dust); and *Class II, group G* (flour, starch, or grain dust). Some manufacturers make motors good for all four classes simultaneously, but if an explosion-proof motor is needed, care must be taken to match the UL rating with the hazard.

All the above are called *Class 1 Division 1* conditions, and require UL-listed motors. *Class 1 Division 2* means an intermittently hazardous location, for example, one that is safe from fumes unless a gasoline line in the area is broken. This condition can be met by any nonsparking motor. The nonsparking requirement eliminates most DC motors and single-phase motors with the common centrifugal governors, but still allows most three-phase or sealed single-phase motors.

Explosion-proof motors are at first glance TEFC motors with UL labels, but a closer look reveals heavier duty fits between body parts, an outlet box with leads potted as they exit the stator, and external through-bolts for more strength. Explosion-proof motors are not guaranteed to never burn out, but rather to fully contain any sparks or flames internally. They are like TEFC motors in all other respects of noise and amperage. The explosion-proof requirement is an instance where a three-phase AC motor is especially preferred, because of the relative ease of manufacture, and therefore its relatively low cost compared to single-phase.

Even after all the above considerations, one of the most important facets of motor selection has not been addressed: matching the design and mounting pattern of the motor to the needs of the pump.

8.9.5   NEMA

The National Electrical Manufacturers Association (NEMA) has established standard mounting dimensions. Frame sizes, mounting dimensions, and letter suffixes are such that motors with identical frame designations are interchangeable in all critical areas such as mounting bolts, shaft sizes, and shaft heights, regardless of manufacturer; motors of a larger size (due to greater horsepower, or because of slower speeds, and therefore greater torque) have higher frame numbers. Pump companies following NEMA standards provide the exact NEMA frames necessary to fit their products. The most common are:

(a)   *Standard Horizontal Motors.* These motors have their own mounting feet, and are designed to be connected via a flexible coupling to the pump, also having its own mounting feet. The typical three-digit frame designation has either a "T" or no suffix at all, such as frames 143T or 184.

(b)   *Close-Coupled Pump Motors.* These motors are intended specifically to have a pump of the proper design mounted to the front face; they therefore have a register machined to guarantee concentricity (a "C" face). The pump impeller (or impeller/shaft assembly) is mounted to the motor shaft. The suffixes designating this type are "C", "TC", "TCZ", "JM", and "JP". Many types of pumps take advantage of close coupling, hence the variety in motors, but all are constructed as described above. Some ("C" and "TC") have short shafts and are used with pumps having their own shaft extensions to slip over the motor shaft and carry the impeller. Others ("TCZ", "JM" and "JP") have longer shafts; the impeller can mount directly on them, saving the cost of a shaft extension. Examples are frames 215TCZ and 254JM.

(c) *Jet Pump Motors*. These are similar to close-coupled pump motors, but they differ in the following respects: they always appear in the same general frame size (designated frame 56), their horsepower rating is fractional or small integral (3 hp or less), and they run at 3600 rpm. They are mass-produced, and are therefore relatively inexpensive. Commonly used by pump companies on their lower-end models, they come in a 56C frame (with a standard shaft) or a 56J model (with a 7/16"-20 threaded end for a threaded-bore impeller), but always with a "C" face for close coupling.

8.10   Troubleshooting

The key to successful troubleshooting is to follow several important steps:
  (1) Make sure you understand the system fully, and have not overlooked anything;
  (2) Use the information you have about the problem to shorten the list of possible sources of your trouble;
  (3) List all possible remaining causes;
  (4) In order of probability, go through them one by one.

System analysis can be tedious. Fear of asking questions that may appear foolish or have obvious answers can tempt one to give this step short shrift. However, if problems began after a change, system analysis usually holds the answer. "Walking through" a typical flow pattern and questioning each step is often helpful. Keep in mind that flow at any division seeks equilibrium; flow divides so that resistance to flow on one path exactly equals the resistance to flow on the other. Double checking that the pump in question can, in theory, meet the system conditions may also be appropriate.

In many cases, you can narrow down whether a problem is in the system or the pump by switching the pump for one in an installation that is working well. If the problem follows the pump, it's in the pump; if it stays with the line, it's in the system. In other cases, you can isolate noises by separating various components of a system and determining if they continue or cease. The motor sometimes can be separated from the pump, for example, or suspect components can be temporarily removed. Sometimes you can check the results by isolating pumps or sections of piping beyond a valve.

There are usually more possible explanations for a problem than manufacturers' lists provide. A sampling of potential causes for a few specific complaints is in Table 8.6. Creativity is essential in this step; never take anything for granted. Often a marginal suggestion, tentatively or casually offered up, is immediately recognized by an engineer and the problem solved.

Depending on the symptoms, some possibilities are always more likely than others. Of these, the most easily checked should be the first inspected, and then the more

Table 8.6 Analysis of Common Problems

I. Low Flow Rate

A. Centrifugal Pumps:
   1. Reverse rotation (Note: a centrifugal impeller "flings" the liquid, rather than "scooping" it, if there is ever a question as to rotational arrows. A centrifugal pump does pump if turning in reverse, but with low flow rate and pressure, and higher horsepower due to lowered efficiency);
   2. More head than originally calculated;
   3. Clogged impeller;
   4. Wrong impeller diameter;
   5. Cavitation (see Section V of this table);
   6. Too-large impeller clearances;
   7. Too-slow motor speed;
   8. Impeller spinning loosely on shaft;
   9. Broken coupling;
   10. Increased viscosity;
   11. Vortexing in source tank {Note: the recommendation is 1 ft of depth per (ft/s) of velocity at outlet pipe from tank to pump};
   12. Air in tank, introduced from splashing or churning, close enough to tank outlet so that the air enters with the stream;
   13. Wrong pump for conditions;
   14. Air leak in suction pipe (if pump is above source tank).

B. Positive-Displacement Pumps:
   1. Excessive wear on components;
   2. Stuck internal check valves, if any;
   3. Torn diaphragms, if present;
   4. Stuck internal (or external) pressure relief valve;
   5. Too-slow speed;
   6. Cavitation (see Section V of this table);
   7. Wrong pump for conditions;
   8. Slipping belt (if belt driven);
   9. Broken coupling;
   10. Speed not corrected for viscosity;
   11. Air leak in inlet pipe (if pump is above source).

Table 8.6 Analysis of Common Problems, *continued*

II.	Motor Overheats

A.	Centrifugal Pumps:
	  1. Too much flow; further out on curve than calculated;
	  2. Wrong rotation;
	  3. Oversized impeller;
	  4. Unbalanced impeller.

B.	Positive-Displacement Pumps:
	  1. Too much back pressure;
	  2. Binding or rubbing components;
	  3. Insufficient motor starting torque to get up to full speed (in case of progressing-cavity or other pump with rubbing components).

C.	All Pumps:
	  1. Voltage outside the ±10% range allowed by manufacturers;
	  2. Bad motor bearings;
	  3. Three-phase motor not getting power to all three lines ("single phasing");
	  4. Overtightened packing;
	  5. Cavitation (see Section V of this table);
	  6. Misaligned motor/pump shaft or misaligned pulleys;
	  7. Undersized cable to motor;
	  8. Loose connection at motor outlet box or control box;
	  9. Worn-out insulation (broken down through time and use);
	  10. Cable too long to handle amperage;
	  11. Too-high ambient temperature;
	  12. Overgreased bearings, causing binding when bearings heat up;
	  13. Broken or slipping fan (on TEFC design);
	  14. Too-high speed for pump;
	  15. Insufficient air movement to keep motor cool (because of confined or poorly vented area);
	  16. Starting windings not disengaging (if single-phase motor);
	  17. Viscosity higher than calculated;
	  18. Pipe torque, causing misalignment and binding;
	  19. Motor stopping and starting too often (consult manufacturer for maximum stops and starts for applicable horsepower);
	  20. Dirty surfaces not allowing efficient cooling;
	  21. Frequency outside the ±5% range allowed by manufacturers;
	  22. Internal diffuser or fans not channeling air flow properly (on open drip-proof design).

Table 8.6 Analysis of Common Problems, *continued*

III. Noise: Different designs generate different levels of noise. The manufacturer can supply an expected noise level if given the operating conditions.

A. All Pumps:
1. Worn motor or pump bearings;
2. Cavitation (see Section V of this table;)
3. Components hitting or rubbing (if a PD pump);
4. Amplification of what might be considered normal noise through lines into abnormal amounts (if rigidly piped);
5. Amplification of noise from pocket of air beneath motor/pump package (if mounted on channel steel base);
6. Coupling hitting coupling guard;
7. Motor fan hitting fan cover guard;
8. Check valves chattering as they stick;
9. Water hammer (valves must close slowly enough for momentum in line to be absorbed). A rough rule of thumb is:
$$t = 2L/a$$
where t is the time for valve to close (minimum), s;
L is the length of pipe via longest route, ft; and
a is the speed of pressure wave, ft/s; (an average for water is 4000 ft/s).
    For example, a 200-ft length of pipe handling water needs a valve that closes in no less than $2 \times 200/4000 = 0.1$ s, to avoid water hammer.

IV. Abnormal Seal Wear: A mechanical seal in a continuous-duty, clean application can be expected to last a minimum of several years, and as many as 10 years or longer if under absolutely ideal conditions.

A. Seat Fracture:
1. Thermal shock from dry running followed by sudden introduction of cooler liquid, or any sudden drastic change in the temperature of the liquid; usually appears as a hairline fracture;
2. Incorrectly installed seal head (causing the head to strike the seat);
3. Physical shock from a blow, such as being dropped;
4. Water hammer;
5. Pipe misalignment, twisting the bracket holding the seat.

B. Prematurely Worn Carbon Washers:
1. Abrasives (evidence of scored faces is clearly visible);
2. Chemical incompatibility between the pumped fluid and either the washer itself, or the binder holding the carbon together;
3. Improper installation, usually with the head cocked.

C. Elastomer Problems:
1. Chemical incompatibility (elastomer may become brittle, softened, swollen, or even eaten away totally);
2. Dry running (elastomer may look burned or cracked, or may become brittle and hard).

Table 8.6 Analysis of Common Problems, *continued*

V. Cavitation: Indicated by crackling noise for centrifugal pumps, banging for most PD pumps, or by finding that TDH has fallen at least 3% below the curve for a centrifugal pump.

A. All Pumps:
  1. Blocked inlet;
  2. Too much friction loss in suction line;
  3. Elbow less than 6× pump inlet diameter from suction port;
  4. Running at extreme left end of the curve, leading to suction or discharge recirculation (for centrifugal pumps only);
  5. Suction line with high spot (as opposed to continuously rising); therefore, air trap to disturb or restrict flow;
  6. Liquid too hot for necessary vacuum to be applied;
  7. Elevation differential too great;
  8. Vacuum level too high, or vacuum level that builds to an unacceptable point only as liquid is pumped out (if drawing from closed tank);
  9. Excess prerotation at inlet of centrifugal pump, causing flow to enter in wrong pattern for design. (Note: rotation against impeller rotation is a positive factor and displaces the curve upward, but the more common prerotation in the same direction as the impeller is harmful and lowers the curve.)

difficult. At times there are two simultaneous flaws in a system, and checking of individual items is fruitless. In these cases logic helps more than experiment.

8.11   Maintenance

Every pump and motor manufacturer provides maintenance data for his product. If these are not already included with a new pump, a phone call to the vendor will bring them. The literature is usually fairly complete and is designed for an informed and motivated audience. It often includes an exploded view, a parts list, and a thorough maintenance manual.

Pump manufacturers are especially professional in this respect, desiring not only that the product carrying their name work well in the application and lead to referrals, but also that parts business will follow. The data should always be requested and kept on file to help deal with the inevitable Friday afternoon failure, and should serve as the definitive word for the pump in question. However, there are a few general suggestions that can be applied to existing equipment as well as to any new purchases.

Close-coupled centrifugal pumps are designed to need little or no regular maintenance. Among the suggestions that may be mentioned in repair manuals are the following:

(a)   If the pump is a mechanical seal pump, monitor leakage periodically to anticipate a complete seal failure. Sometimes weeping corrects itself, especially in a new seal, by the faces wearing in together.

(b)   If the pump is a packing pump, keep packing leakage to 3–5 drops a minute, never less.

(c)   Monitor for noise, especially looking for evidence of cavitation.

(d)   If the pump is a horizontal pump, and if the discharge position is anything other than top horizontal, air is trapped in the top quadrant of the casing. Usually a boss for a pipe plug is present, and a drain cock can be installed to bleed off the trapped air. Anytime air can enter the system it gathers at this point, never purging itself, and therefore needs to be eliminated to keep efficiency at the maximum.

All the above apply also to pedestal-mounted centrifugal pumps, as do alignment requirements. Alignment is not inherent, unlike a close-coupled pump, and is critical. As a rule of thumb, a straightedge placed on the top and side of a coupling should admit no visible light between the straightedge and either coupling hub.

Maintenance of positive-displacement pumps is similar in many ways to that of centrifugal pumps. Alignment is one common concern, since PD pumps are often pedestal-mounted and have close-tolerance components that wear or bind if misaligned. Some key considerations are:

(a)   If the pump is belt driven, the belt tension is important. The rule of thumb is that the belt should be able to be depressed ½" in midlength. The faces of both pulleys also need close alignment; a long straightedge can be used as described for checking a coupling.

(b)   The length of the service life of PD pumps often depends on how clean the pumped liquids are; therefore, filters and strainers should be scrupulously maintained, and the air line feeding double diaphragm pumps monitored for contamination.

Motors rarely need maintenance today. The ball bearings are often double sealed for the life of the bearing; if they are not, care should be taken to avoid the dangers of overgreasing. The section on troubleshooting (Section 8.10) and Table 8.6 may indicate areas in a particular system that could lead to problems and, therefore, require greater attention.

8.12   References

1.   *Hydraulic Institute Standards for Centrifugal, Rotary and Reciprocating Pumps*; 14th edition; Hydraulic Institute, Cleveland, OH, 1983.

2.   INTERNATIONAL ASSOCIATION OF MILK, FOOD AND ENVIRONMENTAL SANITARIANS, INC.; *3-A Sanitary Standards*; International Association of Milk, Food and Environmental Sanitarians, Inc., Des Moines, IA; 1947–present.

3. Current Good Manufacturing Practice in Manufacturing, Processing, Packing or Holding of Drugs, *Fed. Regist. 56* (29), 5671–5674 (Feb. 12, 1991).
4. Regulations Establishing Current Good Manufacturing Practice for the Manufacture, Packing, Storage, and Installation of Medical Devices, 21 CFR Part 820; *Fed. Regist. 43* 31508 (July 21, 1978).
5. Current Good Manufacturing Practice for Finished Pharmaceuticals, 21 CFR Part 211 (specifically Parts 211.63, 211.65, 211.67); *Fed. Regist. 43* 45077 (Sept. 29, 1978).
6. Current Good Manufacturing Practice in Manufacturing, Processing, Packing or Holding of Drugs, General, 21 CFR Part 210; *Fed. Regist. 43* 45076 (Sept. 29, 1978).

Cartridge Filtration for Biotechnology

Jerry M. Martin, A. Mark Trotter, Paul Schubert, and Hyman Katz

Contents

9.1   Introduction

This chapter provides an overview of cartridge-style filters used in the manufacture of biotechnology products. Cartridge filters are widely used in bioprocess manufacturing to clarify, polish, and sterilize almost all fluids (gases and liquids) in both batch and continuous operations. We describe the most commonly applied filter types, cartridge constructions, and housings. The bioprocess engineer should find this information useful in selecting the proper filter assemblies for both pilot and production plants. We include quality requirements associated with pharmaceutical and related biotechnology products. These descriptions can be used to write specifications for various types of filters.

We then describe applications in bioprocessing that benefit from the various types of filters. These filter applications include sterile filtration of gas and liquid feeds to fermentors and bioreactors, exhaust/off-gas containment, cell and cell debris separation, downstream purification, final sterile filtration, and utilities.

We discuss procedures for steam sterilization of process filters, including precautions to avoid damage to filters during the sterilization process. We follow with descriptions of the integrity tests commonly used to ensure proper selection, installation, and integrity of filters prior to and after use. In the concluding section we present an overview of concepts to consider in the validation of filtration systems as part of the overall validation of an FDA–regulated biotechnology manufacturing process.

9.1.1   Filter Ratings

Filter elements are typically rated for removal of particles of a stated size at a stated efficiency. Such ratings can be as absolute (100%) removal, or by percent removal of a specific size fraction of particles on a particle-count basis. The absolute filter rating is the diameter of the largest hard spherical particle that passes through a filter under specified test conditions. This is an indication of the largest opening in the filter element.

Some filter elements may carry only a nominal pore size rating. The nominal pore rating is an arbitrary micron (micrometer) value assigned by the manufacturer, based on removal of some percentage of all particles of a given size or larger. Because of a lack of industry agreement in the designation of nominal ratings, such values are frequently not comparable from one filter manufacturer to another; absolute ratings provide more reliable comparisons [1].

Efficiency ratings may be expressed as percent removal, *beta ratio* (defined as influent concentration over effluent concentration for a given particle size range), or *titer reduction*, T_R (defined as influent concentration or total challenge, over effluent concentration or total recovered, for a given microorganism). The latter rating may also be expressed as a *log reduction value*, *LRV*, which is the titer reduction expressed as a logarithm. Ratings must be determined under defined test conditions.

Specific filter efficiency data are typically available from the vendor. Such data should denote, as a minimum, for final filters, the absolute or 100% removal efficiency value, and, for prefilters, an efficiency for at least one other size fraction.

9.1.2 Specification of Filter Media and Cartridge Design

A broad array of filter media in cartridge form is available for the removal of contaminants from fluids in the manufacture of biotechnology products. The specified process filter cartridge should be designed so that when used in the location and application noted, it performs satisfactorily under the conditions of use. All materials should be compatible with the process fluid over the temperature and pressure range indicated.

To select the appropriate filter medium, one must know the removal efficiency required. This can be based on experience or specification. Having determined the removal efficiency, one must ensure that the fluid is compatible with the filter media; usually more than one medium is suitable for the application.

Filter selection is further based on many variables, including service temperature, flow rate, pressure drop restrictions, etc. The reader should find the information here is also useful for preparation of written filter specifications, after final selection of the filter assembly. Each filter specification should include the filter medium, the absolute rating in micrometers, and appropriate applications, and should provide performance and mechanical criteria.

The specifications provided in the following sections are generic for use in the manufacture of biotechnology pharmaceutical products and can be met by various filter vendors. Filter cartridges are available in a variety of media types, including polypropylene, cotton, resin-impregnated glass fiber, cellulose, nylon, polyamide, cellulose ester, polysulfone, poly(vinylidene fluoride) (PVDF), and polytetrafluoroethylene (PTFE).

9.1.3 Filter Media Design

Filter types that do not exhibit media migration are preferable. *Media migration* is defined as the tendency of particles or fibers of the filter medium to become dislodged, slough off, shed, or in some other way thereby contaminate the filtrate. Media migration may be observed with nonbonded fibrous or particulate filtration media, which should therefore be avoided.

The filter medium should also be of fixed-pore construction and not unload even at maximum differential pressure. *Unloading* is defined as the release of contaminants retained by the filter medium into the effluent as the differential pressure across the cartridge increases. A vendor claiming that a filter medium does not unload should supply data that give the filtration efficiency as a function of increasing differential pressure.

9.1.4 Filter Element Mechanical Design

Process filter cartridges are typically available in 2½" or 2¾" OD (63 or 70 mm), in multiple lengths of 10" (25 cm) up to 40" (100 cm) long (Figure 9.1). Maximum recommended differential pressure and flux for aqueous and air flows should be specified by the vendor.

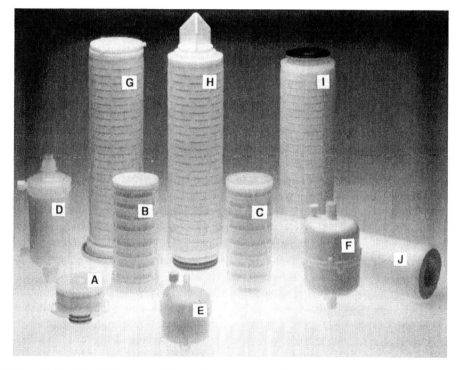

Figure 9.1   Filter Elements.   Filter elements are available in a wide variety of styles, configurations, and sizes, for pilot and production scales.   Filters A–C are small flow replaceable elements and D–F are self-contained assemblies. Filters G and H are sanitary-style 10" process cartridges. Filters I and J are double open-ended industrial-style process cartridges: I a pleated cartridge and J a depth cartridge with integral molded gaskets. (Courtesy of Pall Ultrafine Filtration Co.)

Standard-design depth-filter cartridges are open cylinders of thick filter material (*double open-ended*). Circular knife edges in the housing installation form the seals that isolate upstream from downstream by engaging the filter medium at each end. A more reliable seal is provided in double open-ended industrial-style depth (nonpleated) filter cartridges by end-sealing into thermoplastic gaskets that are heat-welded onto each end of the cartridge. Industrial-style pleated filter cartridges are provided with similar flat gasket seals that are fitted into the end caps at both ends. Gasket materials for industrial-style pleated filter cartridges should be selected to meet process compatibility requirements.

Sanitary-style cartridge configurations feature a finned or flat blind end cap on one end, and an open outlet at the other end. Sealing is accomplished with a piston double O-ring seal at the open end to ensure no fluid bypass. Standard O rings for sanitary-style cartridges are usually of FDA–listed silicone. Other O-ring materials, selected for chemical compatibility, are typically available on request. Filter elements that have locking tabs at the piston double O-ring adapter base, for bayonet locking of the element, seal positively and do not unseat during reverse pressurization cycles.

9.1.5 Optimization for Biotechnology Manufacturing

All materials used in the manufacture of filters intended for processing of biotechnology-derived pharmaceuticals should be traceable. Filter element components are typically made from materials listed for food contact usage according to Title 21 of the U.S. *Code of Federal Regulations* (*CFR*), Parts 170–199 [2a]. These materials should also meet the specifications of biological safety tests listed in the latest revision of the *United States Pharmacopeia* (*USP*) for Class VI plastics at 121 °C [3a].

For sterilizing-grade filters used in manufacturing of pharmaceutical products, including those of biotechnological origin, the filter must be subjected to and required to pass specified Quality Control/Quality Assurance (QC/QA) tests. To verify and document the quality testing of such filters, the vendor should provide a *Certificate of Conformance* listing all such tests and performance.

All sterilizing-grade filter elements must be capable of undergoing nondestructive integrity testing (see Section 9.8, Filter Integrity Testing) to confirm the absolute rating and integrity of the filter element prior to and after use. The integrity test procedure for sterilizing-grade filters must be correlated to a bacterial challenge test. Sterilizing grade filters should also be integrity-tested on a 100% basis by the vendor to ensure integrity prior to shipment. Nonsterilizing-grade sanitary and industrial-style filter elements may be sampled for nondestructive integrity testing.

To assure the highest level of effluent quality, filter samples from each manufacturing lot should also be tested by the filter vendor for cleanliness, oxidizables, pH, and pyrogens (bacterial endotoxins). Effluent cleanliness standards should meet USP limits under Particulate Matter in Injections [3b], with effluent counts determined microscopically. These counts can serve to document conformance with the requirements for non–fiber-releasing filters according to 21 CFR 211.72 and 210.3 (b) [2b].

Levels of effluent oxidizables after flushing should meet the USP requirements under Purified Water. Determination of pH effects on the effluent as a result of filtration is typically made according to the USP requirements under Purified Water [3c]. Influent and effluent samples, taken after the flush for oxidizable substances, should be within 0.5 pH units.

Qualification of the filter as nonpyrogenic can be determined according to the USP requirements under the Bacterial Endotoxins test [3d]. An aliquot from a minimal volume filter soak solution should contain less than 1.0 EU/mL, as determined using the *Limulus amoebocyte lysate (LAL)* test.

9.2   Types of Filters

9.2.1 Polypropylene Depth Filter Cartridges

Absolute-rated polypropylene depth filter cartridges are applicable as prefilters for serum, vaccines, diagnostics, tissue culture products, deionized water, container washing, and final product. These filters are used in fermentation for liquid feeds, makeup water, solvents, and antifoam. In downstream processing they are used for cell and cell debris removal, buffers, cleaning agents, and sanitizing solutions.

These filter cartridges are constructed using nonmigrating continuous filaments of polypropylene filament medium without resin binders. *Positively charged* versions feature a positive zeta potential when immersed in an aqueous solution over a specific pH range. Filters with positive zeta potential offer the same high dirt capacity as the uncharged prefilters, with the added ability to remove organisms and particles, such as bacterial endotoxins (pyrogens), which are much smaller than the absolute pore size rating, by electrostatic binding.

Cartridges are constructed with an inner (downstream) section in which the pore diameter is constant (this section provides absolute-rated filtration), and an outer section in which the pore diameter varies continuously from that of the absolute-rated section to as much as 90 μm or more (Figure 9.2). The regions of different pore size contain filaments of varying diameters, while the void volume is constant throughout the medium. The constant void volume provides the increased dirt capacity and lower clean pressure drop of the filter cartridge.

9.2.2 Polypropylene Pleated Filter Cartridges

Polypropylene pleated depth filters are applicable as prefilters for makeup and rinse water, and serum for culture media; in downstream processing for cell debris removal;

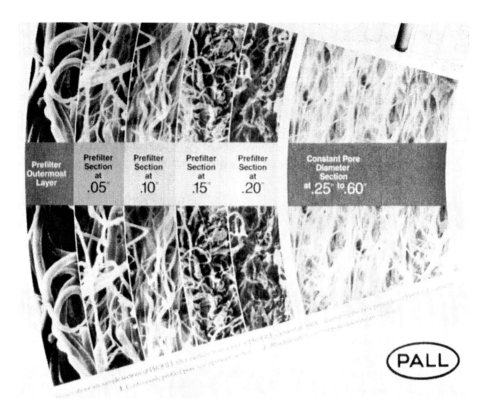

Figure 9.2 Polypropylene Depth Filter Medium. Scanning electron photomicrograph of profiled depth filter element showing the absolute removal rated, constant pore diameter section at the core and the constant voids volume upstream prefilter section of continuously increasing fiber diameter. (Courtesy of Pall Ultrafine Filtration Co.)

and for prefiltration of solvents and buffers. These process filter cartridges are also constructed using nonmigrating continuous strands of nonwoven polypropylene filaments. The medium should have a constant–pore-size section (downstream) for absolute-rated filtration (Figure 9.3) and a continuously graded upstream section for effective prefiltration.

The thin sheet of polypropylene medium is pleated and formed into a cylinder with a longitudinal side seal of melt-sealed polypropylene. The cylinder is then melt-sealed to injection-molded polypropylene end caps to ensure no fluid bypass. Polypropylene hardware components consisting of an inner support core and an external protective outer cage are incorporated. The process filter cartridge with polypropylene hardware should be rated to withstand differential pressures of 80 psi (550 kPa) up to 50 °C (122 °F) and 60 psi (410 kPa) up to 80 °C (176 °F).

 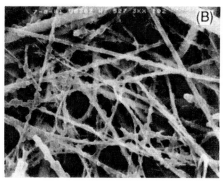

Figure 9.3 Pleated Depth Filter Media. Scanning electron photomicrographs of (A) continuous polypropylene filaments and (B) positive zeta potential glass-fiber filter media. Note small particles collected on the charge-modified fibers in (B). (Courtesy of Pall Ultrafine Filtration Co.)

9.2.3   Resin-Free Cellulose Pleated Filter Cartridges

Resin-free pleated cellulose filter cartridges are applicable as prefilters for makeup and rinse water, for reverse osmosis membranes, for deionized water, and for the inlet air for fermentors and bioreactors. These filter cartridges are constructed of pure cellulose medium (without resin binders) pleated into a high-area cylinder. The longitudinal side seal of the process filter cartridge should be an FDA–listed polypropylene (approved for food contact usage).

Cellulose medium cartridges are assembled with hardware components consisting of a perforated inner support core, an outer support cage, and end caps melt-sealed to embed the medium in the plastic. All hardware components should be of pure polypropylene, without filler or reinforcement, to ensure a high-quality filtrate with a minimum of soluble extractables. The process filter cartridge with polypropylene hardware should be rated to withstand differential pressures of 80 psi (550 kPa) up to 50 °C (122 °F) and 60 psi (410 kPa) up to 80 °C (176 °F).

9.2.4   Resin-Impregnated Glass-Fiber Pleated Filter Cartridges

Resin-impregnated glass-fiber filters are applicable as prefilters for deionized water, solvents, serums, and diagnostic reagents, and for cell debris removal from harvest fluids. The resin impregnation makes available a choice of filters with either a positive or negative zeta potential when used in aqueous service. Filter cartridges with negative zeta potential are constructed of glass fibers that have a negative zeta potential and are reinforced with a resin binder that is also negatively charged when immersed in water. Positive zeta potential filter cartridges are constructed using a resin binder that has a positive zeta potential in the medium. The positive charge enhances electrostatic binding of negatively charged particles much smaller than the filter's absolute removal rating.

The filter cartridges are available with hardware components consisting of end caps and an internal perforated support core made of either polypropylene or stainless steel, and an external polypropylene outer cage or protective net. Stainless steel end caps are attached to the filter by an inert synthetic resin. Polypropylene end caps are melt-sealed to embed the medium in the plastic. Filter cartridges with polypropylene hardware should be rated to withstand differential pressures of 80 psi (550 kPa) up to 50 °C (122 °F) and 60 psi (410 kPa) up to 80 °C (176 °F). With stainless steel hardware, the filter can be rated to 75 psi (520 kPa) up to 135 °C (275 °F).

9.2.5 Porous Stainless Steel Filters

As the name implies, porous stainless steel medium is made by the sintering of small particles of stainless steel or other high-alloy powder to form a porous metal medium (Figure 9.4). Porous stainless steel can be formed as a flat sheet, or, when used in filter elements, preferably as a seamless cylinder. This special manufacturing process produces a high–dirt-capacity medium that is temperature- and corrosion-resistant. The recommended alloy is type 316LB, which has a higher silicon content than type 316L and provides a stronger, more ductile product with better flow properties.

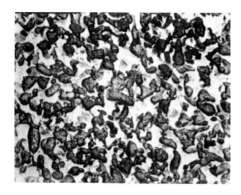

Figure 9.4 Porous Stainless Steel Filter Medium. Scanning electron photomicrograph showing small particles of stainless steel sintered to form a highly porous, temperature- and corrosion-resistant filter medium. (Courtesy of Pall Ultrafine Filtration Co.)

Standard porous stainless steel filters are 2³/₈" (60 mm) OD, with a type 304 or 316 stainless steel flat blind end cap at one end and a 1" or 1½" (25- or 38-mm) NPT connection at the other, or industrial-style flat gasket open end caps welded at each end. Elements have absolute ratings of 0.4–11 μm in gas service applications. Porous stainless steel filter cartridges are chemically or mechanically cleanable, offering economy of reuse.

9.2.6   Membrane Pleated Filters

Membrane filter cartridges incorporate a variety of polymeric microporous media (Figure 9.5) with microbial removal ratings in liquid from 1.2 μm down to 0.04 μm or finer, including the industry standard 0.45-μm and 0.2-μm sterilizing grade. Mechanical design and construction of membrane pleated filter cartridges are similar for most membrane types.

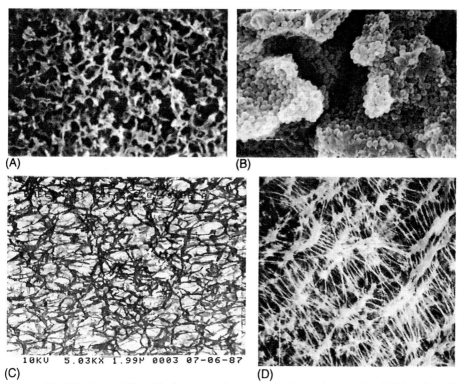

Figure 9.5   Membrane Filter Media.   Scanning electron photomicrographs of (A) nylon 66 membrane, (B) positive zeta potential nylon 66 membrane, (C) poly(vinylidene fluoride) (PVDF) membrane, and (D) polytetrafluoroethylene (PTFE) membrane.   Positive zeta potential membrane (B) shows latex spheres, much smaller than pores, adsorbed to membrane pore walls due to zeta potential effect. (Courtesy of Pall Ultrafine Filtration Co.)

Membranes are typically pleated between layers of nonwoven polyester or polypropylene, which serve as mechanical support and drainage layers.   The process filter cartridge is assembled using hardware components consisting of polypropylene or polyester end caps, an internal perforated polypropylene support core, and an external polypropylene outer cage.   The outer cage provides protection against mechanical damage and ensures retention of structural integrity against accidental reverse pressure up to 50 psi (350 kPa). The longitudinal side seal of the pleated membrane filter cartridge should be an integral, homogeneous melt seal without any additional or extraneous materials, glues, or resins. End caps should be attached by melt sealing to embed the medium in the plastic. Multilength elements are flame-welded together end to end. All hardware should be specified as produced from virgin materials.

Process filter cartridges are available in either single-membrane–layer (prefilter) or double-membrane–layer (sterilizing filter) configurations. Single-layer prefilters are used for bioburden reduction, or upstream of the double-layer final (sterilizing) filters to protect the final filter from premature plugging, thereby prolonging the life of the final filter.

9.2.7 Hydrophilic Membrane Pleated Filter Cartridges

Hydrophilic pleated membrane filter cartridges for process liquid service should be inherently hydrophilic and uniformly wettable when exposed to water or any other liquid. Such *intrinsically hydrophilic* membranes are manufactured from pure nylon 66, hydroxyl-modified polyamide, modified poly(vinylidene fluoride), modified polysulfone, and other modified polymers. Filters constructed with membranes incorporating wetting agents, surface-active additives, or coatings of any kind, which could subsequently leach off, contribute to filter extractables, and adversely affect product quality, should be avoided.

Absolute-rated hydrophilic membrane filter cartridges are applicable to prefiltration and final sterile filtration of parenterals, diagnostic reagents, purified water and Water For Injection, dry gases, organic solvents, buffers, and biological fluids such as serum, plasma, tissue culture media, nondilute protein solutions, and fermentation harvest fluids.

Process filter cartridges are constructed using intrinsically hydrophilic membranes. Casting of the membrane on a nonwoven polyester substrate provides enhanced mechanical properties. For more stringent applications, such as solvent filtration, where a very low level of extractables is required, a pre-extracted version employing a polypropylene substrate can be used. Positively charged hydrophilic membranes contain cationic (positively charged) functional groups that have a positive zeta potential when immersed in an aqueous solution and provide enhanced retention of particles smaller than the absolute rating.

9.2.8 Low–Protein-Binding Membrane Pleated Filter Cartridges

Absolute-rated low–protein-binding pleated membrane filter cartridges are hydrophilic filters applicable to fluids containing dilute protein (typically less than 1 mg/mL). Such fluids include serum-free tissue culture media and protein additives, protein-based therapeutics and diagnostics, diluents and buffers containing dilute protein, recombinant proteins, hormones and growth factors, protein chromatography feeds and eluates, vaccines, and other dilute biologicals.

Hydrophilic absolute-rated low–protein-binding membranes are made of hydroxyl-modified polyamide, modified poly(vinylidene fluoride) (PVDF), modified polysulfone, or other polymers. The low–protein-binding process membrane filter cartridge should bind less than 5 μg of protein per cm^2 of filter area, as determined after throughput of 20 mL dilute bovine serum albumin (at 10 μg/mL) per cm^2 of filter area. Low–protein-binding membrane filter cartridges should be considered for the prefiltration stages as well for service as the absolute-rated final filter. Protein adsorption should be evaluated for an entire filtration system, not just at the final filter. Prefiltration of dilute protein solutions may be a source of significant protein loss.

9.2.9 Hydrophobic Membrane Pleated Filter Cartridges

Absolute-rated hydrophobic membrane pleated filter cartridges are applicable to fermentor and bioreactor sterile inlet air and exhaust gas vents, sterile pressure gas, sterile nitrogen blankets, storage tank sterile vents, formulation tank sterile vents, sterile air for aseptic packaging, and aggressive solvents. Hydrophobic membrane process filter cartridges are typically constructed using two layers of hydrophobic poly(vinylidene fluoride) or polytetrafluoroethylene membrane pleated between two layers of polypropylene support material.

Sterilizing-grade hydrophobic membrane filter cartridges should be inherently hydrophobic and be capable of removing bacteria and bacteriophage from air streams with 100% efficiency under moist or dry conditions. The microbial removal rating typically selected is 0.2 μm absolute in liquids. Such filters can provide a particulate absolute removal rating of 0.01 μm in gas or air service.

These hydrophobic membrane process filter cartridges can be subjected to multiple sterilization cycles and must be designed to be steam sterilized repeatedly in either direction of flow or autoclaved repeatedly. The vendor should provide information on filter cartridge sterilization procedures as part of the proposal, and include operating data limitations (time and temperature). These filters should be able to withstand multiple *in situ* steam sterilizations of at least 165 cumulative hours at up to 142 °C.

9.3   Filter Housings

It is also necessary to specify the performance requirements of replaceable cartridge-type sanitary, air/gas, and industrial service filter housings. Housing size should be adequate for the flow and differential pressure requirements as stated on the manufacturer's data sheets. Filter housings for biotechnology are typically constructed of stainless steel (for example, 304, 316, 316L) or carbon steel, with 316 series stainless steel internal hardware and cartridge seating surfaces. Internal hardware includes tube sheet adapters, tie rods, and seal nuts. Housing closures should utilize quick-release mechanisms such as V-band clamps or fast-action swing bolts to facilitate filter change-outs (Figure 9.6).

Figure 9.6 Sanitary-Style Stainless Steel Filter Housings for large-scale gas and liquid applications. (Courtesy of Pall Ultrafine Filtration Co.)

Design operating pressure of all filter housings should be specified as minimum psig and rated for full vacuum service. Design maximum operating temperature of the housing should also be specified. The housings or pressure vessels that are within the scope of the ASME Boiler and Pressure Vessel Code (BPVC), Section VIII, Division 1 [4], should be designed and U-stamped in accordance with the code. TIG (Tungsten-Inert-Gas) weld construction should be used in sanitary-style housings to minimize weld porosity and ensure high-quality, clean joints, with all internal welds ground smooth and flush. All weld procedures and welders should be qualified to ASME/BPVC, Section IX [5].

Housings should be capable of *in situ* steam sterilization in accordance with the manufacturer's recommended procedures, and housing or system design should provide for condensate drainage. Gasket material and O-ring elastomers must also be capable of withstanding repeated steam sterilization cycles, along with being compatible with process fluids.

Industrial-style housings provide cartridge mounting on a tie rod and sealing to the tie rod assembly by a seal nut at the top of the assembly. Tube sheet adapters should be seal-welded to the tube sheet to prevent fluid bypass. Filter cartridges are thereby sealed in the housing independent of any cover assembly, which ensures positive sealing and no fluid bypass. Filter cartridges should be seated on the tube sheet adapter assemblies above the tube sheet to ensure complete drainage of unfiltered fluid prior to cartridge replacement. This prevents potential contamination of downstream surfaces during change-out of filter elements.

9.3.1 Sanitary Design Housings for Liquid Service

Sanitary design housings are designed to accept sanitary-style filter cartridges. The material of construction should be type 316L stainless steel where the housing comes in contact with product, and should be compatible with the process fluid. Other grades may be used for external hardware and leg stand assemblies, which can be specified where applicable as recommended by the manufacturer.

Internal surfaces are typically finished to a 20–25 microinch Ra (0.51–0.64 μm arithmetic average surface roughness) specification, and external surfaces mechanically polished to a high-quality sanitary finish. Subsequent electropolishing of internal and external housing surfaces provides a smooth, adhesion-resistant surface.

Inlet and outlet sanitary fittings are commonly Tri-clamp® design or equivalent, with vent and drain ports of sanitary stem valve design and no threaded connections in contact with the process fluid. A sanitary Tri-clamp® port for a pressure gauge should be specified where applicable. Jacketed housings used to prevent condensation in vent filters used on holding tanks for hot fluids (such as 80 °C WFI) can have threaded inlet and outlet steam connections, located to facilitate steam entry and exit. Lifting points and handles can be provided on larger multicartridge housings for easy bowl removal.

Sanitary-style housings must be designed so that all product contact areas are readily accessible for thorough cleaning. Housing design should also be optimized for clean-in-place (CIP) processes, with external drains designed and positioned to minimize product holdup.

9.3.2 Stainless Steel Housings for Air/Gas Service

Air/gas service housings should be designed to maximize air/gas flow with minimal pressure drop. The preferred material of construction is again type 316L stainless steel where in contact with the fluid stream. Other stainless steel grades may be used for external hardware. Internal surfaces should be mechanically polished to a uniform 2B finish, and external surfaces can be mechanically buffed to a high-quality finish. After polishing, the entire housing should be passivated.

Inlet and outlet fittings for air/gas service housings should either be 150-lb SO raised flange (RF) or schedule 10S butt weld (American National Standards Institute, ANSI), with appropriate wall thickness corresponding to pipe diameter specifications. Vent and drain ports should be stainless steel fittings with appropriate thread diameters for the housing size.

9.4   Fermentation and Cell Culture—Filter Applications

The term *fermentation* has traditionally been applied to the process of growing yeast or bacteria, either native or genetically engineered, for the biosynthesis of a particular product (for example, cells, antibiotics, amino acids, or recombinant proteins). *Bioprocessing* has more recently been used specifically to refer to large-scale mammalian cell culture production of protein products in modified tank fermentors or other bioreactors such as hollow-fiber modules. Bioprocessing typically also refers to one or more of a sequence of unit operations designed for the manufacture of products of fermentation (such as medium sterilization, fermentation, product recovery, or final product preparation). Nevertheless, the two terms are often used synonymously.

Fermentation and cell culture filtration applications involve preparation of the fermentor or bioreactor liquid and gaseous feeds to support growth of the desired cells. Filtration applications for fermentation are shown in Table 9.1 and illustrated in Figure 9.7. Applications specific to fermentation are described in the immediately following sections, and filtration of utilities such as steam, air, and water is discussed in the Utilities section (Section 9.8) of this chapter.

9.4.1 Prefiltration of Fermentation Air

Compressors are used to generate air flow for the manufacturing facility. There are two types of compressors, oil-free and oil-lubricated. In older facilities, where oil-lubricated air compressors are commonly used, prefiltration of inlet air is necessary for

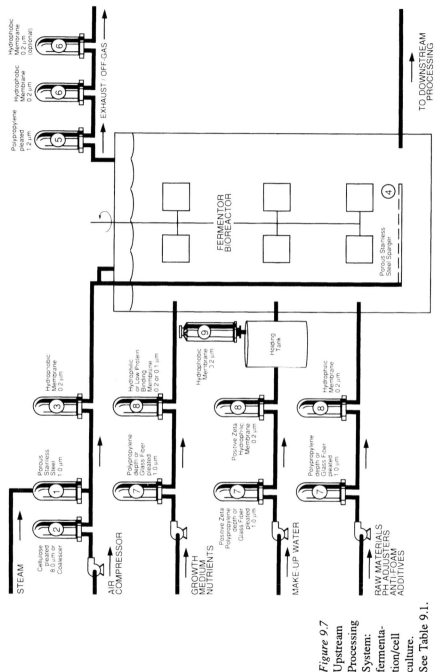

Figure 9.7 Upstream Processing System: fermentation/cell culture. See Table 9.1.

Table 9.1 Filter Recommendations for Fermentation/Cell Culture

| Filter No.[1] | Filter application | Filter type | Absolute rating, μm | Typical flow rates per 10" module | | Typical aqueous clean pressure drop per 10" module[2], psi/gpm |
				Water, gpm	Gas, scfm	
1	Steam for cleaning & sterilizing	Porous stainless steel	1.2	–	130 lb/h (30 psig saturated steam)	–
2	Prefiltration or coalescing of air to sterilizing filters	Cellulosic pleated (oil-free); Coalescer (oil-lubed)	8.0 0.3	– –	75 200–400	– –
3	Sterile filtration of air to fermentor[3]	Hydrophobic membrane	0.2	–	75–100	–
4	Sparging	Porous stainless steel	3.0	–	–	–
5	Prefiltration of exhaust/off-gas	Polypropylene pleated	1.2	–	40	–
6	Sterile filtration of exhaust/off-gas	Hydrophobic membrane	0.2	–	40	–
7	Prefiltration of additives, water, and raw materials	Polypropylene depth (pos. charged); Resin-bonded glass-fiber (pos. charged)	1.0 1.0	1–3 2–4	– –	1.5 0.1
8	Sterile filtration to holding tank, seed tank, or fermentor[4]	Hydrophilic; Low–protein-binding, or pos. charged membrane[5]	0.2 0.1	2–4 1–3	– –	1.2 3–8
9	Tank venting[6]	Hydrophobic membrane	0.2	–	75–100	–

(1) Numbers refer to filter numbers in Figure 9.7.
(2) Liquid pressure drop values are based on water flow rates. For liquids with viscosity differing from water, multiply the pressure drop by the viscosity in centipoise.
(3) Typical clean pressure drop of hydrophobic membrane filters in gas services for 100 scfm (standard ft^3 per min) at 60 psig inlet is 0.8 psi.
(4) Optional 0.1-μm filtration should be used where *Pseudomonas cepacia* and other small organisms, *e.g.*, mycoplasma, have been shown to be a problem.
(5) Low–protein-binding membrane recommended for dilute protein feeds and additives. Positively charged membrane recommended for enhanced removal of endotoxins (pyrogens) from water.
(6) For sizing vent filters for vent service, contact filter manufacturer.

removal of oil droplets. A good coalescing filter, such as a bonded glass fiber filter carrying a positive zeta potential, can provide greater than 99.9% removal of oil and water droplets in the 0.01–0.5-μm range and larger. This also acts as an excellent prefilter for the hydrophobic membrane pleated filters that are commonly used for sterilizing the inlet air to the fermentor.

For oil-free compressors, a prefilter acts to remove dirt in the air system, extending the service life of the final filter. For use with fermentation air, a cellulose pleated filter with an absolute rating of 8.0 μm is normally the filter of choice. Alternatively, polypropylene (2.5-μm rated) or glass fiber (3.0-μm rated) pleated filters also serve as excellent prefilters for this application.

9.4.2 Sterile Air Filtration

Bacteria and bacteriophage, when present in air feeds, can enter fermentation tanks or bioreactors and contaminate the product. Bacteriophage or other viruses can destroy the producing cells and reduce yields.

The recommended filters for sterilization of air feeds to fermentors and bioreactors are the hydrophobic membrane pleated filters. The hydrophobic (water-repelling) nature of these membranes can achieve 100% removal of bacteria and bacteriophage under moist or dry operating conditions. This is an important advantage over glass-fiber towers and cartridges. Filters for sterile air feeds should have a 0.2-μm absolute bacterial rating in liquids and a 0.01-μm particulate rating in air service.

9.4.3 Sparging

Sparging acts to disperse air evenly in the fermentor or bioreactor containing the growth medium and product. The filter of choice for this application is porous stainless steel sparging elements, which can provide an exceptionally uniform and fine aeration gas dispersion. These elements are fabricated with one face of porous metal and one face of solid metal (if both surfaces of the sparging elements were porous, bubbles from the under surface might coalesce with bubbles from the top surface).

Porous stainless steel sparging elements should be positioned horizontally in the tank, with the porous face upward. Fine grades of porous stainless steel (for example, 3.0-μm absolute liquid rated) are ideal even for shear-sensitive mammalian cell cultures because of their high gas transfer and low-shear aeration capability. Elements are available in standard and custom designs.

9.4.4 Exhaust Air/Off-Gas Filtration

Preventing the release of organisms from exhaust and vent lines is a fundamental principle in the control of bioreactors and fermentors. The exhaust filtration system for a recombinant or mammalian cell fermentor/bioreactor must vent sterile air to the environment and provide a sterile barrier to prevent ingress of contaminants. Additionally it must be *in situ* steam sterilizable, and typically has a clean differential pressure of less than 1 psid (7 kPa pressure differential).

The fermentor or bioreactor exhaust gas line can be contaminated with microorganisms or cells, growth medium components expelled from the fermentor/bioreactor as droplets or as solid particles, and aerosol condensate droplets formed during cooling of the gas in the exhaust system. These aerosol droplets, when present, can potentially block the final filter and must be removed before they can reach the final filter. Mechanical separation devices, such as cyclones, condensers, and demisters, may not achieve effective aerosol removal below 5 µm [6]. Removal efficiencies and pressure drops also vary significantly with flow rate in such equipment. Recent studies have shown that aerosols in exhaust lines are predominantly in the very fine 1–5-µm range [7].

The recommended exhaust filtration system design entails two stages using a polypropylene pleated depth filter cartridge as a prefilter to a 0.2-µm absolute rated hydrophobic membrane pleated filter cartridge. The purpose of the prefilter (typically 1.2-µm absolute-rated) is to remove aerosolized particles and liquid droplets containing cells or growth medium from the fermentation off-gas or exhaust air. This serves to extend the service life of the final sterilizing filter. If the medium contains only fully dissolved components, for example, a sterile filtered cell culture medium, and if the fermentation is run at low temperatures (< 30 °C) and low aeration rate (1–1.5 VVM), the prefilter may be optional.

Like the final sterilizing filter, the pleated polypropylene prefilter should be repeatedly steam-sterilizable. An additional function of the prefilter is to retard foamouts from reaching the final sterilizing filter.

Sterilizing-grade 0.2-µm absolute-rated hydrophobic membrane pleated final filters, with PVDF or PTFE membranes, can prevent organisms from entering or leaving the controlled reaction zone, even in the presence of water droplets and saturated gas [8, 9]. Steam sterilizability and integrity test values correlated to microbial retention studies under worst-case liquid challenge conditions provide the highest degree of assurance of satisfactory performance. Redundant systems using a second 0.2-µm-rated sterilizing filter in series are recommended for high-risk recombinant organisms.

If condensate accumulation is excessive, and if the fermentor is operated with overpressure in the head space, positioning the pressure control valve at the fermentor exit upstream of the exhaust gas filters can reduce the amount of condensate accumulating. Formation of condensate can also be prevented if a heating section is fitted into the exhaust gas line upstream of the filters. This can readily be accomplished at the exhaust filter housing if steam jacketing is specified. Elevating exhaust gas temperatures at the filter higher than the temperature at the fermentor exit serves to prevent condensate formation.

9.4.5   Filtration of Fluid Raw Materials

Fermentation and cell culture growth media are frequently sterilized by filtration, especially when heat-labile components are present. During the fermentation process, antifoams, pH adjusters, nutrients, growth factors, and makeup water may also be added. The 0.2-μm absolute-rated hydrophilic membrane pleated sterilizing filter is an ideal choice for these applications. Positive zeta potential hydrophilic membrane pleated filters provide enhanced removal of bacterial endotoxins (pyrogens) from makeup water [10].

Where particle loadings indicate prefiltration, polypropylene depth filters with or without positive zeta potential are recommended as general-service prefilters to 0.2-μm cartridges. Polypropylene pleated filters are also an excellent choice for prefiltration of aggressive solvents and viscous fluids, and for applications requiring exceptionally high flow rates.

Large-scale mammalian cell culture is becoming the most common method of producing therapeutic proteins. Serum content of cell culture medium may range from 10% down to serum-free medium containing low concentrations of specific protein additives. Hydrophilic low–protein-binding filters are recommended for low-serum or serum-free mammalian cell culture medium with dilute protein concentration [11]. Both the membrane prefilters and the 0.2-μm-rated sterilizing filters should be low–protein-binding for optimal protein recovery [12].

9.4.6   Removal of Mycoplasma and Viruses from Medium Feeds

Serum and other biological additives may be potential sources of mycoplasma contamination of the cell culture. Hydrophilic membrane pleated filters rated as 0.1-μm and validated for absolute retention of mycoplasma [13] are widely used to ensure mycoplasma-free serum, serum-containing cell culture medium, and even serum-free medium with growth factor additives.

Among biologicals/plasma fraction producers, the AIDS epidemic, in addition to the risk of hepatitis, has heightened the urgency of preventing viral contamination in serum for mammalian cell culture and plasma fractions. With mammalian cell sources, viral contaminants such as retroviruses must be excluded from the final product. Serum feeds can be a source of viral contamination of the downstream harvest fluid, and use of serum additives free of such adventitious agents is recommended [14].

Elimination of potential virus contaminants from serum in growth media reduces the risk of culture and product contamination. Hydrophilic membrane pleated filters rated at 40 nm (0.04 μm) or less are capable of removing virus particles from serum, plasma, and tissue culture medium. The use of 40-nm filters in series removes viruses in fetal bovine serum [15]. In serum filtration, 0.2-μm sterilizing-grade filters and 0.1-μm absolute-rated filters for removal of mycoplasma can act as prefilters to 40-nm (0.04-μm) filters for viral reduction. The same process is applicable to cell culture medium feed to fermentors and bioreactors.

9.5   Downstream Processing—Filter Applications

The objective of downstream processing is to start with the cells and conditioned broth medium from the fermentor or bioreactor, and to produce a highly purified, biologically active protein product, free of contaminants such as endotoxins, bacteria, particles, or other biologically active molecules. This phase of bioprocessing typically comprises a series of unit operations including cell and cell debris separation, fluid clarification and polishing, concentration, purification, and membrane filtration sterilization of the purified protein product.

The use of through-flow depth and membrane filter cartridges in downstream processing ensures freedom from bacteria and particulate contamination, and helps achieve optimal product purity and yield. Absolute-rated filters can remove contaminants such as cultured cells, bacteria, cell debris, particulates, and viruses from the harvested fluid, or any such contaminants introduced during processing of a fluid stream. In many applications they can be an economical alternative to more costly separation technologies such as centrifuges and cross-flow microfiltration (CFF); in other cases they are complementary to these technologies. Integrity-testable, absolute-rated filters enable aseptic processing conditions that protect concentration and purification systems, while maximizing yield and purity and minimizing the risk of endotoxin contamination.

Cartridge filters are used in many stages of downstream processing, for filtration of both the harvested fluid and the product intermediates, as well as of many additional fluids used throughout the process. These applications for absolute filtration cartridges are described in the following subsections in terms of the benefits offered, typical process streams filtered, and the most commonly recommended filters.

9.5.1 Primary (Bulk) Cell and Cell Debris Removal

The high solids content and viscosity of a fermentation broth or harvested culture fluid often lead process developers to regard this primary solids removal step of downstream processing as an unlikely application for through-flow cartridge filters. This assumption ignores several important facts. Although the attempt to pass a high-solids cell and debris suspension through a membrane filter in "dead-end" mode is very likely to lead to rapid plugging, the use of high-capacity depth prefilters upstream of the membranes can greatly enhance the feasibility, efficiency, and economy of this process.   Absolute-rated cartridge filters can provide single-pass through-flow separation of bulk cell mass from growth media, and are applicable to cultures and suspensions of mammalian, bacterial, or yeast cells.   With extracellular secreted products or disrupted cells, cells and debris can be separated and removed from the broth directly by these filters.

For processing batch volumes up to several hundred liters for bacterial and yeast fermentations, and up to several thousand liters for mammalian cell cultures, the ease of use, simplicity, and very low initial cost of a disposable filter cartridge system are attractive features, compared to installation and operation of centrifugation or CFF systems.   For cell and cell debris removal applications, depth filter cartridges can provide more rapid processing times than CFF, and lower shear than either centrifuges or cross-flow microfilters; easier and more rapid cleaning, sterilizing, and assembly are additional benefits.   Also eliminated are the drawbacks of CFF, such as membrane fouling and inadequate protein transmission.

With centrifuges, and especially with CFF systems, waste solids must be ejected as a slurry, whereas with through-flow filtration, all the liquid containing soluble product is passed through the filter to be collected.   This important additional feature of cartridge filtration provides that liquid volume losses are minimal and therefore yield is maximized.

On large scales, the combination of a centrifuge to remove most of the solids, and subsequent polishing through an absolute-rated depth filter, can often be a better approach than CFF.   It bypasses the hurdle of centrifuges struggling to achieve very high removal efficiencies at particle sizes below 1–2 μm.

The filter types normally recommended for the primary removal of cells and cell debris appear in Table 9.2, along with the absolute removal ratings.   A generic scheme showing filter applications is illustrated in Figure 9.4.   Polypropylene depth filters, with or without positive zeta potential, are recommended for most cell removal applications.   Grades with absolute ratings of 1.0–5.0 μm are typically used with whole mammalian cell suspensions, while 0.3–1.0-μm grades are more applicable for bacteria and cell debris.   High–filter-area, resin-bonded, glass-fiber pleated filters (also with or without positive zeta potential) and polypropylene pleated filters with absolute ratings of 0.7–2.5 μm are recommended for removal of high loadings of cell debris with narrow size distribution.

Specific advantages of using absolute-rated cartridge filters for the applications stated above are:

(a)   100% throughput offered by direct single-pass filtration;

(b)   low shear conditions required for shear-sensitive mammalian cell removal;

(c)   high flow rate required for labile protein products;

(d)   directly scalable from R&D through large pilot-scale operation; and

(e)   potential savings in investment and energy costs when compared with centrifugation and CFF the scales of operation noted above.

Table 9.2 Filter Recommendations for Downstream Processing: Cell/Debris Separation and Clarification

Filter No.[1]	Filter application	Filter type	Absolute rating, μm	Typical flow rates per 10" module		Typical aqueous clean pressure drop per 10" module[2], psi/gpm
				Water, gpm	Gas, scfm	
1	Bulk cell removal from harvested fluid	Polypropylene depth[3]	5.0	1–4	–	0.8
			0.5	1–2.5	–	1.8
2	Secondary cell/debris removal from harvested fluid	Polypropylene depth;	0.5	1–2.5	–	1.8
		Polypropylene pleated; or	1.2	1–3	–	0.2
		Resin-bonded glass-fiber (pos. charged)[4]	1.0	1–3	–	0.1
3	Polishing or sterile filtration of harvested fluid	Hydrophilic; Low–protein-binding membrane	0.2	1–3	–	1.8
			0.1	1–2	–	3.8
			0.04 [5]	1–2	–	4.5
4	Prefiltration of solvents, precipitates	Polypropylene depth	1.0	1–3	–	1.5
5	Sterile filtration of solvents, precipitates, wash buffers	Hydrophilic membrane[6]	0.2	1–3	–	1–8
6	Sterile N_2 blanket	Hydrophobic membrane	0.2	–	75–100	–
7	Tank venting[7]	Hydrophobic membrane	0.2	–	75–100	–

(1) Numbers correspond to filter numbers in Figure 9.9.
(2) Liquid pressure drop values are based on water flow rates. For liquids with viscosity differing from water, multiply the pressure drop by the viscosity in centipoise.
(3) Pore rating of depth filter determined based on cell type and size.
(4) Positively charged prefilters provide enhanced removal of debris finer than absolute rating.
(5) 0.04-μm filters for reduction of potential virus contamination
(6) Positively charged membranes provide enhanced removal of endotoxins (pyrogens) from aqueous buffers.
(7) For sizing filters for vent service, contact filter manufacturer.

9.5.2 Secondary (Residual) Cell and Cell Debris Removal

Preliminary separation processes such as depth prefilters, centrifuges, cross-flow microfilters, settling, or even immobilized cell bioreactors, are typically not capable of producing a sterile or cell- and debris-free effluent. Cartridge filters are invariably employed downstream of this equipment to perform a secondary polishing filtration (Figure 9.8). Secondary filtration serves to further clarify and in some cases sterilize the harvested fluid by removing residual cells, cell debris, bacterial contaminants, and particulate impurities. Absolute removal of particulate solids from the process stream, including sterile filtration, serves as an essential prefiltration/protection step for downstream chromatography and ultrafiltration systems.

Figure 9.8 Cartridge Filters, used for large-scale mammalian cell and debris removal prior to chromatography. (Courtesy of Genetics Institute, Inc.)

The applications most commonly encountered in this category are clarification or sterilization of whole or centrifuged bacterial and mammalian cell supernatants, clarification of homogenized or solvent-lysed centrifuged bacterial supernatants, clarification of precipitated or settled supernatants, and clarification of coarse filtered effluents from filter presses and rotary vacuum filters. More recently, absolute-rated cartridge filters are also being employed to clarify or sterilize effluents from immobilized and continuous mammalian cell cultures grown with microcarriers or hollow fiber bioreactors.

The types and ratings of absolute filters used for polishing, clarification, and sterilization operations are also listed in Table 9.2 and illustrated in Figure 9.9. As discussed earlier, polypropylene depth filters, resin-bonded glass-fiber and polypropylene pleated filters are the absolute cartridge filters recommended for secondary cell and cell debris removal applications, such as downstream of centrifuges and diatomaceous-earth and other coarse prefilters, and settled supernatants. Downstream of these prefilters, or alone if solids loading is already sufficiently reduced, hydrophilic membrane pleated filters are recommended. These filters may be membrane pleated prefilters, or final sterilizing-grade rated at 0.45 or 0.2 μm. Where protein concentrations are already dilute, such as in mammalian cell culture using serum-free culture medium, hydrophilic low–protein-binding membrane pleated filters are recommended.

9.5.3   Reduction of Potential Virus Contaminants

Viral contamination in the final product may come from the cell culture itself, as well as from serum or biological additives. Concern for this possibility is shared by both manufacturers and regulatory agencies, and extends even to undetectable viruses [14, 16]. While several methods of virus reduction or inactivation exist, none has gained widespread acceptance. Integrity-testable filters have been shown to give substantial titer reductions for viruses [15].

In downstream processing, filtration for virus removal using hydrophilic membrane pleated filters rated at 0.04 μm (40 nm) or less is applicable to clarified mammalian cell culture harvested fluids. The recommended position in the purification scheme is immediately after initial clarification and 0.2- or 0.1-μm filtration, or after subsequent concentration by ultrafiltration. Performing virus reduction filtration when protein concentration of the process feed is relatively high reduces loss of product by adsorption on the viral reduction membrane.

Ultrafiltration is often not applicable for virus removal because protein product can be removed along with virus particles in the retentate. Microporous (0.04-μm-rated) membrane cartridge filtration, however, freely passes proteins while providing a sieving mechanism for virus removal [17]. These filters can contribute further assurance of virus-free product when combined with the adsorptive virus removal mechanisms of chromatography during product purification.

Membrane cartridge filtration for virus removal provides many additional advantages over other means of virus removal or inactivation, including:

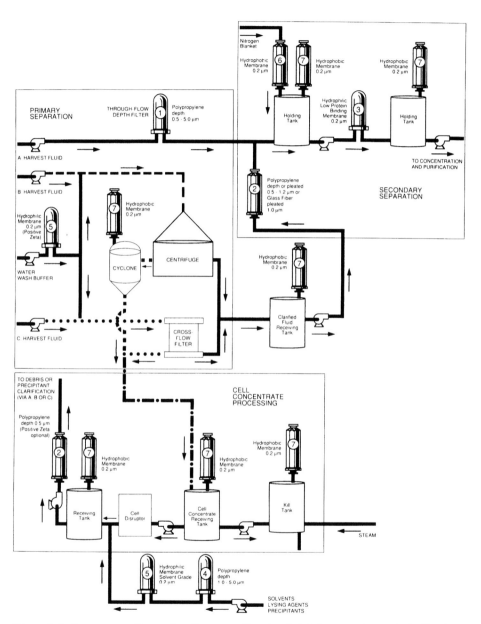

Figure 9.9 Downstream Processing System: cell and cell debris separation and clarification. See Table 9.2.

(a) less product loss than in ultrafiltration,
(b) less activity loss than in heat inactivation,
(c) integrity testing, which can be correlated to virus removal to provide assurance of performance, and
(d) fast and simple processing.

9.5.4 Protection of Concentration and Purification Systems

The specific advantages of using absolute-removal cartridge filters for ancillary process fluids are: (a) protecting downstream concentration and purification systems from premature fouling or clogging due to bacterial, particulate, or endotoxin contamination, and (b) increasing the service life of costly chromatography gels and ultrafiltration membranes, and reducing their cleaning and replacement frequencies.

Process fluids, pumps, piping, and downstream purification equipment themselves can all be sources of microbial and particulate contaminants entering a bioprocess system. The use of absolute filtration serves to eliminate these contaminants from process intermediates and ancillary fluids. Filtration can extend the service life and protect more costly tangential flow microfiltration (TFF) and ultrafiltration (UF) membrane systems and chromatography columns. Solvents, buffer solutions, and other fluids entering a bioprocess must be sterile filtered to maintain aseptic conditions, and particulate impurities must be removed to prevent premature plugging. In most cases, a 0.2-μm-rated sterilizing-grade membrane filter is employed as the fluid filter. Cartridge filtration systems have very low initial cost compared to ultrafiltration systems and chromatography columns, and are therefore more easily justified economically to improve the lifetime, performance, and efficiency of those processes. Where required due to high particulate loadings, the use of an absolute-rated depth filter as a prefilter can have substantial economic advantages in protecting the critical final filter.

The applications where absolute-removal cartridge filters need to be used for prefiltration or sterilization of ancillary process fluids are described in Table 9.3 and illustrated in Figure 9.10. These include prefiltration of chromatography and extraction solvents and buffers with polypropylene depth filters; sterile filtration of solvents and buffers for chromatography, ultrafiltration, diafiltration, and centrifugation washes with hydrophilic membrane pleated filters; and elimination of endotoxins from water and buffers with positively charged hydrophilic membrane pleated filters [18].

An additional application of filters at this stage of downstream purification is the use of porous stainless steel filter disks as support frits in chromatography columns. The mechanical strength, flatness, and controlled porosity with high void volume of this material provides efficient support for the column gel beads with minimal resistance to fluid flow. Porous stainless steel disks are typically supplied as original equipment with process chromatography columns, or can be retrofitted in existing columns.

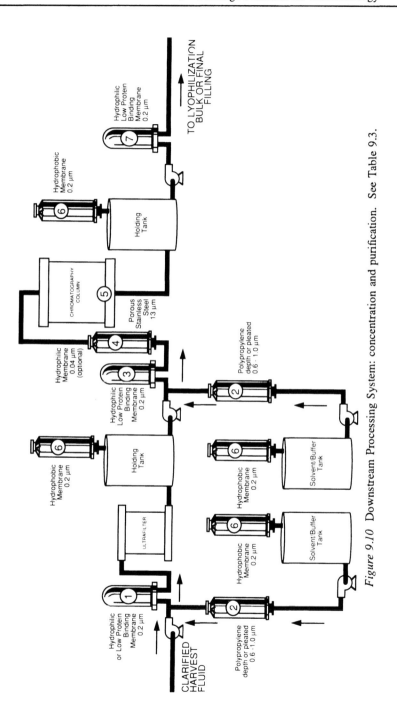

Figure 9.10 Downstream Processing System: concentration and purification.   See Table 9.3.

Table 9.3 Filter Recommendations for Downstream Processing: Concentration and Purification

Filter No.[1]	Filter application	Filter type	Absolute rating, μm	Typical flow rates per 10" module		Typical aqueous clean pressure drop per 10" module[2], psi/gpm
				Water, gpm	Gas, scfm	
1	Polishing or sterile filtration of harvested fluid	Hydrophilic; Low–protein-binding membrane	0.2	1–4	–	1.8
2	Prefiltration of buffers, solvents	Polypropylene depth[3];	1.0	1–3	–	1.5
		Polypropylene pleated[4]	0.6	1–2	–	1.3
3	Sterile filtration of intermediates, buffers, solvents	Hydrophilic membrane[5]	0.2	2–4	–	1.8
4	Virus reduction	Hydrophilic membrane	0.04	1–2	–	4.5
5	Column support	Porous stainless steel	13	–	–	–
6	Tank venting[6]	Hydrophobic membrane	0.2	–	75–100	–
7	Final sterile filtration	Hydrophilic; Low–protein-binding membrane	0.2	2–4	–	1.8

(1) Number corresponds to filter number in Figure 9.10.
(2) Liquid pressure drop values are based on water flow rates. For liquids with viscosity differing from water, multiply the pressure drop by the viscosity in centipoise.
(3) Positively charged prefilters provide enhanced removal of debris finer than absolute rating.
(4) Recommended for filtration of ≥ 1.0 M NaOH for cleaning, sanitizing, and depyrogenation. Use PTFE membrane with fluoropolymer hardware for ≥ 1.0 M NaOH if 0.2-μm filtration is required.
(5) Membranes should be compatible with 0.1 M NaOH. Positively charged membranes provide enhanced removal of endotoxins (pyrogens) from aqueous buffers. Solvent grades provide low extractables for HPLC.
(6) For sizing filters for vent service, contact filter manufacturer.

9.5.5   Final Product Purification and Sterile Filtration

In order to eliminate bacteria and particulate contaminants from purified product, final purification or sterilization must be performed using absolute-removal cartridge filters. Low–protein-binding hydrophilic membrane pleated filters are specifically designed to maximize yield of dilute protein products at these intermediate and final purification or sterile filtration stages. These filters are typically used whenever it is necessary to achieve such objectives as postchromatography filtration of eluted product to remove media fines, final filtration prior to bulk storage, and sterile filtration of unit and bulk fills.

The recommended filter type for dilute final product filtration duties is the 0.2-μm sterilizing-grade hydrophilic, low–protein-binding membrane pleated filter. This and other filters used in final filtration are described in Table 9.4 and illustrated in Figure 9.11.

9.6   Utilities

9.6.1   Water and Chemical Filtration Systems

Manufacturing costs are ultimately lowered, and product quality is improved, when care is taken to process and maintain sterile and particle-free liquids and gases in the production of biotechnology products. A fully integrated filtration system can resolve even the most complex contamination problems to provide economic and efficient delivery of purified liquids and gases for utility applications.

Filters are used in utility services such as water, chemicals, solvents, air/nitrogen or other process gases, and steam (Figure 9.12). Water applications include water for product formulations (WFI), equipment washing (CIP), washing of product containers (WFI and DI), and purified water (PW). Chemical and solvent filtration includes acetone, alcohols, and methylene chloride for crystallization, purification, and use as carrier vehicles. Recommended filters for biotechnology water systems are presented in Table 9.5 and illustrated in Figure 9.13. Recommended filters for chemical filtration systems are presented in Table 9.6 and Figure 9.14.

Table 9.4 Filter Recommendations for Final Purification and Sterilization Systems [1]

Filter No.[2]	Filter application	Filter type	Absolute rating, μm	Typical flow rates per 10" module		Typical aqueous clean pressure drop per 10" module[3], psi/gpm
				Water, gpm	Gas, scfm	
1	Sterile N$_2$ blanket	Hydrophobic membrane	0.2	–	75–100	–
2	Hot WFI filtration to bulk tank	Hydrophilic membrane (PVDF)	0.2	2–4	–	0.6
3	Prefiltration to sterilizing filter for holding tank (optional)	Polypropylene depth	1.0	1–3	–	1.5
4	Sterile filtration into holding tank	Hydrophilic membrane or low–protein-binding membrane	0.2	2–4	–	1.8
5	Tank venting[5]	Hydrophobic membrane	0.2	–	75–100	–
6	Particulate removal[6]	Hydrophilic membrane	5	–	–	–
7	Final filtration at filling machines[6]	Hydrophilic membrane or low–protein-binding membrane	0.2	2–4	–	1.8
8	Hot WFI filtration for container cleaning	Hydrophilic membrane (PVDF)	0.2	2–4	–	0.6
9	Sterile air for container cleaning[4]	Hydrophobic membrane	0.2	–	75–100	–

(1) Includes Small-Volume Parenterals (SVPs) and other small-volume packaging units.
(2) Number corresponds to filter number in Figure 9.11.
(3) Liquid pressure drop values are based on water flow rates. For liquids with viscosity differing from water, multiply the pressure drop by the viscosity in centipoise.
(4) Typical clean pressure drop of hydrophobic membrane filters in gas services for 100 scfm at 60 psig inlet is 0.8 psi.
(5) For sizing filters for vent service, contact filter manufacturer.
(6) Select disk or assembly size to meet flow demand. Low–protein-binding membrane recommended for filtration of dilute protein solutions.

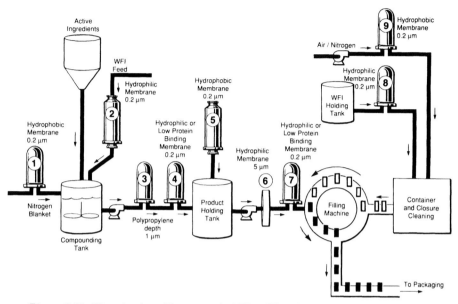

Figure 9.11 Biotechnology Pharmaceutical Final Filtration System. See Table 9.4.

Figure 9.12 Utility Filter Assemblies. Filtration provides contamination control of utility fluids such as water, chemicals, air, steam, and other gases, to lower manufacturing costs and improve product quality. (Courtesy of Pall Ultrafine Filtration Co.)

Table 9.5 Filter Recommendations for Fermentation/Cell Culture

Filter No.[1]	Filter application	Filter type	Absolute rating, μm	Typical flow rates per 10" module		Typical aqueous clean pressure drop per 10" module[2], psi/gpm
				Water, gpm	Gas, scfm	
1	Removal of silica particles	Polypropylene depth	70	10–15	–	< 0.05
2	Protection of ion exchange beds from carbon fines	Polypropylene depth; Resin-bonded glass-fiber (pos. charged)	10–20 10	6–15 3–5	– –	0.3–0.1 0.02
3	Removal of resin fines	Polypropylene depth; Resin-bonded glass-fiber (pos. charged)	1.0 1.0	1–3 2–4	– –	1.5 0.1
4	Deionized water bacteria retentive filtration[2]	Hydrophilic membrane (pos. charged)	0.2 0.1	2–4 1–3	– –	1.2 3.8
5	Tank venting[3]	Hydrophobic membrane	0.2	–	75–100	–
6	Removal of particles generated by pump. Point-of-use filter	Hydrophilic membrane (pos. charged)	0.2 0.45	2–4 3–5	– –	1.2 0.6
7	Removal of particles generated by pump. Point-of-use (WFI)	Hydrophilic membrane	0.2	2–4	–	0.6

(1) Number corresponds to filter number in Figure 9.13.
(2) Optional 0.1-μm filtration should be used where *Pseudomonas cepacia* and other small microorganisms, *e.g.*, mycoplasma, have been shown to be a problem.
(3) For sizing filters for vent service, contact filter manufacturer.

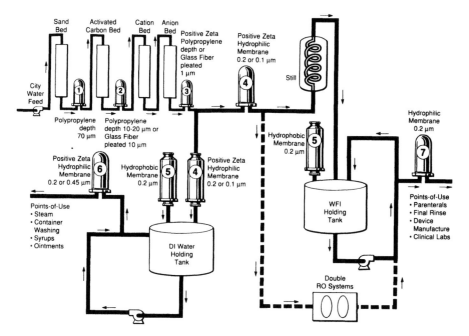

Figure 9.13 Biotechnology Water Filtration System. See Table 9.5.

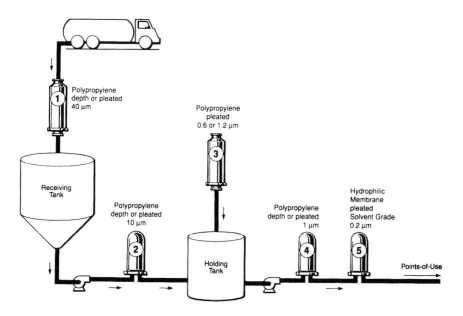

Figure 9.14 Chemical Filtration System. See Table 9.6.

Table 9.6 Filter Recommendations for Fermentation/Cell Culture

Filter No.[1]	Filter application	Filter type	Absolute rating, μm	Typical flow rates per 10" module, gpm	Typical aqueous clean pressure drop per 10" module[2], psi/gpm
1	Coarse prefiltration to remove particles from solvent or chemical	Polypropylene depth;	40	10–15	0.05
		Polypropylene pleated	40	10–15	0.02
2	Fine prefiltration prior to entering a holding tank	Polypropylene depth;	10	6–15	0.3
		Polypropylene pleated	10	6–10	0.04
3	Tank venting[3]	Polypropylene pleated	0.6 1.2	–	–
4	Prefiltration to final filter	Polypropylene depth;	1.0	1–3	1.5
		Polypropylene pleated	1.2	2–3	0.22
5	Final filtration	Membrane pleated (solvent grade)	0.2	2–3	3–4

(1) Number corresponds to filter number in Figure 9.14.
(2) The liquid pressure drop values are based on water flow rates. For liquids with viscosity differing from water, multiply the pressure drop by the viscosity in centipoise.
(3) For sizing filters for vent service, contact filter manufacturer.

9.6.2 Steam Filtration

Process equipment and final filters are frequently sterilized in place by direct steam flow [19]. This eliminates the need for making aseptic connections and risking contamination. Clean steam is required for this *sterilization in place* (*SIP*) of filters, piping, vessels, and filling equipment. Steam is also required for general equipment cleaning and sterilizing. The steam may contain significant amounts of pipe scale and other corrosion products. This particulate material can be removed by filtration to promote overall cleanliness and to avoid burdening the prefilters and final filters.

Particulate contamination in process steam is efficiently retained by porous stainless steel filters with an absolute gas rating of 1.2 μm. Porous stainless steel filter assemblies are typically sized at steam flow rates of 30–40 acfm (actual ft^3 per min) per square foot of filter medium.

9.6.3   Filtration of Air and Gases and Venting Applications

Absolute-rated cartridge filters eliminate contaminants and impurities from air, nitrogen, and other gases used in downstream processing. Vent filtration ensures containment and freedom from product contamination during fluid transfer operations and protects processing equipment during sterilization cycles.

In fermentation, cartridge filters are used to maintain the sterility of the makeup water, feeds, additives, and culture medium in holding tanks, and to decontaminate the fermentor/bioreactor exhaust. Cartridge filters are typically used in downstream processing to filter air and gases; and in venting applications when it is necessary to vent tanks during fluid transfers, to pressurize tanks using inert gases such as nitrogen and argon, to protect vacuum lines, and to vent holding tanks and lyophilizers aseptically. They are also used for gas purging, blanketing, and drying, and when sterilizing equipment by *in situ* steaming or autoclaving.

The recommended filters for nonsterile particulate removal applications are polypropylene pleated filters. Hydrophobic membrane pleated filters such as poly(vinylidene fluoride) (PVDF) or polytetrafluoroethylene (PTFE) are recommended for aseptic processing. The absolute removal rating for the latter filters should be 0.2 μm, determined under liquid flow conditions.

9.7   Steam Sterilization Procedures

Bioprocess equipment is most often sterilized using *in situ* steaming procedures. This section describes the procedures to be adopted for both *in situ* and autoclave sterilization of filter assemblies [16]. First, we make some general recommendations that apply to both methods of steam sterilization. We then present specific recommendations and examples for each method. We also review the most common procedural errors that can result in damage to the filter cartridge during the steaming cycle.

9.7.1   General Recommendations

Sterilizing-grade filters are generally not supplied sterile by the manufacturer. The validation of the steam conditions and sterilization procedure are the responsibility of the process operator. Although time/temperature guidelines can often be cited from the literature [20], these guidelines cannot be substituted for operating experience. As a starting point, a typical guideline of the conditions required to ensure effective sterilization is that the steam be maintained at a minimum pressure of 15 psig (100 kPa) and a temperature of 121 °C for at least 30 minutes. These measurements are taken downstream of the filter housing.

Sterilizing-grade filters should be integrity-tested immediately after steam sterilization. This *pre-use test* is conducted before product is introduced to the filter.

A *post-use test*, conducted after filtration, is also recommended in order to ensure that the cartridge was not damaged during the production run. Often it is also standard practice to conduct an integrity test prior to the initial sterilization. This sterilization test allows early detection of damage to the filter that may have occurred during shipping or handling.

The recommended test methods for verifying the integrity of the filters are the *Forward Flow Integrity test* or the *Pressure Hold Integrity test*. These procedures are described in the Filter Integrity Testing section (Section 9.8) of this chapter.

The filter assembly should be positioned so that condensate from the steam supply cannot accumulate in the housing and the open end of the filter cartridge is oriented in a downward position. The upstream side of the filter system must be equipped with a drain in order to remove condensate formed during the steaming cycle. Piping downstream of the filter should be kept to a minimum.

Steam used for sterilization purposes must be dry and saturated. The steam should be free of particulate matter, such as rust and pipe scale, which can be retained by the sterilizing filter and reduce its service life. A separate steam filter, typically fabricated of porous stainless steel, is recommended for the filtration of steam.

The maximum sterilization temperature and maximum accumulated steam sterilization time are specified by the manufacturer of each type of filter. These recommendations are intended as a guideline and should not be exceeded. The service life of the filter depends on actual process conditions.

We strongly recommend that the filter assembly and downstream equipment be flushed with water, buffer, or product after steam sterilization in order to remove any residues originating from the steam or trace amounts of extractables remaining after sterilization.

9.7.2 *In Situ* Steam Sterilization

A typical filter installation for *in situ* steam sterilization using the recommendations given below is shown in Figure 9.15. Assembly of the filtration system should include installation of pressure gauges upstream and downstream of the filter housing. These are required to monitor both the steam pressure and the differential pressure across the filter assembly during the sterilization cycle. Select pressure gauges that are accurate over the range of 0–45 psi (0–310 kPa).

Condensate that may have accumulated in the steam lines should be drained from the system before steam is allowed to flow into the filter housing. During the steam sterilization cycle, the differential pressure across the filter itself should not exceed the recommendation of the manufacturer. Typically, the stated maximum differential pressure is 5 psi (35 kPa). After steam flow is started, the steam pressure should be increased gradually until the full operating steam pressure is reached and the differential pressure across the filters remains within the maximum specification.

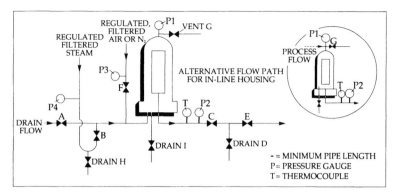

Figure 9.15 Typical Filter Installation for *in situ* steam sterilization. (Courtesy of Pall Ultrafine Filtration Co.) The recommended steaming procedure for the system is:

(1) ensure all valves are closed;

(2) fully open valve C;

(3) fully open condensate drain valve H, housing drain valve I and housing vent valve G;

(4) preset steam pressure (P4) to 5.0 psi above the steam pressure required at the filter assembly, then slowly open steam valve B; after condensate has been expelled from H and I, partially close both valves;

(5) partially close vent valve G when steam flow is evident; when pressure at P2 is within 5.0 psi of pressure at P1, partially open valve D to drain condensate;

(6) permit steam to flow through the system for the required sterilization time, ensuring that no more than 5.0 psi pressure differential (P1 − P2) is developed across the filter;

(7) when sterilization time is complete, close drain valves D, I, and H and vent valve G;

(8) preset pressure (P3) of regulated air or nitrogen at 3.0 psi above the sterilization steam pressure, close steam valve B, and immediately open air or nitrogen valve F;

(9) to flush steam from the assembly, carefully open vent valve G; close valve G after venting;

(10) allow assembly to cool to ambient or to process fluid temperature;

(11) close air or nitrogen valve F; and

(12) relieve the gas pressure in the filter assembly via vent valve G.

   Filter assembly is now ready for use.

During the steaming cycle, all drain and vent valves should be cracked open to ensure that all components of the system have been sterilized. The drain and vent valves should be closed before the steam valve is closed.

   Immediately after the steam valve is closed, nitrogen or air regulated at 3 psi (21 kPa) above the sterilization steam pressure must be introduced at the upstream side of the filter. This prevents a reverse pressure differential from developing across the filter element. To assist cooling, steam may be flushed from the assembly through the opened upstream vent, which should be closed after venting. After the assembly has cooled to ambient temperature, the nitrogen or air valve is closed and the vent valve is opened to relieve the pressure in the assembly.

9.7.3 Alternative Procedures for *In Situ* Steam Sterilization

In situ steam sterilization of water-wetted hydrophilic filters and in-line steaming of filters and downstream equipment require careful modifications of this procedure. These situations are outlined below. Detailed procedures and recommendations should be obtained from the filter manufacturer concerning these modifications.

Steam flow can be readily initiated through a dry filter. When filters have been wetted prior to the sterilization cycle, however, additional steps must be taken in order to establish steam flow but not exceed the maximum differential pressure across the cartridge. The pores of the filter medium must first be cleared of fluid before steam flow can be initiated. In some cases, where sufficient air pressure and flow is available, the pores can be dried if the air pressure is increased above the bubble point of the medium and air is allowed to flow through the assembly.

The filter may also be dried with heat from flowing steam. We recommend that steam be introduced simultaneously on both the upstream and downstream sides of the filter. This procedure permits the steam to be introduced at the sterilization temperature; the maximum differential pressure of 5 psid (35 kPa) across the filter should not be exceeded. The introduction of steam by this procedure greatly reduces the time required to dry the filter.

Some biotechnology processes also require that a positive gas pressure be maintained downstream of the filter after the sterilization is completed. If post-sterilization pressurization is required, an additional hydrophobic vent filter should be installed to maintain the sterility of the downstream section. This hydrophobic filter must also be sterilized during the steaming cycle.

In situ steam sterilization of filters and downstream equipment is a complex process. Sterilization of equipment assemblies requires very careful control and validation, since the required steaming cycle is influenced by the equipment configuration, materials of construction, heat capacity, and volume of the downstream system. In addition, the vent filter must be steamed in the reverse direction.

It is always preferable to sterilize the filter and downstream components separately. However, if the filters and downstream equipment are steamed simultaneously, appropriate vacuum relief safety devices must be fitted on any downstream vessels that cannot withstand negative pressure without collapse.

9.7.4 Elimination of Reverse Pressurization

We recommend that an integrity test be conducted at the conclusion of the sterilization cycle, to ensure that the filter has not been damaged during the steaming procedure. Damage to filter cartridges at this stage has occasionally been experienced. Analyses of such incidents have shown that over 70% of the failures can be attributed to a reverse pressurization condition that developed during the cooling phase of the steam sterilization cycle. The possibility of damage to the filter from exposure to a reverse pressurization condition is increased during steam sterilization, since the element is also subjected to elevated temperature conditions.

Reverse pressurization conditions can be avoided by following recommended procedures [21]. During *in situ* steam sterilization, the steam is introduced into the filter housing in a procedure that displaces the air from the housing and then permits the steam to flow through the filter and exit through the downstream outlet of the system. At the conclusion of the *in situ* steam sterilization cycle, the steam valve is closed, and air or nitrogen is applied to the system at approximately the same pressure as the steam. This procedure is the critical step in preventing reverse pressurization damage.

The most common operator error at this point is to close the steam valve and not to apply the specified air or nitrogen pressure. If the steam valve is shut and a noncondensable gas is not introduced, the housing is essentially an isolated system. Due to differential heat transfer during the cooling cycle, steam condenses to liquid water at different rates upstream and downstream of the filter. As a consequence, pressure differences of 5–15 psi (35–104 kPa) across the wetted filter medium may develop. The condition may exist for only several minutes, but this period is sufficient for permanent damage to be done to the filter.

Filter damage can be completely avoided by maintaining the pressure through the introduction of a noncondensable gas (air or nitrogen) at the conclusion of the steaming cycle. This step eliminates the conditions for reverse pressurization.

9.7.5   Autoclave Sterilization Procedures

During autoclave sterilization, it is essential that air not be trapped in the equipment to be sterilized and that it be readily displaced with steam. If air remains in the system, the time required to effect sterilization is significantly increased. After a cartridge is installed in the filter head, the outlet connection (for example, bell or hose adapter) is also covered with an approved steam-porous covering, which should be loosely sealed or taped to allow for adequate steam penetration. We recommend that the filter head and bowl be separated during autoclaving to facilitate the purging of air. Longer cycle times may result if the assembly is closed. The filter cartridge should not support the head or rest against the bowl during the autoclave cycle.

We suggest a system purge of two vacuum cycles to remove the noncondensable gases, which may interfere with the sterilization. During each vacuum cycle, the internal absolute pressure within the autoclave must be reduced to at least 1 psia. A validation test is required to ensure that the filter assembly and associated items are sterilized by the autoclave steaming conditions.

When a downstream receiving vessel is connected to the filter outlet during the autoclave cycle, the receiver volume should not exceed 6.5 gallons (0.025 m³) nor should the connecting tube be longer than 5 feet (150 cm). The receiver must be fitted with a hydrophobic vent filter. All valves must be left open and outlets covered with an approved steam-compatible porous covering. For larger receivers or for tubing exceeding 5 feet (150 cm), the filter and other components should be autoclaved separately and connected aseptically.

A slow exhaust cycle must be specified at the end of the cycle. Typically, 10–20 minutes are required to fully exhaust the autoclave. A rapid exhaust cycle should not be used. Any obstruction that prevents an equalization of temperature and pressure between the upstream and the downstream sides of the filter during exhausting or cooling should be avoided. In a single filter, such an obstruction may be a fitting on the downstream side of the cartridge, or an improper sanitary wrapping that obstructs the flow. If the filter is attached to a receiving vessel or other processing apparatus, an obstruction at any point along the downstream path may limit the air flow. After the sterilization cycle is complete, filters autoclaved separately from downstream equipment should be installed into the process system using aseptic techniques.

9.8 Filter Integrity Testing

In many critical biotechnology fluid processes involving filtration, there is a requirement to achieve the highest possible assurance of filter integrity and removal efficiency [22]. Such requirements are often at the terminal production stages, but may also be at key intermediate stages. Examples include sterile filtration of parenterals and biological liquids, sterile filtration of air for fermentation, and sterile vent applications.

In such applications, installation of integrity-testable filters and performance of routine integrity testing by the user is essential to demonstrate that the system is performing to specification. The application of nondestructive integrity tests is contingent on the development of validation data correlating nondestructive integrity test values with destructive bacterial challenge testing, and the correlation of filter performance with process conditions.

Filter validation requires challenge testing using a procedure sensitive enough to detect the passage of contaminants of interest. For example, challenges of 0.2-μm sterilizing-grade filter elements are performed under simulated process conditions with *Pseudomonas diminuta* bacteria (ATCC 19146). This is a destructive test and cannot be performed in a production environment to evaluate filter integrity. Therefore, this test must be correlated to a nondestructive integrity test. From the correlation, filter performance can be safely verified in the production environment using a suitable nondestructive integrity test. We discuss filter validation in greater detail in Section 9.9.

The industry-accepted nondestructive integrity tests, such as the *forward flow*, *pressure hold*, or *bubble point* tests, are performed by air pressurization of a wetted filter and observation of air flow through the filter. The forward flow and pressure hold tests quantitatively measure both diffusive and bulk gas flow through all the pores of a wetted filter, while the bubble point test is qualitative and is dependent upon visual observation of bubbling through the largest pores of a wetted filter. These integrity test methods are described in the following sections.

9.8.1 Forward Flow Test

The filter is wetted and a predetermined constant air pressure is applied. A measurement of pure diffusional air flow through the wetted membrane is made. If the diffusional air flow across the membrane is below the maximum allowable value given (originally determined by bacterial challenge and filter validation studies), then the filter is acceptable. Forward flow is an objective and quantitative method of determining filter integrity. Diffusive forward flow can be determined either upstream or downstream of the test filter.

The downstream forward flow test is a measurement of the flow of gas that diffuses across the wetted membrane under the constant test pressure. Because it utilizes a downstream connection for the flowmeter, this test is recommended for use before sterilization of a filter or after the completion of a sterile filtration.

The upstream forward flow test measures the flow of gas required to maintain pressure on the upstream side of the filter as diffusion occurs across the wetted membrane under constant test pressure. It does not require that sterile downstream connections be disturbed, and may be performed *in situ* or off-line, before and after sterile filtration, without compromising the sterility of the downstream system.

The forward flow test is correlated to a bacterial challenge test capable of reliable detection of penetration to one organism in 10^{13}. Membrane pleated filter cartridges can produce a titer reduction value (T_R) greater than 10^{13}. In contrast, glass-fiber cartridges and some PTFE filters used in sterile air applications are validated with a dioctyl phthalate (DOP) 0.3-μm oil-droplet aerosol test with a test sensitivity of only 10^6. Thus the forward flow test method is 10^7 (10,000,000) times more sensitive than the DOP aerosol method, offering significantly greater assurance of bacterial retention.

9.8.2 Pressure Hold Test

The pressure hold test is a modified form of the upstream forward flow test. It can be performed after sterilization of the filter assembly, before filtration, and after sterile filtration, since the downstream sterile connections are not disturbed. Another advantage of the pressure hold test is that it also confirms the integrity of the entire assembly including housing seals.

The filter assembly is pressurized to a predetermined setting and then isolated from the pressure source. Diffusion of gas across the wetted membrane is quantitatively measured as a decay in pressure over a specified period of time. This is a very easy test to perform, but it is important that the upstream volume be known, that temperature in the upstream volume of the filter housing be kept constant, and there be no gas leaks in the upstream system. Published values of pressure decay are based on standard housings with upstream valves located as close as possible to the filter housing.

9.8.3 Bubble Point Test

In the bubble point test, air pressure is applied to a wetted filter disk and slowly increased until water is expelled from the largest pores such that bubbles appear from a submerged tube in a downstream collection vessel. Vigorous bubbling, as opposed to a diffusional air flow or occasional bubbles, is indicative of reaching the bubble point. This visual test can be fairly accurate for low-area filters, such as disks. When used to evaluate high-area filters, however, it is subject to limitations in observation, test time, collection conditions, and pressurization rates. The bubble point test is not recommended for integrity testing of filter cartridges of any style, but should be used for the testing only of small–surface-area disks (for example, up to 142 mm diameter).

9.8.4 Factors Influencing Integrity Testing

Several physical factors can affect integrity test performance. The most important are temperature, surface tension of the wetting liquid, and hydrophobicity of the filter medium.

Testing must be performed as close to constant temperature as possible. The physical characteristics of air and wetting fluid are altered by extremes in temperature. Therefore, temperatures other than ambient may require a set of integrity test values specific for those process conditions. It is good practice to allow the filter system to cool to room temperature after autoclaving or steaming, or to warm after exposure to cold temperatures.

For a given integrity test value, the test pressure varies with the surface tension and wetting characteristics of the product or of the wetting fluid. Since the surface tension of a product may differ from that of water (a standard wetting liquid with a surface tension of 74 dyne/cm), it may be necessary to completely flush all product out of a filter prior to integrity testing with water, or to perform the test using the product or another suitable wetting liquid. Tests performed with liquids other than water require test values based on the physical and chemical characteristics of these liquids.

Hydrophobic membrane pleated filters, such as PVDF or PTFE membrane filters, are inherently nonwettable with water. They should be wetted with an appropriate solvent, such as 60:40 isopropanol:water (v/v) or 25% v/v aqueous solution of *t*-butyl alcohol. Because of the hydrophobic nature of these membranes, it is best to wet the filter by a thorough flush of the wetting liquid through the filter, to ensure complete wetting and ease of integrity testing. Incomplete wetting results in inaccurately high forward flows or rapid pressure drops, thereby providing a false failure or fail-safe mechanism.

The filter manufacturer is responsible for providing the integrity test values for its recommended filter and housing systems. The operator is responsible for obtaining the correct test parameters, recommended wetting agent, test procedures, and values from the filter manufacturer. For multicartridge systems, retrofit filter elements from alternative vendors, special, and other types of housings, the filter manufacturer should be contacted for test values and procedures.

9.8.5 Integrity Testing in Production

Filters must be visually inspected prior to their installation. With sanitary-style sterilizing-grade filter cartridges, an effective cartridge O-ring seal is essential to ensure sterile filtration integrity. O rings should be checked for proper installation and absence of nicks and particulates. Prior to sterilization, water or product may be used to aid insertion of the open-end adapter into the housing. A forward flow or pressure hold test may be performed upon receipt, to verify integrity of the filter after shipping.

Filters for sterilizing applications are installed and sterilized *in situ*, or autoclave-sterilized and installed aseptically. A forward flow or pressure hold test should be performed after sterilization prior to use. If a process liquid other than water is used to flush the assembly and wet the membrane, then an appropriate integrity test value for the wetting liquid must be used.

At the completion of the filtration process, a forward flow or pressure hold test should be performed again to verify that integrity was maintained under the process conditions. Any apparent failure of a pressure hold test should be confirmed by testing for leaks in the upstream side of the system and by a forward flow test. For multicartridge systems, should the forward flow or pressure hold test values exceed the recommended test limits, then the individual cartridges must be tested in a single cartridge housing. If each individual cartridge proves satisfactory, the other source(s) must be investigated.

9.8.6 Automated Integrity Test Instruments

Filter integrity tests performed manually can be labor- and time-intensive, and results may be subjective, open to interpretation, and sometimes dependent on operator technique. Integrity test instruments provide the benefit of automation and simplify these critical filter integrity tests, which are fundamental to a good biotechnology process quality control program. Automated integrity test units quantitatively and reliably perform filter integrity tests recommended by the Food and Drug Administration

(FDA) [23]. Tests include forward flow, pressure hold (decay), and bubble point tests. Instruments should be accurate to 0.1% of the set test pressure over the complete measuring range of 87 psi (600 kPa), with automatic zero calibration before every test.

Automation of integrity testing eliminates the potential of operator error. Electronic circuitry and microprocessors enable precise control and monitoring of test parameters. Sophisticated software can prompt the operator through all functional steps. Automated integrity test instruments with standard software have been considered to be *peripheral devices* by the FDA and not subject to Compliance Policy Guide 7132a.15, May 1987 [24], which requires inspection of software source codes for *application-specific software*. The software should be prepared according to standard operating procedures, standardized, and documented in a recognizable way so that the logic can be readily understood.

Advanced instruments are programmable for permanent memory storage of integrity test parameters for up to 99 different filter assemblies. Hard-copy paper printout of test parameters and results can be directly incorporated into batch records.

9.9   Validation of Filters Used in Biotechnology Manufacturing

No new implementing regulations have been adopted by the United States Food and Drug Administration (FDA) to deal with products of the new biotechnology. This continues to be true despite the diverse nature of these products. Indeed, FDA has stated that the agency's existing requirements are sufficiently comprehensive to apply to the new developments in recombinant DNA research in areas ranging from drugs to biologics, from diagnostics to food additives [25]. Newness has been seen largely as a function of changing cell substrates, and the FDA has long regulated such products from a multitude of manufacturing processes.

Guidelines for process validation [26] exemplify this approach. They are general principles and not a process-by-process compilation of concerns. In these guidelines, validation is defined as "Establishing documented evidence which provides a high degree of assurance that a specific process will consistently produce a product meeting its predetermined specifications and quality attributes". Therefore, FDA guidelines for process validation apply equally to validation of a fermentation process for recombinant DNA drug products [27] and to other processes.

9.9.1   Validation Concerns

Certain common concerns apply to all validation protocols. For filtration processes, these validation concerns are described in Table 9.7.

Table 9.7 Concerns in the Validation of Filters

Concern	Equipment qualification	Process performance	Worst-case test condition	Change control (revalidation)	Documentation
Bacterial removal	Challenge complete assembly with > 10^7 *P. diminuta* per cm^2 of filter surface	Filtrate is sterile after simulated fluid fill	Filter challenged to plugging (> 50 psid)	New conditions or new fluid groups	Test data, conditions, and observations
Integrity testing	Correlate test with removal efficiency for contaminant(s) of interest	Tested prior to and after each use	Safety margins included in minimum bubble point or maximum forward flow/pressure hold values	New filter construction or test pressure	Batch records
Operating conditions	Define maximum temperature and pressure rating	Integrity maintained under extreme pH, viscosity, flow rate, hydraulic shock	Challenged with endotoxin and bacteria while testing to plugging	Reuse, longer service	Batch records
Equipment sterilization	Verify heat distribution	*D* value achieved	Heat-resistant bacterial spore (biological indicator) included	New cycle or new equipment	Validation and processing records
Materials of construction	Demonstrate nontoxicity per biological tests for USP Class VI plastics at 121 °C	Processed drug product tested for biological safety	Drug product processed at upper and lower operating limits	New suppliers or new materials	FDA master file submittal
Particulate cleanliness	Determine counts as a function of flush volume	Meets USP requirements for particulate matter	Effluent counts assayed at upper flow rates, maximum pressure drops	New process conditions	Test and blank counts
Extractables	Determine quantity as a function of flush volume	Meets USP requirements for total solids and oxidizable substances	Samples autoclaved in solvent groups of interest	New solvent group or new procedure	Test data and conditions
Sterilization batch records	Establish serial number traceability	Process variables recorded	Environmental monitoring and product bioburden documented	Interdepartmental review	Approval signatures

These concerns were formulated for filtration validation by a filter manufacturer [28] at a time when the FDA had proposed detailed, process-specific Good Manufacturing Practice (GMP) regulations for large-volume parenteral (LVP) products [29]. Although the LVP GMPs were never finalized, and became outdated within a decade, the proposals in regard to filtration have continued to evolve and subsequently have appeared in Guidelines for the Validation of Aseptic Processing [22] and Guidelines for the Manufacture of In Vitro Diagnostic Products [30].   Manufacturers of pharmaceutical products also formalized and published their approach to filtration validation [31–33].   All of these references, insofar as they deal with validation of filters, may serve as source documents in biotechnology manufacturing.

The FDA expects each regulated manufacturer to establish documentation that the filtration system is appropriate for what is being filtered. This requires monitoring of the operating environment and personnel hygiene as well as the bioburden of raw materials and finished products. It also requires keeping the bioburden as low as possible, regardless of the ability of filtration to remove far greater numbers, in order to assure the greatest safety factor during production.

9.9.2 Validation Programs

The validation program itself often starts with equipment qualification. This consists of performance data in the absence of product.   It provides a baseline level of information for validating equipment, and can be obtained primarily from the equipment manufacturers' published data and test results [28–30]. Where drug product characteristics and process conditions are substantially equivalent to those already tested by a filter manufacturer, additional testing with a simulated product may not be required. If process performance data are necessary, however, then additional testing for at least three runs under actual use conditions may be performed, to ensure that filtration is not adversely affected by the extremes of product characteristics and process conditions that will be encountered. As a result, the validation program in its broadest form may encompass equipment qualification, process performance data, worst-case testing, and periodic revalidation, with proper documentation for each of these.

While there are many correct ways to conduct such comprehensive validation programs, each program dealing with filtration processes needs to address the eight validation concerns listed in Table 9.7. For each of these concerns, certain data are readily available from equipment manufacturers, and other data may be attainable only through use of that equipment [34]. Protocol considerations for each of the eight validation concerns, at different stages in the validation program, are outlined in Table 9.7. A detailed discussion of all eight concerns is beyond the scope of this chapter; however, key points regarding validation of microbial removal ratings are discussed below.

9.9.3 Contaminant Retention and Sterilization

Contaminant removal capability is a primary validation concern in filtration, and must be demonstrated for biotechnology processing. Filters intended to sterilize fluids must be shown to remove all bacteria in a worst-case challenge test. Therefore, filter manufacturers have provided laboratory test data showing that specific types of filters are able to remove bacteria of a certain small size. For example, a 0.2-μm-rated filter is expected to be able to remove all incident *Pseudomonas diminuta* (ATCC 19146), because this bacterium is 0.3 μm in diameter and 0.6–0.8 μm in length. In liquids it penetrates 0.45-μm-rated filter media, but is retained quantitatively by 0.2-μm-rated filter media [35]. Pharmaceutical manufacturers have provided process test data (retrospective and prospective) demonstrating complete removal of bacteria by filtration under their conditions of use, or they have documented the equivalency of their process parameters with the laboratory test parameters used by the filter manufacturer.

While no standard microbial challenge test methodology is applicable to all situations, the American Society for Testing and Materials (ASTM) has published a generalized experimental protocol [36]. This protocol is suitable for validating a sterilizing-grade filter, which is considered at present one that produces a sterile effluent at a minimum challenge concentration of 10^7 *Pseudomonas diminuta* per square centimeter of filter surface area [22]. Filtration efficiency is concentration-dependent, and, although higher challenge concentrations are possible, they are more difficult to prepare and utilize properly. In addition, to obtain a worst-case challenge, testing should be carried out to clogging. This can be defined as the maximum pressure drop across the filter before it is removed from service (for example, 50–60 psi or 350–400 kPa).

Working at such extreme conditions of concentration and pressure drop requires rigorous attention to detail. Proper sampling procedures, blanks, and controls are critical when experimental variables are being set to their highest possible values. Analytical capability can be easily compromised by inadvertent contamination or system bypass. For example, damage to the test microorganisms during testing or recovery can result in failure to recover, while inadequate O-ring squeeze can result in downstream recovery of bacteria; in both instances, false results are obtained.

9.9.4 Mycoplasma and Virus Filtration

In applications of filtration of serum, cell culture media, or deionized water, elimination of mycoplasma or certain small water-line microorganisms may make it necessary to use and validate 0.1-μm-rated filters. A challenge test using *Acholeplasma laidlawii* (ATCC 23206) has been developed for this purpose [13].

Filters rated at 0.04 μm are applicable to reduction of viral contamination from serum, cell culture media, and harvest fluids. For validation of 0.04-μm-rated filters (40-nm filters), challenge testing has been performed using reovirus (an icosahedral virus 75 nm in diameter) and MS 2 bacteriophage (another icosahedral virus 24 nm in diameter) [29]. Viral contaminants are particularly important in biotechnology pharmaceutical manufacturing because even materials that are virus-free when harvested may become contaminated with a retrovirus prior to administration to a human or an animal.

The FDA *Points to Consider* documents specifically indicate incorporation of validated processes to remove or inactivate viruses potentially contaminating rDNA or cell-culture–derived products [37]. They strongly recommend employing validated procedures such as filtration, capable of removing viruses if present, during purification of, for example, hybridoma products [14]. Spiking experiments are suggested for accomplishing process validation to establish exclusion of potential virus contaminants.

9.9.5 Product and Process Simulation

Microbial challenge testing is further complicated by the need to simulate other aspects of fluid and contaminant variation. For example, for certain groups of drug product vehicles, freedom from liquid vehicle effects may need to be validated. The performance of sterilizing filters must be shown not to vary when process solutions contain additives such as surfactants. These are known to enhance filter penetration when the zeta potential of exposed sites throughout the filter membrane microstructure is of like sign to the microbial challenge, resulting in repulsive electrostatic interaction [38]. The removal capability of positively charged filters, therefore, varies with factors such as solution pH, ionic strength, and surface charge density. Validation data for such a limiting process condition must include filter removal ratings derived in the absence of charge effects that enhance contaminant removal capability.

Another set of limiting process conditions comes into play when high-viscosity liquids are filtered. It had been questioned whether the initial elevated pressures often applied in filtration of high-viscosity liquids might enhance penetration by, or extrusion of, contaminants [39]. Test results at 50 psi (350 kPa) with *Pseudomonas diminuta* in glycerol solution adjusted to 160 centipoise, however, demonstrated that 0.2-μm-rated nylon 66 membrane filters can provide sterile effluents as well under these conditions as they do in aqueous fluids of low viscosity [40].

Other standardized challenge tests have been developed. A microbial aerosol challenge test has been described for validating such air and gas applications as vent filters, compressed air used in sterilizers, air or nitrogen used for product and in-process solution transfers and at filling lines, and air or nitrogen used in fermentors

[41]. The efficiency of filters is greater in gases than in liquids, and 0.45-μm filters can be validated as sterilizing-grade in gases. This is due to the relative effectiveness of aerodynamic mechanisms such as inertial effects and Brownian diffusion, which increase the probability of impingement and removal.

Not all tests of contaminant removal capability are concerned with sterilizing-grade filters. If a filter is being used for particulate removal, or in a process where the drug product is terminally heat sterilized, then 0.45-μm filtration may be validated. This is commonly performed utilizing a slightly larger microorganism, *Serratia marcescens* (ATCC 14756). Particle challenge tests also have been developed to qualify removal ratings in various applications [42].

9.9.6 Integrity Testing

Integrity testing is an essential consideration in the validation of all fluids handling equipment, and one that directly influences contaminant removal capability. For fluid filtration, integrity testing serves to qualify removal ratings as well as to confirm system integrity. Because filter manufacturers have access to prototype or worst-case filters produced under the widest range of manufacturing conditions, they can provide empirical correlation between nondestructive integrity test limits and the ability of a filter to remove standardized contaminants such as bacteria or particulates. Integrity testing of filters is discussed in detail in Section 9.8.

9.9.7 Preproduction to Manufacturing

Validation serves as a unique link between basic science and industrial production. In the evolution from laboratory bench-scale operations to routine manufacturing, validation demonstrates the efficacy of the processing techniques. The emergence of the new biotechnology has, therefore, raised new regulatory questions regarding the efficacy of the new bioprocessing equipment and techniques. Validation studies undertaken by equipment manufacturers and users have, by addressing these concerns, continued to advance the state of the art [43].

9.10 References

1. THOMAS, A. J., *et al.*, Validation of Filter Integrity by Measurement of the Pore-Distribution Function, *Pharm. Technol. 16* (2), 32–36 (1992).
2a. Title 21, *Code of Federal Regulations (CFR-21)*, Parts 170–199; Subchapter B—Food for Human Consumption; Food and Drug Administration; Superintendent of Documents, Washington, DC; 1991 ed.

2b. *Ibid.*; Sections 211.72—Equipment, filters; and Section 210.3—Definitions.

3a. *United States Pharmacopeia XXII/National Formulary XVII*; Section 88, Biological Reactivity Tests, *In Vivo*; United States Pharmacopeial Convention, Rockville, MD, 1990.

3b. *Ibid.*; Section 788, Particulate Matter In Injections.

3c. *Ibid.*; Monograph on Purified Water.

3d. *Ibid.*; Section 85, Bacterial Endotoxins Test.

4. *Rules for Construction of Vessels*; Unfired Pressure Vessels, Section VIII Division I—Appendix; ASME Boiler and Pressure Vessel Code; American Society of Mechanical Engineers, New York.

5. *Qualification Standard for Welding and Brazing Procedures, Welders, Brazers, and Welding and Brazing Operators*; Unfired Pressure Vessels, Section IX; ASME Boiler and Pressure Vessel Code; American Society of Mechanical Engineers, New York.

6. PORTER, H. F.; Gas-Solid Systems; Chapter 20 in: *Perry's Chemical Engineers' Handbook*; CHILTON, C. H., PERRY, R. H., Eds.; 5th ed., McGraw-Hill, New York, 1973.

7. JAENCHEN, R., DELLWEG, R.; Process Parameters of an Economic Use of Membrane Filters in Fermentation Exhaust Gas Filtration; *Pall Scientific and Technical Report STR-1149*; Pall Ultrafine Filtration Co., East Hills, NY, 1989.

8. CONWAY, R. S.; Selection Criteria for Fermentation Air Filters; In: *Comprehensive Biotechnology*; COONEY, C., Ed.; Academic Press, New York, 1984.

9. BRUNO, C. F., SZALO, L. A., Fermentation Air Filtration Upgrading by Use of Membrane Cartridge Filters, *Biotechnol. Bioeng. 25*, 1223–1227 (1983).

10. JAMES, J. B., BLANDEN, P. D., KRYGIER, V., HOWARD, G.; Bacterial and Endotoxin Retention by Sterilizing Grade Filters During Long Term Use in Pharmaceutical High Purity Water Systems; *Pall Scientific and Technical Report STR-PUF03*; Pall Ultrafine Filtration Co., East Hills, NY, 1988.

11. MARTIN, J. M., MANTEUFFEL, R. L., Protein Recovery from Effluents of Microporous Membranes, *BioPharm 1* (10), 20–27 (1988).

12. DATAR, R., *et al.*, Dynamics of Protein Recovery from Process Filtration Systems Using Microporous Membrane Filter Cartridges, *J. Parenter. Sci. Technol. 42*, 35–42 (1992).

13. MEEKER, J. T., *et al.*, A Quantitative Method for Challenging 0.1 μm Rated Filters with *A. laidlawii*, *BioPharm 5* (2), 41–44 (1992).

14. *Points to Consider in the Manufacture and Testing of Monoclonal Antibody Products for Human Use*; Center for Biologics Evaluation and Research (CBER), Food and Drug Administration, Rockville, MD, 1987.

15. Greater Retention of Contaminants in Fetal Bovine Serum: 40 nm Filtration; *Special Report Vol. 2* (No. 1); Hyclone Laboratories, Inc., Logan, Utah, 1989.

16. BUILDER, S. E., VAN REIS, R., PAONI, N. F., FIELD, M., OGEZ, J. R.; Process Development in the Regulatory Approval of Tissue-Type Plasminogen Activator; In: *International Biotechnology Symposium*; DURAND, G., *et al.*, Eds.; 8th ed., Vol. 1, Knokke, Belgium, 1988.

17. TOTII, C. S.; Retention of Viral Contaminants by 0.04 μm Nylon 66 Filters; *Scientific and Technical Report STR-PUF14*; Pall Ultrafine Filtration Co., East Hills, NY, 1990.

18. WICKERT, K., Endotoxin Control Using Charge Modified Filter Media, *Pharm. Eng. 11* (4), 25–28 (1991).

19. Steam Sterilization of Pall Filter Assemblies; *Publication PCS-500*; Pall Ultrafine Filtration Co., East Hills, NY, April 1987.
20. STUMBO, C. R.; *Thermal Bacteriology in Food Processing*; 2nd ed.; Academic Press, New York, 1973.
21. SCHUBERT, P.; The Causes of Reverse Pressurization Problems; *Publication PUF-169*; Pall Ultrafine Filtration Co., East Hills, NY, 1986.
22. Food and Drug Administration, Good Manufacturing Processes Used for Sterilization of Liquids, *J. Parenter. Drug Assoc. 38* (1), 37–43 (1984).
23. *Guideline on Sterile Drug Products Produced by Aseptic Processing*; Division of Manufacturing and Product Quality (HFN-320), Office of Compliance, Center for Drugs and Biologics (CDER/CBER), Food and Drug Administration, Rockville, MD, June, 1987; p 31.
24. *Guidance Manual for Drugs and Biologics, Section 7132a.15*; Food and Drug Administration Compliance Program (HFA-250), Food and Drug Administration, Rockville, MD, May 1987 (rev. 1990).
25. MILLER, H. I., FDA Regulation of Products of the New Biotechnology, *Am. Biotechnol. Laboratory 6* (1), 38–43 (1988).
26. *Guideline on General Principles of Process Validation*; Center for Drugs and Biologics, Food and Drug Administration, Rockville, Maryland, May 1987.
27. CHIU, Y. H., Validation of the Fermentation Process for the Production of Recombinant DNA Drugs, *Pharm. Technol. 12* (6), 132–138 (1988).
28. DUBERSTEIN, R., Pall Corporation, personal communication, July 1980.
29. Food and Drug Administration, Proposed Rules, Current Good Manufacturing Practice in the Manufacture, Processing, Packing, or Holding of Large Volume Parenterals, *Fed. Regist. 41* (106), 22202–22219 (June 1976).
30. *Draft GMP Guidelines for the Manufacture of In Vitro Diagnostic Products*; Center for Devices and Radiological Health, Food and Drug Administration, Silver Spring, MD, March 1988.
31. Validation Guide for Pall 0.2 μm Nylon 66 Membrane Cartridges; *Publication No. TR-680*; Pall Ultrafine Filtration Co., East Hills, NY, 1985.
32. Validation Guide for Pall Emflon® II Membrane Cartridges for Air and Gas Filtration; *Publication Nos. TR-870 and TR-880B*; Pall Ultrafine Filtration Co., East Hills, NY, 1987.
33. Validation Guide for Pall Bio-Inert® Membrane Filter Cartridges; *Publication No. TR-890*; Pall Ultrafine Filtration Co., East Hills, NY, 1987.
34. Validation of Sterilizing Grade Filters—A Current Perspective, *Filtration News*, Pall Ultrafine Filtration Company, East Hills, NY, Spring, 1990.
35. BOWMAN, F. W., CALHOUN, M. P., WHITE, M., Microbiological Methods for Quality Control of Membrane Filters, *J. Pharm. Sci. 56*, 222 (1967).
36. *Standard Test Method for Determining Bacterial Retention of Membrane Filters Utilized for Liquid Filtration*; ASTM F838-83; American Society for Testing and Materials, Philadelphia, October, 1983.
37. *Points to Consider in the Production and Testing of New Drugs and Biologicals Produced by Recombinant DNA Technology*; Center for Biologics Evaluation and Research (CBER), Food and Drug Administration, Rockville, MD, April 1985.

38. PALL, D. B., KIRNBAUER, E. A., ALLEN, B. T., Particulate Retention by Bacteria Retentive Membrane Filters, *Colloids Surf. 1*, 29 (1980).
39. RETI, A. R., An Assessment of Test Criteria in Evaluating the Performance and Integrity of Sterilizing Filters, *Bull. Parenter. Drug Assoc. 31* (4), 187–194 (1977).
40. HOWARD, G., MEEKER, J.; Evaluation of the Retention Characteristics of 0.2 μm Rated Membrane Filters When Filtering High Viscosity Liquids; Development of a Microbial Challenge Method; *Pall Scientific and Technical Report STR-PUF06*; Pall Ultrafine Filtration Co., East Hills, NY, 1988.
41. DUBERSTEIN, R., HOWARD, G. Sterile Filtration of Gases: A Bacterial Aerosol Challenge Test, *J. Parenter. Drug Assoc. 32* (4), 192–198 (1978).
42. KRYGIER, V., LATHAM, M., CONWAY, R., Automatic Particle Measurement in Liquids Downstream of Fine Membrane Filters, *Microcontamination 3* (4), 33–39 (1985).
43. STINAVAGE, P., Filter Qualification and Validation: Part II, *Filtration News*, Pall Ultrafine Filtration Company, East Hills, NY, Fall, 1991.

Pressure Relief

Nancy A. D'Elia

Contents

10.1   Introduction

Pressure relief devices are a common part of chemical and petroleum processing plants, since these facilities contain a multitude of pressure vessels that need protection from overpressure.   The development of scaled-up fermentor and bioprocessing systems has made evident the need for similar safety devices in biotechnology plants. There is less chance of explosion or fire in a biotechnology plant in comparison to a chemical plant, but there is the additional danger of releasing infectious or contaminated material to the atmosphere.   Chemical plant relief loads are typically directed to flares for burning.   In contrast, relief waste in biotechnology plants often needs to be contained and treated.   Moreover, most bioprocessing systems require sterility, which further complicates the issue of providing safe and effective pressure relief. Regardless of the application of relief devices, the theory underlying the design of relief systems is the same.   We can use our knowledge and experience from conventional processing plants to develop a method for designing safety relief systems in biotechnology plants.

In this chapter I describe the key points to consider in the design of a pressure relief system for a biotechnology plant.   I provide general guidelines for sizing relief devices, as well as specific considerations for biotechnology systems.   In addition, I give some helpful hints for avoiding installation and maintenance problems.

10.1.1   Codes and Regulations

Relief devices are required by law, according to guidelines set forth in Unfired Pressure Vessels, Section VIII, Division 1 of the ASME Boiler and Pressure Vessel Code [1].   Corporate insurance firms also typically require adherence to the ASME Boiler Code.   Finally, state and local laws, such as administrative codes, impose requirements for the protection of pressure vessels. All vessels operating at more than 15 psig (~100 kPa) and having a diameter greater than 6 inches (15 cm) must have a safety relief device.

The definitive references on pressure relief are the American Petroleum Institute (API) Recommended Practices 520 and 521.   API 520 bears the title *Design and Installation of Pressure-Relieving Systems in Refineries* [2], and API 521 is a *Guide for Pressure-Relieving and Depressuring Systems* [3].   Both references offer extensive information, and I recommend them to anyone who needs to gain thorough understanding of this subject.

10.1.2 Principles

The fundamental purpose of a pressure relief system is to protect pressure vessels from overpressure caused by upsets in mass or energy flow. These upsets are typically the result of some imbalance in the system, which causes heat or material to build up. Determination of relief valve requirements involves a detailed analysis of the heat and mass balances for the processing system.

Before these system balances can be analyzed, the designer needs to understand what is meant by *overpressure*. A vessel needs to be protected from exceeding its *maximum design pressure*, the basis for the pressure at which the vessel is hydrostatically tested in order to obtain its ASME rating. The maximum design pressure is the maximum allowable working pressure of the vessel, and is usually determined by the engineer during its design for its intended use. For vessels operating up to 150 psig (1 MPa), the designer should set the design pressure (or maximum allowable working pressure) at 15 psi (100 kPa) over the maximum operating pressure. Good engineering practice states that the *maximum allowable working pressure (MAWP)* should be at least 10% above the *maximum operating pressure* for all vessels, regardless of pressure rating. The maximum operating pressure is calculated as the *normal operating pressure* plus at least 1% or 5 psi, whichever is greater. A vessel with this leeway in operating and relieving pressure ranges is certain to operate without frequent relief occurrences. An example of this method of determining maximum allowable working pressure is in Section 10.7.1. An exception is sometimes made if a rupture disk is used to protect the equipment. In this case, the operating pressure should never exceed 70% of the burst pressure. The section on rupture disks (Section 10.3.2) of this chapter shows how this requirement can affect the MAWP.

The relief valve set pressure is usually the same as the MAWP, so the sizing of the relief valve depends on the specified MAWP. A single relief valve, set at a pressure not to exceed the MAWP, is generally used to protect the equipment from overpressure. Sometimes more than one valve is used to protect a process vessel. In this case, the set pressure of one of the valves must be at or below the MAWP, and the secondary valve may be set as high as 105% of the MAWP.

Relief loads are calculated at an accumulated pressure based on the rise in vessel pressure during discharge. This *accumulation* allows the pressure relief valve to attain its full lift in order to achieve its rated capacity. The API recommends that an accumulation of 20% over the maximum allowable working pressure be allowed for fire relief cases, 6% for steam boilers, 10% for all other single-valve process relief cases, and 16% for all other multiple-valve relief cases [2, 3]. Figure 10.1 depicts the relationships among these and other pressures.

As is apparent from the preceding discussion, the field of relief system design uses a wealth of terms. These terms are summarized and defined in Table 10.1, and used throughout the remainder of this chapter.

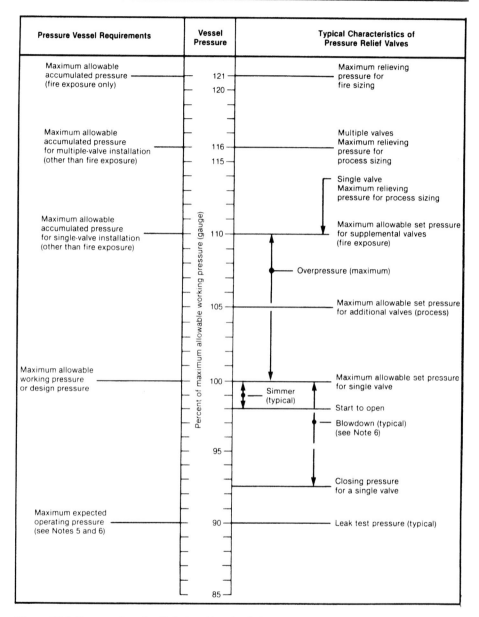

Figure 10.1 Pressure Levels. Relationship of relief pressures to maximum allowable working pressure of a vessel or system. See notes on following page. (Courtesy of API.)

Notes for Figure 10.1 (*on previous page*):
1. This figure conforms with the requirements of Section VIII of the ASME *Boiler and Pressure Vessel Code*.
2. The pressure conditions shown are for pressure relief valves installed on a pressure vessel.
3. Allowable set-pressure tolerances will be in accordance with the applicable codes.
4. The maximum allowable working pressure is equal to or greater than the design pressure for a coincident design temperature.
5. The operating pressure may be higher or lower than 90.
6. Section VIII, Division 1, Appendix M, of the ASME Code should be referred to for guidance on blowdown and pressure differentials.

Table 10.1 Definition of Terms [2, 3, 12]

Accumulation—The pressure increase over the maximum allowable working pressure (MAWP) of the vessel during discharge; expressed as percent of MAWP.

Back Pressure—The pressure on the discharge side of the relief valve. This pressure may be *superimposed* or *built up*.

Blowdown—The difference between the set pressure and the pressure at which the valve reseats, usually expressed as percent of the set pressure (where set pressure is specified as a gauge pressure).

Built-up Back Pressure—The pressure in the discharge line that develops as a result of resistance to flow through the relief valve. It exists only when the valve is open and material is flowing.

Burst Pressure—The pressure at which a rupture disk relieves.

Chatter—Rapid opening and closing of the pressure relief valve.

Cold Differential Test Pressure—The pressure at which the valve is adjusted to open on the test stand, incorporating corrections for service conditions.

Lift—The actual travel of the disk away from the closed position when the valve is relieving.

Maximum Allowable Working Pressure (MAWP)—The maximum permissible pressure at which the vessel may be routinely operated for a designated temperature, typically the same as the design pressure.

Operating Pressure—The pressure at which the vessel is normally operated; usually designed to allow a suitable margin between operating pressure and MAWP to avoid inadvertent relief valve operation; never greater than MAWP.

Overpressure—The pressure increase over the set pressure of the relief device. It is the same as accumulation when the relief valve is set at the vessel's MAWP.

Pressure Relief Valve—A device designed to reclose after normal conditions have been restored; a generic term for either a *relief valve*, a *safety relief valve*, or a *safety valve*.

Relief Valve—A relief device actuated by the inlet static pressure, that opens in proportion to the increase in pressure over the set pressure.

Relieving Conditions—The relief valve inlet pressure and temperature of fluid at a specific overpressure.

Table 10.1 Definition of Terms, *continued*

Rupture Disk—A diaphragm or disk designed to burst when actuated by the inlet static pressure.

Safety Relief Valve—A pressure relieving device suitable for use as either a safety valve or a relief valve. The valve can be adjusted to open rapidly, or in proportion to the pressure increase, depending on its application.

Safety Valve—A pressure relieving device actuated by the inlet static pressure and characterized by rapid full opening (*popping*); typically used in gas, vapor, steam, or air service.

Set Pressure—The inlet pressure at which the pressure relief valve begins to discharge.

Simmer—The audible or visible escape of compressible fluid between the seat and disk at an inlet static pressure below the set pressure and at no measurable capacity.

Superimposed Back Pressure—The static pressure present at the outlet of the relief device when the valve is closed. This pressure is imposed by other sources in a common discharge header. The superimposed back pressure and the built-up back pressure are essentially the same when the device relieves to the atmosphere.

10.1.3 System Design

Designing a pressure relief system involves two steps:
  (a)   calculating the relief loads from individual pieces of equipment for various incidents; and
  (b)   designing the entire system to incorporate loads from each relief valve, and sending the material to disposal.

Before either of these steps can be completed, it is necessary to examine the equipment and systems to be protected. This usually involves collecting pertinent process flow diagrams containing the design and operating parameters of the processing system. Once these data are gathered, the possible causes for overpressure can be determined. Both individual items of equipment and overall systems must be considered. Often one piece of equipment is affected by an adjacent one. The engineer must assess the list of potential causes to determine the most probable or reasonable set of occurrences. These cases are used to design the individual and overall relief systems [4].

10.2   Relief Occurrences

The first step in the design of a relief system is to examine the possible causes for overpressure for each of the vessels to be protected and to determine the relief load associated with each. The underlying cause for overpressure is a disruption of the heat and material balance in a system. This imbalance can be brought about by any of several factors, such as fire, power failure, valve failure, or instrument air failure. The engineer must analyze each of these cases to determine the largest reasonable relief load occurrence, and then size the relief device accordingly. Table 10.2 lists typical relief cases for the various pieces of equipment common to biotechnology. The table can serve as a guide, but the engineer needs to consider all possible causes for the particular system. The focus in this section is on assessing several of the most common relief scenarios.

Table 10.2 Typical Relief Cases for
Biotechnology Equipment

Equipment	Typical Relief Cases
Fermentor	Fire
	Power failure
	Air failure
	Blocked outlet
	Cooling water failure
Auxiliary vessel	Fire
(holding tanks,	Power failure
mixing tanks, prep	Air failure
tanks)	Blocked outlet
Vessel jacket	Thermal expansion
	Blocked outlet
	Abnormal heat input
	Fire
Heat exchanger	Thermal expansion
	Tube rupture
	Blocked outlet
	Fire
	Abnormal heat input
Clean-steam generator	Fire
	Blocked outlet

10.2.1 Fire

Fire must be considered as a potential cause of overpressure in all vessels within 25 feet (7.6 m) of grade. Determine the heat imbalance by calculating the heat absorption due to fire exposure of the wetted surface of the vessel. Then determine the fire relief load by calculating the vapor rate from the latent heat of vaporization of the fluid in the vessel at the relief conditions. Determine the relief conditions by assuming saturated vapor at the relief pressure. The API recommends that an assumption of 20% accumulation over the vessel set pressure be used to determine the relief pressure [2, 3].

Use the following formula to determine the amount of heat absorbed by a vessel exposed to fire:

$$Q = 21,000\ FA^{0.82} \tag{10.1}$$

where Q is the total heat input to the wetted surface, Btu/h;
F is the environmental factor, dependent on the type of installation (see Table 10.3); and
A is the total wetted surface area, ft^2.

Table 10.3 Environmental Factor, F [2]

Equipment	F
Bare vessel	1.0
Insulated vessel: insulation conductance (Btu/h·ft^2·°F)	
4.0	0.3
2.0	0.15
1.0	0.075
Underground storage	0.0
Earth-covered storage above grade	0.03

The environmental factor, F, allows for credit to be taken for insulated vessels. Typically, fire destroys insulation, so credit is not usually considered unless the tanks are steel jacketed. In these cases, a reasonable value of F can be assumed based on the conductance of the insulation.

For relief calculation purposes, the wetted surface area is the area wetted by the liquid contents, and therefore capable of generating vapors when exposed to fire. A 25-ft (7.6-m) height limitation from grade is usually applied. In this case, grade is considered to be ground level or any horizontal surface capable of sustaining a fire. The numerical value assigned to the area of the wetted surface is often somewhat arbitrary, due to the variable liquid level in most vessels. In general, the average liquid inventory should be assumed. For example, the total surface area within the height limitation should be assumed for liquid-full vessels, but surge drums typically operate half full, and therefore half the total surface area should be used. Use the following equations to determine surface area of a standard vertical vessel with dished heads:

$$A_v = \pi dh \tag{10.2}$$

$$A_h = 1.082\, d^2 \tag{10.3}$$

where A_v is the surface area of the vessel cylinder/shell, ft^2;
A_h is the surface area of one dished head, ft^2;
d is the vessel diameter, ft; and
h is the vertical height of the shell, ft.

Fire cases typically cause vapor relief. The vapor requiring relief is that which is in equilibrium with the liquid in the vessel at the relief conditions. Determine the vapor relief rate by assuming it is equal to the rate of vapor formation due to heat absorption in the vessel. Calculate this by dividing the heat input by the latent heat of vaporization of the liquid:

$$W = Q/\lambda \tag{10.4}$$

where W is the vapor formation rate, lb/h;
Q is the heat absorption as calculated by eq 10.1, Btu/h; and
λ is the latent heat of vaporization at relief conditions, Btu/lb.

In some cases, relief valves are located in the liquid zone of vessels that may be exposed to fire. In these instances, the relief valve should be capable of passing a volume of liquid equivalent to the displacement caused by vaporization during fire. When the liquid can flash across the valve, two-phase flow calculations must be done. For cases where the relief conditions are above the critical point of the fluid in the vessel, refer to the API guides for further details [3].

10.2.2 Blocked Outlet

The blocked outlet case is one of the most common relief occurrences. In this case we assume that fluid continues to be fed to the vessel through the inlet valve while an outlet valve is blocked because an outlet control valve has failed or because of operator error. The relief situation can come to pass only if the feed source has sufficient pressure to open the relief valve. For example, the relief valve set pressure must be less than the feed pump shutoff pressure or the air source pressure.

There are usually several cases to consider, involving either vapor or liquid relief. As a rule of thumb, if the vessel has empty space above the high liquid level that can fill in less than ten minutes, the blocked exit case results in liquid relief. The relief load is the flow rate from the pump at the relief conditions. Obtain this value by consulting the pump curves at the accumulated relieving pressure. If gas is the motive fluid source, the flow rate should be calculated using the pressure difference between the gas source and the vessel at the accumulated relief pressure. For vessels that have empty space above the high liquid level that will take ten minutes or longer to fill, only vapor relief needs to be considered. The vapor rate is the vapor generated at the relieving pressure, assuming constant temperature.

10.2.3 Power Failure

Another potential cause of system imbalance is power failure. A crucial part of an overall biotechnology system design is the evaluation of the effect of a power failure on both individual equipment operation and the overall process plant operation. Consideration should be given to placing critical equipment on emergency power. However, in the event that equipment is not on emergency power, or credit cannot be taken for emergency power, this relief scenario should be considered. There are three primary power failure situations: (a) local failure, affecting a single piece of equipment; (b) general power failure, in which the site distribution center is affected; and (c) total power failure, where areawide power is affected and even emergency power is unavailable.

Since local power failure affects individual pieces of equipment, evaluate it by determining the effect of individual component failure on the vessel being protected. For instance, one would analyze the effect of failure of a pump, solenoid valve, or process transmitter on the overall operation of a vessel. With this information in hand, relief sizing is determined as for each of these individual cases.

General and total power failure require much more involved analysis of the complete process system. In these cases, the combined effects of multiple equipment failures must be evaluated. The engineer needs to carefully consider the case when relief material is discharged through a common header system to a disposal area. In this case, simultaneous relief may determine the size of the relief header and waste treatment system.

10.2.4 Instrument Air Failure

As with power failure, several cases of instrument air failure must be considered: individual component, local, and total failure. To determine the potential of a relief situation, the failure positions of each of the individual components must be evaluated. Typically, devices are designed to be fail-safe, but there are often instances where no device can be truly fail-safe. An example of instrument air failure might be local failure of air to an outlet control valve. If the valve fails in the closed position while material is fed to the vessel, then the blocked exit condition exists and can be analyzed accordingly. On the other hand, if the inlet valve fails while the outlet valve continues to function, no relief case occurs. The outlet pump may starve and cut out, but there is no cause for overpressure.

10.2.5 Thermal Expansion

Vessel jackets, heat exchangers, and pipelines typically need relief valves provided for thermal expansion in the event that the inlet and outlet are both blocked while the system is exposed to an external source of heat. The main heat sources are solar radiation, warm ambient air, steam tracing, fire, and adjacent process fluid [5]. Calculate the relief load due to thermal expansion from the following equation:

$$W_Q = 7.48 \; \alpha V \; (\Delta T/t) \tag{10.5}$$

where W_Q is the liquid thermal relief rate, gpm ;
α is the liquid coefficient of expansion, 1/°F (for water, $\alpha = 0.0001/°F$);
V is the volume enclosed between block valves, ft^3; and
$\Delta T/t$ is the rate of temperature rise of liquid in system, °F/min.

Determination of the relief rate is usually difficult because the rate of temperature rise of the fluid in the blocked system is usually unknown. The overall temperature rise necessary to achieve relief conditions can be calculated if the relief pressure of the system is included in the calculations. This pressure must be set at the maximum working pressure of the weakest component of the system being protected.

$$\Delta T = \Delta P/\alpha\kappa \tag{10.6}$$

where ΔT is the temperature rise of the fluid, °F;
ΔP is the pressure difference between normal operating pressure and relief pressure, psi;
α is the liquid coefficient of expansion, 1/°F; and
κ is the liquid isothermal compressibility coefficient, 1/psi.

Note that in both eq 10.5 and 10.6, expansion of the pipe is neglected. This is a conservative approach.

Alternatively, API 521 [3] recommends that for heat exchangers, the thermal expansion relief load can be determined from the following equation:

$$W_Q = \frac{\alpha H}{500 (SG) C_p} \tag{10.7}$$

where W_Q is the liquid thermal relief load, gpm;
α is the liquid cubical coefficient of expansion, 1/°F;
H is the maximum exchanger duty, Btu/h;
SG is the specific gravity of the liquid; and
C_p is the specific heat of the liquid, Btu/lb·°F.

As a general rule, thermal expansion relief valves should be no smaller than ¾" × 1" (18 × 25 mm). Most thermal expansion applications involve relieving liquid only, so the valve size is small [6]. However, for practical purposes a minimum of a ¾" (18-mm) valve is used to prevent accumulation of deposits in the relief valve and ensure adequate mechanical strength.

10.2.6 Summary

In this section I have described the most common relief cases particular to biotechnology processing plants. The engineer should be warned that this listing is by no means complete, and additional cases should be considered as warranted by the system under evaluation. Some examples of other potential relief cases include chemical reaction, heat exchanger equipment failure, abnormal heat input, cooling water failure, and accumulation of noncondensables. Once again, for more in-depth analysis of potential causes for overpressure, the API guidelines should be consulted [3].

10.3   Relief Device Sizing

Once the individual relief loads are known, the difficult analysis has been completed, and a relatively straightforward procedure for relief device sizing can be followed. Before sizing the device, however, the engineer must define the type of device to specify.

10.3.1 Types of Relief Devices

The two types of relief devices are relief valves and rupture disks. A *relief valve* has a spring-loaded disk that actuates at a particular set pressure as its spring is compressed, and a *rupture disk* is a nonmechanical device in which a metal diaphragm bursts within a specified pressure range.

Relief valves are common in most process industries; however, their use becomes less desirable in biotechnology, where sterilizable systems are needed. Most ASME–rated relief valves cannot be cleaned or sterilized, and so are used only on "dirty" lines or nonprocess streams.

Rupture disks are more common in biotechnology because they can be cleaned and sterilized. Many manufacturers offer a sanitary version of the rupture disk. Still, the engineer must be aware of the potential difficulties with these devices before deciding to use a disk over a valve. A rupture disk touted as sanitary can present a cleaning and sterilizing problem unless installed so as to minimize the dead leg associated with its holder design. In addition, unlike relief valves, rupture disks cannot be reseated, so they have a one-time use only.

Another issue is cost. Biotechnology process vessels often contain fluids that may be corrosive to standard relief devices. In these cases, relief valves and disks must be fabricated from stainless steel or other more costly materials. The engineer may often find that a relief valve with stainless steel wetted parts is less expensive than the corresponding rupture disk and necessary support structure.

For these reasons, valves are usually preferred in places where sanitary conditions do not prevail. Sometimes rupture disks are used in series with relief valves when the contents of the vessel need to be contained. In this section I consider the sizing of all such relief devices.

10.3.2   Rupture Disks

Rupture disks are typically sized using manufacturers' literature [7, 8]. The key criterion required to correctly specify a rupture disk is specification of the burst pressure at the maximum allowable working pressure and temperature of the system to be protected. Because the strength of a metal rupture disk varies with temperature [9], the burst pressure must be related to the design temperature in discussions of the application with the disk supplier. The disk is then stamped with its relief pressure at both the design temperature and ambient conditions (72 °F). If only the pressure is supplied to the disk supplier, the disk is fabricated for this pressure at ambient temperature. When the system is operated at a higher temperature, the disk will burst at a much lower pressure than desired.

The nonmechanical operating principle of the rupture disk also creates additional variables in the specification of the device. The burst pressure is not exact, but instead occurs within a range of pressures. The infinite number of possible relief conditions has brought about the establishment of manufacturing ranges. The *manufacturing range* specifies the allowable pressure range within which a rupture disk is rated. Essentially, it defines the maximum permitted variation between the customer-specified burst pressure and the actual burst pressure. Compounding the manufacturing range is the burst or rupture tolerance. The *burst tolerance* represents the maximum variation from the stamped burst pressure. Once the disk has been manufactured, it is tested and stamped with a rated burst pressure. The stamped burst pressure is determined from the average burst pressure observed with comparable disks during destructive testing. The engineer must take into account both the manufacturing range and the burst tolerance to ensure that the relief pressure does not exceed the maximum allowable working pressure of the system. See Section 10.7.2 for an example illustrating this point.

10.3.2.1   Types of Rupture Disks

There are numerous styles and types of rupture disks available. Designs range from conventional prebulged models, to reverse-buckling, knife-cut versions, to specialized composite devices available with Teflon® seals (see Figure 10.2). These designs have

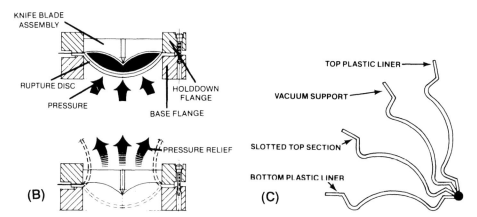

Figure 10.2 Rupture Disk Styles. (A) Conventional, prebulged disk; (B) Reverse-buckling, knife-cut design; (C) Composite device with layers of plastic and metal. (Courtesy of Fike Corp.)

associated with them advantages and disadvantages that distinguish them from one another and make them suitable for a particular application. One specific characteristic that helps to distinguish these designs is the recommended maximum operating pressure of the system to prevent premature failure of the rupture disk [7, 10]:

- Prebulged disks: 70% of the rated disk burst pressure;
- Composite disks: 80% of the rated disk burst pressure;
- Reverse-acting disks: 90–95% of the rated disk burst pressure.

Regardless of the style of disk chosen, sizing rupture disks is the same. A few assumptions are made to eliminate unnecessary complications in the equations. The most important assumption made is that of the disk flow area. The flow through the device is calculated as 62% of the flow through a theoretical nozzle of the same area, to take into account disk material that protrudes into the flow stream after rupture.

Table 10.4 Gas Flow Constant, C_1, for Subsonic Flow [7]

k \ Pe/Po	0.95	0.95	0.85	0.80	0.75	0.70	0.65	0.60	0.55	0.50	0.45
1.05	0.0440	0.0598	0.0703	0.0777	0.0829	0.0864	0.0884	0.0891	0.0887	0.0871	0.0844
1.10	0.0441	0.0600	0.0707	0.0783	0.0837	0.0874	0.0897	0.0906	0.0904	0.0891	0.0867
1.15	0.0441	0.0602	0.0711	0.0788	0.0844	0.0884	0.0908	0.0920	0.0920	0.0909	0.0888
1.20	0.0442	0.0604	0.0714	0.0793	0.0851	0.0892	0.0919	0.0933	0.0935	0.0927	0.0907
1.25	0.0442	0.0606	0.0717	0.0798	0.0857	0.9000	0.0929	0.0945	0.0950	0.0943	0.0926
1.30	0.0443	0.0607	0.0719	0.0802	0.0863	0.0908	0.0938	0.0956	0.0963	0.0958	0.0943
1.35	0.0443	0.0609	0.0722	0.0805	0.0868	0.0915	0.0947	0.0967	0.0975	0.0973	0.0959
1.40	0.0444	0.0610	0.0724	0.0809	0.0873	0.0921	0.0955	0.0977	0.0987	0.0986	0.0975
1.45	0.0444	0.0611	0.0726	0.0812	0.0878	0.0927	0.0963	0.0986	0.0998	0.0999	0.0990
1.50	0.0445	0.0612	0.0728	0.0816	0.0882	0.0933	0.0970	0.0995	0.1009	0.1011	0.1003
1.55	0.0445	0.0613	0.0730	0.0819	0.0886	0.0938	0.0977	0.1003	0.1019	0.1023	0.1017
1.60	0.0445	0.0614	0.0732	0.0821	0.0890	0.0944	0.0983	0.1011	0.1028	0.1034	0.1029
1.65	0.0446	0.0615	0.0734	0.0824	0.0894	0.0948	0.0989	0.1019	0.1037	0.1040	0.1041
1.70	0.0446	0.0616	0.0735	0.0826	0.0897	0.0953	0.0995	0.1026	0.1045	0.1054	0.1053
2.00	0.0448	0.0620	0.0743	0.0839	0.0915	0.0976	0.1024	0.1061	0.1088	0.1104	0.1111
2.10	0.0448	0.0621	0.0746	0.0842	0.0920	0.0982	0.1032	0.1071	0.1100	0.1119	0.1127
2.20	0.0448	0.0622	0.0748	0.0845	0.0924	0.0988	0.1040	0.1080	0.1111	0.1131	0.1142

Table 10.5 Gas Flow Constant, C_2, for Sonic Flow

k	C_2	k	C_2	k	C_2	k	C_2
1.00	0.0876	1.20	0.0937	1.40	0.0989	1.60	0.1035
1.02	0.0883	1.22	0.0942	1.42	0.0994	1.62	0.1039
1.04	0.0889	1.24	0.0948	1.44	0.0999	1.64	0.1043
1.06	0.0895	1.26	0.0953	1.46	0.1003	1.66	0.1048
1.08	0.0902	1.28	0.0959	1.48	0.1008	1.68	0.1052
1.10	0.0908	1.30	0.0964	1.50	0.1013	1.70	0.1056
1.12	0.0914	1.32	0.0969	1.52	0.1017	2.00	0.1120
1.14	0.0920	1.34	0.0974	1.54	0.1022	2.10	0.1129
1.16	0.0925	1.36	0.0979	1.56	0.1026	2.20	0.1145
1.18	0.0931	1.38	0.0984	1.58	0.1031		

10.3.2.2   Vapor Relief Sizing

This section covers sizing a rupture disk for vapor relief. Here, we determine the area required for vapor flow.

For subsonic flow through the device:

$$A_r = \frac{W(T)^{0.5}}{K C_1 P (M)^{0.5}} \tag{10.8}$$

and for sonic flow through the device:

$$A_r = \frac{W(T)^{0.5}}{K C_2 P (M)^{0.5}} \tag{10.9}$$

where A_r is the required relief area, in²;

W is the mass flow rate, lb/s;

T is the relief temperature, °R;

K is the capacity correction factor = 0.62 (to account for disk material protruding into the flow path);

C_1 is the gas flow constant for subsonic flow (see Table 10.4);

C_2 is the gas flow constant for sonic flow (see Table 10.5);

P is the accumulated relief pressure, psia (set pressure + accumulation); and

M is the molecular weight of relieving vapor, lb/lb mol.

To determine if sonic flow exists, calculate the flow pressure ratio (FPR) and compare it to the critical pressure ratio (CPR).

$$\text{FPR} = P_e/P \tag{10.10}$$

where P_e is the exit pressure, psia (usually atmospheric); and
P is the accumulated relief pressure, psia.

$$\text{CPR} = [2/(k + 1)]^{k/k-1} \tag{10.11}$$

where k is the gas constant = C_p/C_v (see Table 10.6);
  C_p is the specific heat of the relieving material at constant pressure; and
  C_v is the specific heat of the relieving material at constant volume.
If FPR > CPR, flow is subsonic.
If FPR ≤ CPR, flow is sonic.

Table 10.6 Gas Constants [7]

Gas or Vapor	$k = C_p/C_v$	Gas or Vapor	$k = C_p/C_v$
air	1.40	ethylene	1.26
acetic acid	1.15	helium	1.66
acetylene	1.26	hydrochloric acid	1.41
ammonia	1.33	hydrogen	1.41
argon	1.67	hydrogen sulfide	1.32
benzene	1.12	methane	1.31
n-butane	1.09	methyl alcohol	1.20
isobutane	1.09	natural gas	1.27
butene	1.10	nitric acid	1.40
carbon monoxide	1.40	nitrogen	1.40
carbon dioxide	1.30	nitrous oxide	1.30
chlorine	1.36	oxygen	1.40
cyclohexane	1.09	propane	1.13
ethane	1.22	sulfur dioxide	1.29
ethyl alcohol	1.13	water vapor	1.32

10.3.2.3   Liquid Relief Sizing

The sizing equation for liquid flowing through a disk [7] is:

$$A_r = \frac{W_Q\,(SG)^{0.5}}{23.1\,(P)^{0.5}} \tag{10.12}$$

where A_r is the required relief area, in^2;
  W_Q is the liquid relief load, gpm;
  SG is the specific gravity of liquid; and
  P is the accumulated relief pressure, psig.

10.3.2.4   Steam Relief Sizing

Sizing rupture disks for steam loads depends on the type of steam being relieved [7]. For our purposes, we assume that the vessel is relieving dry, saturated steam:

$$A_r = 70 \ W/KP \tag{10.13}$$

where A_r is the required relief area, in^2;
  W is the relief load, lb/s;
  K is the capacity correction factor = 0.62; and
  P is the accumulated relief pressure, psia.

To determine the actual steam conditions, the system relief conditions can be compared to those published in steam tables. Steam relieving at a higher temperature than the saturation temperature at the accumulated relief pressure is superheated; steam relieving at a lower temperature is considered wet steam. In each of these cases, different sizing equations are used. Refer to manufacturers' literature for these conditions.

For each of the sizing equations above, determine the rupture disk size from the area by using the formula:

$$A_r = \pi d^2/4 \tag{10.14}$$

where A_r is the required relief area, in^2; and
  d is the rupture disk diameter, in.

10.3.3   Relief Valves

Although relief valves present many problems for clean, sterile bioprocesses, they are still quite common throughout much of the processing plant. There are two general types of relief valves: conventional and balanced.

Conventional relief valves, shown in Figure 10.3, have a larger disk area than nozzle seat area. This difference in size causes the valve opening pressure and flow rate to be affected by the system back pressure [11]. Since the valve opens and closes in response to forces acting on the disk and the nozzle seat, if the exposed disk area differs from the nozzle seat area, the difference in back pressure imposes differing forces on the disk. It is this lack of equilibrium that causes the valve to open at pressures other than the set pressure [2]. The valve bonnet may be vented to either the atmosphere or the discharge side of the valve. In either case, the valve operation is affected: when venting is to the atmosphere, the relief valve back pressure causes the opening pressure to be less than the set pressure; when venting is to the discharge line, back pressure causes the opening pressure to be greater than the set pressure. Obviously, this inconsistency in valve operation makes the conventional valve somewhat undesirable.

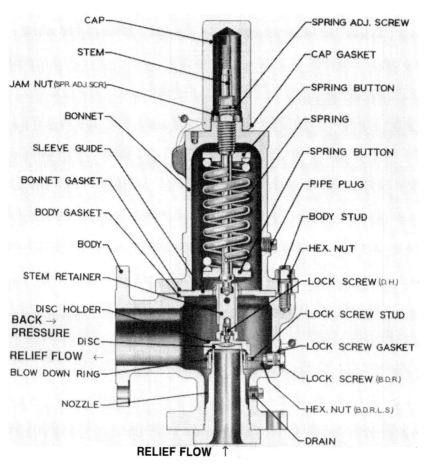

Figure 10.3 Conventional Relief Valve. A larger disk area than nozzle seat area causes this valve to be affected by back pressure. (Courtesy of Teledyne Farris Engineering.)

The *balanced valve*, on the other hand, is not affected by back pressure, since it is designed so that the valve actuation area is the same as the nozzle seat area. As shown in Figure 10.4, a bellows is used to isolate the top of the disk from back pressure. The effective area of the bellows becomes the valve actuation area. The bellows is also connected to the valve bonnet so that the bonnet and valve spring are isolated from the process fluid. The bonnet is then vented to the atmosphere. In this way, the forces resulting from back pressure cancel each other, while the relieving forces due to vessel pressure are always acting against atmospheric pressure. For these reasons, back pressure does not affect the operation of the valve.

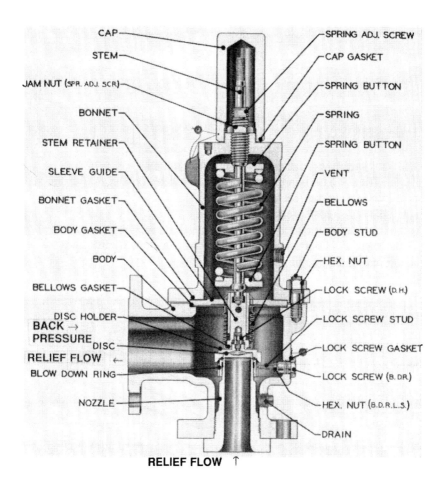

Figure 10.4 Balanced Relief Valve. The valve actuation area and the nozzle seat area are the same, so back pressure does not affect valve operation. (Courtesy of Teledyne Farris Engineering.)

Conventional valves are normally used when back pressures do not exceed 10% of the set pressure. Balanced valves are used when the back pressure is 40–50% of the set pressure. For biotechnology, balanced valves are recommended in all cases because of their cleaner nature and consistent operation. The techniques used to size these valves differ only in the determination of the capacity correction factor. A lifting device should be specified on all pressure relief valves not relieving hazardous material. The lifting lever is helpful during inspection and maintenance. In addition, sealed guide stems are specified, to prevent leakage of fluid during relief.

10.3.3.1   Vapor Relief Sizing

Use the following equation to determine the effective discharge area for relief of vapor:

$$A_d = \frac{W(TZ)^{0.5}}{CKP K_b (M)^{0.5}} \qquad (10.15)$$

where A_d is the effective discharge area, ft^2;
W is the relief load, lb/h;
T is the inlet temperature of vapor at relief conditions, °R;
Z is the compressibility factor ($Z = 1$ is conservative);
C is a coefficient based on the ratio of specific heats ($k = C_p/C_v$) of the vapor at standard conditions (see Tables 10.6 and 10.7);
K is the discharge coefficient (normally 0.97);
P is the upstream pressure at relief conditions (i.e., accumulated pressure), psia;
K_b is the capacity correction factor (see Figure 10.5A or 10.5B); and
M is the molecular weight of the relieving vapor.

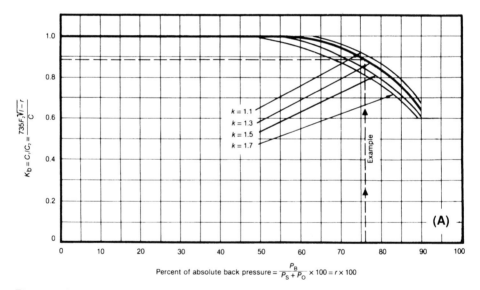

Figure 10.5A Correction Factor, K_b, for Vapor and Gas Capacity Sizing of Conventional Relief Valves. (Example is worked out on facing page.)

C_1 = capacity with back pressure;
C_2 = rated capacity with zero back pressure;
P_B = back pressure, psi;
P_S = set pressure, psi;
P_O = overpressure, psi.

Example shown in Figure 10.5A:
Set pressure (MAWP) = 100 psig;
Overpressure = 10 psig;
Superimposed back pressure (constant) = 70 psig;
Spring set = 30 psi;
Built-up back pressure = 10 psi;
Percent absolute back pressure = (70 + 10 + 14.7)/(100 + 10 + 14.7) × 100 = 76;
K_b (follow dashed line) = 89 (from the curve);
Capacity with back pressure = 0.89 × (rated capacity without back pressure).

Note: This chart (Figure 10.5A) is typical and suitable for use only when the make of the valve or the actual critical flow pressure point for the vapor or gas is unknown; otherwise, the valve manufacturer should be consulted for specific data. This correction factor should be used only in the sizing of conventional (nonbalanced) pressure relief valves that have their spring setting adjusted to compensate for the superimposed back pressure. It should not be used to size balanced-type valves. (Courtesy of API).

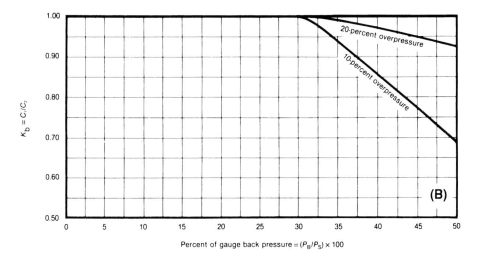

Figure 10.5B Correction Factor, K_b, for Vapor and Gas Capacity Sizing of Balanced Relief Valves.
C_1 = capacity with back pressure;
C_2 = rated capacity with zero back pressure;
P_B = back pressure, psi;
P_S = set pressure, psi;

Note: The curves above represent a compromise of the values recommended by a number of relief valve manufacturers and may be used when the make of the valve or the actual critical flow pressure point for the vapor or gas is unknown. When the make is known, the manufacturer should be consulted for the correction factor. These curves are for set pressures of 50 pounds per square inch gauge and above. They are limited to back pressure below critical flow pressure for a given set pressure. For subcritical flow back pressures below 50 pounds per square inch gauge, the manufacturer must be consulted for values of K_b. (Courtesy of API.)

Table 10.7   Values of Coefficient C, based on $k = C_p/C_v$ [7]

k	C	k	C	k	C	k	C	k	C
0.45	228.30	0.85	296.43	1.25	342.19	1.65	376.37	2.05	403.37
0.50	238.83	0.90	303.04	1.30	346.98	1.70	380.08	2.10	406.37
0.55	248.62	0.95	309.35	1.35	351.60	1.75	383.68	2.15	409.29
0.60	257.79	1.00	315.38	1.40	356.06	1.80	387.18	2.20	412.15
0.65	266.40	1.05	321.19	1.45	360.38	1.85	390.59		
0.70	274.52	1.10	326.75	1.50	364.56	1.90	393.91		
0.75	282.20	1.15	332.09	1.55	368.62	1.95	397.14		
0.80	289.49	1.20	337.24	1.60	372.55	2.00	400.30		

10.3.3.2   Liquid Relief Sizing

Use the following equation to determine the effective discharge area for relief of liquids:

$$A_d = \frac{W_Q (SG)^{0.5}}{27.2 \, K_p \, (P_d)^{0.5}} \tag{10.16}$$

where A_d is the effective discharge area of the valve, ft^2;
   W_Q is the liquid relief load, gpm;
   SG is the specific gravity
   K_p is the capacity correction factor based on overpressure (see Figure 10.6); and
   P_d is the pressure differential (relieving pressure – back pressure and discharge piping pressure drop), psi.

Once the effective discharge area is known, manufacturers' literature can be consulted to determine the relief valve size (see Table 10.8).

Table 10.8   Sizes of Relief Valves and Effective Areas of Nozzles*

Effective Area of Nozzle, in^2	Normal Size Designation	Effective Area of Nozzle, in^2	Normal Size Designation	Effective Area of Nozzle, in^2	Normal Size Designation
0.110	1D2	1.287	2J3 or 3J4	6.379	4P6
0.196	1E2	1.838	3K4	11.045	6Q8
0.307	1.5F2	2.853	3L4 or 4L6	16.000	6R8 or 6R10
0.503	2G3	3.000	4M6	26.000	8T10
0.785	2H3	4.340	4N6		

* Size designation is expressed as the inlet diameter, in inches, followed by a letter designating the orifice, followed by the outlet diameter, in inches. For example, 1D2 describes a valve with a 1" inlet, size D orifice, and 2" outlet.

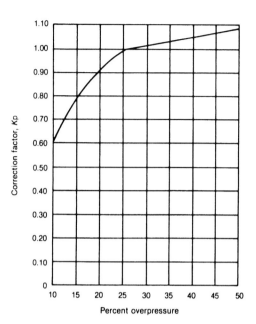

Figure 10.6 Correction Factor, K_p, for Liquid Capacity Sizing. This is the liquid sizing factor based on over-pressure.

Note: The curve shows that up to and including 25% overpressure, capacity is affected by the change in lift, the change in the orifice discharge coefficient, and the change in overpressure. Above 25%, capacity is affected only by the change in overpressure. Valves operating at low overpressures tend to chatter; therefore, overpressures of less than 10% should be avoided. (Courtesy of API.)

10.4   Design of the Relief Header System

With the individual relief loads calculated and the relief valves sized, the overall relief discharge piping system can be designed. The header design depends on the type of process and the nature of the material being processed. In general, there are three means for disposal in biotechnology plants: (a) relieve to atmosphere, (b) relieve to biohazardous waste system, and (c) relieve to incinerator. The method and destination of disposal of the relieved material depend on the material itself. Biotechnology plants typically have two reasons for processing in a sterile environment: either to protect the product from contamination, or to protect the operator and environment from a viable organism.

If materials are processed under sterile conditions solely to protect the product from contamination, there is usually no viable organism present. In this case, relief material is usually vented to the atmosphere. The reason for this is twofold. Firstly, it is more desirable to vent to atmosphere to avoid the risk of back pressure effects on the relief discharge flow rate. Secondly, a relief occurrence is an emergency measure in the event of overpressure, so the engineer needs to weigh the risk of vessel overpressure against the risk of exposure to an undesirable organism. If the likelihood that viable organisms are present is sufficiently low, the waste material can be vented to atmosphere in an emergency situation. If, on the other hand, a viable organism is possibly present, venting to a waste header should be considered.

In operations where potentially hazardous materials are processed, the utmost care must be taken to contain the waste. Such waste material should be discharged to a biohazardous waste disposal system, an incinerator, or both. The system should be designed and validated to render the material nonhazardous prior to discharge to the environment. It is not uncommon for the waste system to include an incinerator for treatment of exhaust gases from the system. Whether to dispose of the relief material to a waste system or an incinerator alone is a decision that depends on the type of relief and the potential for back pressure from the waste system. If the relief is all vapor, it may be sent to an incinerator alone. If, however, it contains liquid or can condense, it should be sent to the waste treatment system or a knockout drum that drains to the waste system.

Relief valves need an unobstructed path for flow of fluid during relief occurrences. Excess back pressure in the relief system can restrict flow. The relief header and waste treatment system should be designed using the standard practices of pipe sizing. The engineer must establish the destination of the relief fluid and then work backwards from this location to the point of relief. He or she can determine the line size by performing back-pressure calculations at the relief flow rates. Multiple relief occurrences should also be considered. In cases such as power failure, instrument air failure, or fire, where plantwide failure can trigger multiple relief valves, the relief header should be sized to handle the aggregate of all relief occurrences. The example in Section 10.7.3 demonstrates the procedure used to design a complete relief system.

10.5   Installation

The installation of relief devices should ensure safe, trouble-free performance and allow access for maintenance and inspection. A device can be maintained properly only if it is properly located and oriented.

10.5.1 Valve Position and Location

Relief devices are required on all equipment that has the potential for overpressure. Usually they are installed on vessels, reactors, vessel jackets, long piping runs, compressors, and heat exchangers. Just as relief valves are installed on vessel jackets, they are also usually required on the cold liquid side of heat exchangers, to protect against thermal expansion or abnormal heat input. This doesn't, however, eliminate the need for a relief device on the hot liquid side; the entire potential scenario should be reviewed.

A valve should be mounted with the valve stem upright and plumb, to minimize the accumulation of particles and the formation of deposits as a result of poor drainage. If mounting a valve in the vertical position requires an elbow, use a long-radius elbow to minimize the pressure drop.

In general, the relief device should be installed at the highest point and as near to the pressure source as possible. It should be mounted on a port or coupling specifically designated for it. Long inlet lines and excessive pressure drop must be avoided if possible. An exception to this rule may be made in installations where inlet pressure fluctuations may prematurely pop the relief device. In these cases, the relief device should be located farther from the pressure source, to damp the pressure fluctuations.

Valves should also be installed with care to minimize vibrational damage or thermal stress. Specifically, proper support and expansion joints should be used.

When a rupture disk is used in combination with a relief valve, a pressure gauge and a bleed valve must be installed between the disk and valve. This is required by ASME code so that operators can tell whether the rupture disk has burst [1]. Figure 10.7 illustrates this type of installation.

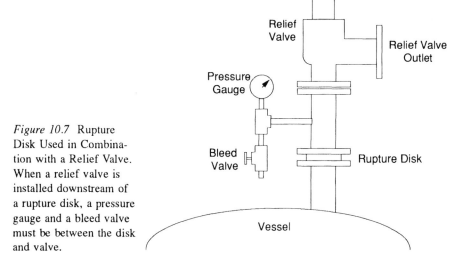

Figure 10.7 Rupture Disk Used in Combination with a Relief Valve. When a relief valve is installed downstream of a rupture disk, a pressure gauge and a bleed valve must be between the disk and valve.

10.5.2 Inlet Piping

The relief device inlet piping should be free and clear of obstructions. The connection to equipment should be at least as large as the largest inlet line to the system. The API suggests that the inlet line pressure drop be less than 3% of the valve's set pressure [2]. This is a good rule of thumb for the establishing of initial line sizes, but it should not be used as a definitive rule for inlet line sizing. The 3% criterion is based on the assumption that valve chatter can be avoided if the inlet line pressure drop is less than the valve's blowdown. In reality, valve chatter depends on the combination of inlet pressure drop and outlet back pressure [12]. I discuss this further in Section 10.6 (Maintenance and Troubleshooting).

In general, inlet pressure drop should be as low as feasible. Short pipe runs of the same diameter as the relief device are recommended. Long, small-diameter inlet lines and those with multiple turns should be avoided.

10.5.3 Discharge Piping

As with inlet piping, the discharge piping should provide an unobstructed path for the relief fluid. The discharge piping should be at least as large as the pressure relief device outlet, and there should be a minimum of bends. The effect of back pressure on the relief device must be considered. In atmospheric venting applications, the outlet back pressure is rarely a problem. However, in systems where the relief devices are manifolded, the discharge piping must be so sized as to limit the superimposed back pressure to that allowed by the device with the lowest back-pressure tolerance.

The discharge piping must be directed to prevent injury to personnel. When a rupture disk bursts, particles may be discharged. The engineer must make sure that flow is directed from the relief device to a safe location, so that no injury can result from flying particles or the relieving vapor or liquid. Discharge of nontoxic, nonhazardous material can be vented to the atmosphere. Typically, the relief is vented high enough to avoid risk, or it is pointed downward to the floor. When it is vented high, the piping usually extends a minimum of 6 feet (2 m) above the roof level, or at least 12 feet (4 m) above ground level. Vent piping directed to the floor should terminate about 1.5 pipe diameters from the floor, to provide an unobstructed path for flow of relieving fluid, with minimum potential for harm to personnel. All discharge piping should be sloped to drain or provide a means of draining the valve body. Most valves are equipped with drainage ports in the valve body. This port should be piped to a safe location, to prevent accumulation of material that could damage the valve [6]. Spool pieces (sections of pipe or tube with end connections that allow connection from the rupture disk to another location) are often used as discharge piping from relief devices. This allows access to the discharge side of the relief device with no need to remove the device from the system it is protecting.

Because of the potential for cross contamination, relief discharge lines should not generally be manifolded. An exception is in cases where the system contains hazardous materials, so all discharges must be manifolded to a treatment system. In these cases, special care must be taken to size the discharge lines to take multiple loads and back pressure into account.

To prevent leakage of the relief device, the discharge piping must be supported independently of the relief device. During relief, reactive forces cause vibration in the discharge piping. This vibration must be controlled by the use of flexible supports and vibration dampeners. The API gives detailed calculations for the determination of the reactive force from the relief device [2].

10.6   Maintenance and Troubleshooting

Relief valve malfunction can take three primary forms: premature actuation, chattering, and leaking. *Premature actuation* occurs when the valve or rupture disk relieves at a pressure less than the intended set pressure. *Chattering* is the rapid opening and closing of a valve during relief conditions. The most common reason for valve chattering is oversizing of the valve. Finally, *leaking*, as the name implies, occurs when fluid or gas finds its way through or around the valve or rupture disk. The most common causes are operation too near the set pressure, and particulates between the seat and disk that prevent the device from sealing. Causes, explanations, and solutions for each of these problems are presented in Tables 10.9–10.11.

Table 10.9 Troubleshooting: Premature Actuation [17]

Causes	Explanations and Solutions
Lack of temperature correction during valve setting.	• Relief devices are set at ambient temperature, so when they are used at elevated temperatures, the manufacturer's cold setting correction factor must be used. At higher temperatures the valve body and bonnet expand, while the spring force is reduced. These factors lead to a reduction in the set pressure and can lead to premature popping. To resolve the problem, provide the vendor with the maximum relief temperature and make sure that the temperature correction factor is used during valve setting.
Improper test apparatus calibration.	• The gauge used for setting the relief device must be properly calibrated, so that the set pressure is not set low. (Conversely, if the gauge reads high, the system pressure may exceed the allowable limit.)
Vibration in the system.	• Just as vibration can lead to misalignment of the valve and cause leaking, it can also cause premature popping. Provide appropriate valve and piping support.

Table 10.10 Troubleshooting: Chattering [17]

Causes	Explanations and Solutions
Valve oversized.	• A valve that is oversized by 25% has a tendency to chatter. If the volume of fluid to be relieved is less than 25% of the rated capacity, there is insufficient force to lift the disk fully off the seat, and the valve opens and closes only partially. It is difficult to choose a valve for all potential relief conditions, but it is important to carefully size the valve for the worst-case scenario, without oversizing. Sometimes two valves are used with staggered set points to accommodate widely differing relief loads.
Excess pressure drop in the valve inlet piping.	• Pressure losses in the relief valve inlet piping can restrict relief flow significantly and result in insufficient lift. A typical rule of thumb is for the inlet pressure drop to be less than 3% of the valve set pressure. The ultimate rule is for the pressure drop to be less than the valve's blowdown. To prevent excess pressure drop, the inlet piping should be at least as large as the valve orifice diameter, there should be a minimum number of bends, and the valve should be located as close as possible to the pressure source. (See Section 10.5.2 for proper inlet piping methods.)
Pressure variations in the inlet or outlet piping.	• Unavoidable pressure variations in the inlet piping can be resolved by the use of two relief valves sized and set to accommodate minor upsets in pressure. One valve is sized smaller and at a lower pressure to relieve the minor pressure variations, and the second is set at no more than 105% of the MAWP and sized to relieve the remainder of the relief load. • Pressure variations in outlet piping occur only when relief valves are manifolded into a common discharge header. Often the relief valve discharge line can be sized and designed to minimize variations in pressure (see ASME Code, Section VIII, Appendix M). However, if the pressure variations are still a problem, a balanced bellows valve should be used to isolate the valve bonnet from the discharge piping.

Table 10.11 Troubleshooting: Leaking Valves [17]

Causes	Explanations and Solutions
Operating pressure too near the set pressure.	• Design vessel according to ASME code and API recommendations. Allow 5 psi above normal operating pressure, and set pressure an additional 15 psi above that. • Allow for manufacturing and burst tolerances in rupture disks.

Table 10.11 Troubleshooting: Leaking Valves, *continued*

Causes	Explanations and Solutions
Particulates between seat and disk.	• Replace metal-to-metal seat with resilient O-ring seat. • Use bellows to isolate the valve stem from the relieving fluid. • Use a specially designed knife-edge seat that cuts through particles.
Unsupported outlet piping.	• Be sure to support the relief valve and outlet piping adequately to prevent the valve internals from becoming misaligned. (See Section 10.5.3.)
Excessive popping tolerance.	• All valves and disks have a popping tolerance associated with their set pressure. ASME code requires that the popping tolerance for valves not exceed ± 2 psi for set pressures up to 70 psi, and 3% for pressures above 70 psi. Rupture disk burst tolerance varies with manufacturer. Ensure that the popping tolerance is minimal and operate the system well outside this range.
Thermal stress.	• Large shifts in temperature of the operating system can affect the internal alignment of the valve and its associated piping. Be sure to provide appropriate expansion joints, loops, and flexible supports to accommodate thermal expansion.
System vibration.	• Just as thermal expansion can cause relief device misalignment, system vibration can lead to leaks. Support the relief device and piping with flexible supports and use resilient O-ring seals and seats.
Relief valve is not installed upright.	• A relief valve that is not installed vertically can accumulate particles and form deposits because of poor drainage. In addition, the valve may become misaligned. The result is poor sealing and, therefore, leakage. All valves should be mounted upright and vertical.
Valve misadjusted for long blowdown.	• The valve blowdown is the pressure at which the valve reseats after popping (expressed as % of set pressure, usually 5–7%). If the blowdown is set too high, the valve stays open for an excessively long time. The reseating pressure can fall into the operating pressure range and the valve never fully reseats. Check that the blowdown setting does not put the reseating pressure too near the operating pressure.
Valve corrosion and erosion.	• The process fluid in contact with the relief valve may be corrosive or erosive (clean steam, or fluid containing particulates). Choose the relief device materials carefully to ensure maximum life under conditions of corrosive and erosive attack. Stainless steel is the most likely material choice for most biotechnology processes, but if stainless steel is incompatible or the fluid is excessively erosive, more exotic materials, such as Monel®, should be considered.
Incorrect seat lapping.	• For the valve to seal tightly, seats must be smooth, flat, and highly polished. If the seat surfaces don't have a fine finish, they should be lapped and polished using the manufacturer's recommended procedure.

10.7   Example Problems

10.7.1 Calculation of Maximum Allowable Working Pressure

An engineer needs to design a vessel to contain sterile medium for automatic feeding into a fermentor. The engineer's first job is to determine the design pressure of the vessel.

This vessel must be capable of being sterilized in place with 15 psig steam at 121 °C. Although the typical operating pressure of the vessel is much less than the sterilization pressure, the designer should assume for design purposes that 15 psig is the normal operating pressure of the vessel. Since steam needs to be supplied at a higher pressure to achieve sterilization conditions in the vessel, it makes sense to set the maximum operating pressure at 5 psi higher than the normal operating pressure. The designer should therefore set the maximum operating pressure of the vessel at 20 psig. The maximum allowable working pressure or design pressure of the vessel is then 35 psig. This is shown below:

- Normal operating pressure = 15 psig;
- Maximum operating pressure (add 1% or 5 psi) = 20 psig;
- Maximum allowable working pressure (add 15 psi) = 35 psig.

Note that if a rupture disk is used in this application, the MAWP would probably be set at 50 psig to take burst tolerance requirements into account (see Section 10.7.2).

10.7.2 Calculation of Burst Pressure for a Rupture Disk

A vessel operating at 15 psig has a MAWP of 50 psig at 72 °F (22 °C). A rupture disk must be chosen to protect this vessel from overpressure. The engineer looks at manufacturers' data for standard manufacturing ranges and burst tolerances of rupture disks at the specified burst pressure [13–16]. Typical data are as follows:

- 50 psig at 72 °F corresponds to a specified rupture pressure range (psig @ 72 °F) of 46–90 psig;
- In this range, manufacturing range (% @ 72 °F) = +12 to −6 [12, 13]; and
- burst tolerance (% @ 72 °F) = ±5 [12, 13].

Therefore, a rupture disk specified to burst at 50 psig has the following range of stamped burst pressures:

- +12% of 50 psig = +6 psig => **56 psig** and
- −6% of 50 psig = −3 psig => **47 psig**.

Thus, the stamped burst pressure can be anywhere between 47 and 56 psig. The burst tolerance is then used to calculate the actual burst pressure range:

- −5% of 47 psig = −2.35 psig => **44.65 psig** and
- +5% of 56 psig = +2.8 psig => **58.8 psig**.

Notice that the manufacturing range and burst tolerance are essentially additive. If the engineer does not specify any further requirements on the rupture disk, the disk could have a burst pressure range of 44–59 psig. This is not acceptable for a vessel whose maximum allowable working pressure is 50 psig. To take into account this range of burst pressures, the designer can specify that all of the manufacturing tolerance is on one side of the burst pressure.

For example:

- manufacturing range = +12 to –6 => 18% total;
- burst tolerance on the high side = 5%; so
- total tolerance = 23%.

Rather than split this tolerance on each side of the specified burst pressure, the engineer can specify that it all go on the low side:

- 18% of 50 psig = 9 psig, and
- –23% of 50 psig = –11.5 psig => **38.5 psig**, so
- maximum stamped pressure = 38.5 + 9 psig = **47.5 psig**.

The stamped burst pressure is then between 38.5 and 47.5 psig. The burst tolerance range is:

- –5% of 38.5 psig = –1.9 psig => **36.6 psig** and
- +5% of 47.5 psig = +2.4 psig => **49.9 psig**.

In this case, the rupture disk burst pressure does not exceed the MAWP of the vessel. However, the low side of the pressure range is significantly reduced. Instead of relieving at a minimum of 44 psig, the disk may now relieve at 36 psig. This points out the importance of providing sufficient leeway between the operating pressure and the set pressure to prevent premature popping. Typically, the system operating pressure should not exceed 70% of the stamped burst pressure, although this depends on the rupture disk design; manufacturers' literature should be consulted for actual operating pressure limits.

In this example, the stamped burst pressure range is 38.5–47.5 psig.

- 70% of 38.5 psig = **27 psig** > 15 psig.

The vessel normal operating pressure is 15 psig, so the specified disk should be acceptable for this application. The engineer can specify the following:

"Rupture disk with burst pressure Not To Exceed 50 psig @ 72 °F; all manufacturing tolerance on lower end of pressure range."

In this example, the vessel operates at ambient conditions. If, instead, the vessel were used in a steam sterilization application, the operating pressure would approach 20 psig and the design temperature would be near 250 °F. Since the burst pressure decreases as the operating temperature increases, specification of the rupture disk should include the maximum operating temperature. Rupture disk data are always tabulated at 72 °F, so the disk manufacturer must assist with this problem. Sometimes it becomes more expensive to provide a vessel with a pressure rating to cover the required rupture disk tolerance than to use a relief valve.

10.7.3 Design of a Complete Relief System

Figure 10.8 illustrates a typical biotechnology processing system, consisting of a medium holding tank fed by a mixing vessel, a fermentor, and a product holding tank that feeds a continuous-flow centrifuge for product clarification. The medium mixing station is on a mezzanine, and the medium holding tank is elevated approximately 12 feet. The fermentor is fed by gravity flow. The fermentation product is transferred by air pressure to the product holding tank, where it is held at 2–8 °C (by a propylene glycol–chilled jacket) until ready for clarification. The product from this system contains potentially viable organisms that must be contained. The entire system must be sterile. The engineer involved in designing this facility must develop a relief system capable of protecting the equipment and personnel.

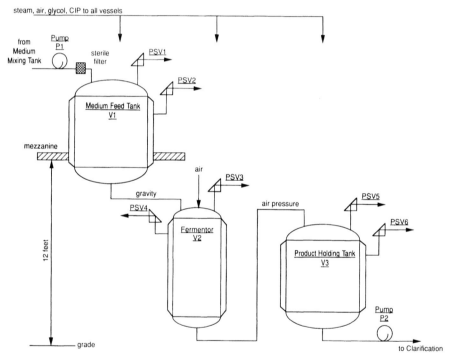

Figure 10.8 Typical Biotechnology Processing System, consisting of a mixing vessel, a media holding tank, a fermentor, and a product holding tank. This system is used as the basis for the design of a complete relief system (see Section 10.7.3).

Pertinent Design Data:
(1)   Media Holding Tank, V1:
      Design pressure = 50 psig/full vacuum;

Design temperature = 300 °F;
Jacket design pressure = 90 psig;
Jacket design temperature = 400 °F;
Mezzanine height = 12 feet;
Vessel dimensions = 42" dia. × 24" tangent-to-tangent (TT);
Nominal vessel volume = 600 L;
Jacket heat transfer medium: propylene glycol.

(2)   Media Feed Pump, P1:
Pump type: positive-displacement;
Maximum discharge pressure = 150 psi;
Flow @ 50 psig = 18 gpm.

(3)   Fermentor, V2:
Design pressure = 50 psig/full vacuum;
Design temperature = 300 °F;
Jacket design pressure = 50 psig;
Jacket design temperature = 300 °F;
Vessel dimensions = 24" dia. x 48" TT;
Nominal vessel volume = 300 L;
Jacket heat transfer medium: water or steam;
Jacket maximum water flow = 2 gpm.

(4)   Product Holding Tank, V3:
Design pressure = 50 psig/full vacuum;
Design temperature = 300 °F;
Jacket design pressure = 90 psig;
Jacket design temperature = 400 °F;
Vessel dimensions = 42" dia. × 24" TT;
Nominal vessel volume = 600 L;
Jacket heat transfer medium: propylene glycol.

(5)   Product Transfer Pump, P2:
Pump type: positive-displacement;
Maximum discharge pressure = 150 psi;
Flow @ 50 psig = 18 gpm.

(6)   Utilities:
Steam: 45 psig maximum, regulated to 22 psig;
Air: 65 psig, ambient, maximum flow = 35 scfm (standard ft^3/min);
Glycol: −2 °C, 25 psig, 40% solution.

The system consists of three vessels that need to be relieved: the media holding tank, the fermentor, and the product holding tank. Each of these also has a jacket that must be relieved.

With pertinent design data, process flow diagram, and detailed piping and instrument diagrams in hand, we can begin design of the relief system. The first step is analysis of potential relief cases for each vessel and sizing of each relief device.

(1)   Media Holding Tank, V1, relieved by PSV1:
Potential relief cases—
    (a)   fire;
    (b)   blocked exit;
    (c)   blocked exit during sterilization;
    (d)   power failure/instrument air failure; and
    (e)   steam regulator failure.

P_{set} = MAWP = 50 psig.
P_{relief} = 50 × 1.1 = 55 psig (nonfire),
      = 50 × 1.2 = 60 psig (fire).

(a)   Fire—
    • Vessel is on a mezzanine, 12 ft above grade.
    • Assume vessel liquid-full to high liquid level (HLL).
    • Vessel overall height above mezzanine is about 3 ft.
    • Entire vessel is within 25 ft of grade, so entire vessel must be considered.

From eq 10.2,
$$A_v = \pi dh = \pi(42/12)(24/12) = 22 \text{ ft}^2$$
From eq 10.3,
$$A_h = 1.082(42/12)^2 = 13.25 \text{ ft}^2$$
Therefore,
$$A_{total} = 22 + 13.25 = 35.25 \text{ ft}^2$$
From eq 10.1, the heat absorbed by a vessel during fire, assuming no credit for insulation, is:
$$Q = 21,000\, A^{0.82} = 21,000\, (35.35)^{0.82} = 389,843 \text{ Btu/h}$$
To determine the vapor relief rate during fire, assume the medium is similar to water. The value of the latent heat of vaporization of water at the relief pressure of 60 psig can be found in steam tables:
$$\lambda = 904 \text{ Btu/lb @ } T = 308 \text{ °F}$$
From eq 10.4,
$$W = 389,843/904 = \mathbf{431 \text{ lb/h}} = W_{fire}$$

(b)   Blocked exit—
    • All outlets closed.
    • Medium is fed from mixing vessel through pump P1, until vessel is full; therefore liquid relief.
    • Determine maximum flow from pump when pumping is against relief pressure of 55 psig.

From design data (pump curves), maximum flow @ 55 psig is 18 gpm, therefore,
$$\mathbf{W_{BO} = 18 \text{ gpm}}$$

(c)   Blocked exit during sterilization—
* Vent valve fails closed and steam valve remains open.

Steam flows into vessel, but only against a pressure less than 22 psig. The pressure in the vessel does not reach 50 psig; therefore, **no overpressure case** exists.

(d)   Power failure/instrument air failure—
* All valves fail closed (fail-safe).
* Therefore, there is no potential for fluid entry to overpressure.

No case

(e)   Steam regulator failure—
* Pure steam is fed to system at 45 psig maximum.
* Again, steam can't overpressure vessel designed for 50 psig.

No case

(Note: Good design practice in biotechnology is to design vessels to prevent overpressure where economically reasonable.)

Sizing basis—

There are two possible sizing cases: fire and blocked exit. Since this is a sterile vessel, assume we use a rupture disk.

For fire, from eq 10.9:

$$A_r = \frac{(431)(308 + 460)^{0.5}}{(3600)(0.62)(0.0969)(60 + 14.7)(18)^{0.5}} = 0.17 \text{ in}^2$$

For blocked exit, from eq 10.12 for liquid flow:

$$A_r = \frac{(18)(1)^{0.5}}{(23.1)(60)^{0.5}} = 0.10 \text{ in}^2$$

Therefore, the sizing basis for PSV1 is fire, with $A_r = 0.17 \text{ in}^2$.

(2)   Media Feed Tank Jacket, V1, relieved by PSV2:

Potential relief cases—
(a)   fire;
(b)   blocked exit;
(c)   abnormal heat input during vessel sterilization (thermal expansion); and
(d)   power failure.

$P_{set} = \text{MAWP} = 90$ psig.
$P_{relief} = 90 \times 1.1 = 99$ psig (nonfire),
$\qquad = 90 \times 1.2 = 108$ psig (fire).

(a)   Fire—
* Same assumptions as for PSV1.

From eq 10.2,

$$A_v = \pi d h = 22 \text{ ft}^2 \text{ (from PSV1 calculations) (no jacket on head).}$$

From eq 10.1,

$Q = 21,000 (22)^{0.82} = 264,854$ Btu/h;

and therefore, from eq 10.4,

$W = 264,854/359 = 737$ **lb/h** $= W_{fire}$

Here the latent heat of vaporization for pure glycol at 1 bar (359 Btu/lb) is used. In the absence of data, this is a conservative estimate for the latent heat of vaporization of a 40% glycol solution at 99 psig. Since water boils first, we look at steam tables to determine the temperature at which this fluid vaporizes (again a conservative approach). At 108 psig, the boiling point of water is 343 °F.

(b) Blocked exit—

The maximum pressure from the glycol system is 25 psig, so **no relief case** exists.

(c) Abnormal heat input—
- During sterilization of a vessel, both the inlet and outlet of the jacket may be blocked.
- Steam is sent to the vessel shell at 121 °C.

The heat transfer equation is:

$H = U A \Delta T_{ln}$

Assuming 0 °C as an average glycol temperature, and that we can heat the glycol to within one degree, that is, 120 °C,

$$\Delta T_{ln} = \frac{\Delta T_2 - \Delta T_1}{\ln (\Delta T_2 / \Delta T_1)} = \frac{121 - 1}{\ln (121/1)} = 25.02 \ ^{\circ}C = 76.4 \ ^{\circ}F$$

where ΔT_1 is the cold side temperature difference, here (temperature of steam in − temperature of glycol out) or (121 °C − 120 °C);

ΔT_2 is the hot side temperature difference, here (temperature of steam out − temperature of glycol in) or (121 °C − 0 °C); and

ΔT_{ln} is the log mean of ΔT_1 and ΔT_2 (also called the *LMTD*).

Estimate $U = 150$ Btu/h·ft²·°F, so

$H = (150)(22)(76.4) = 252,120$ Btu/h.

Since the boiling point of the glycol solution is 177 °C, and that of water at 99 psig is 164 °C, the solution does not boil. Therefore, we need consider only liquid relief. In order to determine the liquid relief, we look at eq 10.7 for thermal expansion:

$$W_Q = \frac{\alpha H}{500 \ (SG) \ C_p} = \frac{(0.00054/^{\circ}F) \ (252,120 \ Btu/h)}{(500) \ (1.0514) \ (0.89)} = \mathbf{0.29 \ gpm} = W_{thermal}$$

Values for α, SG, and C_p were taken from physical data tables for propylene glycol solutions.

(d) Power failure—
- If equipment is on emergency generator, there is **no case**.
- If equipment is not on emergency generator, then glycol pumps fail, as do valves, so still **no case**.

Sizing basis—
Since this is a jacket, we decide to use a relief valve.
For fire, from eq 10.15,

$$A_d = \frac{(737 \text{ lb/h}) \left[(343 + 460) (1) \right]^{0.5}}{(348) (0.97) (108 + 14.7) (1) (18)^{0.5}} = 0.12 \text{ in}^2$$

For thermal expansion, from eq 10.16,

$$A_d = \frac{(0.29) (1.0514)^{0.5}}{(27.2) (0.62) (99 + 14.7)^{0.5}} = 0.002 \text{ in}^2$$

The sizing basis for the jacket is the fire case, with a required area of 0.12 in^2. This requires a valve of approximately 1" $\times$ 2". Typically, only ¾" $\times$ 1" valves are needed on jackets. We recognize that we were very conservative in calculating the fire case, so we may want to reevaluate the assumptions and determine if a 1" $\times$ 2" valve is truly necessary. This is particularly pertinent since thermal expansion is a more likely relief occurrence than fire. The valve size required for thermal relief is significantly smaller than that for fire. For maintenance reasons, we won't usually specify a valve smaller than ¾" $\times$ 1". This is already oversized for thermal relief, and we must remember that oversized valves cause problems (see Section 10.6).

We've now seen the basics of valve sizing. For the remainder of the relief vessel cases, I note only those occurrences that differ significantly from the cases already described. Table 10.12 summarizes the potential relief cases.

(3)   Fermentor Vessel, V2, relieved by PSV3:
Potential relief cases—
  (a)   fire;
  (b)   blocked exit;
  (c)   abnormal heat input;
  (d)   power failure;
  (e)   instrument air failure; and
  (f)   steam regulator failure.
P_{set} = MAWP = 50 psig.
P_{relief} = 50 $\times$ 1.1 = 55 psig (nonfire),
       = 50 $\times$ 1.2 = 60 psig (fire).
(a)   Fire—
  • Assume vessel contents are similar to water/steam and size as for PSV1.
(b)   Blocked exit—
  • No liquid case since gravity fed.
  • Can have air relief because air pressure to discharge.
  • Determine the air mass flow by assuming maximum volumetric flow at the relief pressure.

(4)   Fermentor Vessel Jacket, V2, relieved by PSV4:
Potential relief cases—
  (a)   fire;
  (b)   blocked exit; and
  (c)   abnormal heat input (thermal expansion).
P_{set} = MAWP = 50 psig.
P_{relief} = 50 × 1.1 = 55 psig (nonfire),
       = 50 × 1.2 = 60 psig (fire).
These cases are straightforward.

(5)   Product Holding Tank, V3, relieved by PSV5:
Potential relief cases—
  (a)   fire;
  (b)   blocked exit;
  (c)   power failure; and
  (d)   steam regulator failure.
P_{set} = 50 psig.
P_{relief} = 50 × 1.1 = 55 psig (nonfire),
       = 50 × 1.2 = 60 psig (fire).
These cases are identical to PSV1 for the media holding tank, except that no blocked exit case exists since the vessel is fed by air pressure (not a pump) from the fermentor. If the pressure exceeds 50 psig, the fermentor relief device bursts, not the product holding tank.

(6)   Product Holding Tank Jacket, V3, relieved by PSV6:
Potential relief cases—
  (a)   fire;
  (b)   blocked exit; and
  (c)   abnormal heat input (thermal expansion).
These cases are identical to those for PSV2.

Table 10.12 summarizes the relief valve sizing cases for each of the vessels in Figure 10.8.

With these data in hand, we may now design the relief header and containment system to handle the relief scenarios. To size the relief header, we must decide which relief devices to tie into a common header, and which relief cases cause a multiple relief occurrence.

Since the vessel jackets contain only glycol or steam, these are relieved independently to the atmosphere. We should design relief outlet piping to minimize safety hazards. Therefore, the relieving fluid should leave the system near grade. The piping should also be designed to minimize back pressure on the relief valves.

Table 10.12 Summary of Relief Case Analysis

Cause	Individual Relief Loads					
	PSV1	PSV2	PSV3	PSV4	PSV5	PSV6
Fire	431 lb/h	737 lb/h	372 lb/h	372 lb/h	431 lb/h	737 lb/h
Blocked exit	18 gpm	NC	12.75 lb/min	2 gpm	NC	NC
Abnormal heat input	NC	0.29 gpm	NC	NC	NC	0.29 gpm
Power failure	NC	NC	NC	NC	NC	NC
Instrument air failure	NC	NC	NC	NC	NC	NC
Regulator failure	NC	NC	12.75 lb/min	NC	NC	NC
Device size, in^2	0.17	0.12	0.21	0.096	0.17	0.12
Sizing basis	fire	fire	blocked exit	fire	fire	fire

NC: no case

The media holding tank does not contain anything that could harm the environment, so it may also be relieved to the atmosphere. Again, we should consider safety when directing the outlet piping.

The fermentor and product holding tank may each contain viable organisms. Relief from these vessels should be contained. The most appropriate discharge location is the plant's biodisposal system. Assuming this is a gravity-flow system, the relief piping may be directed to a gravity-flow waste header.

To size the common relief header system, we need to consider which relief occurrences can cause multiple device discharge. For instance, a blocked exit on vessel V2 does not cause a relief occurrence on vessel V3. However, fire, power failure, or instrument air failure can affect more than one vessel. In sizing the relief header, we should add the relief loads from both the fermentor, V2, and the product holding tank, V3, for each of the multiple relief cases. Since fire is the only case that causes relief in both vessels, we use fire as the basis for manifold sizing. The total relief load during fire is (372 + 431 lb/h) = 803 lb/h.

Once the header is sized, we should check the sizing of the outlet piping for the relief device, to ensure that excessive back pressure is not present at the relief device. We do this by starting with the pressure in the biodisposal system and doing pressure-drop calculations back along each leg of the relief system to the device outlet.

10.8   References

1. *Rules for Construction of Pressure Vessels*; Unfired Pressure Vessels, Section VIII; ASME Boiler and Pressure Vessel Code, Division 1; American Society of Mechanical Engineers, New York, 1980.
2. *Recommended Practice 520 for the Design and Installation of Pressure-Relieving Systems in Refineries*; American Petroleum Institute, Washington DC, 1976.
3. *Recommended Practice 521 Guide for Pressure-Relieving and Depressuring Systems*; American Petroleum Institute, Washington, DC, 1982.
4. JENKINS, J. H., KELLY, P. E., COBB, C. B., Design for Better Safety Relief, *Hydrocarbon Process. 56* (8), 93–97 (1977).
5. WONG, W. Y., Safer Relief Valve Sizing, *Chem. Eng. 96* (May), 137–140 (1989).
6. *Safety Relief Valves—Recommendations*; In: Manual FE-80-100, Section 7; Teledyne Farris Engineering, Palisades Park, NJ, 1986.
7. *Catalog No. 7387*; Fike Metal Products, Division Fike Corporation, Blue Springs, MO, 1992.
8. *Rupture Disc Sizing*; Bulletin 1-1110, Continental Disc Corporation, Kansas City, MO, 1991.
9. MYERS, J. F.; *Overpressure Protection from Metal Disks Designed for Failure*; Publication 77-721, BS&B Safety Systems, Tulsa, OK, 1972.
10. *Rupture Discs/ASME Code*; Bulletin 73-3002, Continental Disc Corporation, Kansas City, MO, 1980.
11. PAPA, D. M., How Back Pressure Affects Safety Relief Valves, *Hydrocarbon Process. 62* (5), 79–81 (1983).
12. MUKERJI, A., How to Size Relief Valves, *Chem. Eng. 87* (June 2), 79–86 (1980).
13. *Standard-Type Rupture Disc Catalog*; No. STD-1184; Continental Disc Corporation, Kansas City, MO, 1984.
14. *Composite-Type Rupture Disc Catalog*; No. CMP-185; Continental Disc Corporation, Kansas City, MO, 1986.
15. *Rupture Disc Data Sheet 3010*; Fike Metal Products, Blue Springs, MO, February, 1986.
16. *Rupture Disc Data Sheet 3030*; Fike Metal Products, Blue Springs, MO, September, 1985.
17. SCULLY, W. A., Safety Relief Valve Malfunctions: Symptoms, Causes and Cures, *Chem. Eng. 88* (August 10), 111–114 (1981).
18. COX, O. J., WEIRICK, M. L., Sizing Safety Valve Inlet Lines, *Chem. Eng. Prog. 76* (November), 51–54 (1980).

Instrumentation and Control of Bioprocesses

Thomas Hartnett

Contents

11.1   Fundamentals of Process Control

11.1.1   The Basic Control System

Instrumentation is essential for the control of any process. A visual inspection can be enough to check the status of a mechanical process such as a beverage bottling operation, but a visual examination of an ongoing bioprocessing reaction such as that taking place within a fermentor usually means nothing. Instrumentation is considered our window to the process, and by utilizing instrumentation in conjunction with process control, we can keep the difference between measured and desired values of variables to a minimum.

A bioprocess is significantly different from a standard synthetic chemical reaction in many ways that affect its instrumentation, both negatively and positively. One way that bioprocessing reactions are said to be easy to control is that they are fairly stable, and most major variables tend to change slowly over time in the absence of major equipment failures. A significant problem with bioprocesses, however, is that once the cell population has died, the process is over with no hope of restarting, and the product yield up to that point may be unacceptable. This could mean the loss of a substantial amount of money and time. Another way that bioprocesses pose problems for instrumentation is that they must be absolutely sterile. As any biotechnologist knows, the environment established for the production of a desired cell line can also support the growth of unwanted cell types. All sensors measuring process variables must be installed and maintained in a sanitary manner.

Overall, the objective of a good measurement and control strategy is to provide optimal conditions for growing the culture and producing the desired product. The well-instrumented bioreactor can therefore be considered a reactor supporting another reactor, the cells themselves. By initially providing and then maintaining an environment that encourages growth, with the proper temperature, pressure, nutrient feeding, substrate composition, etc., we obtain the desired product at optimum titer.

Figure 11.1 shows a well-instrumented fermentor. Although many industrial fermentors get by with only temperature, pressure, and nutrient flow control, the level of control illustrated is not considered excessive. In an economic sense, it is a modest portion of the total installed cost for a new facility at the industrial scale. At smaller scales the cost of an instrumentation system becomes a larger portion of the total cost. However, smaller scale installations can usually justify a well-instrumented fermentor by the need for data collection in a research setting or by the production of a high–value-added product.

In the fermentor example shown in Figure 11.1, several individual classes of instrumentation must be defined. Of primary importance are the measuring devices. These include the sensors measuring a process value, such as temperature, pressure,

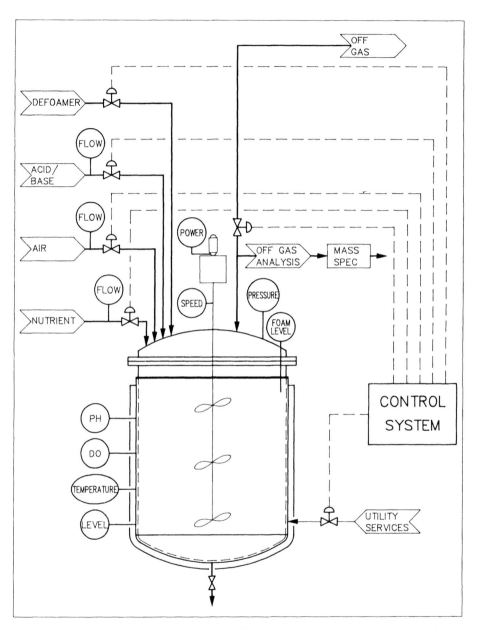

Figure 11.1 Well-Instrumented Fermentor. While such complete instrumentation is not necessary in all applications, it provides a good basis for a flexible bioreactor installation.

level, flow, pH, dissolved oxygen, or power. These sensors transmit a scaled pneumatic or electrical signal of the variable to either an indicator, a recorder, or a controller. These indicators, recorders, and controllers can range from individual units, to combination units that do all three functions for one or more variables, to complex distributed control systems with almost limitless capabilities. As the terms imply, *indicators* simply read out the process variable or a scaled value of it; *recorders* log a hard copy of it in addition to indicating the value; *controllers* compare the measured variable to a desired value, the *set point*, compute the *error*, and calculate an output based on the controller algorithm. The output of the controller usually is transmitted to a final control element: the signal usually actuates a control valve, but it can be an input to a speed controller or the set point of another controller, or it can be used in other ways to control the process. The final control element acts on the manipulated variable to produce the desired results.

The combination of a sensor, an indicator, recorder or controller, and a final control element (when a controller is used) constitutes a *control loop*.

11.1.2   Control Loops

11.1.2.1   Open Control Loop

The term *open control loop* is a misnomer, since these loops do not provide automatic control. An example of an open control loop is shown in Figure 11.2. The sensor detects the batch temperature and transmits this value to a controller. The controller compares this value to the set point. An error is calculated, but since the controller is not associated with a control valve, the batch temperature is not adjusted. Human intervention based on an indicated value is considered an open control loop, but if it is left alone, an open control loop acts as shown in Figure 11.3. This example shows the temperature rising because the supply of cooling water is not adequate for removing the heat of reaction. Examples of an open control loop are indicators, recorders, or any situation where constant human mediation is applied if necessary. Open-loop control is often used to determine dynamics of the control system. This is useful when the controller is being tuned for a closed-loop mode.

11.1.2.2   Closed Control Loop

The closed control loop is the heart of a well-designed control strategy. An example is shown in Figure 11.4. The control system compares the measured variable to the set point, giving an error that is supplied to the appropriate control algorithm, and an output is calculated. The output acts on the final control element, here a control valve, which adjusts the cooling water flow as needed to act on the manipulated variable. A graphical representation of this is shown in Figure 11.5. A process disturbance occurs that increases the error in the loop. The controller changes the position of the

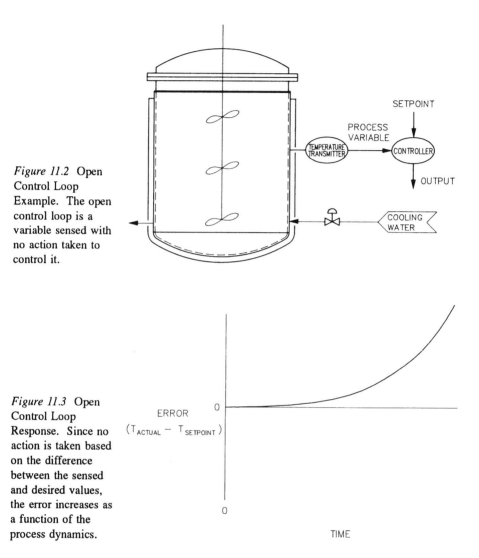

Figure 11.2 Open Control Loop Example. The open control loop is a variable sensed with no action taken to control it.

Figure 11.3 Open Control Loop Response. Since no action is taken based on the difference between the sensed and desired values, the error increases as a function of the process dynamics.

control valve and the error is eventually removed. The amount of time that a control loop takes to remove the error depends on the constants of the control algorithm and the dynamics of the process.

The most common control algorithm in the process industry is the *proportional/ integral/derivative* (*PID*) control algorithm. The error is supplied to the PID algorithm and an output is calculated that depends on the controller constants.

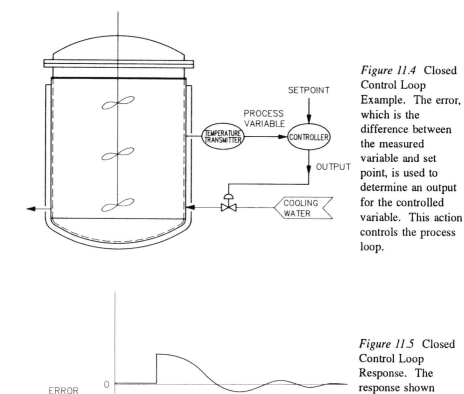

Figure 11.4 Closed Control Loop Example. The error, which is the difference between the measured variable and set point, is used to determine an output for the controlled variable. This action controls the process loop.

Figure 11.5 Closed Control Loop Response. The response shown indicates quarter-amplitude decay. In each successive cycle after the first, the error in a loop decreases to one-quarter of its previous level.

The response due to the *proportional* component of the control algorithm is based on the size of the error, and is described by the following equation:

$$O = K_c e \tag{11.1}$$

where O is the output, K_c is the proportional gain, and e is the error. Note that no units are given. The error and output are scaled, and range from 0 to 1. The gain can be any value in the range the controller accepts. If an output for a control valve is greater than 1 or less than 0, the valve is fully open or fully closed (depending on the action). The proportional gain therefore determines how sensitive the control loop is. An alternative form of proportional gain is the *proportional band*, defined as:

$$P_b = K_c/100 \tag{11.2}$$

where P_b is the proportional band.

The *integral* or *reset* element of the PID controller calculates an output based on the length of time the error is present. It is described by the following equation:

$$O = K_c/T_1 \int e \, dt \tag{11.3}$$

where T_1 is the integral time, and t is time.

The third and final component of the control algorithm is the *derivative* or *rate* action. This mode responds to the rate of change of the error, and is described by the following equation:

$$O = K_c T_2 \, (de/dt) \tag{11.4}$$

where T_2 is the derivative time.

It is possible to have any combination of the three control algorithm components (proportional, integral, and derivative); use of all three gives the following control equation:

$$O = K_c [e + 1/T_1 \int e \, dt + T_2 \, (de/dt)] \tag{11.5}$$

In a bioprocessing environment, where the lag of a control loop is usually not a major concern, the most common form of control is the proportional/integral or PI.

11.1.3   Additional Forms of Control

11.1.3.1   On/Off Control

In addition to the PID control algorithm, many other mathematical models can be used to control a system. One of the most universal is the *on/off controller*. This works by having an allowable range about a measured variable. The manipulated variable is turned fully on when one extreme of the range is reached, and off when the other extreme of the range is achieved. An example of this action is batch temperature control in which the cooling water is turned on at 26 °C and off at 28 °C to maintain an average temperature of 27 °C. The response of this system is shown in Figure 11.6; we can see an unacceptable oscillating temperature pattern that is detrimental to the growth of a culture. The use of on/off control should be confined to manipulated variables that do not have a major effect on the process.

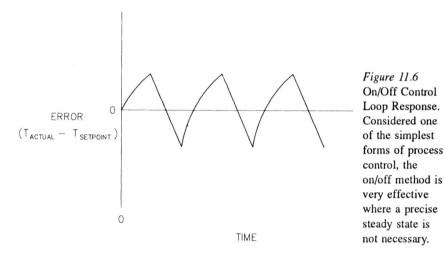

Figure 11.6
On/Off Control
Loop Response.
Considered one
of the simplest
forms of process
control, the
on/off method is
very effective
where a precise
steady state is
not necessary.

11.1.3.2   Cascade Control

Cascade control is the type of control where the output of one controller is the set point of another. A schematic example of this is shown in Figure 11.7, which illustrates a dissolved oxygen controller sending its output to a speed controller, which regulates the agitator speed in a fermentor. This allows the fast-acting speed controller to maintain the value of the set point of the speed, but still respond to the need to raise or lower speed to adjust the dissolved oxygen concentration in the broth. Cascade control is applied so as to have the slave loop (here the speed control) be fast acting and control most of the possible disturbances, and the master loop (the dissolved oxygen control) respond to the lag in the system. A good guideline for stable cascade control is to have the slave loop response time 5 to 10 times (or more) faster than the master loop response time.

11.1.3.3   Ratio Control

Ratio control is applied when the proportion of two or more variables is to be maintained relative to each other. An example is shown in Figure 11.8. The flow of nutrient 1 is to be maintained at a preset ratio to the uncontrolled flow of nutrient 2. As the flow rate of nutrient 2 varies with the changing process conditions in the line, the signal is compared to the flow signal of nutrient 1. Their ratio is the measured variable and is compared to the set point ratio. The controller determines the error and calculates an output.

Figure 11.7
Cascade Control
Loop Example.
A cascade loop
is useful for
improved control
response in a
loop where the
measured
variable is slow
to respond to
changes in the
manipulated
variable. The
figure shows a
speed control
slave loop (fast)
and a dissolved
oxygen control
master loop
(slow).

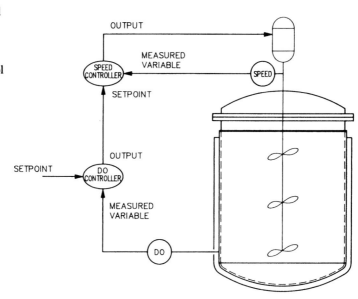

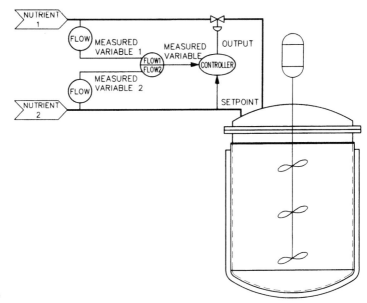

Figure 11.8
Ratio Control
Loop Example.
In this example,
the ratio of flow
rates of two
feeds to a
bioreactor is
being controlled.

11.1.3.4   Feedforward Control

Feedforward control is a control method that measures disturbances before they affect the process, and changes the manipulated variables as needed to maintain the set point. This requires an exact mathematical model of the process relative to the set point, manipulated variables, and measured variables. An example of this scheme is the control of the temperature of a fed-batch system that relies on measurements of the flow and temperature of the incoming and outgoing streams and an accounting for heat losses. Although in theory this is an ideal way to control a process, in practice it has limited application. Loop error, model inadequacies, or unmeasured disturbances such as heat loss to the atmosphere can cause cumulative drift in the desired process condition in the reaction vessel. If processes require fast response and tight control of process variables, feedback control is usually combined with feedforward control. Feedforward control is not normally used in bioprocessing applications, since the variables are not fast acting and the process dynamics are somewhat stable.

11.1.3.5   Multivariable Control

Multivariable control is known by several names, including *computer control*, *high-level control*, and *supervisory control*. Conceptually it is relatively simple to implement, as long as it is possible to both send and receive signals from a control computer. Multivariable control, as the name implies, is the control of at least one manipulated variable by two or more (usually several) measured variables. The measured variables are supplied to a computer programmed with a predetermined specific mathematical algorithm, which calculates a process value and compares it to the set point in the controller. The loop functionality from the controller is the same, in that an output is calculated and supplied to the final control element. This is an extremely useful tool and has many applications in biotechnology. An example is in Section 11.2.6 on dissolved oxygen instrumentation for the calculation of $K_L a$.

11.2   Measurement Elements

11.2.1   Temperature Measurement

Of all the variables sensed in the production of natural products, the most important is usually temperature. The precise and accurate measurement and control of temperature is essential to the viability of the organism being cultured. As discussed in Section 11.5.2, Calibration Methods and Accuracy Determination, an accuracy of 0.5 °C or less can be considered reasonable with currently available and well-calibrated *resistance temperature devices* (*RTDs*). In most fermentations, a variance of 1.0–1.5 °C can mean the difference between an acceptable and an unacceptable run.

The margin between the best accuracy possible and an undesirable temperature reading is 0.5–1.5 °C, so a thorough knowledge of temperature-sensing control loops is necessary for optimal production of any culture.

11.2.1.1   Resistance Temperature Devices

RTDs are the preferred method of temperature measurement in bioprocesses. Their inherent accuracy and stability make them ideal for temperature measurement. One disadvantage is their moderately high cost, but that is unimportant for such a critical variable. Their somewhat high time constant (5 seconds) is insignificant for a growth cycle of several days.

The theory behind RTDs is simple: the resistance of a metal conductor changes with temperature. The four most common metals used in the manufacture of RTDs are nickel, nickel alloy, copper, and platinum. Platinum has the most linear characteristic, with the broadest range and highest stability, and is therefore the most widely used. Figure 11.9 shows a cross-sectional view of an RTD. Platinum RTDs are, in fact, the international standard of temperature measurement, as stated in the International Practical Temperature Scale of 1968 (IPTS-68) [1].

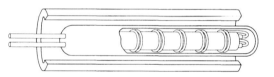

Figure 11.9 Resistance Temperature Device Cross-Sectional View, showing a resistance element.

The relationship of temperature and resistance is described by the CALLENDAR–VAN DUSEN equation:

$$R_T/R_0 = 1 + \alpha\ [T - \sigma\ (T/100 - 1)\ (T/100)] \tag{11.6}$$

where R_T is the resistance at temperature T, Ω;
   R_0 is the resistance at 0 °C, Ω;
   α is a constant (0.00382–0.00393 for platinum);
   T is the temperature, °C; and
   σ is a constant (1.48–1.51 for platinum).
   The equation applies in the range 0–630 °C [2].
Several different standards besides IPTS-68 exist for the manufacture and use of RTDs, including those of the Deutsches Institut für Normung E.V. (DIN 43760-1968) from Germany, and the British Standards Institute (BSI 1904-1964). The resistance versus temperature chart shown in Table 11.1 conforms to the IPTS-68, and is within the tolerances of BSI 1904-1964 and DIN 43760-1968. This standard chart is based on a resistance of 100 Ω at 0 °C and an α of 0.00385. The temperature versus

Table 11.1 Temperature versus Resistance for Resistance Temperature Devices

Temp (Deg C)	Resistance (Ohms)	Temp (Deg C)	Resistance (Ohms)	Temp (Deg C)	Resistance (Ohms)	Temp (Deg C)	Resistance (Ohms)
0	100.00	40	115.54	80	130.89	120	146.06
1	100.39	41	115.93	81	131.27	121	146.44
2	100.78	42	116.31	82	131.66	122	146.81
3	101.17	43	116.70	83	132.04	123	147.19
4	101.56	44	117.08	84	132.42	124	147.57
5	101.95	45	117.47	85	132.80	125	147.94
6	102.34	46	117.85	86	133.18	126	148.32
7	102.73	47	118.24	87	133.56	127	148.70
8	103.12	48	118.62	88	133.94	128	149.07
9	103.51	49	119.01	89	134.32	129	149.45
10	103.90	50	119.39	90	134.70	130	149.82
11	104.29	51	119.78	91	135.08	131	150.20
12	104.68	52	120.16	92	135.46	132	150.58
13	105.07	53	120.55	93	135.84	133	150.95
14	105.46	54	120.93	94	136.22	134	151.33
15	105.85	55	121.32	95	136.60	135	151.70
16	106.24	56	121.70	96	136.98	136	152.08
17	106.63	57	122.09	97	137.36	137	152.45
18	107.02	58	122.47	98	137.74	138	152.83
19	107.40	59	122.86	99	138.12	139	153.20
20	107.79	60	123.24	100	138.50	140	153.58
21	108.18	61	123.62	101	138.88	141	153.95
22	108.57	62	124.01	102	139.26	142	154.32
23	108.96	63	124.39	103	139.64	143	154.70
24	109.35	64	124.77	104	140.02	144	155.07
25	109.73	65	125.16	105	140.39	145	155.45
26	110.12	66	125.54	106	140.77	146	155.82
27	110.51	67	125.92	107	141.15	147	156.19
28	110.90	68	126.31	108	141.53	148	156.57
29	111.28	69	126.69	109	141.91	149	156.94
30	111.67	70	127.07	110	142.29	150	157.31
31	112.06	71	127.45	111	142.66	151	157.69
32	112.45	72	127.84	112	143.04	152	158.06
33	112.83	73	128.22	113	143.42	153	158.43
34	113.22	74	128.60	114	143.80	154	158.81
35	113.61	75	128.98	115	144.17	155	159.18
36	113.99	76	129.37	116	144.55	156	159.55
37	114.38	77	129.75	117	144.93	157	159.93
38	114.77	78	130.13	118	145.31	158	160.30
39	115.15	79	130.51	119	145.68	159	160.67

resistance relationship allows calibration of an RTD against a standard set of values. For greater accuracy, the temperature versus resistance relationship for a specific RTD can be determined against an NIST–traceable RTD and in a constant-temperature bath.

Four types of wiring arrangements are available for RTDs. They are the *two-wire, two-wire with compensation loop circuit, three-wire,* and *four-wire* configurations, and are shown in Figure 11.10. The two-wire configuration is adequate to make an RTD work, but does not provide for lead-wire resistance compensation. Therefore it should be used only in cases where the sensor is integral with the transmitter. The three-wire

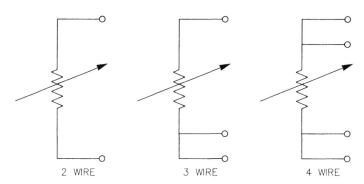

2 WIRE   3 WIRE   4 WIRE

Figure 11.10 Resistance Temperature Device Wire Configurations. Two-, three-, and four-wire RTD configurations are shown. While the accuracy increases from two- to three- to four-wire sensors, three-wire RTDs are the most common and have nearly the accuracy of four-wire RTDs.

circuit has a third lead connecting the RTD to the transmitter assembly that, when balanced by a Wheatstone bridge and of the same length and in the same run as the other leads, cancels out the undesired resistance effects of the wire. The two-wire with compensation loop circuit has a second loop of the same length, material, and gauge as the RTD loop and is subject to the same conditions. The two pairs are connected opposite each other in the Wheatstone bridge, and they cancel out the lead-wire resistance. The four-wire is the most accurate of these circuit configurations. Current is passed through opposing leads of the RTD, and the voltage is measured across the remaining two leads. The four-wire and two-wire with compensation loop provide the best accuracy, but the additional cost is not generally justified, and the three-wire configuration is standard [3, p 61].

11.2.1.2   Thermocouples

Although available with accuracies that equal or exceed those of RTDs, commercial thermocouples are generally considered to be less accurate than their RTD counterparts. RTDs are preferred for critical temperature measurements, such as culture batch temperature, but thermocouples are excellent low-cost, rugged alternatives for other services, such as utilities, sterilization, and HVAC applications. They can be configured similarly to RTDs, in that they can be supplied with metal-sheathed tip ends, or they can be provided bare to take advantage of their inherent tip sensitivity (in contrast to the average temperature measurement over the sensitive area that RTDs give). Figure 11.11 shows both a bare-wire and a metal-sheathed thermocouple. In practice, thermocouples are almost always used in conjunction with a thermowell.

Figure 11.11 Bare-Wire and Sheathed Thermocouples.

The theory behind a thermocouple is also simple. A closed electrical circuit is made between two dissimilar metals. If a temperature difference between the two junctions is present, a current flows. One of the junctions is placed in the sensing region, and the other junction, considered a reference, is handled in one of several ways. The reference junction can be placed in an ice bath, or it can have its temperature measured. When an ice bath is used, the measuring junction temperature can be determined from the current of the thermocouple loop, because the reference junction is at the ice point. If the temperature at the reference junction is measured, the reference junction temperature and loop current are used to determine the measuring junction temperature.

Several types of thermocouples are standard; Table 11.2 lists them, with their associated range and tolerance, based on the ISA standard [4]. Table 11.3 gives the voltage values for J-, K-, and T-type thermocouples over the temperature ranges of interest to the biotechnologist, so that a particular thermocouple can be calibrated against a standard set of values.

Table 11.2 Initial Calibration Tolerances for Thermocouples;
Reference Junction at 0 °C

Thermocouple type	Temperature range, °C	Standard tolerances*	Special tolerances*
J	0–750	±2.2 °C or ±0.75%	±1.1 °C or ±0.4%
K	0–1,250	±2.2 °C or ±0.75%	±1.1 °C or ±0.4%
T	0–350	±1.0 °C or ±0.75%	±0.5 °C or ±0.4%

* Use whichever tolerance is greater.

Table 11.3A Temperature versus EMF for Type J Thermocouples

Temp (Deg C)	Voltage (mV)	Temp (Deg C)	Voltage (mV)	Temp (Deg C)	Voltage (mV)	Temp (Deg C)	Voltage (mV)
0	0.000	40	2.058	80	4.186	120	6.359
1	0.050	41	2.111	81	4.239	121	6.414
2	0.101	42	2.163	82	4.293	122	6.468
3	0.151	43	2.216	83	4.347	123	6.523
4	0.202	44	2.268	84	4.401	124	6.578
5	0.253	45	2.321	85	4.455	125	6.633
6	0.303	46	2.374	86	4.509	126	6.688
7	0.354	47	2.426	87	4.563	127	6.742
8	0.405	48	2.479	88	4.617	128	6.797
9	0.456	49	2.532	89	4.671	129	6.852
10	0.507	50	2.585	90	4.725	130	6.907
11	0.558	51	2.638	91	4.780	131	6.962
12	0.609	52	2.691	92	4.834	132	7.017
13	0.660	53	2.743	93	4.888	133	7.072
14	0.711	54	2.796	94	4.942	134	7.127
15	0.762	55	2.849	95	4.996	135	7.182
16	0.813	56	2.902	96	5.050	136	7.237
17	0.865	57	2.956	97	5.105	137	7.292
18	0.916	58	3.009	98	5.159	138	7.347
19	0.967	59	3.062	99	5.213	139	7.402
20	1.019	60	3.115	100	5.268	140	7.457
21	1.070	61	3.168	101	5.322	141	7.512
22	1.122	62	3.221	102	5.376	142	7.567
23	1.174	63	3.275	103	5.431	143	7.622
24	1.225	64	3.328	104	5.485	144	7.677
25	1.277	65	3.381	105	5.540	145	7.732
26	1.329	66	3.435	106	5.594	146	7.787
27	1.381	67	3.488	107	5.649	147	7.843
28	1.432	68	3.542	108	5.703	148	7.898
29	1.484	69	3.595	109	5.758	149	7.953
30	1.536	70	3.649	110	5.812	150	8.008
31	1.588	71	3.702	111	5.867	151	8.063
32	1.640	72	3.756	112	5.921	152	8.118
33	1.693	73	3.809	113	5.976	153	8.174
34	1.745	74	3.863	114	6.031	154	8.229
35	1.797	75	3.917	115	6.085	155	8.284
36	1.849	76	3.971	116	6.140	156	8.339
37	1.901	77	4.024	117	6.195	157	8.394
38	1.954	78	4.078	118	6.249	158	8.450
39	2.006	79	4.132	119	6.304	159	8.505

Reference junction at 0 deg. C

Table 11.3B Temperature versus EMF for Type K Thermocouples

Temp (Deg C)	Voltage (mV)	Temp (Deg C)	Voltage (mV)	Temp (Deg C)	Voltage (mV)	Temp (Deg C)	Voltage (mV)
0	0.000	40	1.611	80	3.266	120	4.919
1	0.039	41	1.652	81	3.307	121	4.960
2	0.079	42	1.693	82	3.349	122	5.001
3	0.119	43	1.734	83	3.390	123	5.042
4	0.158	44	1.776	84	3.432	124	5.083
5	0.198	45	1.817	85	3.473	125	5.124
6	0.238	46	1.858	86	3.515	126	5.164
7	0.277	47	1.899	87	3.556	127	5.205
8	0.317	48	1.940	88	3.598	128	5.246
9	0.357	49	1.981	89	3.639	129	5.287
10	0.397	50	2.022	90	3.681	130	5.327
11	0.437	51	2.064	91	3.722	131	5.368
12	0.477	52	2.105	92	3.764	132	5.409
13	0.517	53	2.146	93	3.805	133	5.450
14	0.557	54	2.188	94	3.847	134	5.490
15	0.597	55	2.229	95	3.888	135	5.531
16	0.637	56	2.270	96	3.930	136	5.571
17	0.677	57	2.312	97	3.971	137	5.612
18	0.718	58	2.353	98	4.012	138	5.652
19	0.758	59	2.394	99	4.054	139	5.693
20	0.798	60	2.436	100	4.095	140	5.733
21	0.838	61	2.477	101	4.137	141	5.774
22	0.879	62	2.519	102	4.178	142	5.814
23	0.919	63	2.560	103	4.219	143	5.855
24	0.960	64	2.601	104	4.261	144	5.895
25	1.000	65	2.643	105	4.302	145	5.936
26	1.041	66	2.684	106	4.343	146	5.976
27	1.081	67	2.726	107	4.384	147	6.016
28	1.122	68	2.767	108	4.426	148	6.057
29	1.162	69	2.809	109	4.467	149	6.097
30	1.203	70	2.850	110	4.508	150	6.137
31	1.244	71	2.892	111	4.549	151	6.177
32	1.285	72	2.933	112	4.590	152	6.218
33	1.326	73	2.975	113	4.632	153	6.258
34	1.367	74	3.016	114	4.673	154	6.298
35	1.407	75	3.058	115	4.714	155	6.338
36	1.448	76	3.100	116	4.755	156	6.378
37	1.489	77	3.141	117	4.796	157	6.419
38	1.529	78	3.183	118	4.837	158	6.459
39	1.570	79	3.224	119	4.878	159	6.499

Reference junction at 0 deg. C

Table 11.3C   Temperature versus EMF for Type T Thermocouples

Temp (Deg C)	Voltage (mV)	Temp (Deg C)	Voltage (mV)	Temp (Deg C)	Voltage (mV)	Temp (Deg C)	Voltage (mV)
0	0.000	40	1.611	80	3.357	120	5.227
1	0.039	41	1.653	81	3.402	121	5.275
2	0.078	42	1.695	82	3.447	122	5.324
3	0.117	43	1.738	83	3.493	123	5.372
4	0.156	44	1.780	84	3.538	124	5.420
5	0.195	45	1.822	85	3.584	125	5.469
6	0.234	46	1.865	86	3.630	126	5.517
7	0.273	47	1.907	87	3.676	127	5.566
8	0.312	48	1.950	88	3.721	128	5.615
9	0.351	49	1.992	89	3.767	129	5.663
10	0.391	50	2.035	90	3.813	130	5.712
11	0.430	51	2.078	91	3.859	131	5.761
12	0.470	52	2.121	92	3.906	132	5.810
13	0.510	53	2.164	93	3.952	133	5.859
14	0.549	54	2.207	94	3.998	134	5.908
15	0.589	55	2.250	95	4.044	135	5.957
16	0.629	56	2.294	96	4.091	136	6.007
17	0.669	57	2.337	97	4.137	137	6.056
18	0.709	58	2.380	98	4.184	138	6.105
19	0.749	59	2.424	99	4.231	139	6.155
20	0.789	60	2.467	100	4.277	140	6.204
21	0.830	61	2.511	101	4.324	141	6.254
22	0.870	62	2.555	102	4.371	142	6.303
23	0.911	63	2.599	103	4.418	143	6.353
24	0.951	64	2.643	104	4.465	144	6.403
25	0.992	65	2.687	105	4.512	145	6.452
26	1.032	66	2.731	106	4.559	146	6.502
27	1.073	67	2.775	107	4.607	147	6.552
28	1.114	68	2.819	108	4.654	148	6.602
29	1.155	69	2.864	109	4.701	149	6.652
30	1.196	70	2.908	110	4.749	150	6.702
31	1.237	71	2.953	111	4.796	151	6.753
32	1.279	72	2.997	112	4.844	152	6.803
33	1.320	73	3.042	113	4.891	153	6.853
34	1.361	74	3.087	114	4.939	154	6.903
35	1.403	75	3.131	115	4.987	155	6.954
36	1.444	76	3.176	116	5.035	156	7.004
37	1.486	77	3.221	117	5.083	157	7.055
38	1.528	78	3.266	118	5.131	158	7.106
39	1.569	79	3.312	119	5.179	159	7.156

Reference junction at 0 deg. C

11.2.1.3   Thermowells

Thermowells are mechanical protection devices for separating the temperature sensing device from the process environment. They shield the sensor from contamination, corrosion, and impingement, and allow maintenance while the system is in production. Several standard configurations are shown in Figure 11.12.

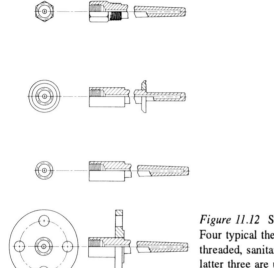

Figure 11.12 Standard Thermowell Configurations. Four typical thermowell configurations are shown: threaded, sanitary clamp, welded, and flanged. The latter three are used when avoiding contamination is important.

No firm rules govern the sizing and conformation of thermowells. In a bio-processing environment, however, two important requirements must be met. First, the thermowell should be secured to the sensing area in such a manner that sites for contamination are not present. This is best accomplished by a welded, sanitary-type connection, a threaded and seal-welded assembly, or a flange-mounted assembly. Of the three, the welded-in-place assembly is preferred. The other concern in thermowell construction and installation is the insertion length. It must be long enough so that conduction of heat from the outside environment has no effect on the temperature reading, with 1.5" a minimum, but it must not be so long that agitation of broths, especially viscous broths, causes mechanical fatigue. A formula for estimating the length of thermowells is:

$$L = 7 / \sqrt{d_o U / 3 K (d_o^2 - d_i^2)} \tag{11.7}$$

where L is the immersion length of the thermowell, in;

d_o is the outer diameter of the thermowell, in;

U is the overall heat-transfer coefficient, Btu/h·ft^2·°F;

K is the thermal conductivity of the well material, Btu/h·ft^2·°F·ft; and

d_i is the inner diameter of the thermowell, in [5].

11.2.1.4 Miscellaneous Temperature Measuring Devices

Of the many standard methods of temperature sensing besides RTDs and thermo-couples, those of interest to the biotechnologist are thermistors, thermal filled systems, bimetal thermometers, liquid-in-glass thermometers, and temperature-sensitive surface materials. A description of the principles and utility of each follows.

Thermistors are thermally sensitive resistors that exhibit a decrease in resistance with an associated increase in temperature. They are manufactured from metal oxides and come in a variety of shapes and sizes for many applications. Although they are somewhat accurate, with a fast response, they see limited use in bioprocessing applications due to their limited range of about 100 °C [3, p 62] and their unavailability in standard industrial configurations.

Thermal filled systems and *liquid-in-glass thermometers* operate in a similar fashion. They contain either a liquid or a gas that expands with increasing temperature, and produce a mechanical movement proportional to temperature. Four classes of thermal filled systems are defined by the Scientific Apparatus Makers Association (SAMA) [6], and are: Class I, liquid-filled; Class II, vapor-filled; Class III, gas-filled, and Class V, mercury-filled.

Liquid- and mercury-filled systems relate temperature to a volume change of an incompressible fluid, and vapor- and gas-filled systems relate temperature to a pressure change. Although available with standard accuracies of 0.5% of span, this is still not as good as the accuracy offered by a good-quality resistance temperature device. Also, filled systems do not lend themselves to electrical or pneumatic signal transmission. For these reasons, filled systems should be considered only when other methods of temperature sensing are not available.

Bimetal thermometers are a useful temperature measuring device for almost any industrial application, including biotechnology. They are made from two metal strips with different coefficients of thermal expansion, fastened to each other usually in a helical fashion. As temperature varies, the metals expand or contract to different extents; an indicating needle is displaced linearly, proportionally to temperature. They are best used where measurements do not require a high degree of accuracy, and where numerous, low-cost displays are necessary. An example is a long transfer line that must be maintained at the proper temperature for sterility.

Liquid-in-glass thermometers are a standard laboratory instrument that can be used to measure any liquid or vapor stream when a portable, easily inserted, low-cost, temperature-sensitive indicator is required. These should be used only as temporary devices in an industrial environment, since they are very easily broken by the slightest rough handling. Their most useful application is as comparative instruments for calibration of other temperature sensing devices. Liquid-in-glass thermometers for this application can be purchased with high accuracies and individual calibrations traceable to the National Bureau of Standards.

Temperature-sensitive materials include tapes, paints, markers, and crayons that change physical properties such as color or consistency at predetermined temperatures. They are most useful as a gross measurement for checking sterility, as long as their inherent poor accuracy is kept in mind. For example, if a process line is considered sterile at 123 °C, it is best to check the line with a material that melts or changes color at 125–130 °C.

11.2.1.5   Applications

Applications for temperature measurement are relatively standard, and include batch temperature control, utility jacket temperature measurement for control and heat balance information, sterility verification, and utility temperature measurement.

11.2.1.6   Manufacturers

Manufacturers of temperature sensing equipment include Ametek, Ashcroft, Bailey Controls, Combustion Engineering, Degussa, Foxboro, Heraeus, Honeywell, Leeds and Northrup, Omega, Robertshaw, Rosemount, Signet, and Thermo Electric. Many more suppliers of good-quality devices are listed in the latest Instrument Society of America Directory of Instrumentation.

11.2.2   Pressure Measurement

Pressure is measured in units of force per unit area. It is commonly referred to in one of three ways: *absolute pressure*, which is the pressure above a complete vacuum; *gauge pressure*, which is the pressure above atmospheric pressure; and *differential pressure*, which is the difference between two pressure levels. The application constraints for a pressure sensor in a bioprocessing environment are that the sensing area be easily sterilized and be able to withstand exposure to sterilization temperatures that can reach 140 °C for long periods of time, yet the sensor should still be able to provide acceptable accuracy. For most pressure sensors, this is readily accomplished. Pressure sensors are of two general types, *mechanical* and *electronic*.

11.2.2.1   Mechanical Sensors

The three predominant types of mechanical pressure sensors are the *bellows, bourdon tube,* and *diaphragm.* Although it is possible to purchase pressure transmitters that use this principle to create an analog signal proportional to pressure, these gauges are best used to provide local readout of pressure. For local readout, the mechanical action of increasing or decreasing pressure moves a pointing device on the gauge face to indicate pressure. The three types of element are all relatively equal in terms of accuracy. To choose a mechanical gauge it is only necessary to specify an instrument that is isolated from the process by a diaphragm, is capable of withstanding sterilization temperatures, has the necessary end connections and materials of construction, and gives the required accuracy throughout the range.

11.2.2.2   Electronic Sensors

Electronic pressure sensors are used in applications where an analog signal proportional to pressure must be supplied. Pressure transmitters can be of the *strain gauge, piezoresistive,* or *capacitive* type. Newer technologies better suited for digital circuitry include the *double-ended tuning fork* and the *resonant wire pressure sensors* [7], which are also acceptable for pressure sensing. As with mechanical pressure sensors, the primary considerations when electronic pressure sensors are specified include accuracy, sterility, ability to withstand a sterilization atmosphere, and configuration. Numerous manufacturers supply sensors to meet these needs. An example of a pressure transmitter with associated tank spud is shown in Figure 11.13.

Figure 11.13 Sanitary Clamp-Type Pressure Transmitter. (Courtesy of Rosemount, Inc.)

11.2.2.3   Applications

All reaction vessels and any other vessels not open to the atmosphere must be equipped with a sensing device that monitors the pressure in the vapor space. Pressure information, considered along with the dissolved oxygen content and the off-gas composition, describes the gassing characteristics of the culture. Beyond process pressure considerations, pressure measurement is necessary for safety reasons. It is not a good idea to operate a process at a pressure too close to the set point of the pressure relief device, because relief of pressure is so likely to lead to contamination of the culture.

11.2.2.4   Manufacturers

Manufacturers of mechanical pressure sensing gauges include Ametek, Anderson Instrument, and Dresser Industries Ashcroft Division. Manufacturers of electronic pressure transmitters include Ametek, Anderson Instrument, Bailey Controls, Taylor Division of Combustion Engineering, Dresser Industries Ashcroft Division, Fischer and Porter, Foxboro, Honeywell, ITT Barton, Leeds and Northrup, Rosemount, and Yokogawa. Many more suppliers of good-quality instruments are listed in the latest Instrument Society of America Directory of Instrumentation.

11.2.3   Level Measurement

Although level is not a critical variable, the measurement of level in a charging or reaction vessel is important so that proper amounts of medium, nutrient materials, and any other necessary components are supplied, and also to avoid excessive foaming. Many different types of level instrumentation are available, but, because sterility is usually a concern, only *differential pressure cells*, *load cells*, and *conductance probes* see widespread use in biotechnology applications. Other types of level instrumentation either have numerous moving parts that are difficult or even impossible to sterilize, are not useful for process fluids of changing or undefined composition, or are not suited to the measurement of level in an agitated vessel. These include displacement floats, torque tubes, nuclear-based sensors, sonic measuring devices, and resistance, thermal, infrared, and microwave probes.

11.2.3.1   Differential Pressure Cells

To sense level, a differential pressure (dP) cell measures the difference between the pressure at the bottom of the liquid in the vessel and that in the vapor space, and converts the difference in readings to a height of liquid. The principle of operation is the same as that of pressure sensors, with the primary distinction between the two being the location of the reference. For a pressure indication, the sensing point is referred to external pressure. For liquid level measurement, the bottom of the liquid column is referred to the vapor space, unless it is open to the atmosphere. Since the device is actually measuring a pressure head, location of the sensor at or near the bottom of the liquid is critical. An example of a dP cell is shown in Figure 11.14.

Figure 11.14
Sanitary Clamp-Type
Level Transmitter.
(Courtesy of
Rosemount, Inc.)

A dP cell transmits a signal proportional to pressure, and is therefore affected by changes in density. The following equation applies:

$$H = 2.307 \; \Delta P/\text{SG} \tag{11.8}$$

where H is the height of liquid above the pressure sensor, ft; ΔP is the differential pressure, lb/in^2; and SG is the density, expressed as specific gravity. If water is used to calibrate a differential pressure cell, then a correction factor must be used if the fluid in the vessel has a specific gravity other than that of water at the calibration temperature. For example, if a solids-laden medium stream with a specific gravity of 1.2 is charged to a fermentor and gauged by a dP cell calibrated with water, and the reading is not corrected, the true level is 83% ($1/1.2 \times 100\%$) of the indicated level. A dP cell should therefore be calibrated with the process fluid it will be sensing. If this is not possible, or the specific gravity of the process stream is constantly changing, the transmitter output must be compensated, either manually or automatically.

11.2.3.2   Load Cells

Load cells differ from dP cells in that they measure mass directly. If the geometry of the vessel and the density of the process fluid are factored in, load cells can indicate level. Changes in density affect the level reading just as they do with dP cells. Load cells also differ from dP cells in that they are external to the process. They are mounted on the load-bearing surface of the vessel; the total mass, less the mass of the vessel and fittings, is the mass of the process fluid.

Load cells are simply strain gauges mounted in compression and wired in a bridge arrangement to compensate for temperature. They are made from either piezoelectric or semiconductor material, much like a pressure transmitter. Numerous specific concerns need to be considered in the design of a load cell for level indication. These include, but are not limited to, the selection of a specific load cell, the installation of connecting piping and other fittings, the structural design of the vessel and support structure, and wiring considerations [8].

11.2.3.3   Conductance and Capacitance Probes

Conductance, capacitance, and admittance level probes, although utilizing different principles of operation, are similar in configuration and operating characteristics. All three are electronic sensors that employ a chemically resistant probe, usually inserted from the top of the vessel and protruding down over the range of level to be measured. They are used for both point and continuous measuring applications.

Conductance probes operate on a very simple principle: an open electrical circuit is closed when both terminals are in contact with a common conductive fluid. A low voltage is established between the vessel and the probe tip. The circuit is closed when the liquid level reaches the probe tip, and this actuates a relay to either the open or the closed position.

Capacitance is the ability to store an electric charge. When two conductors are placed adjacent to and insulated from each other, the result is a capacitor. In a process vessel, the capacitance probe is inserted into a vessel and isolated from its ground. The probe is one conductor, and the vessel is the other. An insulator is necessary between the two conductors if the level of conductive fluid is to be sensed. This insulator is the dielectric of the system, and is usually a PTFE or polypropylene coating on the probe. The change in capacitance is directly proportional to the liquid height; the value is converted into an analog electrical signal that is transmitted to an indicating device.

11.2.3.4   Applications

Level instrumentation can be used to measure fluid level, foam level, on/off point sensing, and density. In pilot- to production-scale level sensing, the easiest and most economical way to sense level is with dP cells. As previously mentioned, their main drawback is the inability to correct automatically for density. This is of little concern if the process fluids are always at or near the density of water, or if the density is known and adjusted for as necessary. It is possible, however, to both read out and

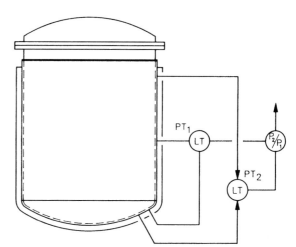

Figure 11.15 Dual Differential Pressure Cell Configuration. With a dual differential pressure cell configuration, the level can be sensed and corrected for density to provide a true level measurement that is independent of density.

correct for density with a dP cell system. Figure 11.15 shows a dual dP cell arrangement for the measurement of liquid head pressure and density that gives a true level reading. Note that the standard level dP cell is present, and the additional dP cell is for measuring density. The dP cell for density must have both its sensors submerged. The following equations give the values of density, expressed as specific gravity, and level:

$$SG = 2.307\ \Delta P_1/L \qquad\qquad (11.9)$$

where ΔP_1 is the pressure differential between density dP cell sensors, lb/in^2; and L is the length between density dP cell sensors, ft.

$$H = 2.307\ \Delta P_2/SG \qquad\qquad (11.10)$$

where H is the height of liquid above the pressure sensor, ft; and ΔP_2 is the pressure differential between the level dP cell sensors, lb/in^2.

If eq 11.9 and 11.10 are combined and simplified, the true height of the liquid level is obtained:

$$H = \Delta P_2\ L/\Delta P_1 \qquad\qquad (11.11)$$

It is therefore not necessary to use an external device to calculate the required values, since the equations are linear and can be readily scaled at the transmitter. It is necessary, however, to install a circuit to divide the level signal by the density signal.

11.2.3.5   Manufacturers

Manufacturers of electronic pressure transmitters for dP cell applications include Ametek, Anderson Instrument, Bailey Controls, Taylor Division of Combustion Engineering, Dresser Industries Ashcroft Division, Fischer and Porter, Foxboro, Honeywell, ITT Barton, Leeds and Northrup, Rosemount, and Yokogawa. Manufacturers of load cells include Action Instruments, BLH Electronics, Cole-Parmer, Entran, Fisher Scientific, HITEC, Kulite, and Transducers Inc. Suppliers of conductance and capacitance probe level sensing equipment include Drexelbrook, Endress & Hauser, Great Lakes Instruments, Penberthy, and Robertshaw. Many more suppliers of good-quality instruments are listed in the latest Instrument Society of America Directory of Instrumentation.

11.2.4   Flow Measurement

Numerous types of flow metering devices exist, covering a wide range of applications and economic considerations. In any bioprocessing facility there are many applications besides those directly in contact with the process, so it is necessary to have a well-rounded knowledge of the metering devices available to meet all processing needs, not just the meters for use on sterile streams. A description of each type of flowmeter follows, with emphasis on the meters most suitable for use in bioprocessing applications.

11.2.4.1   Variable Area Meters

The two predominant types of variable area flowmeters are the *rotameter* and the *vane-type meter*. I focus my discussion on the rotameter, since it has the widest use and the operation is the same for both. When a fluid enters the meter, its force acts upon a float. The force moves the float up an increasingly wider glass or metal tube. The orifice equation holds:

$$W_m = C_o \sqrt{\Delta P} \qquad (11.12)$$

where W_m is the mass flow rate; C_o is the orifice coefficient; and ΔP is the differential pressure. At constant pressure drop, the flow rate is proportional to the orifice opening, which is proportional to the movement of the float. If the tube is glass, the position of the float can be read by an operator; if the tube is metal, the float may contain a magnet that causes motion of an indicator, or float position may be conveyed to an electrical transducer/transmitter. A glass-tube and a metal-tube rotameter are shown in Figure 11.16.

Figure 11.16 Glass-Tube and Metal-Tube Rotameters.
(Courtesy of Brooks Instruments.)

Variable area flowmeters can be calibrated to read either mass flow rate or volumetric flow rate, of either liquids or gases. When the operating conditions differ from the calibration conditions, it is necessary to correct for the deviations [9]. For liquids that need to be corrected for changes in density, the following equation applies:

$$L_a = L_i \, \rho_c \, / \rho_a \tag{11.13}$$

where L_a is the actual flow rate; L_i is the indicated flow rate; ρ_c is the density of the calibration liquid; and ρ_a is the actual density. The specifications for a particular rotameter include a maximum viscosity. If this viscosity is exceeded, the rotameter must be calibrated for the specific process viscosity. No practical conversion formula exists.

For gases flowing through a variable area flowmeter, it is necessary to correct for variations in density, pressure, and temperature from those of the calibration fluid. The corrections for temperature and pressure are as follows:

$$L_a = L_i \, T_c \, / T_a \tag{11.14}$$

where T_c is the absolute temperature of the calibration liquid, and T_a is the actual absolute temperature; and:

$$L_a = L_i \, P_c \, / P_a \tag{11.15}$$

where P_c is the pressure of the calibration liquid, and P_a is the actual pressure. It is important to know and keep in mind the calibration basis for each particular variable area flowmeter. *Actual*, *standard*, and *calibrated* flow conditions are used, and they are not interchangeable; furthermore, mass flow rates can be used instead of volumetric flow rates. *Actual flows*, as the name implies, are the true flow rates. *Standard flows* are the rates at standard conditions, usually 70 °F and one atmosphere. The *calibrated conditions* are the conditions that the process stream is expected to be at during processing. A variable area flowmeter is usually supplied with the calibrated conditions and either a standard, actual, or mass flow scale. If the processing conditions differ from the calibrated conditions, the value needs to be adjusted, as discussed above.

11.2.4.2   Magnetic Meters

Magnetic flowmeters, since they are unobtrusive and easily sterilized, are ideal for use in liquid streams that have some electrical conductivity (that is, above 2.0 microsiemens/cm) and require a sterile environment. An example is shown in Figure 11.17. Their operation is based on FARADAY'S law, which states that a voltage is induced when a conductor moves through a magnetic field. A schematic depiction of how this applies to a magnetic flowmeter is in Figure 11.18. The electrical voltage generated, E, is a linear function of velocity, according to the following equation:

$$E = \varepsilon_r \cdot B \cdot v \cdot d \tag{11.16}$$

where ε_r is the dielectric constant (relative permittivity), B is the strength of the magnetic field, v is the velocity, and d is the diameter of the pipe. The electronics in the transmitter assembly convert voltage into flow units. The flow is indicated locally or transmitted remotely as a volumetric rate.

Figure 11.17 Magnetic Flowmeter. (Courtesy of Rosemount, Inc.)

11.2.4.3   Mass Flowmeters

Two types of mass flowmeters are common in the process industry: *thermal* and *Coriolis-based* meters. *Thermal mass flowmeters* are useful for measuring the flow of gases, and operate under one of several principles. The simplest involves a gas

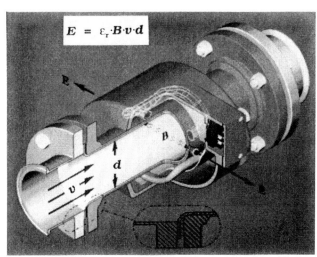

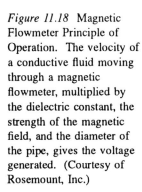

Figure 11.18 Magnetic Flowmeter Principle of Operation. The velocity of a conductive fluid moving through a magnetic flowmeter, multiplied by the dielectric constant, the strength of the magnetic field, and the diameter of the pipe, gives the voltage generated. (Courtesy of Rosemount, Inc.)

flowing over a heating element maintained at a constant temperature. The amount of energy necessary to maintain the constant temperature is proportional to the mass flow. Another type of meter utilizes two temperature sensors and a heater. The gas flows over the first temperature sensor, then the heater, and finally the second temperature sensor. The temperature difference, caused by heat going toward only the downstream temperature sensor, is proportional to the flow. A combination mass flow sensor and controller is shown in Figure 11.19.

The *Coriolis-based mass flowmeter* relies on NEWTON'S law, $F = ma$. As a liquid flows through an S- or U-shaped tube vibrating at a particular frequency, the tube tends to bend to an extent that depends on the force exerted by the liquid. The amount of bending is measured and is proportional to flow. Proper mounting is essential for accurate readings.

11.2.4.4   Displacement Meters

A *displacement meter* is one in which the liquid flow displaces a particular measuring device, and the rate of displacement is proportional to flow. Many types exist, which include but are not limited to the *turbine, undulating disk, oval gear, rotary piston*, and *vane meters*. Mechanical or electrical pulses are produced by the sensor, and are mechanically or electrically converted to flow units. Since the fluids in a bioprocess often contain solids. the use of displacement-type meters is usually not a recommended practice. They are best used for low-viscosity clean fluids at moderate temperatures and pressures.

Figure 11.19 Thermal Mass Flow Sensor with
Controller. (Courtesy of Brooks Instruments.)

11.2.4.5   Differential Pressure Meters

Differential pressure flowmeters, such as the *venturi* and *orifice meters*, are among the
most common in the processing industries. Their inherent lack of accuracy and the
fact that they are somewhat difficult to clean make them of limited utility in a sterile
environment. They are based on the pressure and flow relationship of a fluid flowing
through an orifice. As with a rotameter, the flow is computed as the square root of
the pressure, multiplied by an orifice coefficient. However, instead of the orifice
changing in proportion to the flow, the fluid passes through a venturi or an orifice and
the pressure is sensed by a dP cell. The value is then squared and scaled to a flow
rate [10]. The most common equation used for sizing is:

$$W_{\mathrm{m}} = K \cdot A \cdot Y \cdot F_a \sqrt{2 g_c \, \Delta P \cdot \rho} \qquad (11.17)$$

where W_{m} is the flow rate, lb/s;
  K is the flow coefficient;
  A is the cross-sectional area of the orifice, ft^2;
  Y is the gas expansion factor;
  F_a is the area factor for expansion of the orifice;
  g_c is NEWTON'S law conversion factor, 32.17 lb mass·ft/lb force·s^2; and
  ρ is the density, lb/ft^3.
  This equation applies to a sharp-edged, concentric orifice. Besides orifice meters,
there are also *venturi, target, pitot, elbow, area, weir*, and *flume pressure differential
flowmeters*.

11.2.4.6   Miscellaneous Meters

In addition to the aforementioned flowmeters, many other types of flow measuring devices exist. Although they may not be well-suited to flow measurement in a bioprocessing environment, they may see application in a support service such as steam, cooling water, or other utilities. Among these are ultrasonic and vortex shedding meters. *Ultrasonic flowmeters* work on the Doppler principle, transmitting sound into a fluid and computing the flow from the change in return frequency. *Vortex shedding meters* have a stationary bluff body perpendicular to the fluid stream being measured. As the fluid passes the stationary body, vortices are formed that shed at a frequency proportional to the flow rate.

11.2.4.7   Applications

Applications for flow measurement are many, but of primary concern are additions to a bioreactor. To measure moderate flow rates of air and other gases to a reaction vessel, a rotameter or other variable area meter is often the best choice. If the flow rate is small, a thermal mass flowmeter may be the recommended choice. A sterile, clean flow is required, since variable area flowmeters are not easily sterilized on a regular basis.

The measurement of nutrient addition and medium feed rates is somewhat difficult, since readings must be accurate and rates can be very low, for example in bioconversions. When flow rates are substantial, a Coriolis-based mass flowmeter is a good choice, albeit an expensive one. Another alternative is a magnetic flowmeter that can handle a wide range of flows, down to very low velocities, at moderate cost. Coriolis-based and magnetic flowmeters are easily sterilized. Table 11.4 gives application information for a number of flowmeters, with the caveat that specific applications always need to be considered case by case.

11.2.4.8   Manufacturers

Manufacturers of variable area flowmeters include Brooks, Erdco, Fischer and Porter, Universal, and Wallace and Tiernan. Manufacturers of differential pressure flow transmitters include Anderson, Bailey Controls, Taylor Division of Combustion Engineering, Fischer and Porter, Foxboro, Honeywell, ITT Barton, Leeds and Northrup, Rosemount, and Yokogawa. Manufacturers of mass flowmeters include Brooks, K-Flow, Exac, Fischer and Porter, Fluid Components, Micro Motion, Molytek, and Neptune. Manufacturers of displacement-type flowmeters include Badger, Brooks, Foxboro, Haliburton, Smith Meter, and Tokheim. Magnetic flowmeter manufacturers include Endress and Hauser, Foxboro, Rosemount, and Yokogawa. Many more suppliers of good-quality devices are listed in the latest Instrument Society of America Directory of Instrumentation.

Table 11.4 Flowmeter Applications

	Clean liquid	Dirty liquid	Corrosive liquid	Viscous liquid	Slurry liquid	Clean vapors	Dirty vapors	Corrosive vapors	Steam
Variable area	E	F	P	G	F	E	F	G	F
Magnetic[1]	E	E	E	E	E	N	N	N	N
Mass flow	E	E	E	G	E	P	P	P	P
Displacement	E	P	F	G	P	F	N	F	P
Differential pressure	G	F	G	F	F	E	F	G	G
Ultrasonic	F	G	F	F	G	P	P	P	P
Vortex shedding[2]	F	F	F	F	F	E	E	E	E

E: Excellent; application sensor was designed for.

G: Good; application sensor was designed for but limitations apply.

F: Fair; not application sensor was designed for but will work in some applications.

P: Poor; not application sensor was designed for and will work only in few instances.

N: Not recommended; not recommended under any circumstances.

Approximate accuracies:
   Variable area: ±0.5% of rate to ±5% of full scale.
   Magnetic: ±0.5% of rate to ±2% of full scale.
   Mass flow: ±0.25% of rate to ±2% of full scale.
   Displacement: ±0.5% of rate to ±2.5% of full scale.
   Differential pressure: ±2% of rate to ±10% of full scale.
   Ultrasonic: ±2% of rate to ±10% of full scale.
   Vortex shedding: ±0.5% of rate to ±1% of full scale.

Notes: (1) within the limits for conductivity; (2) with adequate velocity.

11.2.5   pH Measurement

11.2.5.1   Theory of Operation

The pH is a measure of the activity of the hydronium ion in an aqueous solution, and is defined by the following equation:

$$pH = -\log_{10} [H^+] \tag{11.18}$$

where $[H^+]$ is the hydronium ion concentration.

   In an industrial bioreactor, pH is an important variable, since excursions of as little as 0.5 pH units can quickly kill a culture, and optimum production is often confined within pH ranges of 0.1 units. The pH can be an arduous variable to sense and control, but the obstacles to pH control in a bioprocessing environment are not traditional. Under most circumstances, difficulties in controlling pH are caused by the logarithmic relationship between pH and the addition of controlling reagent. This can cause extremely large and rapid changes in pH with the addition of small volumes of acid or base. In cell-based product production, however, the addition of small amounts of acid or base about a neutral pH of 7 does not cause wide swings in the solution pH. In fact, the titration curve for many cultures is relatively linear in the pH range of 5 to 9, and not very steep. Also, the inherently slow nature of a large-scale fermentation, and the absence of a need for fast response, results in pH control being a manageable process. Figure 11.20 shows the control of pH in an *E. coli* fermentation over its production cycle. Most pH control problems in a bioprocessing environment are related to difficulties in making the measurement.

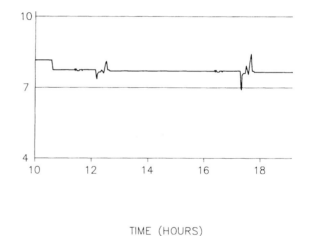

Figure 11.20 Response of pH over Time for a Typical Fermentation. Response of pH over time can be a discontinuous function. Usually a non-PID control algorithm is needed for acceptable control.

TIME (HOURS)

The theory of pH measurement is that a galvanic cell is formed when two metal conductors are connected via one or more electrolytes. As long as the current is small, the composition of the broth remains unchanged. A pH electrode consists of two half cells: a pH sensing element and a reference element. The sensing element consists of a silver chloride coated silver wire immersed in a buffer solution, usually at pH 7, in a sealed tube with a pH sensitive glass tip. The reference element contains silver chloride coated silver wire in a potassium chloride solution, and a porous diaphragm. A pH electrode is commonly constructed as a single unit, with the reference electrode built in an annular fashion around the pH sensing unit. A schematic depiction of a pH electrode is in Figure 11.21, and an industrial probe is shown in Figure 11.22.

Figure 11.21 (at left) Schematic of a pH Electrode. The reference electrode is the annular region, and the sensing electrode is the bulb and stem element.

Figure 11.22 (at right) pH Electrode. (Courtesy of Ingold Electrodes, Inc.)

As a result of the construction of the electrode, the potentials developed are all constant except for the potential between the outer gel layer of the sensing element and the sample solution. This potential is related to the pH by the NERNST equation [11]:

$$E = 0.1984 \ (T + 273.2) \ (7 - \text{pH}) \tag{11.19}$$

where E is the potential between sensing element gel layer and solution, mV; and T is the temperature, °C.

11.2.5.2   Operational Characteristics

Several important concepts in the application of pH emerge from the NERNST equation:
- (a)   millivolt output is positive below pH 7 and negative above pH 7;
- (b)   at a standard temperature of 25 °C the output per pH unit is 59.16 mV; and
- (c)   the temperature effect at a pH of 7 is 0, but as the pH moves away from this point, the effect increases linearly.

The value of pH is temperature dependent. This follows from the fact that the pH of the solution varies due to the change in pK_a. If the batch temperature deviates significantly from 25 °C, the signal to the indicating device must be corrected.

A pH loop is calibrated by a three-point check, to correct for differences between the theoretical NERNST equation and actual conditions. First, the probe is immersed in a pH 7 buffer and the standardization is adjusted until the transmitter produces the desired output. The probe is immersed in a second buffer, at a pH other than 7, such as pH 4, and the reading is adjusted (and compared to the first reading, as a slope) until the transmitter again produces the desired output. A third buffer, commonly at pH 10, is then used as a check on linearity. It is best to utilize a transmitter that allows calibration of the pH probe separately from the calibration of the transmitter. This permits a change in desired range without effect on the probe calibration, and it allows installation of a new probe without effect on the transmitter calibration.

Once a probe is calibrated, a number of problems can arise that give erroneous readings of pH. Most often these problems result from neglect or aging, and their symptoms include pH fixed at 7, slow response, erratic response, drifting, and a reduced slope of millivolts versus pH. The most common problems encountered in a bioprocessing environment and their associated symptoms are as follows:

(a)   Medium has coated the pH electrode: Response is slow, and the pH reading is low for a pH below 7 and high for a pH greater than 7. When the probe is completely coated the response is erratic.

(b)   Electrode is broken: Often reads close to pH 7, or response is erratic.

(c)   Probe terminals are shorted, or an open wire to the probe exists: A constant pH reading of 7 or an erratic response results.

(d)   The glass bulb layer has dried out or glass has been etched: High readings of pH result, and attempts at calibration show shortened span.

When a pH probe is clearly no longer giving accurate readings and attempts at calibration are unsuccessful, several procedures can be used to extend the life of the electrode [12]. If the electrode is a filled system, flush out the old solutions and replace with fresh ones. Several treatments can then be tried to make the electrode usable. First, clean it in a commercially available pepsin/HCl or 2% sodium hypochlorite solution for several hours, to remove any protein coating. Next, if any discoloration of the reference junction exists, clean it in thiourea/HCl. Finally, in cases where the electrode is old and responds sluggishly due to aging of the glass, rejuvenate the pH sensitive glass by immersing the electrode in a dilute HF solution for one minute and then soaking overnight in the reference electrode solution (3M KCl). The electrode should now respond properly. If it does not, contamination, a cracked pH membrane, loss of electrolyte, or an open circuit is likely, and the probe must be discarded.

Probes should be stored in their reference electrolyte solution when not in use.

Since bioprocesses are usually buffered about the region of pH control, an on/off process control strategy can be employed. This involves creating a dead band about the desired pH range. Whenever the pH drifts out of the dead band, either acid or base is fed to the process via pressure flow, pumping, or gravity flow. The flow ceases after a preset time, and stays off for a (different) preset time, to allow complete mixing of the acid or base. When the mixing timer stops, the sequence repeats as necessary, until the pH is again within the dead band range.

For sterile applications, pH probes come in a number of useful configurations. They generally have a 25-mm diameter and an *Ingold*-type mounting connection.

Varying lengths are available, to suit the application. Two general types are in common use: gel filled (permanently sealed and disposable), and pressurized (with renewable electrolyte). Retractable, sterilizable assemblies are available that allow on-line changing of a pH probe under sterile conditions.

11.2.5.3 Manufacturers

Manufacturers of pH sensors or transmitters specifically for use in a sterilizable environment include Bradley-James, Ingold, ISI, Rosemount, and TBI. Other suppliers of good-quality sensors and other instruments are listed in the latest Instrument Society of America Directory of Instrumentation.

11.2.6   Dissolved Oxygen Measurement

The measurement of dissolved oxygen in a bioprocessing reaction is of major importance to the development and production of valuable natural products. If the partial pressure of oxygen is known, it is possible to calculate the oxygen transfer coefficient, $K_L a$. Knowledge of $K_L a$ is essential for the scale-up of any large-scale fermentation.

Dissolved oxygen is measured amperometrically, by the reduction of oxygen at the cathode and the formation of silver chloride at the anode. An electrolyte solution connects the anode and cathode, and a polarizing voltage is established across them. When a fluid containing oxygen is in contact with the probe, oxygen diffuses across a gas-permeable membrane, and the following reactions result:

$$\text{Cathode reaction:} \qquad O_2 + 2H_2O + 4e^- \;\rightarrow\; 4OH^- \tag{11.20}$$

$$\text{Anode reaction:} \qquad 4Ag + 4Cl^- \;\rightarrow\; 4AgCl + 4e^- \tag{11.21}$$

The result is an electric current proportional to the oxygen partial pressure.

Dissolved oxygen probes come in several lengths and configurations, similar to the pH probe assemblies, that fit standard Ingold-type 19-mm and 25-mm fittings. The probes and amplifier/transmitter assemblies differ from supplier to supplier, so they are not interchangeable, other than for identical units. The probes themselves are usually reconstructible, so the housings can be used repeatedly. Rebuilding can be as simple as a membrane replacement, or, if necessary, a replacement of the anode/cathode assembly. Two-point calibration is simple, with a zero point in an oxygen-free atmosphere (easily created by nitrogen passing over the probe), and the second point in normal air. The reading can be in pressure units, reflecting absolute partial pressure, or it can be a relative measure of dissolved oxygen. An example of a dissolved oxygen probe and amplifier is shown in Figure 11.23. Suppliers of steam-sterilizable polarographic dissolved oxygen probes and amplifiers include Ingold and Orbisphere.

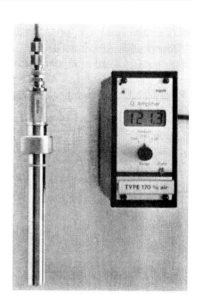

Figure 11.23 Dissolved Oxygen Probe and Amplifier/Transmitter. (Courtesy of Ingold Electrodes, Inc.)

11.2.7   Carbon Dioxide Measurement

It is now possible to measure the partial pressure of carbon dioxide with a steam-sterilizable probe. This probe measures the carbon dioxide evolution from the cells; off-gas data can thus be verified, and the carbon dioxide evolution rate (CER) is obtained without the lag associated with calculations from the output of a mass spectrometer.

The measurement of the partial pressure of carbon dioxide is, in fact, a measurement of pH. Carbon dioxide diffuses from the process fluid, across a semipermeable membrane and into a standardized bicarbonate electrolyte solution. The formation of hydrogen ions is proportional to the dissolution of carbon dioxide in an electrolytic solution, according to the following reaction:

$$CO_2 + H_2O \ \rightarrow \ H^+ + HCO_3^- \tag{11.22}$$

The output, in millivolts, is thus directly proportional to carbon dioxide partial pressure. Since this is a pH measurement, changes in temperature affect the reading, as shown by the NERNST equation. Also, temperature fluctuations affect the dissociation constant, which reduces accuracy. Readings of carbon dioxide should therefore be made with a probe calibrated at the process temperature, and data obtained at other temperatures should be ignored.

Carbon dioxide probes come in configurations similar to pH probe assemblies, and fit standard Ingold-type 25-mm fittings. The probes themselves can have their membrane replaced (like the dissolved oxygen probe). The pH probe is built differently from the ones described in Section 11.2.5.2, and therefore cannot be cleaned and regenerated. To calibrate a probe, remove the electrolyte solution with a syringe and add a standard buffer with another syringe. Then remove the buffer and replace the electrolyte. An example of a dissolved carbon dioxide probe is shown in Figure 11.24. A supplier of industrial-grade steam-sterilizable carbon dioxide probes is Ingold.

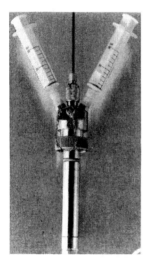

Figure 11.24 Carbon Dioxide Probe. (Courtesy of Ingold Electrodes, Inc.)

11.2.8   NADH and NADPH Measurement

With fluorescence measurement techniques, it is possible to measure concentrations of the nicotinamide adenine dinucleotides (NADH and NADPH) in a cell and provide an on-line indication of cell activity. These compounds are present in all living cells and are involved in numerous reactions. NADPH in cells absorbs light at 340 nm and emits at 460 nm, so a properly calibrated spectroscopic sensor can be used on-line to measure cell viability and biomass [13]. This technique is extremely useful, since it works for yeast, bacterial, fungal, and even high–cell-density mammalian cell cultures.

A fluorescence detector is shown in Figure 11.25. It has an amplifier assembly, and a probe with both the 340-nm light source and the sensor that measures fluorescence at 460 nm with a coaxial beam. It is inserted into a vessel through a standard 25-mm Ingold port, and can be provided steam-sterilizable and autoclavable. Output is a 4–20-mA signal that can be scaled to normalized fluorescence units (NFU) if corrections are made for pH and temperature influences external to the detector. This unit is supplied by BioChem Technology.

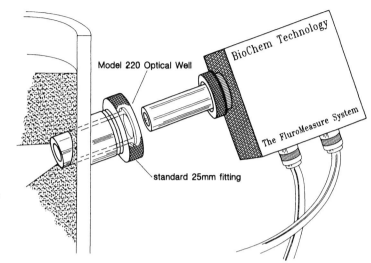

Figure 11.25
Fluorescence
Detector.
(Courtesy of
BioChem
Technology.)

11.2.9   Cell Density Analyzers

The optical density can be used as a measure of the cell density of a culture. This measurement, in contrast to that of fluorescence probes, gives no information on cell viability. The readout can be as optical density, or as cell count if a standard curve has been produced. A Monitek cell density analyzer is shown in Figure 11.26.

A near infrared light is passed through a fiber optic connection to the probe. A silicon detector measures the radiant energy lost to cells in the light path. The probe is constructed such that it is steam-sterilizable and fits into a 25-mm Ingold connection. It must have a signal from the agitation source sent to the unit microprocessor, to correct for variations due to aeration. Specific software has been developed that can be used on a host computer for data conditioning. An off-line sample, read as optical density or other units, can be used to calibrate the signal.

11.2.10   Viscosity Sensors

Although on-line viscosity sensing is not a common measurement in a bioprocessing environment, it is useful in certain applications. An example is the monitoring of viscosity to estimate the growth rate of a mycelial fermentation. Two types of viscometers are common: the oscillating viscometer, and the rotational viscometer.

An example of an *oscillating viscometer* is shown in Figure 11.27. The spherical element, in contact with the fluid, vibrates at constant amplitude. The electrical power

Figure 11.26 Cell Density Monitor. (Courtesy of Monitek.)

required to maintain this constant amplitude is proportional to the viscosity. Since the vibrating element is not coated, the unit provides accurate values of viscosity. Steam-sterilizable industrial-grade oscillating viscometers are supplied by Nametre.

The *rotational viscometer* is shown in Figure 11.28; it is similar in configuration to the oscillating viscometer, but the moving element rotates instead of oscillating. The torque necessary to maintain a preset speed is proportional to the viscosity. Steam-sterilizable industrial-grade rotational viscometers are supplied by Brookfield.

Both units are steam-sterilizable and require a larger than normal aseptic connection to the vessel. It is not easy to retrofit these instruments into an existing vessel, so if use of a viscometer is envisioned, it is necessary to include the required fitting in the specifications.

11.2.11   Biosensors

Biosensor is the term used for a whole class of sensors that utilize a biochemical reaction to determine a specific compound. An immobilized enzyme or cell is combined with a transducer to monitor a specific change in the microenvironment. The probe tip is immersed in the liquid phase and is in contact with the process either directly or through a membrane. To date these instruments have not seen widespread use since, as a class, they exhibit many disadvantages. These include an inability to

Figure 11.27 (at left) Oscillating-Type Viscometer. (Courtesy of Nametre.)

Figure 11.28 (at right) Rotational-Type Viscometer. (Courtesy of Brookfield.)

be steam-sterilized, interactions with media and substrate components that lead to failure, membranes that frequently become impermeable, sensitivity to variations in the physical environment, and an inability to be utilized *in situ* [14].

Although the problems exist, a number of sensors have been developed for specific industrial applications; they are listed in Table 11.5. Glucose sensors are the most common, since glucose is a common constituent of fermentation substrates. As the technology advances, more biosensors are likely to see industrial use, since the direct reading of a biologically active ingredient is a useful indication and control point.

Table 11.5 Applications for Biosensors

Measured variable	Biocatalyst/Enzyme	Application
glucose	glucose oxidase	general fermentation control
sucrose	glucose oxidase	general fermentation control
starches	glucose oxidase	general fermentation control
ethanol	ethanol oxidase	alcohol and yeast production
penicillin	β-lactamase	antibody production
amino acids	amino acid oxidase	general food production
phenols	polyphenol oxidase	chloroform production
pyruvate	pyruvate oxidase	milk production

11.2.12   Analytical Instrumentation

11.2.12.1   Mass Spectroscopy

A mass spectrometer is the best instrument for off-gas analysis from a biochemical reaction. Although the initial expense is high, a single mass spectrometer can be connected via a manifold to several fermentors to determine the off-gas composition of several streams. The relative amounts of nitrogen, methane, argon, carbon dioxide, water, and oxygen are most commonly determined. A Perkin Elmer mass spectrometer is shown in Figure 11.29.

Figure 11.29 Mass Spectrometer. (Courtesy of Perkin Elmer.)

In a mass spectrometer, a sample stream is ionized and then directed into a magnetic field. The ions are deflected to an extent that depends on their ratio of charge to mass, and strike one of several Faraday collectors. The signal from each collector is proportional to the mass of the particular ion. A microprocessor then computes the composition of the sample stream. Figure 11.30 is an illustration of this principle.

Many practical considerations go into the successful application of on-line mass spectroscopy. The main concern is the exclusion of liquids and particles from the instrument feed stream. Proper system design goes a long way to meeting this objective. The instrument feed system should be designed without any pockets, be of minimum distance, and be free-draining back to the fermentor. Since the off-gas is water-saturated, the lines should be heated to avoid condensation and filtered to

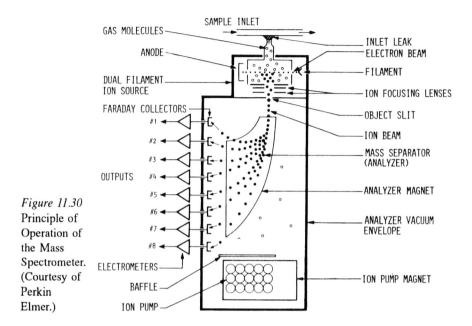

Figure 11.30
Principle of
Operation of
the Mass
Spectrometer.
(Courtesy of
Perkin
Elmer.)

remove medium. Once the system is set up it should be inspected and calibrated with reference gases frequently.

11.2.12.2   Ultraviolet Spectroscopy

Ultraviolet spectroscopy is used to do multicomponent analysis of the effluent stream from a protein separation. The UV spectrometer senses the absorbance of UV light at different wavelengths and uses the data to compute the composition of the stream.

Ultraviolet spectroscopy is based on the spectrographic principle that when UV light shines on a sample stream, particular chemical bonds absorb light at specific wavelengths. This wavelength absorption is a function of the concentration of molecules having the characteristic structure. The original instruments were used for single components, but newer technology allows ultraviolet spectroscopy to be applied to multicomponent systems without the need for changes in the light source. This type of ultraviolet detection system is available from Du Pont.

11.2.12.3   On-Line High Performance Liquid Chromatography

On-line high performance liquid chromatography is a useful process analysis tool for the determination of many complex molecules encountered during protein separation.

A sample stream is drawn from the process, mixed with a carrier solvent, and passed over an adsorbent packing. The influent components are selectively adsorbed within the bed to extents that depend on their affinity to it. After a predetermined cycle time the sample stream is shut off. Another solvent, one in which the sample components are soluble, is passed over the bed. The components are separated on the basis of their affinity for the bed. The solvent displaces and carries the separated components individually, and this effluent stream is then sent to a detector where the amount of each component is determined.

11.3   Instrumentation for Indicating, Controlling, and Recording

11.3.1   Indicating and Recording Elements

Any number of instruments are available to indicate the signal transmitted from a process measurement device. The signal can be viewed as an instantaneous value, as with an indicator, or a recorder can be used to create a hard copy or electronic record of the reading. Indicators and recorders are similar in that they are part of an open loop and perform no control function. They only provide an indication of the measured variable in question.

Indicators are available in many configurations. The most common are single-loop units, and distributed control systems with a video display terminal for indication of several process variables. An example of a double-loop controller is shown in Figure 11.31. Double-loop indication points on a distributed control system are used when it is necessary to monitor a process variable but no record or automatic control is required.

Recorders are used when, as with an indicator, no automatic control is required and an open loop is sufficient. They provide a record of the process variable in one of several forms, the most common being a circular chart, a strip chart, or as a display on a video display terminal. An example of a strip chart recorder with controlling capabilities is shown in Figure 11.32.

11.3.2   Controlling Elements

Although several types of controller algorithms are available, the traditional proportional/integral/derivative (PID) controller is the instrument of choice for a bioprocessing environment. It is well suited for the steady-state gain, dead-time, and lag dynamics inherent in a cell-based process. It is widely used, relatively simple to apply, and obtainable in many useful configurations. As previously discussed, a PID controller works by making a change in the controller output based on the difference

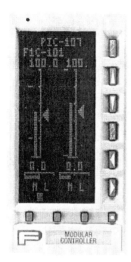

Figure 11.31 Double-Loop Controller. (Courtesy of Fischer and Porter.)

Figure 11.32 Multifunction Strip Chart Recorder and Controller. (Courtesy of Kaye Instruments, Inc.)

between the measured variable and the set point (the error). The components of the control equation are the magnitude of the error (the proportional mode), the duration of the error (the integral mode), and the rate of change of the error (the derivative mode). The most common controller is the proportional/integral (PI) type. The derivative mode is usually necessary on only the fastest of control loops.

Controllers are available in many configurations, including the circular chart and strip chart recorder/controller, the programmable controller, the single-loop or multiloop rack-mounted controller, and the controller as part of a distributed control system.

11.3.3   Microprocessor Based Control Systems

11.3.3.1   Distributed Control Systems

In a *distributed control system*, the individual indicators, recorders, and controllers are not used as the operations interface. Instead, a video display terminal is used to view and manipulate process variables. An example of this type of display is shown in Figure 11.33. The term *distributed* comes from the system configuration, which, as the name implies, has the operator interface distributed from the input and output

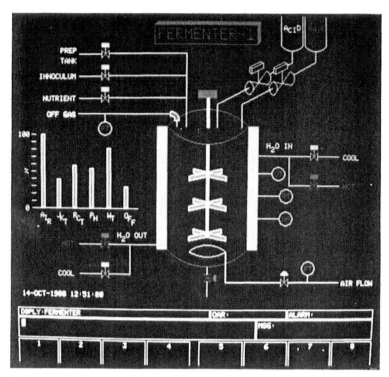

Figure 11.33 Example of a Fermentor Display on a Control System Video Terminal. (Courtesy of Fisher.)

electronics. All the individual analog and discrete signals are transmitted from the individual sensors and switches to the respective remote electronics. From there, the signals are converted into digital form and transmitted over the data highway to the various interfaces, such as operator consoles, engineering consoles, data collection computers, mass storage devices, and printer and plotter output devices. A schematic of a distributed control system is shown in Figure 11.34.

The relative cost of a distributed control system varies with scale. For a facility with 50–100 sensors and switches, the cost of distributed control can be extremely high compared to that of individual instruments. Although costs are falling for small-scale distributed control systems, this still holds. However, as the number of instrument points approaches 1000 sensors and switches, distributed control can be much less expensive than the alternative of individual instruments. If a system is between these extremes, it is necessary to add up detailed costs for an accurate comparison.

Cost aside, distributed control systems are superior to individual instruments in many respects. The advantages include the ability to have redundant electronics for

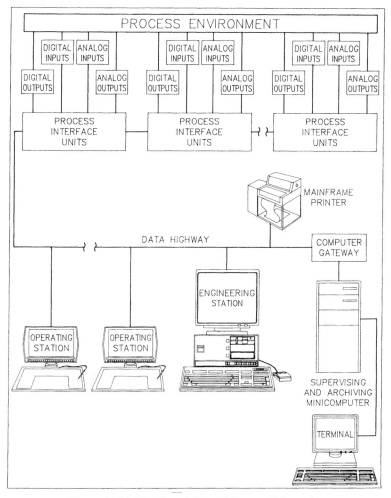

Figure 11.34 Distributed Control System Schematic.

critical components, the use of electronic media or hard copy output for data storage and presentation, much improved reliability, system flexibility, and easier repair and maintenance. Distributed control can seem esoteric to the uninitiated because it is set up in a specialized configuration and initial programming is needed. This drawback at the beginning of a system's life cycle becomes an advantage, as modest changes to the control scheme (in both size and methodology) require only software changes and not hardware changes. Be aware that different vendors have vastly different features in their control systems. When a control system is to be installed, the requirements must be specified clearly and in detail. Several vendors should then be compared before a decision is made.

11.3.3.2   Programmable Logic Controller Systems

Programmable logic controller (*PLC*) based control systems are very similar to distributed control systems. A schematic depiction of a PLC system is in Figure 11.35.

The main differences, when PLC systems are compared to a vendor-specific distributed control system, are that PLC systems can consist of components from several different manufacturers, and also that PLCs are somewhat standardized. Several advantages and disadvantages are apparent. The chief advantages of a PLC type of system are lower cost and facile system upgrades. Since it consists of PLCs, mini- or microcomputer-based interface hardware, and control system software, any of these components can be changed without the need to replace any other components. The main drawbacks include the need to program the system configuration at both the PLC and operator interface software level, a loss of redundancy capabilities, and the lack of a central supplier responsible for all the components.

As with a distributed control system, the functional specification should be specified clearly and in detail before a vendor is selected. Not only must the right equipment be selected (it is likely to include components from several manufacturers) but also a system integrator must be found. This job can be done either by an in-house expert or by an outside firm with the appropriate background. In any case, if a PLC–based system is specified, someone needs to know details of all the elements and how to make them work as expected.

11.3.3.3   Hybrid Microprocessor-Based Control Systems

Hybrid microprocessor-based control systems are a class of instrumentation between, on the one hand, distributed and PLC systems, and, on the other hand, individual instruments. They take many forms, including the strip chart recorder shown in Figure 11.32.

They can also be purchased with small display screens, LED, or LCD operator interfaces. They are well suited for installations with a low point count, and are ideal as a control system on individual fermentors. Some manufacturers can supply this equipment with digital communication capabilities so that the control system can be monitored with a central computer, if that is required.

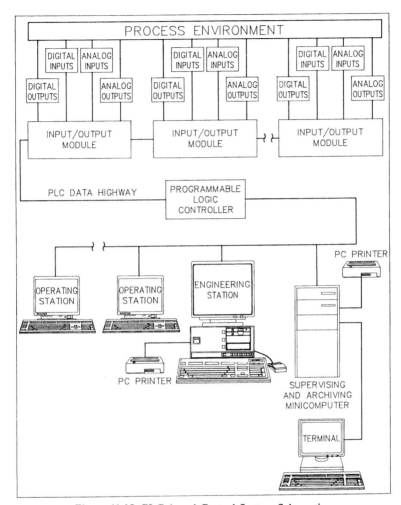

Figure 11.35 PLC–based Control System Schematic.

11.3.3.4   Manufacturers

Manufacturers of microprocessor-based control systems equipment include Allen Bradley, Bailey Controls, Combustion Engineering, Fischer and Porter, Fisher Foxboro,

General Electric, Honeywell, Leeds and Northrup, Moore, Rosemount, Square D, Texas Instruments, and Yokogawa. Many more suppliers of good-quality equipment are listed in the latest Instrument Society of America Directory of Instrumentation.

11.4   Signal Transmission

11.4.1   Analog Signals

Analog signals can be transmitted either pneumatically or electronically. Although different ranges continue in use, the standard for the pneumatic transmission of instrument process variables is 3 to 15 psig of air. This means that a pneumatic instrument senses the variable in question, scales it, and transmits by means of an air pressure value between 3 and 15 psig. Pneumatic instrumentation typically requires a regulated air supply that meets certain criteria for pressure and quality. A low moisture content is necessary, to prevent both corrosion and condensation in the system. A dew point of 0 °C is a good target for moisture content. The oil content of the air supply should be below 5 ppm, with 1 ppm as a preferable maximum, also to prevent corrosion. To minimize plugging, the air should be filtered at the source and locally to remove particles larger than 5–10 μm. Although different instruments require different supply pressures, a regulated supply pressure of 20 psig is suitable for most applications.

Electronic transmission of analog instrument signals also has a standard, which is a current signal of 4–20 mA. A current instrument senses the variable, scales it, and transmits by means of an electronic current of 4–20 mA. Two types of signal transmission are used: two-wire and four-wire. A two-wire transmitter is powered by the DC voltage that exists over the signal transmission lines. A four-wire transmitter has two leads for signal transmission and an additional two leads for AC or DC power.

Besides the 4–20-mA standard, there is also a European standard, which is a voltage transmission of 0–5 V. This presents a minor problem when European instruments are mixed with instruments built to the 4–20-mA standard, since two conversions must be made: the first is from 0–5 V to 1–5 V, which is done with an amplifier; and the second is to 4–20 mA, which is done with a 1000-Ω resistor. Although less common, some instrumentation is available with a 1–5-V voltage range, in which case only a resistor is needed to make the loop a 4–20-mA current loop.

11.4.2   Digital Signals

Numerous methods exist for transmitting data digitally. From a general perspective, this involves the process of taking an analog signal, converting it into an absolute value, encoding it in binary form, and transmitting this value over a data transmission linc for conversion back into an analog value for indication, control, or further transmission. Although the Instrument Society of America is attempting to establish a standard for digital data transmission based on the Modified Automation Protocol (MAP), for use in process control, no accepted method prevails in industry. At present, digital communication is done between instruments from the same vendor or from different vendors that have established a standard for communication between them. An alternative is to have custom software written for digital communication between devices that do not have the requisite software available. An example of this is a custom communications software driver that allows a distributed control system to communicate with a minicomputer for data archiving.

11.5   Maintenance and Calibration

11.5.1   Maintenance Procedures

Maintenance procedures in a biotechnology production environment are based on both standard maintenance procedures and the requirements of Current Good Manufacturing Practices (CGMPs). A facility whose maintenance is derived from these two sets of guiding principles can be maintained to provide the highest and most efficient levels of reliability and accuracy from the instrumentation.

Standard maintenance procedures are based on the simple premise that instrumentation can and will fail, and that data from instrumentation can be erroneous, even if a scheduled calibration is not due. Repair of a failed instrumentation component is straightforward. The subject instrumentation is either tested in place, or removed to a shop environment for troubleshooting, and fixed or replaced as necessary. It is important to be ready for this eventuality with the required tools and electronics for each component. This could range from a voltage-ohm meter (VOM) or a variable resistance box for the analysis of an RTD temperature loop, to an esoteric electronic service device for a distributed control system. A listing of the required service tools for instrumentation troubleshooting and repair is extensive, changing as technology changes, and is somewhat manufacturer-specific. It is therefore imperative that consideration be given to maintenance when the instrumentation is installed, so that the proper procedures and the items necessary for implementation are available when required. An alternative is to contract for maintenance service with an organization familiar with the instrumentation in the facility.

CGMP–based instrumentation needs are an extension of standard calibration requirements and good record-keeping. Standard maintenance procedures call for routine calibration of instrumentation at fixed intervals to assure that the readings obtained are as accurate as necessary. This may entail only the calibration of individual components of the loop that are liable to drift over time, such as the transmitter. Other components, such as the sensor itself, may be neglected since their tendency to drift is negligible. For a bioprocess, here assumed to be for production of a product for human or animal consumption, the FDA mandates that instrumentation reflect the true condition of the process with a great deal of certainty. This translates, in a maintenance-related practical sense, to the calibration of each element of an instrument loop, and the calibration of the entire loop at once. In addition, this work must be traceable to an acceptable standard, such as the National Bureau of Standards.

Besides calibration, CGMPs also call for complete and accurate record-keeping. This includes both the calibration records and the maintenance records. Supporting documentation, such as Standard Operating Procedures and CGMP protocols, are also to be implemented and recorded.

11.5.2   Calibration Methods and Accuracy Determination

Calibration can take one of several forms, depending on the required accuracy of the instrumentation under consideration. As a rough guide, for example, a temperature instrument can be calibrated by simply placing the sensor in boiling water. If it reads at or near 100 °C, it is considered somewhat accurate. If more accurate results are desired, the sensor must be compared to a sensor considered reliable. An example of this is to compare the reading of a pressure transmitter to the reading from an analog pressure indicator. While both of these methods are commonly used and even sometimes appropriate, they do not assure accuracy.

To assure accuracy, it is necessary to compare the sensor to be calibrated to a known standard, such as one from the National Bureau of Standards (NBS). Although it is not possible to make a comparison directly to the NBS standard measurement, it is possible to obtain a sensor with a calibration traceable to the NBS. Since temperature is a critical variable in the production of cell-based products, it is a good example. If a temperature sensor is immersed in a bath equipped with an NBS–traceable sensor, a comparison can be made between the two, and accurate characteristics of the sensor to be calibrated can be obtained. Doing this at three points supplies the data necessary to develop an individual calibration curve for the sensor. If more than one NBS–traceable sensor is used for comparison and they are rechecked or replaced frequently, a good standard calibration reference can be maintained.

In determinations of accuracy, the error source of a control loop must be known. These errors can be related to the sensor, the transmission wiring, the controlling and indicating device, and any other element of the loop. Once all the individual errors are determined, the total probable error is the geometric average of the individual errors. The following formula applies:

$$E = \sqrt{\sum E_i^2}$$

(11.23)

where E is the total probable error, and E_i are the individual errors.

Error analysis involves knowing not only the absolute value of accuracy for a particular component of a control loop, but also the basis of an error. *The basis of an error* can be the percentage of overall range of calibration available, the actual calibrated range, the maximum value, or the scaled measured variable. The error can also be an absolute value relative to the process variable. The following is an example of an accuracy calculation for an RTD temperature control loop at 25 °C with a range of 0 °C to 45 °C, and indicated at a distributed control system:

- sensor accuracy: ±0.06 °C
- transmitter accuracy: ±0.14 °C
- transmission accuracy: ±0.1% of reading
- A to D conversion at control system: ±0.15% of scale.

The calculated error, E, is as follows:

$$E = [(0.06)^2 + (0.14)^2 + (0.001 \times 25)^2 + (0.0015 \times 45)^2]^{0.5} \qquad (11.24)$$

or $E = 0.15$ °C.

Note that the likely error of 0.15 °C is less than the maximum possible error of 0.21 °C.

11.6   References

1. *The International Practical Temperature Scale of 1968 (IPTS-68)*; HMSO, London, 1969.
2. *PRT Handbook*; Rosemount, Inc., Burnsville, MN, 1986; pp 2–3.
3. KENNEDY, R. H., Selecting Temperature Sensors, *Chem. Eng. 90* (16), 54–71 (August 8, 1983).
4. *Temperature Measurement Thermocouples*; ANSI/MC96.1; Instrument Society of America, American National Standards Institute, Research Triangle Park, NC, 1982.
5. RICHMOND, D. W., Selecting Thermowells for Accuracy and Endurance, *InTech 27*, 59–63 (February, 1980).
6. *Filled Thermal Thermometers*; SAMA Standard RC; Scientific Apparatus Makers Association, Washington, DC, 1963.
7. LEWIS, J. D., Pressure Sensing: A Practical Primer, *InTech 35* (12), 44–77 (Dec., 1988).
8. *Electronic Weigh System Handbook*; BLH Electronics, Canton, MA, 1986; pp 5–12.
9. *Basics of Variable Area Flowmeters*; Brooks Handbook HT-1100; Hatfield, PA, 1983; pp 5–7.
10. *Handbook, Flowmeter Orifice Sizing*; Fischer and Porter Publication 19713a; Fischer and Porter, Warminster, PA, 1978; p 2.
11. MCMILLAN, G. K.; *pH Control*; Instrument Society of America, Research Triangle Park, NC, 1984; p 55.
12. BUHLER, H., BAUMANN, R.; *pH Electrodes: Storage, Aging, Testing, and Regeneration*; Ingold Publication E-TH-7; Ingold Electrodes, Inc., Wilmington, MA, 1988; pp 12–13.
13. MACMICHAEL, G. J., ARMIGER, W. B., LEE, G. F., KRAUSE, R. J., Fluorescent Measurement of Biomass in Mammalian Cell Cultures, *BioPharm 1* (May), 48–52 (1988).

14. CLARKE, D. J., CALDER, M. R., CARR, R. J. G., BLAKE-COLEMAN, B. C., MOODY, S. C., CILLINGE, T. A., The Development and Application of Biosensing Devices for Bioreactor Monitoring and Control, *Biosensors 1*, 231–320 (1986).

PART III

SUPPORT SYSTEMS

Cleaning of Process Equipment: Design and Practice

P. William Thompson

Contents

12.1 Introduction

Cleaning of a bioprocess plant is necessary, whatever the industry or application, for two main reasons. First, it is important for process reasons that equipment be clean so that potential contamination is minimized. For example, a residue of soil remaining in a medium preparation tank can quickly become a substrate for bacterial growth; the bacteria rapidly begin to consume substrates in medium that is made up for a subsequent batch, thus reducing the nutritional value of that medium prior to sterilization. In addition, it is important that there be no fouling buildup that may compromise performance. For example, if a chromatography column is not thoroughly cleaned, residual molecules bound to the matrix can build up and gradually soak up the available capacity of the column. Second, in many industries, such as food and pharmaceuticals, it is a statutory requirement that adequate levels of hygiene be maintained. Both the FDA (Food and Drug Administration) and European regulatory authorities for pharmaceuticals demand that cross contamination be eliminated from production systems; in a multipurpose production facility, this can only be ensured by adequate cleaning between batches. Some materials (such as chromatography matrices) cannot be cleaned sufficiently or reliably enough to satisfy some regulatory authorities, and therefore their use must be limited to a single product.

The types of soil that may be encountered in a bioprocess plant clearly depend on the use to which the plant is put, and the type of fouling that is generated by the process. The range of bioprocesses that exist makes it difficult to generalize, and the problems in cleaning a large-scale penicillin fermentation plant are very different from those encountered in a small-scale mammalian cell culture process. However, one may identify certain types of soil associated with individual sections of the plant.

Medium makeup facilities become fouled with medium components such as proteins, lipids, and sugars. These may be densely accumulated at the surface of the medium in a tank, and at the vortex point in an agitated vessel, if complex media containing suspended solids are used, or thinly distributed if the medium components are completely soluble. If medium is heat-sterilized, denatured protein and caramelized sugars may be added to the background soil and may be still more difficult to remove. It is a sensible precaution to attempt to prevent the formation of these persistent forms of soil at the design stage. Likewise, the use of hard water for medium makeup results in a layer of inorganic scale buildup, which can be avoided if softened or demineralized water is used.

Fermentation vessels can also suffer from fouling with medium components, especially if medium is steam-sterilized in the vessel. In addition, if foaming has occurred, there are deposits of denatured protein around the top of the vessel, and residual biomass left over from harvesting the fermentor. In some instances, such as highly aerated microbial culture broths with high cell concentrations, and highly viscous fungal fermentations, the cultured organisms demonstrate a strong affinity for the internal surfaces of the fermentor, resulting in wall growth that is sometimes several centimeters thick. This can occur around the liquid surface, on impeller blades, and around protrusions into the vessel, such as addition pipes and sensor probes, possibly impairing their function.

Once the fermentor has been harvested, the main source of fouling becomes the biomass and the product or any by-products. In a harvesting centrifuge there are many places that biomass can accumulate, and it is important to select such equipment with cleanability in mind, especially if the biomass concerned is in any way sticky. Even low levels of biomass, such as appear in animal cell cultures, can accumulate in centrifuges and can be consequently difficult to remove.

Other harvesting devices that concentrate biomass also generate high levels of fouling. Plate-and-frame filters, rotary filters, and cartridge filters all accumulate biomass. Cross-flow microfiltration or ultrafiltration systems tend to be of inherently cleaner design, permitting simpler cleaning procedures if appropriate on other grounds.

If biomass is not the product, the supernatant contains the product and impurities. Typical soils from this point on consist of these, together with residues of reagents added in subsequent purification steps. Of the impurities associated with protein products, albumin is a common source of soil, but fortunately is not difficult to remove, assuming that no denaturation has taken place. In many downstream processing operations for products such as low molecular weight antibiotics, quite high levels of solvents or inorganic salts may be added in processes such as solvent extraction, HPLC, and chromatography. It is often important to ensure complete removal of one reagent from a piece of equipment before processing on the next batch is begun. Rinsing is often sufficient in these cases.

Residual product buildup is usually undesirable because it poses the threat of cross contamination between batches of different products in the same plant. Thorough cleaning of all plant equipment and materials is necessary to eliminate this possibility.

The types of soil most commonly encountered in a facility outside a piece of process plant equipment are usually spillages. Generally, good housekeeping must be maintained to remove such spillages before they can create major sources of contamination and increase the overall level of bioburden in the facility. Typical waterborne organisms such as *Pseudomonas* are the principal causes of many contaminated fermentors, and are endemic in many fermentation facilities.

12.2   Hygienic Plant Design

12.2.1   Materials

For those parts of the equipment not in contact with the product, less emphasis need be placed on the cleanability of the material. However, as it is still generally desirable to maintain a hygienic appearance, it is preferable to use materials that can be cleaned without causing overt corrosion. For example, the external covering of cladding on piping and vessels should generally be aluminum with well-sealed joints. Spillages such as sodium hydroxide should be removed as soon as possible from aluminum to prevent corrosion. Stainless steel 304 gives a superior finish, especially on vessels, but is not strictly necessary. Plastic coverings for insulation cladding give a usually inferior finish and should be confined to those areas where hygiene is less critical. Where the cleanability of equipment is concerned, the materials of construction in direct contact with the raw materials and products merit the most attention. These include the materials for the main structure of the equipment, jointing materials, coating materials, and sealing compounds.

Austenitic stainless steel is the most commonly used material, in the form of AISI 304 or 316. Type 304 stainless steel is quite suitable for most applications where there are no harsh process conditions. Corrosive environments include those with high temperatures, acid conditions, or high chlorine concentrations, including high salt concentrations. These environments can cause pitting in 304 stainless steel, which results in surfaces that are difficult to clean. Therefore, the 316L grade of stainless steel, containing a proportion of molybdenum, is used in many cases, so long as chlorine levels are not excessive or associated with high temperature or low pH.

Where conditions are highly corrosive it may be necessary to use rather more exotic materials, such as titanium or Hastelloy®, but such materials are very rarely required.

Plastic materials may also be used for some plant items such as piping. Acrylonitrile-butadiene-styrene (ABS) or polyvinylidene fluoride (PVDF) is commonly used for deionized water pipework, and for very small tubing that connects small additive vessels to larger systems, silicone rubber tubing is usually acceptable. If such flexible tubes need to tolerate higher pressures than silicone tubing can stand, hoses lined with PTFE (polytetrafluoroethylene) reinforced with stainless steel may be used. Plastics may also be used in vessels, such as storage vessels, if glass-reinforced plastic is acceptable, or for internal coatings for vessels if corrosive conditions prevail. In this case epoxy resin coatings or PTFE linings are preferred. The use of a plastic lining should always be accompanied by a suitable (probably stainless steel) base material, so that in the event of a failure of the lining there is no hazard from toxicity or marked chemical activity.

Materials such as PTFE and PVDF are commonly used in pumps and valves, both for structural items such as impellers and bodies, and for valve diaphragms. Plastics are also extensively used as seals and gaskets. PTFE is a common sealing material, and gaskets make use of polymers such as butadiene, silicone, and EPDM (ethylene-propylene-diene monomer). Glass is a good material for its cleanability and for the ease of inspection, and is generally acceptable for small or highly corrosive applications.

The plastics and polymers to be avoided include those that have a porous surface, such as natural rubber, because they are almost impossible to clean, and those that tend to leach out additives (such as plasticizers) that can potentially contaminate the product. In this respect, materials to avoid include low density polyethylene, neoprene, and PVC. Nonferrous metals are not generally acceptable, with the exception of titanium, nickel, and some nickel alloys, all of which are high-cost materials. Materials to specifically avoid include zinc, cadmium, lead (even in solder if pharmaceutical or food products are to be produced), and plastics containing free phenol, formaldehyde, and plasticizers [1].

12.2.2   Surface Finish

All surfaces in contact with the product should be smooth, nonporous, and free from pits and crevices. All of the suitable materials described above are nonporous, and, where there is no corrosion, free from pitting.

The main concerns are those of surface smoothness and weld finish. Mechanical polishing or electropolishing is required, to generate a suitably smooth finish on the internal surface of most vessels. For nonsterile storage vessels, a polish quality of 120–180 grit is adequate. For sterile storage vessels, 180–240 grit may be specified, and for fermentor vessels in which soil adhesion to the walls is to be minimized and cleaning performance enhanced, a surface finish of 240–360 grit with a mirror polish on top is common [2].

The quality of the surface finish may be more accurately defined (though not so easily measured) by referring to the *surface roughness* or *Ra* value. In this case, a 120-grit polish can result in an *Ra* of between 1 and 3 μm, a 240-grit produces a finish between 0.5 and 1.5 μm, and a 320-grit with mirror polish should result in less than 0.2 μm *Ra*. These are only guidelines, since the *Ra* values are influenced by material quality and care of polishing as much as by the specification of the polishing tool [3, 4].

The accepted internal finish for pipework for nonsterile duties is usually *descaled* only; this quality of finish has been used for some time within the food and drink industries. It appears that although the surface of a descaled sheet may contain irregularities, they are generally rounded irregularities that are not difficult to clean. Polishing with a coarse grit may create an improved appearance, but is likely to degrade the cleanability of the surface because sharp, jagged edges are created, behind which soil can accumulate and not be easily dislodged. Polishing usually creates a grain in a particular direction, so in a vessel that grain should be oriented vertically, to facilitate good draining from the surface [5]. Nevertheless, for duties in which hygiene is more critical, and for sterile systems, a polish of 180 grit or even 320 grit is often specified [4, 6] as it has been found that a lower *Ra* value can reduce the time required to clean a surface to a specified standard [7].

Untreated welds can be a particularly troublesome source of crevices and hence, if accessible, should be ground smooth and polished to the same standard as the rest of the surface. In pipework where the internal weld is inaccessible, it is important that the initial quality of the weld be of a high standard. Automatic orbital welding is usually the most reliable method of ensuring a smooth weld, although good manual welding can match those standards. Thorough inspection is necessary to ensure that welding is up to quality and that poor quality welds are remade.

12.2.3   Vessels

There are a number of features in all vessels, whether sterile or nonsterile, pressurized or open ,that require attention to ensure cleanability. All vessels should drain from the lowest point, preferably from the center of the bottom. If this is not possible, the bottom of the vessel should be sloped towards the outlet valve to ensure complete drainage. Outlets should finish flush with the wall of the vessel.

Sensor pockets for pH and dissolved oxygen electrodes should slope downwards into the vessel to ensure drainage of liquids. Pockets should be designed so that either the probe seals flush with the vessel and leaves no crevice or pocket, or the pocket is of sufficient size that it self-drains. Generally the length of a pocket should be no greater than twice its width.

Inlets to the upper side of the vessel should protrude at least 50 mm into the vessel and slope slightly downwards, to ensure that incoming process streams leave no deposit down the side of the vessel. If the material in the vessel is likely to suffer by foaming if the inlet stream is allowed to fall directly into the bulk liquid, an extended inlet tube leading to near the bottom of the vessel should be provided.

The top of the vessel should permit a person to enter for inspection and manual cleaning if necessary. This may be via a removable top, in the case of open vessels, or an 18" (46-cm) manway in the top dishing of a closed vessel. To assist in inspection, closed vessels should normally incorporate an inspection lamp on the top; the arrangement can consist of a flush-mounted sight glass with integral lamp, or two sight glasses, one with a lamp mounted on it and the other for viewing.

Agitators should be mounted so that the shaft is sealed against leakage. For sterile duties a double mechanical seal is normally adequate. Recently developed magnetically coupled agitators may also be appropriate and foster inherently clean conditions. Agitators should be easily removable for inspection and manual cleaning if necessary. Much of the design of an agitator is dictated by factors other than its cleanability, but a number of features can be highlighted. Frequently an agitator includes movable impellers secured by screws tightened against the shaft. These create inherently unhygienic crevices. The free end of a high-speed turbine agitator (the end opposite the point of entry to the vessel) may be steadied by a bearing on a securing bracket. This bearing needs to be sealed against leakage into the bearing as effectively as the vessel entry seal.

If internal fittings such as agitators and pipes exist, attention should be paid to the location of internal spray balls for cleaning duties, so that any shadowing of internal surfaces by fittings is minimized. If this may be a significant problem, rotating spray balls may provide a solution. Ultimately, manual cleaning must be considered if all else fails.

The geometry of a vessel can affect its cleanability. A vessel of low aspect ratio (short and fat) has surface faces that are more accessible to a top-mounted spray ball than one that is tall and thin. Tall vessels, such as bubble columns and airlift fermentors, may need an extra spray ball mounted at the bottom if significant soil is expected there. Otherwise, care in devising the cleaning methods and their validation will be necessary.

12.2.4   Pipework

The design of pipework is one of the most critical factors in the cleanability of a plant. A number of standards exist to guide the designer, including the American 3-A standards [6] and British Standard BS 5305 [8]. However, these standards were developed for the food and dairy industry and may not in every case be appropriate for bioprocess applications. A useful discipline currently emerging is that of hazard analysis of critical control points (HACCP), which uses modified hazard analysis techniques to identify potential sites where bacterial contamination can grow and survive in a processing system, thus enabling corrective action at the design stage [9].

For most bioprocessing applications, pipeline pressures are quite low and permit the use of the "OD tube" or ASTM A269 (American Society for Testing and Materials standard A269) type of piping, as used in the food industry. Many prefer the bright annealed ASTM A270 standard of piping for its higher weld specification. If pressures are higher, such as in some circulating water systems, schedule standard pipework (ASTM A312) is necessary [4].

The preferred method of joining pipe is welding. Even if plastic pipe is used, it is preferable to heat-weld wherever possible. However, for maintenance and inspection purposes, it is frequently necessary to dismantle piping. In this situation there are different types of connection that may be used.

Flanges are most common if high temperatures and pressures are to be contained. Flanges may employ point contact O-ring gaskets, with which there is a minimal gap between the gasket and the internal pipe surface. The preferable arrangement for flanges is with full-face soft food-grade gaskets (silicone or butadiene rubber) that fit flush against the inside of the pipe. Care must be taken in fitting the gasket and tightening the flange, to ensure that the gasket does not overly protrude into the flow space.

Various unions originally developed for the food and dairy industry may be used successfully in a bioprocessing plant. Such unions are available in sizes from 1" (2.5 cm) (occasionally ½") up to 4" (10 cm), as illustrated in Figure 12.1. The ISS (International Sanitary Standard) or IDF (International Dairy Federation) joint consists of a nut and liner to join the pipe, with a flush-fitting T-section gasket between the

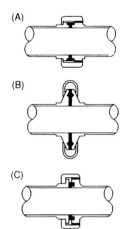

Figure 12.1 Hygienic Pipe Unions. (A) International Sanitary Standard (ISS) union; (B) clamp union; (C) Deutsches Institut für Normung (DIN) 11851 union. All these unions are designed to provide a flexible sealing ring that fits flush against the tube wall so as to allow an uninterrupted flow path and avoid crevices. The main drawback of such unions is that people overtighten them, thus causing the gasket to bulge into the pipe void.

pipe ends. The Tri-clamp® type of union developed in the U.S.A. comprises a full-face gasket gripped between two profiled flanges that are held together by a hand-tightened clamp. The DIN (Deutsches Institut für Normung) 11851 union also consists of a nut and liner with a C-section gasket held between the pipe ends. Several other proprietary unions exist, but they may not be universally available.

All of these unions can be easily cleaned, and if installed correctly, generate no crevices where soil can accumulate. They are easily assembled and disassembled for inspection. However, I point out that an excessive number of these unions in any one section of piping can decrease the rigidity of the system because of the cumulative flexibility of the elastomeric gaskets. They should therefore be kept to the minimum.

Piping bends should have a radius not less than the outside pipe diameter. Where the pipe diameter changes, the decrease or increase should be effected by a smooth taper with no sudden steps. The taper is preferably eccentric, with the lower surface remaining unchanged. There should be no sumps where liquid cannot freely drain. All pipework should be installed with a fall of approximately 1:100 in the direction of a drain point. Pipework should be adequately supported to prevent sagging [2].

If possible, dead legs should be eliminated in pipework. However, if these are unavoidable, the length of the dead leg should not exceed 2 to 3 pipe diameters. If possible, the direction of fluid flow should be into the dead leg rather than away from it. This increases the turbulence in, and hence cleanability of, the dead leg. All dead legs should be drainable and should slope towards the main pipe [7].

Piping design should ensure that there is no possibility of cross mixing of product streams with cleaning reagents that may be used in the same plant at the same time. The *block-and-bleed* layout of pipework achieves segregation by ensuring that there are two valves and an open drain line between different streams, as illustrated in Figure 12.2.

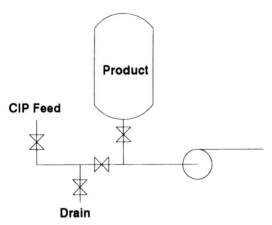

Figure 12.2 Block-and-Bleed Piping. The block-and-bleed piping arrangement ensures that there are always two valves between incompatible fluids. Any leakage through the valve is diverted to drain. The system can be made more secure by appropriate use of automated valves.

Separation of streams can also be ensured by a swing bend, as shown in Figure 12.3. This mechanism uses a single swept bend connected to a tank. If the union at the tank end is partially released, it can be swiveled to connect to either of two pipes; hence, only one pipe can be connected to the tank at any one time.

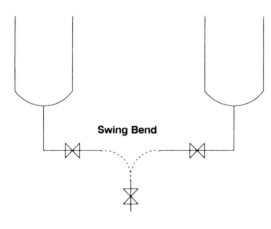

Figure 12.3 Swing Bend Arrangement of Piping. The swing bend is a version of the "key piece", intended to make it physically impossible for a supply pipe to be connected to two tanks at the same time. However, repeated tightening and loosening of the unions can damage the sealing rings, and frequent replacement is necessary.

A more sophisticated version of this is the flow plate. Several pipes can terminate open-ended, with connectors (and perhaps valves) at a flat plate at a precise distance from each other, so that a U bend can make a connection between two pipes. The arrangement of the pipe ends depends on the variety of interconnections that need to made. For example, if a single tank needs to supply its material to any one of six different destinations, then a suitable flow plate consists of a central inlet pipe surrounded by a concentric ring of outlet pipes; the U-bend connector fits between the center and the ring but not between any two pipes that are both in the ring.

If the system is automated, microswitches can be mounted on the plate, to be tripped when the U-bend connector is in place, thus allowing identification of which pipes (if any) are connected.

Flow plates are of necessity manual devices. Fully automated systems must use the block-and-bleed layout of pipework, and the controlling program must be fully validated to compensate for the safety of the swing bend or flow plate. Flow plates can be used in automated systems if prompts for manual intervention are included.

12.2.5   Valves

For sterile systems, membrane or diaphragm sealed valves have a completely flushed flow path, are free from crevices, and hence are easily cleaned. These are generally the valves of choice for sterile duties. In horizontal pipes it is important that diaphragm valves be installed at an angle of about 15° so that the weir does not impede free draining through the valve. If assurance of the diaphragm integrity is important, it is possible to have a leak indicator discharge fluid from the nonsterile side of the diaphragm when leaks occur. The grade of diaphragm to be used should be robust enough to withstand the expected chemical conditions and temperatures, yet pliable enough to promote a good seal between the diaphragm and the weir.

If pipeline sizes exceed the normally available diaphragm valve sizes, and the flow path through the valve is smooth, a globe valve with a bellows-sealed stem or steam-traced stem is usually acceptable.

For hygienic but nonsterile duties, the choice of valves is wider. In addition to the diaphragm valve, acceptable valves include the butterfly valve, the ball valve, and special hygienically designed globe valves [3, 6].

Both the butterfly valve and the ball valve depend on a rotating action to actuate the valve, and each has a smooth flow path. The ball valve contains a potential crevice between the rotating ball and the valve body, but, provided that the valve is accurately machined and lined with PTFE on each face, this crevice seldom causes a problem in practice. The butterfly valve suffers from a small sump at the bottom of the void, which does not easily drain in a horizontal line, so it is advisable always to install this valve in a vertical line. Both these valves utilize a rotating stem seal and so are generally unsuitable for sterile duties.

The traditional drawback of the globe valve is that the rising stem travels from a hygienic to a nonhygienic environment each time the valve is actuated. Recent developments have produced globe valves in which the stem is sealed either with a flexible metal bellows or a diaphragm. Close attention to the hygienic flow path for the fluid results in a valve that can be used for hygienic and even sterile duties.

A special type of globe valve is usually employed as a tank outlet valve, with the globe flush with the inside surface of the tank when closed. When open, the globe may either rise into the tank or descend into the body of the valve. Elaborate stem seals are usually incorporated into tank outlet valves.

Pressure relief valves usually consist of a spring-loaded globe valve mounted flush against the pipe or vessel to be relieved. Stainless steel relief valves are freely available, but if hygiene is important a stainless steel bursting disk is preferable.

12.2.6   Pumps

Various pumps suitable for hygienic duties are available, the most commonly used pumps being the diaphragm, centrifugal, rotary lobe, peristaltic, and helical displacement pump. (Specially designed versions are required, as it is possible to obtain nonhygienic versions of all of these pumps.)

Most pumps are equipped with mechanical seals between the impeller and the gear box; this path may be steam-traced for sterile duties or, alternatively, replaced with a magnetic coupling. For the latter it is important to ensure that the power transmission through the magnetic coupling is sufficient for the duty. In many cases, fluid in sterile systems is transferred by application of pressure to the top of the source vessel, which blows the fluid over to its destination. However, in situations such as chromatography or diafiltration, pumps are indispensable.

Ideally, pumps for hygienic processing are self-draining. In practice this is not easy, as most pumps have some form of sump within the pump body that can be drained only if a special bleed port has been provided at the lowest point. However, pumps such as the lobe pump, helical displacement diaphragm pump, and peristaltic pump have impellers that completely sweep the whole of the internal volume. Thus, when the fluid stream ends the pump almost completely empties itself. Centrifugal pumps, which lack this property, tend to be reserved for high flow rate duties, so that product may be flushed by water and there is no temptation to run the pump dry.

The internals of a hygienic pump should be smooth, with minimal crevices. Impellers should be secured with flat-faced cap nuts, as exposed threads are unacceptable. If threads are unavoidable, they should be male, coarse, and shallow, preferably not less than $60°$ and not more than 8 threads per inch [1].

The externals of pumps should ideally be shrouded, so that the traditional ribbed motor body is not available as an unhygienic surface.

12.2.7   Utilities

A number of plant utilities require regular cleaning, including deionized water systems, pyrogen-free water systems, and clean-steam systems.

Since deionized water is not intended to be sterile, it is acceptable to install ABS piping, although stainless steel may be preferable. Deionized water should recirculate around a pressurized loop, to prevent stagnant water from breeding contamination. If filters are included in the loop they should be regularly changed, as grow-through of bacterial organisms occurs over time.

Deionizers should be duplicated, so that each bed may be regenerated while the other is online, and also so that any contaminating organisms that have accumulated in the bed may be removed during the regeneration process. If such contamination is persistent, the regeneration process may also include a cleaning step.

The hygiene of pyrogen-free water systems is critical, and all components of such systems should be selected as for sterile systems. Inline filters in the recirculation system may be replaced or supplemented by inline ultraviolet sterilizers.

Clean-steam systems should maintain similar hygiene levels to pyrogen-free water. However, the temperatures and pressures involved dictate the appropriate equipment.

12.2.8 Facility Design

For pharmaceutical and food applications, it is important that there be a clean environment in which the bioprocess plant may operate. For other applications, such as chemicals and effluent treatment, the plant is traditionally exposed to the weather, making any control of external hygiene impractical. Even in the pharmaceutical industry there are situations where the process plant is too large to be enclosed in a building, and, therefore, external hygiene is difficult to ensure.

For those applications in which the process is to be indoors, there are gradations of cleanliness. For food applications the conditions should conform generally to food industry standards. However, if products such as vaccines and therapeutic proteins are to be manufactured, the standards tend to be more stringent.

The following is therefore a general guide; specific applications may require specific features of facility design that are not covered here.

Plant layout should allow sufficient space around the equipment to permit adequate cleaning of plant items. Equipment should be raised off the floor if possible, to permit cleaning underneath. All parts of the plant should be accessible for cleaning; it is desirable to eliminate ledges that may accumulate dust and dirt, and any remaining ledges should certainly be accessible [10].

Floors should consist of a continuous sheet of material, such as vinyl or epoxy resin. Tiles are acceptable only if regularly scrubbed to remove contamination from the grout. Floor coverings should extend up a wall curb, and the floor itself should slope downwards towards drain points to avoid puddling. Drains at the highest point of the floor are not acceptable. Drains should be trapped, and long gullies with overlying metal grilles should be avoided. Process drains should discharge directly to a floor drain and not onto the floor.

Walls should be painted with an impervious material that withstands washing. It is advantageous to incorporate a fungicide into the wall covering. It is sometimes appropriate to cover walls with sheet plastic material, to avoid the prospect of having to repaint walls at intervals. It may also be appropriate to install crash rails onto walls and doors to prevent damage. There should be no dust traps, such as window ledges or door ledges. If horizontal surfaces do exist, they should be profiled to a steep slope of at least 45°.

Ceilings should be washable. This means that fibrous ceiling tiles are not acceptable, and solid ceilings with hatches to access above-ceiling equipment such as ventilation are preferable. Ceilings should be coved so as to make a smooth juncture with the walls. Light fittings should be sealed to allow washing.

Ventilation grilles should be flush with the ceilings. The quality of ventilation is dependent on the situation, but there is generally little benefit in building a complex process plant in class 100 laminar flow rooms, as the disruption to flow patterns and the difficulty in bringing such equipment to the relevant standard of hygiene are incompatible with that level of cleanliness. Class 10,000 or 100,000 is probably the most appropriate standard for bioprocess plant.

Utilities pipework should ideally be concealed behind walls or in the ceiling void, and enter the processing area only at the required position. If such arrangements are not possible, pipework should be accessible for cleaning, clad in a continuous impermeable covering, and correctly labeled.

Protection against infestation by insects, rodents, and birds is important whatever the application. Good housekeeping is as important as good facility design. Good ventilation should replace openable windows, and doors should be self-closing and preferably paired into air locks. Drains should be disinfectable, and pipework that runs through external walls should be carefully sealed [11].

12.3 Cleaning Reagents

Most cleaning reagents are prepared as mixtures or solutions in water. It follows that the water should be of a certain minimum quality. Water for cleaning purposes should be at least potable mains water of acceptable bacteriological standards and should conform to the following analysis [12]:

Total hardness	< 50 ppm expressed as $CaCO_3$
Chloride	< 50 ppm
Chlorine	> 1 ppm (for bactericidal effect)
pH	6.5–7.5
Suspended solids	Substantially free

In many cases it is preferable to use deionized water instead of mains water, provided that sufficient quantity is available for use. A compromise may be to use deionized water only for final rinsing.

12.3.1   Detergents

The ideal detergent dissolves organics, and is good at wetting, dispersing solids, rinsing, and sequestering, as well as being a powerful bactericide. Unfortunately no such single detergent exists, which is why most available preparations consist of a mixture of reagents such as alkalis, surfactants, phosphates, acids, and sequestrants.

Bioprocessing plants involve a great need to solubilize proteins and lipids, which is best satisfied by sodium hydroxide (among the alkalis). However, sodium metasilicate is also a powerful alkali with good dispersant properties, which makes it very suitable for loosening and removing heavy soil such as cellular debris.

The most common phosphate is trisodium phosphate, which is often included for its free-rinsing properties, as well as because it is a good dispersant and emulsifier.

Of less importance in the cleaning of bioprocessing equipment are the acids, which are used to dissolve carbonate scales and certain mineral deposits. So long as the process uses deionized water, acid cleaning should not be necessary. However, nitric acid is used in the initial passivation process to impart an oxide layer on weld surfaces. This should not need to be a regular operation, and is usually limited to equipment that is to contain corrosive liquids such as purified water.

Sequestering agents such as EDTA and sodium gluconate are primarily used to prevent water hardness from precipitating, and therefore are seldom necessary in bioprocesses.

To allow the detergent to effectively contact the soil to be removed, it is frequently necessary to add a surfactant that reduces the surface tension of the aqueous medium. Surfactants also contribute emulsification and dispersant properties to the detergent. They can be classified as either anionic, cationic, nonionic, or amphoteric (anionic at high pH and cationic at low pH), and are selected according to the type of soil to be cleaned.

Typical formulas that have been used in cleaning vessels and piping are given in Table 12.1. The concentrations can be varied, but the total weight of the ingredients should be in the range of 0.2–0.5%. Occasionally some plant items such as membranes cannot tolerate harsh cleaning reagents. In these cases it may be feasible to use enzymatic detergents, usually based on alkaline proteases. Clearly, if protein products are being processed, great care must be taken to ensure the complete removal of these proteolytic detergents [13].

12.3.2   Sanitizers

Chemical sanitizing is seldom required in most bioprocessing plants, since the equipment is subsequently steam sterilized. However, sanitizing may occasionally be considered desirable if the equipment to be sterilized (such as many ultrafiltration membranes) cannot tolerate high temperatures.

Table 12.1 Formulas for Cleaning Solutions [12]

	g/L
For cleaning vessels by CIP:	
sodium hydroxide (0.1 N)	4
trisodium phosphate	0.2
For pipeline cleaning:	
surfactant	0.1
sodium tripolyphosphate	1
sodium metasilicate	0.4
sodium carbonate	1.2
sodium sulfate	1.2

The most common chemical sterilant is sodium hypochlorite, which will break down to release free chlorine, a powerful bactericide. The available chlorine from hypochlorite reacts rapidly with and is inactivated by residual organic matter, and so its potency decreases. However, when chlorine solution is used as recommended, the volume and concentration are such that the quantity of residual soil present on equipment does not seriously impair the sterilizing efficiency. Used solutions can deteriorate markedly in strength.

Hypochlorite solutions are corrosive to most metals, including stainless steel. However, at low concentrations (50–200 mg/L) of available chlorine, in alkaline (pH 8.0–10.5) solutions, at low temperature, and with short contact times, these corrosion hazards are minimized, and the sterilizing action remains effective.

Quaternary ammonium compounds can be safer to handle and less corrosive than hypochlorite, but under certain conditions at low concentration and low temperature their effectiveness can be less. For example, some *Pseudomonas* species can survive in quaternary ammonium preparations.

Iodine may be used in the form of iodophors, with effectiveness similar to that of quaternary ammonium compounds. However, iodophors have a tendency to stain rubber gaskets a brown color [12].

12.3.3   Special Cleaning Agents

Sometimes it is necessary to remove tightly bound proteins from organic surfaces, such as chromatography resins, which cannot themselves tolerate harsh cleaning agents such as sodium hydroxide [14]. In such cases it may be possible to use chaotropic agents, such as urea or guanidine hydrochloride, at high concentrations (in the region of 6 M) to release the protein. Protein A affinity chromatography columns are frequently cleaned with 6 M guanidine; the contaminating immunoglobulins and albumins are removed without damage to the protein A that is bound to the matrix [15].

12.4   Cleaning Methods

Traditional methods for cleaning processing equipment consist of stripping down the equipment and cleaning manually. This has the disadvantages of inconsistent cleaning, lack of safety (both for the operator and for the product), and excessive downtime before the equipment can be brought back into operation. As a result, most methods now consist of *cleaning in place* (*CIP*), in which cleaning reagents are circulated through the equipment. The cleaning cycle may be controlled manually or automatically. However, there still remain isolated instances for which this is not effective, and equipment must be cleaned by hand.

12.4.1   Pipework and Valves

A typical sequence of operations for cleaning pipework in place is given in Table 12.2. The initial water rinse is intended to remove loose soil from the pipe, and may be discharged directly to drain. The detergent removes the remaining soil from the surface. This material may be recycled to the detergent tank to conserve materials, as the solids present in the return flow do not significantly affect the efficiency of the detergent. Alternatively, the detergent may be discharged to drain. The intermediate water rinse may be discharged to drain, or recycled; if deionized water from a limited supply is used, recycling may be preferable. Sanitizer may also be recycled, but the final water rinse should always be discarded directly to drain to ensure that the process equipment is clear of all detergents or sanitizers once cleaning is complete.

Table 12.2 Sequence of Operations for Cleaning Pipework

Operation	Time	Temperature
Water rinse	5–10 min	ambient
Detergent wash	15–20 min	ambient to 75 °C
Water rinse	5–10 min	ambient
Sanitizer wash	15–20 min	ambient
Water rinse	5–10 min	ambient

Generally a fluid velocity of about 1.5 m/s is sufficient for satisfactory cleaning. Cleaning is often better at high Reynolds numbers, but in a comparison between two pipes at identical Reynolds numbers, the pipe with the higher velocity was cleaned more effectively, as measured by reduction in bioburden. Velocities higher than 1.5 m/s have little additional effect. Likewise, cleaning times in excess of 20 minutes bring little improvement [12]. Turbulent flow does not appear to necessarily improve cleaning effectiveness, compared with laminar flow, so long as the same velocity is maintained [16].

It is also important not to have too high a temperature during the detergent wash, because sugars can caramelize, lipids can polymerize, and proteins can be denatured, and the products of these reactions are more difficult to remove from the soiled surface. A temperature of about 75 °C is a practical maximum. Cleaning should begin immediately after the process is complete, because dried-on soil significantly reduces the efficiency of the cleaning procedure.

Once cleaning is complete, equipment not to be used immediately should preferably be drained and allowed to dry. This prevents stagnant water in the pipework from becoming a breeding ground for microorganisms.

12.4.2   Vessels

The obvious way to clean vessels, by filling them with detergent and leaving them to soak, is rarely effective, and, except for small vessels, extremely wasteful of materials.

The generally accepted method is to spray the various cleaning reagents through a spraying device in the top of the vessel. Cleaning therefore takes place through a combination of direct impingement of a fluid jet onto the soil and irrigation of reagents across the surface. Direct impingement is so effective at removing soil that the concentration of detergent can be lower than with other methods.

Two types of spraying device are used in vessels: the static spray ball and the rotating jet. Spray balls are inexpensive, simple, and effective. They have no moving parts and are virtually maintenance-free. They are self-draining and self-cleaning, and provide continuous surface coverage during cleaning. Spray balls are virtually unaffected if a single hole becomes blocked. The flow rates tend to be very high and pressures relatively low. They have a limited reach, so most parts of the surface are contacted by irrigation rather than impingement. However, each spray ball can be individually machined to direct jets at particular parts of the vessel where soil accumulates.

In contrast, rotating jets have a long radial throw at lower flow rate, with much higher impingement velocities than spray balls. Rotating jets can also overcome problems of shadowing more effectively than can spray balls. However, nozzle blockage can be serious, reliability is lower, maintenance requirements are higher, and they are more expensive. They are not always self-draining and self-cleaning.

A typical cleaning procedure for vessels is similar to that for pipework, and the same arguments for recycling cleaning reagents apply. Some sensors, such as pH and dissolved oxygen probes, are sensitive to aggressive detergents, and should be removed before cleaning.

Spraying systems need to be operated with care, to avoid some potential hazards. Manways should never be opened during cleaning with aggressive chemicals, to prevent burns to the head and arms. This danger can be forestalled by interlocks (such as microswitches) that trigger valve closure, or by more physical barriers such as interlocking keys.

Another potential hazard is that of the vessel collapsing during the cold water rinse after a hot detergent wash. Some nonsterile vessels may have a vacuum relief system, but fermentors and other sterile tanks should be stressed to accept full vacuum. All pumps should have emergency stop buttons.

12.4.3   Downstream Processing Equipment

Cell harvesting and other solids clarification methods often use continuous disk-stack centrifuges, of either the nozzle or desludging variety. These machines are easy to clean quite successfully, if they have not been used to separate sticky biomass. A sticky biomass sludge tends to collect around the sludge outlet and the collection bowl, where, because of the dimensions, detergent has trouble wetting all surfaces, and certainly does not reach them if fluid velocity is low. Frequently, the only remedy is to strip down the components and clean manually. In many situations it is possible to CIP quite successfully as long as frequent desludging operations are performed. However, this consumes large quantities of detergent.

Cross-flow microfiltration or ultrafiltration systems can usually be cleaned by the conventional CIP methods used for vessels and pipework. However, with long processing times, a persistent gel layer of large molecules can build up on the membrane surface, and it appears that these molecules can insinuate themselves into the pores of the membrane. In this situation it is necessary to cycle through a number of detergent washes and water rinses, finishing off with several water rinses to adequately purge all remaining detergent. Depending on the characteristics of the membrane, a combination of forward and reverse flushing can be effective at removing soil from first one end of the membrane and then the other.

If it is appropriate to backflush the membrane, cleaning reagents are pumped from the permeate side to the retentate side to remove soil from the pores. However, the effectiveness of backflushing depends on the ability of the membrane to withstand reverse transmembrane pressures high enough to achieve the necessary cleaning velocity. Some membranes cannot withstand aggressive chemicals or cleaning temperatures, although the newer varieties available are more robust. In this respect, ceramic cartridges are extremely resistant to cleaning conditions, but this does not necessarily guarantee their processing performance.

The cleaning of chromatography columns can create special difficulties. The silica-based packings commonly found in HPLC systems are generally sensitive to high pH, and thus do not tolerate highly alkaline detergents such as sodium hydroxide [14]. In these cases, alternatives such as sodium orthosilicate have to be employed. Fortunately, HPLC systems tolerate the high pressures and velocities needed for cleaning.

Chromatography systems that use softer matrices and operate at low pressures and velocities can generally withstand sodium hydroxide, but only at low flow rates, so cleaning time tends to be long. Rinsing off the detergents is therefore also a much more time-consuming operation than for robust systems. In some circumstances where these restrictions prevent adequate cleaning, it may be necessary for operators to dismantle the column and clean the matrix manually, mixing it with detergent in a separate vessel.

The aspect ratio of columns also affects their cleanability. Long thin columns are cleaned up more effectively than short, wide-diameter columns.

Most matrices cannot withstand high cleaning temperatures. However, the soil to be removed from them is rarely solid material; rather, it tends to consist of soluble product and impurities adsorbed onto the surface of the matrix and in solution between the matrix particles.

If therapeutic products that include proteins are purified in a multiproduct facility, it is difficult to guarantee that all traces of the preceding product have been completely cleared from the column, especially if that product is highly active. In this situation it is best to dedicate a column to each product; the columns still need regular cleaning, of course. Some practitioners also use this strategy with ultrafiltration membranes.

12.4.4   Ancillary Equipment

The cleaning of ancillary equipment such as pumps, filters, and heat exchangers is generally quite straightforward. There are, however, occasional problems of which to be aware.

Air sterilizing filters can occasionally become soiled with foam from the fermentor, which subsequently dries on the surface of the housing as air flows through. Normal cleaning flow rates correspond to very low velocities in a filter housing, so the dried-on soil is difficult to remove. Manual cleaning may be necessary. Liquid filters with accumulated solid deposits may also suffer from the low-velocity problem.

Where medium is sterilized in continuous heat exchangers of either the plate or spiral variety, there is a constant risk of medium burning onto the heating surface, especially if the correct flow rates are not observed. In such heat exchangers there are also a number of contact points whose purpose is to maintain the correct distance between the heat exchanger surfaces. These can collect solid particles from the medium, again if incorrect flow rates are used. It is extremely difficult to remove this type of burnt-on soil; prevention is preferable (cure may entail stripping down the heat exchanger and removing the soil by hand).

12.4.5   Depyrogenation

Pyrogens and endotoxins cannot be tolerated in pharmaceutical applications of biotechnology. Pyrogens are generally held to be the outermost cell wall layer of Gram-negative bacteria, although streptococcal exotoxins and staphylococcal enterotoxins can also produce a pyrogenic response (a temperature rise) in rabbits. They may come from either air or water, although water is the more common source.

Pyrogens are quite difficult to remove from a product; it is far better to ensure that a plant remains free of pyrogens from the start. Therefore, it is important to depyrogenate the equipment before raw materials are admitted.

After normal cleaning, the equipment must be fully flushed with a solution of 0.1 M sodium hydroxide made up in pyrogen-free water, and left to soak for 30 minutes. The sodium hydroxide is then drained and the equipment is thoroughly flushed with pyrogen-free water [17].

If laboratory glassware is to be used to prepare process additives, it may be depyrogenated in special hot air ovens at 180 °C for 3 hours. Besides the initial depyrogenation, all process materials should be prepared with pyrogen-free water and clean steam.

12.4.6 Facility Cleaning

The traditional methods of cleaning facilities (with mop and bucket) are still frequently followed. However, innovations include mobile wet vacuum cleaners, piped vacuum systems, and long-reach spray lances for inaccessible areas.

Antibacterial, antifungal, antiviral detergents, such as various Tego® products, are also now available; they may be used on virtually all hard surfaces.

Occasionally a gross contamination occurs that requires a complete room to be decontaminated. This necessitates sealing off the inlet and outlet ventilation and all doors, windows, and ventilation systems, then fumigating the area with a fumigant such as formaldehyde. This should be an infrequent occurrence if normal hygienic practices are correctly followed.

12.5   CIP Systems

CIP systems can take various forms. Traditionally, single-use systems were used: the detergent was used once and then discarded. With the development of more complex process plants, there are advantages to recovering the detergents for reuse if there is no risk of cross contamination. It is also possible to install multi-use systems, which combine features of single-use and reuse systems [12]. A recent study suggests that a reuse system employing chemical sterilants rather than hot water results in the lowest operating costs, if chemicals, utilities, energy, labor, and depreciation are all considered; however, these relative costs change from region to region and should not be relied upon [18].

Single-use systems are appropriate if the desired detergents have a short shelf life and deteriorate rapidly, or if a plant contains a high level of soil loading that grossly contaminates detergents and renders them unfit for reuse. Single-use systems are typically small local units, situated adjacent to the individual process plant; there may be several such installations on site. With short pipework runs, the result is reduced holdup and detergent losses.

A typical single-use system is illustrated in Figure 12.4. It consists of a tank with level probes and water inlet, a centrifugal pump for distributing the cleaning reagents around a CIP feed and return system, and an injection point, through which steam for heating is admitted and CIP additives are injected by diaphragm metering pumps.

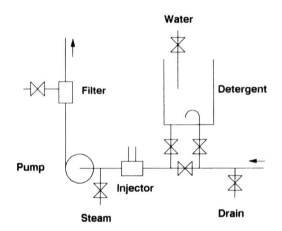

Figure 12.4 Single-Use CIP System. The simple single-use CIP system consists of a single tank to contain the cleaning reagents, which are made up as required. The detergents may be supplied from a separate preparation tank. Such a system is wasteful of both chemicals and heat energy, but is relatively inexpensive to install.

If a bioprocess plant is running on a single product or on a campaign basis, it may be possible to reuse the detergent, to save raw material and to minimize the discharge of waste detergents to drain. Typically, cleaning prerinsed equipment does not pollute detergents very much, so they can be reused until a gross contamination threshold is crossed.

Detergents can be wasted if operation is manual instead of automatic, if self-desludging centrifuges are cleaned, and if detergents are dosed in excess into contaminated solutions.

Since recovered detergents are still hot, they may be stored in insulated tanks and for a short time need little reheating to raise them to operating temperature for another cycle.

A typical reuse system is illustrated in Figure 12.5. It consists of a separate tank for each cleaning reagent, and a water recovery tank if required. Water can be conserved if the recovered final rinse water from one cycle becomes the prerinse water for the next cycle. An in-tank coil heats the detergent, and the fluids are distributed by feed and return pumps and piping. The detergent concentrate is metered from a central reservoir to ensure the correct concentration of cleaning reagent in each tank. Neutralizing tanks, with acid metering pumps to neutralize effluents, can be provided.

Figure 12.5 Reuse CIP System. Reuse systems are most common where fouling is light; cleaning reagents can be reused only if there is no fear of cross contamination. Clearly these recovered reagents cannot be used after accumulated soil reaches unacceptable levels. These systems are relatively economical to operate.

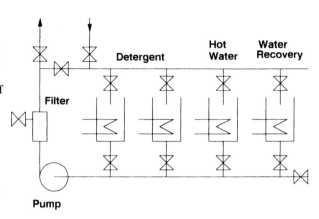

Among recent developments are the combined systems, termed multi-use because they combine features of both recovery and single-use systems. These units are designed for CIP of tanks and pipelines. They work with automatically controlled programs, and consist of different combinations of cleaning sequences in which cleaning reagents are circulated through the circuits for varying lengths of time at varying temperature.

Figure 12.6 shows the simplified flow diagram of a typical multi-use system. It consists of tanks for chemical and water recovery, an associated single pump, recirculatory pipework, and a heat exchanger, mounted on adjacent subframes.

Figure 12.6 Multi-Use CIP System. Multi-use systems are a compromise, intended to provide some recovery of energy and materials at an acceptable capital cost. Typically, the chemical solutions are recovered by the local CIP system and may be sent back to a central recovery tank for reuse as required, with resultant savings in pipework and storage tanks.

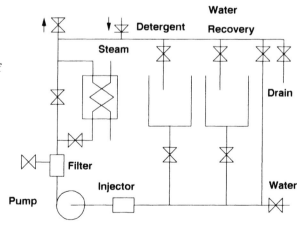

The prerinse water comes either from the recovery tank or from the mains, and can be either directed to drain or diverted for a timed period, if any intermediate washes are required, and then sent to drain. During washing, the detergent can circulate through either the detergent tank or the bypass loop. Recirculation times are variable, and any combination of chemicals can be injected as needed. Temperature can be controlled as required; either the whole system, including the tank, may be heated, or just the loop may be heated, with the tank remaining at ambient. Chemicals can either be recovered or sent to drain. Times and temperatures for the final water rinse are variable, and the rinse water may be discharged to drain or recovered.

CIP systems are ideally suited to automation. The entire process and CIP operations are automatically controlled, and electromechanical interlocks render an error in valve operation virtually impossible. Automation can reduce operator error, cleaning time between batches, and waste of cleaning reagents.

12.6   Validation of Cleaning

Like every process over which there is some regulatory control, the cleaning procedures need to be validated to demonstrate their efficacy for maintaining a hygienic plant and to prevent cross contamination. The validation of cleaning systems and methods generally consists of Installation Qualification, Operational Qualification, and Performance Qualification.

Performance Qualification demands that the equipment actually perform the task for which it was designed, either individually or as a complete system. For a CIP system this means that it should be capable of removing a worst-case soil from the process plant both efficiently and repeatably. Tests are usually carried out in triplicate, and the equipment should meet specifications each time.

The methods of validation for the installation and operational phases are common to other processes, but those for the performance qualification must necessarily be relatively specific. To validate a particular cleaning operation, the cleaning sequence must be performed in the manner that would normally be used (either manual or automatic), to determine whether the operation passes certain predefined criteria of success. This involves depositing a worst-case soil in a variety of locations around the plant, running the cleaning sequence, and then analyzing those areas of soil to determine how much of the various components remains.

The following are criteria for a clean surface:
(a)   It must be free from residual film or soil;
(b)   No contamination is visible under good light, with the surface wet or dry;
(c)   No objectionable odor is apparent;
(d)   The surface does not feel rough or greasy to clean fingers rubbed on it;
(e)   A new white paper tissue wiped over the surface picks up no discoloration;
(f)   The surface shows no sign of excessive water break as water drains from it;
(g)   No fluorescence is detectable when the surface is inspected with long wavelength (340–380 nm) ultraviolet light.

However, since many of these tests are somewhat subjective, it may be necessary to perform some more quantitative tests on the major components of the soil. Most common tests are for residual protein and microbial bioburden. Examples of assays for these are given in Tables 12.3 and 12.4. More specific assays may be developed to detect individual proteins, such as the immunoperoxidase test for whey proteins [19]. More quantitative tests for the amount of protein adhering to a surface may be conducted using equilibrium contact angle analysis [20] or ellipsometry [21].

Table 12.3 Assay to Validate Cleaning of Protein [26]

1.	Prepare a sample surface (perhaps a sensor port blank) with a standard quantity of a protein solution and allow to dry;
2.	Perform the cleaning operation;
3.	Remove the surface from the vessel or pipe and shake off excess water;
4.	Test for residual protein by impressing against a sheet of nitrocellulose that adsorbs protein;
5.	Stain the nitrocellulose with Coomassie Blue and leave overnight in acetic acid to remove excess stain. A blue color remaining on the nitrocellulose indicates the presence of protein on the sample. A control should always be run to ensure that the test is valid.

Table 12.4 Bioburden Assay to Validate Cleaning [7]

1.	Load a sample surface with known organisms and allow to dry;
2.	Perform the cleaning operation;
3.	Remove the surface from the vessel or pipe and shake off excess water;
4.	Swab the surface and culture the swab in petri dishes for 24–48 hours;
5.	Count the colonies on the plate.

Another way to measure bioburden is to mix a known number of test organisms, such as *Bacillus stearothermophilus*, with a typical soil, apply it to a surface, and conduct a cleaning operation. The organisms remaining (if any) are counted after the surface has been coated with a nutrient agar medium and incubated. A comparison of the final and initial counts gives a measure of the efficiency of cleaning [22]. Tests developed more recently include direct epifluorescent fluorometry [23] and ATP bioluminescence [24].

Pyrogen testing is also necessary, to verify that depyrogenation is effective. Tests for pyrogens have traditionally involved injecting a sample into rabbits and measuring their rise in temperature, if any. This may be compared to a standard curve of known pyrogen concentration versus temperature rise, and the level of pyrogens calculated. More recently the *Limulus amoebocyte lysate* (*LAL*) test has become available; it detects endotoxins, a common source of pyrogenic activity [25], by exploiting the ability of an aqueous extract of blood cells (amoebocytes) from the horseshoe crab (*Limulus polyphemus*) to form a firm clot when incubated with endotoxins at 37 °C and pH 7.0. The degree of clotting may be read by a spectrophotometer at 360 nm. Typically, the LAL test can detect endotoxin down to approximately 10^{-10} g/mL. Firm gelation can require between 0.075 and 0.3 ng/mL, depending on the source of the pyrogen. Interference in the assay can be caused by inadequately depyrogenated glassware, SDS, urea, EDTA, heparin, and penicillin G, so care is required in the preparation of the sample.

Lastly, it may be desirable to demonstrate the effectiveness of the final rinsing stage and to test for residual detergent. The most common test is to apply a drop of phenolphthalein to the sample surface and look for the color change to purple that indicates the presence of sodium hydroxide.

12.7   Conclusions

Cleaning is a vital part of any biochemical manufacturing process. It improves the performance of the plant and satisfies regulatory requirements. The design of a bioprocess plant should incorporate provisions for hygiene at a fundamental level, and plant function should include those cleaning operations essential for the maintenance of hygiene.

Modern CIP systems can greatly assist in the running of the plant, by reducing downtime and ensuring quality and consistency. CIP systems can be specified to match the requirements of almost any installation.

The validation of cleaning systems and their operation is frequently necessary, and the analysis methods are usually simple.

12.8   References

1. *Hygienic Design of Food Plant*; Joint Technical Committee of Food Manufacturers' Federation and Food Machinery Association; London, UK, 1967.
2. TROLLER, J. A.; *Sanitation in Food Processing*; Academic Press, London, UK, 1983.
3. CARVELL, J. P., Sterility and Containment Considerations in Valve Selection, *Pharm. Eng.* *12*, 1 (1992).
4. GOODFELLOW, H. R.; *Tubing Systems for Pharmaceutical and Biotechnological Applications*; presented at Bioprocess Systems Design Conference; Institution of Chemical Engineers, Rugby, Warwickshire, UK, 1992.
5. MILLEDGE, J. J.; Fundamentals and Applications of Surface Phenomena Associated with Fouling and Cleaning; In: *Food Processing*; HALLSTRÖM, B., LUND, D. B., TRAGARDH, C., Eds.; Lund University Press, Lund, Sweden, 1981; pp 152–170.
6. INTERNATIONAL ASSOCIATION OF MILK, FOOD AND ENVIRONMENTAL SANITARIANS, INC.; 3-A Accepted Practices for Permanently Installed Product and Solution Pipelines and

Cleaning Systems Used in Milk and Milk Product Processing Plants, *Dairy Food Environ. Sanit. 9*, 416 (1989).

7. TIMPERLEY, D. A., LAWSON, G. B.; Test Rigs for Evaluation of Hygiene in Plant Design; In: *Hygienic Design and Operation of Food Plant*; JOWETT, R., Ed.; Ellis Horwood, Chichester, UK, 1980; pp 79–108.

8. *Recommendations For the Sterilization of Plant and Equipment Used in the Dairying Industry*; BS5305; British Standards Institute, London, 1976.

9. VAN SCHOTHORST, M., Hazard Analysis in Hygienic Engineering, *Food Control 1*, 133 (1990).

10. GRAHAM, D. J., Sanitary Design—A Mind Set (Part V), *Dairy Food Environ. Sanit. 11*, 669 (1991).

11. *Guide to Good Manufacturing Practice*; HMSO, London, 1983.

12. TAMPLIN, T. C.; CIP Technology, Detergents and Sanitisers; In: *Hygienic Design and Operation of Food Plant*; JOWITT, R., Ed.; Ellis Horwood, Chichester, UK, 1980; pp 183–226.

13. SMITH, K. E., BRADLEY, R. K., Evaluation of Efficacy of Four Commercial Enzyme-Based Cleaners of Ultrafiltration Systems, *J. Dairy Sci. 70*, 1168 (1987).

14. PRENETA, A.; Separation on the Basis of Size: Gel Permeation Chromatography; In: *Protein Purification Methods*; HARRIS, E., ANGAL, S., Eds.; IRL Press, Oxford, UK, 1989; pp 293–306.

15. ANGAL, S., DEAN, P.; Use of Affinity Adsorbents; In: *Protein Purification Methods*; HARRIS, E., ANGAL, S., Eds.; IRL Press, Oxford, UK, 1989; pp 245–261.

16. BERGMAN, B. O., TRAGHARDH, C., An Approach to Study and Model the Hydrodynamic Cleaning Effect, *J. Food Process Eng. 13*, 135 (1990).

17. ROE, S.; Separation Based on Structure; In: *Protein Purification Methods*; HARRIS, E., ANGAL, S., Eds.; IRL Press, Oxford, UK, 1989; pp 175–244.

18. MARTENSSON, I., The Hidden Costs in a Cleaning System, *Dairy Ind. Int. 55*, 47 (1990).

19. PARTRIDGE, J. A., FURTADO, M. M., Immunoperoxidase Detection of Whey Protein Soils on Ultrafiltration Membranes, *J. Food Prot. 53*, 484 (1990).

20. YANG, J., MCGUIRE, J., KOLBE, E., Using the Equilibrium Contact Angle as an Index of Contact Surface Cleanliness, *J. Food Prot. 54*, 879 (1991).

21. MCGUIRE, J., AL-MALAH, K., BODYFELT, F. W., GAMROTH, M. J., Application of Ellipsometry to Evaluate Surface Cleaning Effectiveness, *J. Food Sci. 55*, 1749 (1990).

22. LECLERQ-PERLAT, M. N., LALANDE, J. M., TISSIER, J. P., A Method for Assessment of the Cleanability of Equipment in the Food Processing Industry, *Sci. Aliments 10*, 17 (1990).

23. PONTEFRACT, R. D., Bacterial Adherence: Its Consequences in Food Processing, *Can. Inst. Food Sci. Technol. J. 24*, 113 (1991).

24. MACKINTOSH, R. D., Hygienic Monitoring in the Food Processing Industry, *Eur. Food Drink Rev. 55*, 8 (1990).

25. SULLIVAN, J. D., VALOIS, J. W., WATSON, S. W.; Endotoxins: the *Limulus* Amoebocyte Lysate System; In: *Mechanisms in Bacterial Toxicology*; BERNHEIMER, A. W., Ed.; John Wiley & Sons, New York, 1976; pp 383–410.

26. DUNN, M. J.; Determination of Total Protein Concentration; In: *Protein Purification Methods*; HARRIS, E., ANGAL, S., Eds.; IRL Press, Oxford, UK, 1989; pp 10–17.

Sterilization of Process Equipment

Tim Oakley

Contents

13.1   Introduction

The importance of proper sterilization of bioprocess equipment and solutions has long been known [1]. Without the ability to sterilize process equipment and solutions, few of the major advances in biochemical engineering would have been possible. It is important to stress that the term *sterile* is an absolute one; a piece of equipment either is sterile or it is not. With mammalian cell culture fermentations lasting up to several weeks, the presence of even one contaminating microorganism can be disastrous [2].

At present, by far the most common technique for sterilization of large-scale process equipment is heat sterilization using steam, and it is on this method of sterilization that this chapter focuses. Heat sterilization has to kill both live microorganisms and their spores. Of the two, the spores have a thermal resistance to moist heat several thousand times greater than that of live microorganisms [3]. The resistance of bacterial spores to dry heat is even greater; the use of steam for heat sterilization is therefore not only convenient but also more effective, as all surfaces are in contact with hot condensate at steam temperature.

The primary challenge with steam *sterilization in place* (*SIP*) as a process is that it must be designed into the equipment. Processes that have been developed using autoclaved equipment may have to be radically changed if they are to be scaled up and steam sterilized in place. The necessity of being able to SIP every component (vessel, filter, transfer line, etc.) means that the ability to do so must be considered from the very first.

Many of the books on biotechnology and biochemical engineering make passing reference to this, but few [4] actually show how to arrange a process flow sheet for steam sterilization. It is on this aspect that I concentrate in this chapter: the production of flow sheets for steam sterilization in place.

For successful steam sterilizing of process equipment, the time and temperature of the sterilization process are very important. The Medical Research Council (MRC) Working Party [5] proposed the values for saturated steam in Table 13.1, and PERKINS [6] recommends the values in Table 13.2:

Table 13.1

T, °C	t, min
121	15
126	10
134	3

Table 13.2

T, °C	t, min
116	30
118	18
121	12
125	8
132	2

The usual time/temperature combination chosen is 121 °C/15 min. In practice these values are taken as a minimum and the time period of steaming is often extended to ensure a good margin of safety.

The margin of safety chosen depends on the piece of equipment being sterilized. For a simple short pipe section a time of 30 minutes is adequate, for small vessels 45 minutes, and for large complex items 1 hour is sufficient. These times are drawn from experience, but have a basis in logic. The larger an item is, the greater its thermal capacity and the longer it takes for the heat to be transferred into all its extremities.

The steam pressure required for *in situ* sterilization can be obtained from Table 13.3, which refers to saturated steam:

Table 13.3

P, bar	T, °C
1.0	120.4
1.2	123.5
1.4	126.3
1.6	128.9
1.8	131.4
2.0	133.7

The value usually chosen is around 1.5 bar (22 psi) gauge pressure, which gives a good margin of safety and ensures that the equipment being steamed rapidly attains sterilizing temperature.

The steam supply must be saturated at 1.5 barg (1.5 bar, gauge pressure). In practice, this means that it must be adequately trapped and drained. Also, to avoid superheated steam, the take-off point should not be too close to the steam pressure reducer. If the steam is superheated, the conditions inside the equipment may well be those of dry heat.

For mammalian cell culture, the steam supply should be from a pure steam generator, distributed along stainless steel pipework. Plant steam has many impurities, which may either contaminate the product or inhibit growth of the organism being cultured. For large-scale antibiotic production, plant steam containing FDA–approved boiler additives is commonly used.

As previously mentioned, equipment must be designed for steam sterilization from the start. In order to do this there are several cardinal rules that should be followed:

(a) Be sure that all parts of the equipment are able to withstand sterilizing temperatures of up to 130 °C.

(b) Where possible, use welded connections. Any joint is a potential weakness.

(c)   Where joints are necessary, use sanitary fittings (for example, IDF, ILC, Tri-clamp®). *IDF* (*International Dairy Federation* BS 4825, sometimes known as *ISS* or *International Sanitary Standard*) is very common in the UK and Europe; *ILC* (*In Line Cleaning*) is less common, but is available in ½" and ¾" sizes, which IDF isn't. Tri-clamp® is readily available in the U.S.A.

(d)   Avoid dead spaces, crevices, and the like. Any pipe dead leg, if unavoidable, should be shorter than six times the pipe diameter. If dead legs do exist, they should be steamed through using a valve with a steam supply or trap beyond.

(e)   If possible, avoid having only one valve between sterile and nonsterile areas. Bacteria can grow through previously sterilized valve seats. On transfer lines that are sterilized and used immediately, this is not so important.

(f)   Use only valves that are easy to clean, maintain, and sterilize. The biotechnology industry typically uses diaphragm valves for sterile service.

(g)   Design the equipment so that it can be sterilized in sections, and so that each part of the section has a unique direction of steaming.

(h)   Be sure the steam supply is dry and saturated and free from particles and gases.

(i)   Introduce the steam at the highest point(s) and remove condensate at the lowest point(s).

(j)   Design lines for complete drainage, without pockets where condensate can accumulate. This requires sloping of lines and installation of extra traps.

The sterilization of media is well documented elsewhere [1, 4], so in this chapter I do not deal with that aspect of sterilization. I assume in the following pages that empty equipment is being sterilized by direct steam injection, and medium is then introduced in an aseptic manner.

13.2   Equipment Sterilization

13.2.1   Vessels

There are several prerequisites in the design of a vessel for steam sterilization. The first is that the vessel must be able to withstand the required steam pressure of 1.5 bar. The vessel must therefore be designed to an appropriate pressure vessel code and fitted with appropriate relief devices for safety. The vessel jacket must also be designed as a pressure vessel to at least the same standard as the vessel itself.

Glass vessels are not generally desirable for *in situ* steam sterilization because damaged, chipped, or scratched glass may result in an explosion. Small glass vessels can be autoclaved, but if a vessel is too big for autoclaving it should not be glass. If, however, it must be glass for a particular reason, it can be protected either by one of the plastic films available or by being shielded.

The vessel jacket of an empty stainless steel vessel being sterilized by *in situ* steaming should be empty, so provision must be made for draining and venting of the jacket. If this is not done, the vessel takes longer to get up to temperature, and may have cold spots.

The vessel must be pressure-hold tested with air before use, to detect minor leaks or pinholes. Normally a 24-hour test should be performed after installation and after any mechanical work done on the vessel. The vessel should hold the applied pressure for 24 hours after the temperature has stabilized. This cannot be done before every sterilization, so normally a ½-hour test is done to detect gross leaks due to incorrect fitting of probes or filters, or valves left open. It is highly desirable to extend this ½ hour as much as possible. For the pressure hold test, the vessel must be fitted with a sanitary-type pressure gauge capable of being steam sterilized.

Leak detection is a rapidly developing technology. The simplest and cheapest method is to use a soap solution, which as a technique still has much to recommend it. However, there are now several new techniques available that look promising. The most appropriate one is probably the hand-held ultrasonic detector, which can detect the sound of air escaping at high velocity through tiny leaks.

The process of steam sterilization is very simple. The equipment to be sterilized is first leak-tested. It is set up for the sterilization when all the condensate outlet valves are opened, and lastly the steam inlet valves are opened. As the steam goes into the equipment, the internal temperature and pressure rise. At this point, filter bleed valves should be opened to vent air. Since almost all vessels are fitted with exhaust filters, this is also a good way of venting air from the vessel. Vacuum cycling is an alternative method of air removal in large or complex pieces of equipment. When the temperature has reached 121 °C and the pressure has reached 1.1 barg, the timing of sterilization can begin. The condensate is typically eliminated using steam traps fitted with near-to-steam elements.

At the end of the sterilization, the drain valves are shut first, and then the steam inlet valves. When the pressure inside the vessel has dropped to 1 barg (with all valves closed), sterile air at a pressure of 1 barg should be introduced through the air inlet filter. This ensures that as the vessel cools it does not develop a vacuum and hence pull in contaminants or damage itself. It also allows a slow replacement of the condensing steam with sterile air at a higher than ambient pressure as the vessel cools. In this way the risk of poststerilization contamination is greatly reduced.

The usual piping arrangement for steam sterilization of any piece of equipment is simple: steam in at the top, condensate out at the bottom. However, there are usually many connections to vessels, which need careful consideration when steam sterilization is designed (these include spargers, dip pipes, side inlets, spray balls, air filters, and secondary outlets).

Spargers should have a valve arrangement like that shown in Figure 13.1. The sparger should have drain holes as well as the normal upward-facing holes. When the

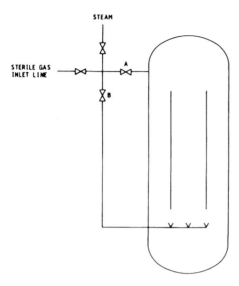

Figure 13.1 Valve Configuration for
Steaming a Vessel Sparger. Initially valve
A is closed, valve B open. When the
pressure in the vessel reaches approximately one bar, valve A is opened.

vessel is steamed, the sparger leg is steamed first, until the vessel pressure reaches
about 0.5 barg. The direct horizontal line is then steamed as well. Both valves A and
B (Figure 13.1) then stay open for the duration of the sterilization. Dip pipes, as
illustrated in Figure 13.2, are often used for filling vessels when splashing is to be
avoided. They should be steamed downwards, and a steam supply needs to be
available both inside and outside the pipe. It should be possible to feed steam into the
vessel directly, as well as down the dip pipe. Both steam valves should be open
throughout the sterilization.

Side inlets can be steam sterilized in one of two ways, depending on whether the
side inlet goes up or down in the vessel (Figure 13.3). The difference depends on how
the side inlet line is sterilized apart from the vessel. In the "up" arrangement, valves
A and B are open during steaming and valve C is shut. Steam then exits through A
and B to the trap, which opens and shuts as condensate contacts it. When the line
itself is steamed, A and C are open, B is shut, and the trap functions as a condensate
drain for the line.

In the "down" arrangement, as the vessel is steamed, valves D and F are open, E
is shut, and the side inlet acts as another steam supply point into the vessel. When the
line itself is steamed, F acts as a steam supply for the line.

Spare ports on vessels should be avoided. A new pressure vessel should be
designed with just the right number of ports for the function required. If the vessel
is second-hand, every port into or out of the vessel should have either a valved steam
supply or a valve and trap arrangement. This is because entry ports are often longer
than six times their diameter, which is the maximum allowed for dead legs.

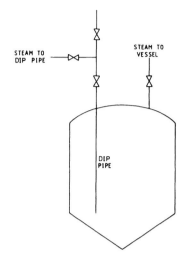

Figure 13.2 Valve Configuration for Steaming a Vessel Dip Pipe. Throughout steaming, both steam supply valves are open.

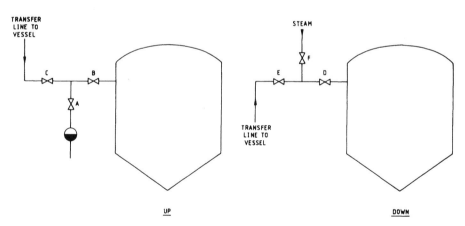

Figure 13.3   Valve Configurations for Steaming Side Inlets to Vessels.   In the "up" configuration, when the vessel is steamed, valves A and B are open, and valve C is closed. The transfer line is then steamed with valves A and C open, B closed. In the "down" configuration, valves F and D are open, E closed during vessel sterilization. Then valve F acts as a steam supply for sterilization of the transfer line.

The vessel outlet should be at the very lowest point of the vessel, otherwise it does not drain properly, which prevents proper cleaning and sterilization. As most vessel outlets need to serve a triple purpose, the detailed design of the valve cluster underneath a vessel is very important. The vessel outlet serves as a product outlet, a condensate outlet during sterilization, and a drain during cleaning in place (CIP). The vessel outlet should be sized to meet drainage requirements during cleaning, not sterilization requirements.

A good valve layout is shown in Figure 13.4. During sterilization of the vessel, valves A, C, and F are open, and B, D, and E are shut. During cleaning of the vessel, valves A, C, and E are open, and B, D, and F are shut. This valve layout allows the vessel to be sterilized and filled (from the top) with sterile liquid. At this point there are at least two valves between the sterile vessel contents and the outside world, which is good practice. The area between valves A, B, and C has also been steamed, so that bacteria have to come through two valve seats to get to the vessel. It may appear that the arrangement shown in Figure 13.5 is equally satisfactory, and it saves a valve. However, in this situation, the valve on the drain/CIP return line forms a dead leg and there could be contamination left under the valve seat. This means that the line from the vessel to elsewhere may not be satisfactorily sterilized.

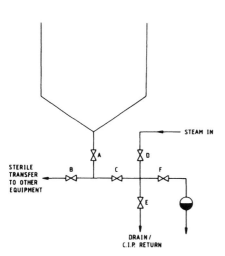

Figure 13.4 A Recommended Vessel Outlet Valve Configuration for Vessel Sterilization, CIP, and Sterile Transfer.

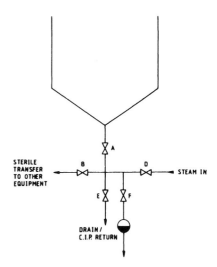

Figure 13.5 A Poor Configuration for Combined Vessel Sterilization, CIP, and Sterile Transfer that is not recommended, due to the likelihood of dead legs and cold spots.

One arrangement that is nearly as good as that shown in Figure 13.4 is shown in Figure 13.6. While the drain valve, E, does not get completely steamed through, the presence of the condensate valve stabbed into the side of the valve (this can be done to most diaphragm valves) means that the whole of valve E does in fact get up to

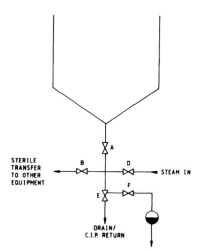

Figure 13.6 An Acceptable Alternative to the
Configuration Shown in Figure 13.4 for combined
vessel sterilization, CIP, and sterile transfer. This
minimizes dead legs and hence possible cold
spots.

temperature, due to the steam going through the upper half of it. This arrangement
can be recommended where space demands preclude the bulkier arrangement in Figure
13.4.

Most vessels and fermentors are now designed so that *CIP* (*cleaning-in-place*)
methods can be used. This invariably means that some sort of spray ball or nozzle is
built into the top of the fermentor or vessel. In installations where this spray ball is
fixed, it must also be sterilized with the vessel. In practice this means a supply of
steam must pass through the spray ball during the sterilization cycle. On smaller,
more manual systems, it is sometimes better to arrange the CIP spray ball to be
removable; it is put in position only for the cleaning cycle.

The ideal valve arrangement is shown in Figure 13.7. During sterilization, valves
B and C are open and A is kept closed, so the whole of the spray ball assembly is
sterilized. Note that the CIP line normally also has a flow plate or similar arrangement to positively disconnect the pressurized CIP line downstream of valve A.

The arrangement shown in Figure 13.8 is simpler, but should not be used if the
vessel needs to remain sterile for more than six or seven days. This is due to the
possibility of bacterial contamination across a single valve.

Agitator seals are a weak link on any sterile vessel. Fermentor manufacturers all
have their own systems for ensuring that the agitator seal remains lubricated and
sterile. These are usually quite complex. However, there are many occasions when
a project engineer has to install a sterile vessel with agitators and devise some simple
method of maintaining seal lubrication and vessel sterility.

Most agitator seals now are double mechanical seals. The simplest method for
sterilizing them is to put live pure steam into the seal chamber, with a steam trap on

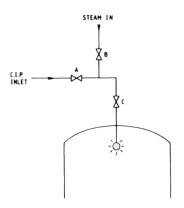

Figure 13.7 Ideal Valve Configuration for Steaming a Fixed CIP Spray Ball.

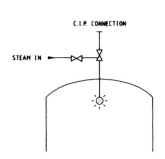

Figure 13.8 Acceptable Alternative to Valve Configuration Shown in Figure 13.7 for steaming a fixed CIP spray ball.

the downstream side. This is effective, but crude. The whole of the agitator tends to run at 121 °C, which is itself a safety concern; mechanically, all components have to be specified to run at this temperature, in particular, bearings and plastics. While effective, this method cannot be generally recommended.

The second method is nearly as simple as the first, with one added valve downstream of the seal assembly. During vessel sterilization this valve is opened, so the seal and chamber are sterilized with the vessel. Afterwards this valve is closed, and the seal chamber fills with cold sterile condensate at 1.5 barg. This approach is equally effective, and has none of drawbacks of the first method. I recommend it for keeping seals sterile for several days; I have no experience with longer periods, although it is a simple matter to resterilize the seal every few days. The drawbacks to this method are that it needs a constant uninterrupted steam supply to maintain the pressure in the seal chamber, and that it is not recommended by seal manufacturers to operate the vessel at a pressure above that of the steam supply.

The third method (shown in Figure 13.9) is much more complex, but is capable of maintaining sterility for very long periods (up to 40 days), and the vessel pressure is limited only by the available air pressure. The air filter and seal are steam sterilized at the start of the run, and then air pressure is used to maintain the pressure behind the sterile condensate in the seal. The sight glass is used to check the condensate level during operation, and more condensate can be produced as required.

The use of magnetically coupled stirrers is becoming more common on laboratory equipment. They have much to recommend them, but may have some drawbacks. The main one is their tendency to shed particles into the process fluid, especially if Teflon® bushings are used.

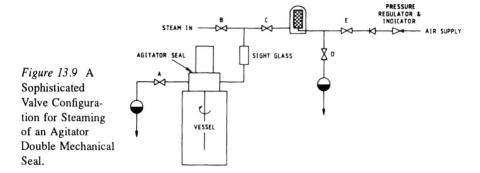

Figure 13.9 A Sophisticated Valve Configuration for Steaming of an Agitator Double Mechanical Seal.

The overall approach to steam sterilization of vessels should incorporate steaming into all additional ports and out through the exhaust filters. For the sterilization of vessels full of medium, such as in bacterial fermentation, the protocol must be modified to take into account the liquid contents.

13.2.2   Air Filters

Air sterilizing filters are designed to be installed on the air inlet and vent ports of sterile vessels, with the sterile side on the inside of the filter element. This is the correct way, as then the filter housing joint and vent and bleed valves are all on the nonsterile side of the filter element. However, if the filter membranes are designed to be bidirectional, the filter housing can be installed and used either way.

Most modern air sterilizing filters are of the membrane type. These membranes are hydrophobic: they repel and shed water. This characteristic means that their steam sterilization requires care. If during the sterilization process they become wet (*blind*), steam cannot pass through the membrane, and the filter membrane and downstream piping are not sterilized satisfactorily. Therefore, the filter housing and the pipework to and from it must be arranged for satisfactory draining of the condensate, and the steam supply to the filter must be dry.

The other factor that must be taken into account is whether the air filter is to be sterilized with the vessel or separately. If the air filter is likely to need changing during a fermentation, it must be able to be sterilized separately from the fermentor. Figure 13.10 shows a simple air filter arrangement for sterilizing with the vessel. Note that the air filter housing is installed the wrong way around, which means that the joint of the filter housing and the condensate valve are both on the sterile side of the filter; if there is a leak at either of these points the vessel may get contaminated. However, the simplicity of this configuration has the advantage that the filter can be sterilized with the vessel.

It is very important to have a steam trap or vent on both sides of the filter membrane. This ensures that condensate is removed as effectively as possible. It is a mistake to combine the steam traps (Figure 13.11). If this is done, the steam short circuits the filter, through both open valves, since there is no driving force to push steam through the membrane, and hence effective sterilization cannot be guaranteed.

One might conclude after an inspection of Figure 13.10 that there are four possible arrangements of filter installation and direction of steaming (Figures 13.10 and 13.12). Only the arrangement in Figure 13.10 can be guaranteed to work repeatably. In Figure 13.12, all are incorrect; neither A or C works because the steam is likely to push a slug of condensate up the air filter line into the middle of the air filter. Once in, there is no way out and the filter may blind. B is not recommended because steaming the vessel via the air filter strains the filter membrane enormously and may rupture it.

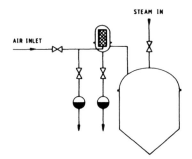

Figure 13.10 Valve Configuration for Steaming a Vessel Air Filter. In this configuration, the air filter is steamed with the vessel.

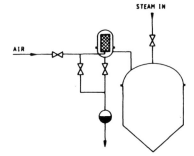

Figure 13.11 Incorrect "Commoning up" of Steam Traps on an Air Filter. The steam pressure rapidly equalizes across the filter membrane. There is no driving force pushing sterilizing steam through it.

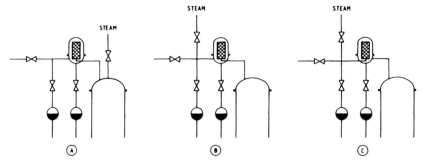

Figure 13.12 Three Incorrect Ways of Sterilizing Air Filters. The filter membrane either blinds with condensate (A, C), or may possibly rupture (B).

The ideal way to steam sterilize air filters [7], which is worth doing in large installations and with large filter housings, is shown in Figure 13.13. In this case the air filter is installed the correct way, with all the joints on the nonsterile side. This can be done because two steam supplies are used, at different pressures, and so the driving force of 0.25 bar is towards the vessel, which ensures that the steam is going the correct way through the filter membrane.

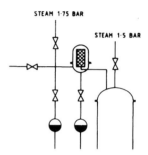

Figure 13.13 Ideal Valve Configuration for Steaming an Air Filter.

In installations where the air filter may need changing during a fermentation, either of the valve arrangements shown in Figure 13.14 is possible. The arrangement on the left has the advantage that the air filter is installed the correct way, but also has the drawback that filter and vessel have to be sterilized separately unless steam at two different pressures is supplied. The arrangement on the right has the advantage that the steam needs to be supplied at only one pressure, and the vessel and filter can be sterilized together. However, the air filter is installed the wrong way around.

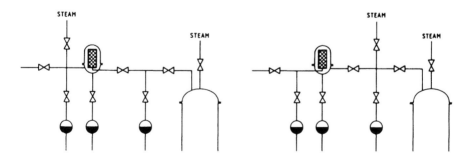

Figure 13.14 Valve Configurations for Steaming Air Filters Separately from a Vessel.

13.2.3 Medium Filters

The sterilization of medium-sterilizing filters is less of a problem than that of air filters because they are not hydrophobic. Medium-sterilizing filters typically are installed in trains of two or three, depending on the duty. In a three-filter train these could be 1.0-μm, 0.2-μm, and 0.1-μm filters, with the first 1.0-μm prefilter being a depth filter. The two-filter train typically dispenses with the coarse prefilter.

Medium filter trains able to withstand integrity testing if required can also be designed. However, the process of integrity testing may not be reliable, and the quality of filters is so high that integrity testing can often be dispensed with. In many cases, however, GMPs require integrity testing.

A cell culture medium filter train, like that shown in Figure 13.15, has to be steam sterilized with the final filter in position, but the coarse and fine prefilters can be left out and put into position later if necessary. The housings should all have vent valves and pressure gauges. When the system is sterilized, the vent valves should be left open for the first 2–3 minutes. This ensures all air is thoroughly purged from the filter housings. Each low point should have a valve and steam trap to drain the condensate.

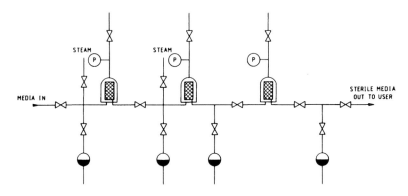

Figure 13.15 A Valve Configuration for Steaming of Medium Sterilizing Filters.

One factor that needs consideration is the order of sterilization of the medium filters, the transfer line from them, and the vessel into which the medium is being introduced. It is generally best to sterilize the vessel first and then sterilize the transfer line and filter system together. In this way, at the end of the filter sterilization sequence, the system steam pressure can be released into an already sterile vessel. This eliminates the possibility that the condensing steam might develop a vacuum and pull microbial contamination into the pipe across a closed valve seat. For further information on this subject, see Reference 7.

13.2.4   Valves and Piping

The diaphragm valve is the most commonly used valve in biotechnology, because the actuating mechanism is completely sealed from the process solution, and it is considered the most easy to clean and sterilize. However, it does have the disadvantage of being anything but crevice-free. There is the continuous crevice all around the body/diaphragm joint, and, when the valve is closed, there is a crevice between the central weir and the diaphragm.

The correct way of sterilizing these valves is the subject of much debate. There are three possible modes of sterilization. I believe the first is totally satisfactory, the second is an acceptable alternative, and the third one is to be avoided if at all possible.

The first mode is to steam completely through the valve to a drain or condensate trap on the far side of the valve from the steam supply. In this mode, the whole of the valve sees full steam, and the pipework beyond the valve also gets steamed.

The second mode depends on using the sampling or drain connection provided by most diaphragm valve manufacturers. This is a boss to one side of the weir, included as part of the forging. If this boss is drilled out, a small valve can be welded directly onto the body of the main valve. This small valve can be used either to put steam directly into the body of the valve or to take condensate away. In this way the critical half of the body can be adequately sterilized, as can the area under the valve seat.

The third mode relies on the rule of thumb that dead legs should not exceed six pipe diameters. The problem with this approach is that unless one is very careful with slopes, etc., the dead leg can fill with condensate and thus not attain sterilizing temperatures.

In the example shown (Figure 13.16), when steam sterilization starts, the first nonsterile condensate gets pushed past the trap valve and never gets hot because it is too far down the pipe. All the steam goes down the trap/valve. In all these arrangements, the valves and tees should be welded together as closely as possible in order to keep within the six-pipe-diameter rule. This rule is a very important one, and the design of all valve clusters should be based on it. Sometimes it even requires that either the commercially available fittings or the valve stubs must be shortened.

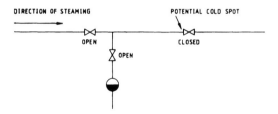

Figure 13.16 Diagram Showing How a Dead Leg Can Produce a Cold Spot.

The other important feature in the design of piping systems for steam sterilization is that they should slope, so that the condensate can drain away. Nominally horizontal pipes must be given a slope because they tend to sag between supports, and low places can act as collection points for cool condensate. The distance between pipe supports

must be adequate to prevent sagging, even when pipes are sloped. Piping constructed
of PVDF requires continuous support to prevent sagging. Having pipes slope does
create additional work, as standard 90° fittings have to be modified to suit the required
angles. A minimum slope of 1:100 is recommended.

Lines for the sterile transfer from one vessel to another should be designed to be
separately sterilizable as far as possible, as this gives an extra degree of operational
flexibility.

Shown in Figure 13.17 is an ideal configuration. It is a combination of a vessel
outlet cluster and a side inlet. Before transfer from vessel 1 to 2 is begun, vessel 2
is sterilized. During this time, valves E and F are open, and D is closed. As the line
is sterilized, valves A and F are closed, and valves E, D, B, C, and the steam inlet
valve are opened. At the end of sterilization, valves E and then C are closed, and
valve F is opened to release the steam pressure and to avoid the possibility of the
condensing steam pulling a vacuum in the transfer line. Vessel 2 and the transfer line
are now sterile and ready for use.

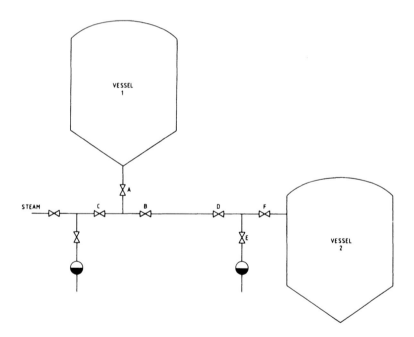

Figure 13.17 Valve Configuration for a Transfer Line Between Two Vessels.

Transfer lines can be quite long; it is best to keep the route as simple as possible,
with a minimum number of high or low points. Every low point should have a drain,
and high points must have bulk steam flow.

13.2.5   Elimination of Condensate

When pipework or vessels are steam sterilized empty, the condensate produced must be removed. There are three ways of doing this. *Free steaming* relies on having all the drain valves cracked open slightly, to permit the condensate to drain away. This is a ticklish procedure and is used only in certain circumstances. One instance is for air filter housings, as it ensures that the air is purged at the same time as the condensate. The valves are left fully open for the first few minutes to remove the air and the start-up condensate, and then "cracked back". Free steaming cannot be recommended as it is unsatisfactory, unreliable, wasteful of steam, and introduces a large humidity load on the HVAC system.

Steam (or condensate) traps are devices designed to remove condensate from steam lines automatically. They must be sized to deal with the required flow of condensate, which is at its peak during the start of sterilization. Thermostatic steam traps should be fitted with near-to-steam elements. Steam traps should be installed according to the manufacturer's instructions, and care should be taken to ensure that they are installed so as to ensure that all the necessary parts of the pipework system reach sterilizing temperature. In particular, if thermostatic-type traps are used, a cooling leg is required before the steam trap to ensure the condensate cools to below steam temperature.

On automatic systems, it is possible to dispense with steam traps entirely and use computer control to open particular drain valves for short periods at regular intervals. This saves on steam traps but increases the computer programming effort. This approach is particularly useful when solids may be present in the lines, such as in large-scale antibiotic production.

In general, I far prefer the use of steam traps. They are well known, inexpensive, robust, reliable, and specifically designed for the job.

13.2.6   Schematic of Complete Vessel

The vessel arrangement in Figure 13.18 shows how a complete vessel can be built up from the items described in the previous pages. However, it is still important to perform a sterility audit on the finished design, taking particular account of the operating philosophy to be used to run the finished plant.

13.3   Monitoring of Sterilization

During routine production operation there are only two parameters that can be measured to give an indication of the progress of sterilization, the temperature and the pressure. Most vessels have both a temperature and a pressure gauge, and it is important that these be calibrated regularly. On simple equipment where a single vessel is being sterilized, these two gauges are usually adequate to give a routine

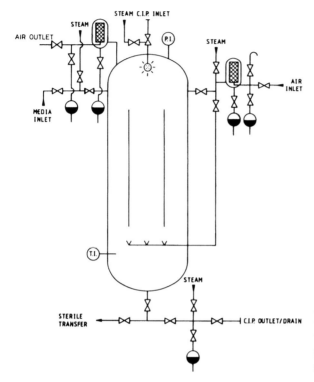

Figure 13.18 Valve Configuration for Sterilization of a Whole Vessel, showing how the system can be built up as a collection of parts.

method of sterilization monitoring. However, it is an important benefit if the temperature can be recorded on a chart recorder, so any dips in temperature below 121 °C can be spotted on the recording. The two types of chart recorder, continuous-pen and dotting-type, are both satisfactory for this purpose. The continuous-pen type gives a much more legible recording, while the dotting type allows many more channels and information (date, channel identification, scale, etc.) on the chart.

On complex pipework systems, this approach is not particularly effective, as there are practical difficulties of fitting sanitary and sterile pressure and temperature gauges to the pipework, especially if it is small-bore piping.

In these situations, one solution that is very effective is to use stainless-steel–sheathed thermocouples fixed to the outside of the pipework by specially designed clamps. (A permanent attachment by welding, brazing, or silver soldering is not recommended; the thermocouples cannot then be removed for calibration or replacement.) This provides excellent thermal contact and allows the probe to register temperatures very close to those inside the pipework. They are best located near steam traps, on the underside of the pipe on long horizontal runs, on bends, or near large heat sinks. These thermocouples are then connected into one of the many sophisticated multichannel chart recorders on the market, which can take direct thermocouple inputs, display the temperature digitally on the front panel, and print out the actual values, as well as record them.

Where the capital cost of permanent thermocouples is too great, thermal labels, temp sticks, or hand-held thermometers are very useful. *Thermal labels* are small stick-on labels that show the maximum temperature they have attained. They are commonly available with a 116 °C to 126 °C range. One drawback is that they are not reversible; if the item being sterilized reaches temperature briefly but then cools, they do not record the fact. *Temp sticks* are very similar. Hand-held surface contact thermometers are essential in any installation, as they can measure the surface temperatures of any part of the process. It is good practice to use them routinely as part of normal production sterilizations. Thermal labels and temp sticks can be used in a similar way.

Another technique that is becoming useful is *infrared imaging*, both for overall thermal imaging of the plant and for noncontact surface temperature thermometry. The thermal imaging cameras can be used on a campaign basis during commissioning or troubleshooting, as they give an instantaneous temperature picture of the plant and hence allow rapid identification of problem areas.

13.4 Validation of Sterilization

The validation of a steam sterilization process requires careful planning. It is possible to validate in two ways, either directly, using media, or indirectly, using temperature and pressure requirements. Each has certain advantages. Validating sterilization directly involves sterilizing the equipment, followed by sterile filling with a standard medium such as Tryptone Soy Broth (TSB), and then incubating this for 7 to 14 days. If the TSB is sterile at the end of this period, the initial vessel sterilization can be said to have been successful.

This method has the advantages that it is close to a real situation, and it directly tests the equipment for its ability to be sterilized and its ability to maintain sterility. It has the drawback of being very time-consuming and expensive. In addition, the sterile filtration system may not have been sized to sterilize TSB, which can be quite difficult to filter compared to simple media.

It is, of course, possible to validate using the actual fermentation medium, and this has been done successfully. However, many such media are carefully tailored to the requirements of the specific organism or cell being cultured, and contaminating microorganisms may not grow well.

Validating steam sterilization indirectly involves ensuring that all parts of the equipment reach sterilizing temperature and remain there for the required time period. There are three ways of measuring the temperature, by using either internal temperature probes, external thermocouples fixed to the surfaces, or a hand-held surface contact thermometer. A combination of all three should be used.

The number of external thermocouples to be used should be appropriate to the size and complexity of the item being sterilized. The number of these thermocouples is best kept to a minimum, as they all must be calibrated before and after the validation. The number of points using the surface contact thermometer can be totally flexible. The surface contact thermometer used must be the fast-response type, otherwise good temperature measurements cannot be obtained. A rule of thumb for external temperatures is that if the required internal temperature is 121 °C, the external temperature should be greater than 115 °C. However, it is advisable to do some tests on typical areas of the plant to check this. Insulation of the line affects the differential. The pressure gauge, if fitted, provides a useful cross-check to the temperature data.

All equipment used for validation should be calibrated to a traceable standard before and after the validation exercise, and these calibration certificates should be enclosed with the validation report. The general acceptance criteria are: internal temperature, greater than 121 °C; external temperature, greater than 115 °C; and pressure, greater than 1.1 barg. These figures must be maintained for the whole of the sterilization period, and no drop must be seen.

13.5   Troubleshooting/Problem Solving

Problems with sterilization are of two types. The first occurs when the equipment fails to reach sterilizing temperature, and the second occurs when microbial contamination is found during a production run.

Failure to achieve temperature can be either a general failure (that is, all points fail to get up to temperature), or failure to achieve temperature in a specific area. There are also differences between automated and manual systems. If the failure is general:

(a)   check that steam pressure is greater than 1.5 barg;

(b)   check the regulator or strainer for blockages;

(c)   if there is only one temperature probe, check the calibration;

(d)   check the main incoming steam valve—it may be failing to open fully.

If the failure is specific to one area, check the following possibilities:

(a)   valves failing to open properly, either on the actuator, the diaphragm, or the solenoid;

(b)   steam trap failure/blockage. Most steam traps fail open rather than shut, but this may starve other areas of steam. A blocked steam trap must be cleared out;

(c)   failure of the specific local thermocouple, resulting in a false reading;

(d)   on air filter arrangements, blinding of the filter with condensate (very small filters can be prone to this);

(e)   cooling of uninsulated pipe runs if they run close to air conditioning grilles or vents. Pipework should therefore be insulated if reasonably possible (this is also a safety factor).

In all these investigations, the use of a hand-held surface contact thermometer or temp stick is invaluable, as it allows the problem areas to be pinpointed very quickly. Any areas that do not reach 115 °C externally should be investigated.

When a system is first installed, a spirit level, to check pipe slopes, is also a very useful tool. All pipes, without exception, must slope down from the steam inlet to the condensate drain points. These slopes should be 1:100 (1/8" per foot) or better. In certain situations, although all pipes slope as they should, and all valves and traps function perfectly, it is still possible to run into problems due to the arrangement of the direction of steaming (see the end of this section).

If microbial contamination occurs, the first action is to get samples of the contamination for identification, to indicate whether the culprit is airborne or waterborne. While this investigation is going on, many checks can be done:

(a)   Did the vessel get sterilized properly? Did it achieve temperature?
(b)   Had the vessel been standing empty for a period of time? If so, it may be necessary to double sterilize. It is good practice to sterilize immediately after use (postcleaning) and just prior to the next use.
(c)   Had the validated procedures been followed?
(d)   Had the vessel passed its pressure test? Recheck this, and, if necessary, repeat a full 24-hour air-pressure test. At this point check all joints, gaskets, and fittings for leaks. If there is a leak, but it cannot be found, a hydraulic test can be done and the tell-tale seep of water locates the leak immediately.
(e)   Are all the valve seats in good order? A pressure test may continue to fail, but not identify the leak. Check all valve diaphragms.
(f)   Are all O rings and Tri-clamp® seals undamaged?

13.6   Sterility Audit

In the case of a new plant, or an old one that has repeated contaminations, a sterility audit must be done. On a new plant it should be done at the design stage. There are three requirements for a sterility audit: an up-to-date flow sheet, a check of pipe slopes, and a set of up-to-date procedures.

The whole plant must then be examined in detail, with checking of the operational procedures against the flow sheet and the slopes. Various tasks emerge:

(a)   Identify any areas where sterile is separated from nonsterile by only one valve.
(b)   Identify any long dead legs, including blanked-off ports on vessels, etc.
(c)   Identify any valves that do not get steamed through.
(d)   Identify any adverse slopes.

(e)   Identify any undrained low points.

(f)   Identify any places where there is conflicting direction of steaming. Figure 13.19 illustrates an example. Around point A it is possible to get a dead spot where the condensate coming down one leg gets held by the steam pressure from the other leg. This tends to happen only on small-bore lines (½" or 13-mm OD), but is a very real occurrence. If this does happen, it is necessary to add another valve on either side of the trap and to steam the problem area in two parts.

(g)   Identify any places where a single steam trap is doing the duty of two.

(h)   Identify any pipe networks that after sterilization can develop a vacuum.

Such a sterility audit should always be done at the design stage of a new plant.

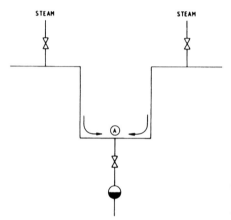

Figure 13.19 Conflicting Direction of Steaming can lead to a possible holdup of cool condensate at or around point A.

13.7   Automation

On a big piece of equipment that is being steam sterilized, a large number of valves need opening and shutting in a specific sequence for sterilization. To do this repeatedly and reliably on a manual system is possible, but it needs highly skilled, trained, and motivated staff, and is also very time-consuming. The chances of error are significant. Because of this, many large plants are automated.

The automation of process equipment is expensive, and is time- and space-consuming. A good rule of thumb is that automated equipment costs twice as much and takes twice as long to build and commission as manually operated equipment. Even so, it is often economically viable to automate, as it reduces significantly the routine labor required, and also improves the process reliability.

The control of the system can be performed by computer, with operator interface by video screen and keyboard. Mimic panels showing the status of valves and other equipment are also highly desirable. Valve actuators should be fitted with feedback switches for both positions. The computer must check that when it asks a valve to open or close, the valve actually does so. Because of the general need to monitor sterilization, there should be at least one temperature probe on the item being sterilized, and the data should be fed into the computer. The computer then can time its sterilization sequence from when the probe reaches 121 °C.

This computer can also be used for control of process variables (pH, temperature, etc.) and also for sequencing of cleaning operations and vessel-to-vessel transfers. Once a computer has been installed, it makes sense to automate as much as possible.

13.8   Conclusions

The principles of steam sterilization are well known, and the methods have been established for more than 20 years [8–10]. However, the increased use of mammalian cell cultures and recombinant organisms has greatly increased the importance of achieving complete sterilization 100% of the time. With a good initial design, a careful process validation, and attention to calibration and maintenance during operation, a high degree of successful operation can be achieved.

13.9   References

1. AIBA, S., HUMPHREY, A., MILLIS, N.; *Biochemical Engineering*; 2nd ed.; Academic Press, New York, 1973.
2. BIRCH, J. R., LAMBERT, K., THOMPSON, P. W., KENNEY, A. C., WOOD, L. A.; Antibody Production with Airlift Fermentors; In: *Large Scale Cell Culture Technology*; LYDERSEN, B. K., Ed.; Hanser Publishers, New York, 1987; pp 1–20.
3. RAHN, O., Physical Methods of Sterilization of Micro-organisms, *Bacteriol. Rev. 91*, 1–47 (1945).
4. SOLOMONS, G. L.; *Materials and Methods in Fermentation*; Academic Press, London, 1969.
5. KELSEY, J. C.; *Methods of Testing the Bactericidal Efficiency of Steam Sterilizers*; Symposium, The Operation of Sterilizing Autoclaves; The Pharmaceutical Press, London, 1959; p 22.
6. PERKINS, J. J.; *Principles and Methods of Sterilization*; Thomas, Springfield, IL, 1956.
7. *Steam Sterilization of Pall Filter Assemblies*; Reference SD805F; Pall Process Filtration Ltd., East Hills, NY.
8. WANG, D. I. C., COONEY, C. L.; *Fermentation and Enzyme Technology*; John Wiley and Sons, NY, 1979.
9. RICHARDS, J. W.; *Introduction to Industrial Sterilization*; Academic Press, London, 1968.
10. *Practical Steam Trapping*; Spirax Sarco, Ltd., Cheltenham, UK.

Pharmaceutical Water Systems: Design and Validation

Rostyslaw Slabicky

Contents

14.1   Introduction

Water is the most widely used raw material in the pharmaceutical industry, and the most misunderstood. In this chapter I provide an overview of the different types of water, guidance on design and operational concerns, and a perspective on the validation of pharmaceutical water systems.

Different grades of water are used as a utility, for drinking and sanitation, as a raw material, as an adjunct for the process or product, and as a finished product. The use of the water dictates the quality attributes, the design of the system, and the level of control in the system.

Water used for utilities such as process heating or cooling needs to be within the parameters that may affect the intended use, such as the concentration of corrosion reduction chemicals, mineral content, or freeze protection.

Water for drinking, bathing, and food preparation must meet the requirements established for potable water. The quality attributes are established by national and international organizations, such as the U.S. Environmental Protection Agency (US EPA), the World Health Organization (WHO), and the European Community (EC). The regulations identify organic and inorganic chemicals and microorganisms that may have an adverse health effect on humans.

Water used for the preparation or processing of pharmaceutical raw materials, drug products, or biopharmaceuticals must meet the requirements for the appropriate grade of pharmaceutical water or the needs of the process. The quality attributes for pharmaceutical waters are established by the various national pharmacopeias. Major pharmacopeias include the British Pharmacopeia, the European Community Pharmacopeia, the Pharmacopeia of Japan, the Nordic Pharmacopeia, and the U.S. Pharmacopeia. The pharmacopeias establish a minimum requirement for chemical, microbial, and physical attributes that may affect the properties of the ingredient or preparation, or adversely affect patients.

14.2   Grades of Water

14.2.1   Potable Water

The quality of *potable water* is established in the United States by the U.S. Environmental Protection Agency and is published in the Code of Federal Regulations. The maximum levels for regulated contaminants are identified in the Primary Drinking Water Regulations [1]. The goals for the reduction of other contaminants are identified in the Secondary Drinking Water Regulations [2]. The major contaminants and levels for the United States, the European Community, Japan, and Canada are listed in Table 14.1A and 14.1B.

Table 14.1A Water Quality Standards for the United States, European Community, Japan, and Canada: Inorganics

Parameters	Units	USA Primary [1]	USA Secondary [2]	European Community [3]	Japan [4]	Canada [5]
Aluminum	mg/L			0.2		
Ammonium	mg/L			0.5	0.05	
Antimony	mg/L	0.006		0.01		
Arsenic	mg/L	0.05		0.05		0.05
Barium	mg/L	1		0.1		1.0
Beryllium	mg/L	0.004				
Boron	mg/L					5.0
Bromide	mg/L					
Cadmium	mg/L	0.005		0.005	0.01	0.005
Calcium	mg/L			100		
Chlorides	mg/L		250	25	200	< 250
Chromium	mg/L	0.05		0.05		0.05
Copper	mg/L	1.3		3.0	1	< 1.0
Cyanide	mg/L	0.2		0.05	0.01	0.2
Fluorides	mg/L	4.0		0.7–1.5		1.5
Heavy metals (Pb)	mg/L				1	
Iron	mg/L		0.3	0.3	0.3	0.3
Lead	mg/L	0.015		0.05	0.1	0.05
Magnesium	mg/L		0.05	30		
Manganese	mg/L			0.05		0.05
Mercury	mg/L	0.002		0.001		0.001
Nickel	mg/L	0.1		0.05		

Table 14.1A Water Quality Standards for the United States, European Community, Japan, and Canada: Inorganics, *continued*

Parameters	Units	USA Primary [1]	USA Secondary [2]	European Community [3]	Japan [4]	Canada [5]
Nitrates (N)	mg/L	10		50	10	10.0
Nitrites	mg/L	1		0.1	no color	1.0
Phosphorus	mg/L			0.5		
Potassium	mg/L			10		
Selenium	mg/L	0.01		0.01		0.01
Silver	mg/L	0.05		0.01		0.05
Sodium	mg/L			150–175		
Sulfates	mg/L		250	25		500
Sulfide (H_2S)	mg/L					0.05
Surfactant	mg/L		0.5		0.5	
Thallium	mg/L	0.002				
Total dissolved solids	mg/L		500		500	< 500
Uranium	mg/L					0.02
Zinc	mg/L		5	5.0	1	< 5.0

Table 14.1B Water Quality Standards for the United States, European Community, Japan, and Canada: Microbial, Inorganic, and Physical Parameters

Parameters	Units	USA Primary [1]	USA Secondary [2]	European Community [3]	Japan [4]	Canada [5]
Coliform	#/100mL	note 1		0	0	10
Microorganisms	CFU/mL				100	500
Color	CU		15		clear	< 15
Hardness (CaCO₃)	mg/L				300	500
Odor	TON		3		odorless	
pH			6.5–8.5	6.5–8.5	5.8–8.6	6.5–8.5
Turbidity	NTU	1		0–5		1–5
Organics	mg/L	note 2				
Pesticides (individual)	mg/L			0.0005		
Pesticides (total)	mg/L			0.0001		
Total trihalomethane	mg/L	0.1				

Note 1: No more than one positive sample per month, if fewer than 40 samples collected.
Note 2: U.S. Primary Drinking Water Regulations establish limits for over 60 organic chemicals.
CFU = colony-forming unit; CU = Color Units; TON = Threshold Odor Number; NTU = Nephelometric Turbidity Units.

14.2.2   Pharmaceutical Water

The grades for pharmaceutical waters are established in the pharmacopeias as monographs. The monographs detail the method of preparation, packaging, labeling, storage conditions, and chemical, microbial, and physical quality attributes. The edition of the pharmacopeia current at the time of the preparation of water identifies the specific quality attributes to be maintained.

The U.S. Pharmacopeia (USP) XXII has six monographs for water [6]. The most widely used grades are *Purified Water* (*PW*) and *Water For Injection* (*WFI*). These are used for cleaning and compounding, and as the ingredients for the remaining grades of water: Sterile Water For Injection, Bacteriologic Water For Injection, Sterile Water For Irrigation, and Sterile Water For Inhalation. The major quality attributes defined in the monographs are listed in Table 14.2.

Table 14.2   Quality Attributes of Purified Water and Water For Injection, USP [6]

	Purified Water	Water For Injection
Treatment method	suitable	distillation, RO
Source water	drinking water	
Ammonium	< 0.3 ppm	< 0.3 ppm
Bacterial endotoxin		< 0.25 EU/mL
Calcium	no turbidity	no turbidity
Carbon dioxide	remains clear	remains clear
Chloride	no opalescence	no opalescence
Heavy metals	less color than control	less color than control
Microbiological	US EPA (no coliforms)	
Oxidizable substances	remains pink with $KMnO_4$	remains pink with $KMnO_4$
pH	5.0–7.0	5.0–7.0
Sulfate	no turbidity	no turbidity
Total solids	0.001%	0.001%

EU = endotoxin units

The quality attributes in the monographs in the British Pharmacopeia, European Pharmacopeia, and Japanese Pharmacopeia, as well as the USP, are in Table 14.3.

Table 14.3 International Quality Attributes for Purified Water and Water For Injection

Parameter	Units	USA		Britain		Europe		Japan	
		PW	WFI	PW	WFI	PW	WFI	PW	WFI
Acidity/alkalinity	mg/L			1	2	1	2	0.1	0.1
Ammonium	mg/L	0.3	0.3	0.2	0.2	0.2	0.2	0.05	0.1
Bacterial endotoxin	EU/mL		0.25		Pyrogen		Pyrogen		0.25
Calcium	mg/L	0.5	0.5	1	1	1	1		
Carbon dioxide	mg/L	4	4						
Chlorides	mg/L	0.5	0.5	0.5	0.5	0.5	0.5	0.5	0.5
Heavy metals (Pb)	mg/L	0.5	0.5	1	1	0.1	0.1	1	
Magnesium	mg/L			0.6	0.6	0.6	0.6		
Nitrates (N)	mg/L	0.2	0.2	0.2	0.2	0.2	0.2	0.2	0.2
Oxidizable substances	mg/L	5	5	1	0.5	5	10	5	
pH		5–7	5–7						
Sulfates	mg/L	0.5	0.5	0.5	0.5	0.5	0.5	0.5	0.5
Total organic carbon	mg/L								0.500
Total solids	%	0.001	0.001	0.001	0.003	0.001	0.003	0.001	0.003

The values in the table are the calculated endpoints for wet chemistry tests, at the point where the test shows a positive result.
Refer to the current Pharmacopeia for the actual test method and limit.
PW = Purified Water; WFI = Water For Injection; EU = endotoxin units; Pyrogen = Pyrogen Test or Rabbit Test (qualitative).

14.2.3   Clean Steam

Steam used in the sterilization of components and process equipment needs to be considered as a potential source of waterborne product contamination. No definitive monographs or guidelines are available for additive-free or clean steam. The steam must be free of boiler treatment additives. The quality of the steam should be consistent with the quality requirements of the water used for other pharmaceutical processing, cleaning, or compounding steps.

14.2.4   Microbial Attributes

The monographs for Purified Water and Water For Injection do not adequately address the concerns for microbial control in pharmaceutical waters. The microbial attributes, relative number of organisms, and types of organisms present, must be weighed as functions of the manufacturing processes and the products in which the water is to be used [7]. The processing steps may further reduce the microbial level, or the product may inhibit microbial growth. Therefore, the microbial attribute levels must be determined by the user for the particular application.

Suggested guidelines for an upper microbial level for water are in Table 14.4. The associated test methods for these levels are also given.

Table 14.4   Microbial Levels and Test Methods for Grades of Water [8]

Grade	Acceptance Level	Sample Preparation Method	Sample Volume	Growth Medium	Incubation Conditions
Purified Water	100 CFU/mL	Pour Plate	1.0 mL	plate count agar	48 h minimum; 30–35 °C
Water For Injection	10 CFU/100 mL	Filtration (0.45 μm)	100 mL	plate count agar	48 h minimum; 30–35 °C

CFU = colony-forming unit

14.2.5   Revisions in the USP

Revisions to USP XXII have been proposed [9] that would update the quality attributes from the existing qualitative tests to tests for conductivity and total organic carbon. Minor changes are being evaluated for total solids and pH. No changes are being proposed for the microbial and endotoxin attributes. The inorganic chemical attributes would be replaced with a proposed test for conductivity [10]. To determine the conductivity of Purified Water and Water For Injection, the water would be equilibrated with atmospheric carbon dioxide at 25 °C, the conductivity and the pH would be measured, and the values would be compared to those in Table 14.5. The pH value would have to be within the listed range of 5.0–7.0, and the conductivity would have to be no greater than the corresponding listed value.

Table 14.5 Proposed Values for Conductivity and pH of Pharmaceutical Water [10]

1. Water meets test if the conditions in Stage 1 are met.					
2. If Stage 1 is exceeded, equilibrate with atmospheric CO_2, and test for the conditions in Stage 2.					
3. If Stage 2 is exceeded, determine pH and use table below.					
	Temperature	Conductivity		Resistivity	
	° C	μS/cm		MΩ·cm	
Stage 1	≥ 25	≤ 1.3		> 0.77	
Stage 2	25 ± 1	≤ 2.4		> 0.42	
Stage 3	25 ± 1	Determine pH and use table:			

pH	Conductivity μS/cm	Resistivity MΩ·cm	pH	Conductivity μS/cm	Resistivity MΩ·cm
5.0	≤ 4.7	> 0.21	6.1	≤ 2.4	> 0.42
5.1	≤ 4.1	> 0.24	6.2	≤ 2.5	> 0.40
5.2	≤ 3.6	> 0.28	6.3	≤ 2.6	> 0.38
5.3	≤ 3.3	> 0.30	6.4	≤ 2.8	> 0.36
5.4	≤ 3.0	> 0.33	6.5	≤ 3.1	> 0.32
5.5	≤ 2.8	> 0.36	6.6	≤ 3.4	> 0.29
5.6	≤ 2.6	> 0.38	6.7	≤ 3.8	> 0.26
5.7	≤ 2.5	> 0.40	6.8	≤ 4.3	> 0.23
5.8	≤ 2.4	> 0.42	6.9	≤ 5.0	> 0.20
5.9	≤ 2.4	> 0.42	7.0	≤ 5.8	> 0.17
6.0	≤ 2.4	> 0.42			

Refer to the USP [6] for the actual test method and limits.

Total organic carbon (TOC) is being proposed as a replacement for the test for oxidizable substances. This quality attribute provides information on the performance of several unit operations, determines the level of organic contaminants, and is an indicator of the available nutrients for microbial growth. The quality attribute and the test method for TOC proposed for Purified Water and Water For Injection in the USP are as follows [11]:

Note: Organic contamination of sampling containers and glassware can result in falsely high TOC results; therefore, use scrupulously clean glassware and sample containers. This method measures TOC as nonpurgeable organic carbon. Since purgeable organic carbon is not a significant contributor to TOC in pharmaceutical water, the instrument differences may be disregarded. The use of a conductivity detector may introduce bias during instrument calibration and system suitability assessment. This bias must be corrected.

Standard Preparation: Dissolve an accurately weighed quantity of powdered potassium biphthalate, previously dried at 120 °C for 2 h, in high purity water (see reagents under Containers <661>), to obtain a solution having a known concentration of 1.06 ppm (0.50 µg of carbon).

System Suitability, preparation: Dissolve an accurately weighed quantity of sodium dodecylbenzenesulfonate in high purity water (see reagents under Containers <661>), to obtain a solution having a known concentration of 0.80 ppm (0.50 µg of carbon).

System Suitability: Determine the TOC (as nonpurgeable organic carbon) concentration of the system suitability preparation as directed under Procedure. A suitable system has a recovery of not less than 85% carbon.

Procedure: Determine the TOC (as nonpurgeable organic carbon) concentration of the sample using the standard preparation for calibration of the apparatus and the High Purity Water used in the standard and system suitability preparations as the blank, and make any necessary corrections. It contains not more than 0.50 ppm.

14.2.6   Reagent Water

The quality of water used for laboratory analysis is specified by several organizations that establish test methods or laboratory practices.   These include the National Committee for Clinical Laboratory Standards (NCCLS), the College of American Pathologists (CAP), and the USP. The water used for USP test methods must be at least Purified Water or as specified. The major quality attributes for reagent waters are listed in Table 14.6.

14.3   Sanitary Design

In order to provide the appropriate quality of water, all of the components in the water system should meet sanitary design standards.   These standards are prepared by industry and government agencies. The International Association of Milk, Food, and Environmental Sanitarians [13], U.S. Public Health Service, and Dairy Industry Committee have prepared standards for various components. The components meeting these standards are designated with the "3-A" symbol.

Sanitary design practices for components minimize the potential for contamination by chemical or microbial sources that may exist as a result of crevices, rough finish, poor welds, threaded connections, poor gasketing, areas of low flow, and poor cleanability. The practices call for smooth pipe material with interior polish to remove seams, no dead legs, flush gaskets, smooth transitions, sweep elbows, self-draining valves with no voids, and ease of cleaning.

Table 14.6 Quality Attributes for Reagent Waters [12]

Attribute	Units	CAP I	CAP II	CAP III	NCCLS I	NCCLS II	NCCLS III
Ammonium	mg/L	0.1	0.1	0.1			
Carbon dioxide	mg/L	3	3	3			
Hardness		negative	negative	negative			
Heavy metals (Pb)	mg/L	0.01	0.01	0.01			
Microorganisms	CFU/mL	note 1	note 1	note 1	10	1000	
Organics					note 2		
Oxidizable substances	mg/L	60	60	60			
Particulate matter					note 3		
pH		6–7	6–7	6–7			5–8
Resistivity	MΩ·cm	10	0.5	0.2	10	1.0	0.1
Silicate (SiO_2)	mg/L	0.01	0.01	0.01	0.05	0.1	1
Sodium	mg/L	0.01	0.01	0.01			

Note 1: Minimum microbial growth observed.
Note 2: Treat water with a carbon filter.
Note 3: Treat water with a 0.22-μm filter.
CAP = College of American Pathologists; NCCLS= National Committee for Clinical Laboratory Standards; CFU = colony-forming unit.

14.3.1   Turbulent Flow

To prevent the development of favorable growth conditions for microorganisms, water must flow within the piping, components, and equipment of a pharmaceutical water system. In stagnant water or in an environment where they may adhere to surfaces, organisms attach to a surface and start to grow. Conditions favorable for organisms to adhere include rough surfaces, areas of low flow rate, or piping or equipment that contains stagnant water (*dead legs*).

In turbulent flow, elements of fluid move all over the cross section, from the interior of the stream to the pipe wall. This provides a scouring action on the pipe and aids in reducing the chemical and microbial contamination of the surfaces [14]. In turbulent flow, the *Reynolds number*, N_{Re}, is greater than 3000 [15]:

$$N_{Re} = vd/\nu \qquad (14.1)$$

where v is the average velocity, ft/s; d is the internal diameter, ft; and ν is the kinematic viscosity, ft^2/s.

The minimum flow rate is calculated for a design condition of full flow with no use of water. In a properly designed distribution system, the intermittent reduction of flow due to use is limited, so that turbulent flow is maintained. The design velocity for the piping should be greater than 5 ft/s (1.5 m/s), significantly exceeding that required to achieve turbulent flow, to ensure that a fully developed turbulent condition exists. Table 14.7 gives the flow rates to achieve minimum turbulent flow and fully developed turbulent conditions.

Table 14. 7 Flow Rates for Turbulent Flow

Pipe diameter, inches	Reynolds number	Temperature, ° C	Velocity, ft/s	Flow rate, gpm
0.5	3,000	10	1.4	0.5
	3,000	20	1.0	0.4
	3,000	60	0.5	0.2
	3,000	80	0.4	0.1
0.5	50,000	10	22.9	7.7
	50,000	20	17.5	5.8
	50,000	60	8.2	2.7
	50,000	80	6.5	2.2
0.5	Flow rate to maintain 5 ft/s velocity > 1.8 gal/min			
1	3,000	10	0.6	1.1
	3,000	20	0.4	0.8
	3,000	60	0.2	0.4
	3,000	80	0.2	0.3
1	50,000	10	9.7	18.0
	50,000	20	7.4	13.8
	50,000	60	3.5	6.5
	50,000	80	2.7	5.1
1	Flow rate to maintain 5 ft/s velocity > 10 gal/min			
1.5	3,000	10	0.4	1.7
	3,000	20	0.3	1.3
	3,000	60	0.1	0.6
	3,000	80	0.1	0.5
1.5	50,000	10	6.2	28.4
	50,000	20	4.7	21.7
	50,000	60	2.2	10.2
	50,000	80	1.7	8.0
1.5	Flow rate to maintain 5 ft/s velocity > 23 gal/min			
2	3,000	10	0.3	2.3
	3,000	20	0.2	1.8
	3,000	60	0.1	0.8
	3,000	80	0.1	0.7
2	50,000	10	4.5	38.7
	50,000	20	3.5	29.6
	50,000	60	1.6	13.9
	50,000	80	1.3	10.9
2	Flow rate to maintain 5 ft/s velocity > 45 gal/min			

14.3.2   Surface Finish

Rough surfaces provide a large area for adhesion of organisms. They also shelter the contaminants from sanitizing agents. The condition may exist on the interior of piping, at welds, as a result of mismatched gasketing, on instrument surfaces, and on valves or pumps. Inherently rough surfaces in a water system exist on ion exchange resin, activated carbon, depth filter media, and filters.

Surface finishes are specified by the number of scratches per inch or *grit* of the polishing agent used during the mechanical polishing. Grit finishes range from 60 to 500. A higher grit number indicates a more polished surface. A more accurate method of indicating surface irregularity involves the use of a profilometer that measures the height of the peaks along the surface. This measure is an indication of the average deviation of the peak from the surface. This measure is generally represented as a roughness average, Ra, with units of microinches ranging from 4 to 250. A lower number indicates a lower profile and frequency of surface irregularities. Table 14.8 provides a rough correlation between grit number and surface irregularity.

Table 14.8 Surface Finish Measurements

Ra, Roughness, arithmetic average, microinches	Grit number
250 maximum	60
125 maximum	120
85 maximum	180
15–63	240
10–32	320
4–16	500

Electropolishing is the process of removing a layer of metal from the pipe or component using electric current. The process removes the peaks and overhanging surfaces that remain after mechanical polishing. This reduces the depth of the surface irregularities and exposes the surface to full fluid flow. The additional smoothness of the pipe reduces the surface area and irregularities available for adhesion by microbial or organic contaminants and improves cleanability [16]. Electropolishing provides cleaning characteristics comparable to a mechanically polished surface that has a surface roughness approximately ten times lower than the electropolished surface [17].

The specification for an interior finish for sanitary tubing should include the maximum surface roughness and details of the mechanical polishing or finish, since the electropolishing process does not remove surface irregularities.

14.3.3   Dead Legs

The number and length of dead legs within a piping system should be minimized, to prevent stagnant conditions. These legs occur as a result of tees, valves, instruments, sampling ports, manifolds, and equipment connections. When legs from a flowing pipe occur, the length of the branch should be maintained at less than six times the diameter of the branch. This is measured from the center line of the flowing pipe to the end of the branch. This is also expressed as:

$$L / d < 6 \qquad\qquad (14.2)$$

where L is the interior length of the branch measured from the center line; and
   d is the interior diameter of the branch.

The length of a branch for common diameters of sanitary tubing is listed in Table 14.9.

Table 14.9 Maximum Length to Prevent Dead Legs in Sanitary Tubing

Diameter, inches	Wall, inches	Length, inches, measured from	
		center line	outer wall
0.5	0.065	2.0	1.8
0.75	0.065	3.4	3.0
1	0.065	4.8	4.3
1.5	0.065	7.5	6.8
1.5	0.12	6.9	6.2
2	0.065	10.3	9.3
2	0.12	9.7	8.7

L/d = 6, calculated from center line of main branch

14.3.4   Slope

Distribution systems for water or steam are designed to be self-draining. This is to prevent stagnant water from accumulating during periods of system shutdown. Steam distribution systems are additionally pitched to drain condensate within the pipe, preventing conditions such as water hammer and slugs of water. The nominal slope of pipes in a water system should be 1/4" per foot of horizontal pipe, with a minimum slope of 1/8" per foot. The slope must be toward a low point, such as a use point or valve that may be drained during a shutdown of the flow, and may be cocurrent or countercurrent to the direction of flow.

The minimum slope for a steam system should be 1/8" per foot of horizontal run for a pipe pitched to provide condensate drainage in the direction of the steam flow, and 1/4" per foot for countercurrent condensate drainage. The condensate flows by gravity against a steam flow gradient. The slope must be directed to a low point that has a trap to separate condensate from steam.

14.3.5   Flow Regulators

A consideration in the design of sanitary distribution systems is the prevention of contamination of the water by back siphoning of fluid or air. This may occur when demand for water is greater than what is being supplied. Water may flow backwards downstream of the high demand, providing no water to subsequent use points or creating a vacuum in the pipe. An open downstream use point permits air or fluid to enter and contaminate the distribution system. A method for providing sufficient flow to the use points is to restrict the amount of water returning from the distribution loop by the use of a back pressure regulator. The regulator continuously throttles the flow in the pipe to maintain a constant pressure. As the return flow is reduced because of consumption, the valve closes, increasing the resistance in the pipe and increasing the loop pressure. This makes additional water available to the use points. Back pressure regulators are available as self-contained, spring-controlled valves (Figure 14.1), or as a pressure sensor/controller with a control valve (Figure 14.2).

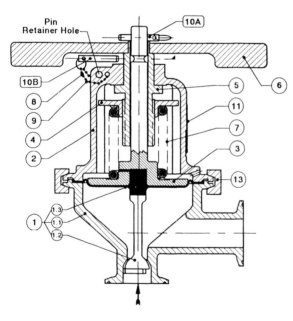

Figure 14.1 Spring-Balanced Back Pressure Regulator. Distribution system back pressure is regulated by a spring that applies constant pressure to the valve diaphragm and seat. Key to numbered items: (1) body assembly; (1.1) body; (1.2) plug; (1.3) diaphragm/stem assembly (2) spring chamber; (3) pressure plate; (4) spring button; (5) adjusting screw; (6) handle; (7) spring; (8) connector; (9) ball chain; (10A) pin (quick-release); (10B) pin (lock-open); (11) nameplate; (13) clamp. (Courtesy of Cashco Co.)

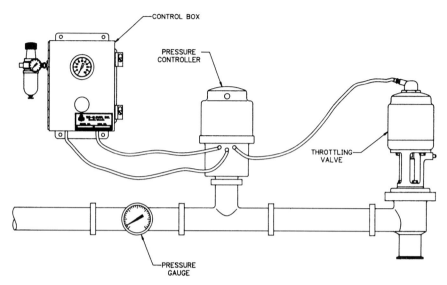

Figure 14.2 Pressure-Balanced Back Pressure Regulator. Distribution system back pressure is monitored upstream of the control valve by a pressure sensor; a pneumatic signal is applied to the valve diaphragm and seat. (Courtesy of Tri-Clover.)

14.3.6 Materials of Construction

Pharmaceutical water systems are fabricated of materials that are nonreactive with water and do not leach or contaminate the fluid stream. The material must be resistant to sanitization regimes, including heat and chemicals such as peroxide, hypochlorite, ozone, and quaternary ammonia. Leaching of metal ions such as lead, copper, zinc, iron, and aluminum needs to be considered in the design of the pretreatment equipment and piping, to minimize the potential for contamination of the water. If plastics are used, consideration must be given to leaching of organic plasticizers and fillers, fusing solvents, and the absorption of contaminants from the water.

Typical materials for pharmaceutical water systems include stainless steel and plastics. The use of plastics is increasing as the technology to fabricate and fuse the materials is advancing. Table 14.10 lists various materials and their properties.

The specifications for stainless steel tubing are in ASTM A270, Seamless and Welded Austenitic Stainless Steel Sanitary Tubing [18]. The composition of common types of stainless steel alloys are in Table 14.11. The actual composition of the material used for the tubing is available from the fabricator or mill. This is provided in the form of a certificate of analysis for the bulk material, and is designated by a *heat number*.

Table 14.10 Comparison of Materials of Construction

Material	Strength	Operating temperature, °C	Extractables
Stainless steel	high	> 250	low
Polyethylene	low	< 60	moderate
Polypropylene	low	< 90	moderate
Poly(vinyl chloride)	moderate	< 60	high
Chlorinated poly(vinyl chloride)	moderate	< 90	high
Poly(vinylidene fluoride)	moderate	< 140	low
Teflon® (PTFE, TFE)	low	< 250	low

Table 14.11 Composition of Austenitic Stainless Steel Alloys [18]

Alloy	Composition, %							
	C	Mn	P	S	Si	Ni	Cr	Mo
304	0.08	2.00	0.040	0.030	0.75	8.00–11.00	18.00–20.00	–
304L	0.035	2.00	0.040	0.030	0.75	8.00–13.00	18.00–20.00	–
316	0.08	2.00	0.040	0.030	0.75	10.00–14.00	16.00–18.00	2.00–3.00
316L	0.035	2.00	0.040	0.030	0.75	10.00–15.00	16.00–18.00	2.00–3.00

14.3.7   Inspection

Materials used for pharmaceutical water systems should be treated as critical components. Pipes, valves, fittings, heat exchangers, and tanks need to be identified and inspected for defects. Inspections should include the interior surfaces of the distribution piping and tanks. Conformance to specifications, materials of construction, finishes, and cleanliness are major points that should be verified and documented prior to use of the item in the installation.

14.4   Water Treatment System

14.4.1   Treatment Equipment

Pharmaceutical water systems are designed to deliver an acceptable quality of water in a consistent and reproducible manner to the point of use. There are many unit operations and treatment sequences that may be specified and maintained to achieve the acceptable quality. The design of each component must take into consideration areas that can affect the quality of the water. Each step in the process must be evaluated for its contribution, expected results, and potentials for reducing the desired quality. I discuss in subsequent sections of this chapter the major design considerations for individual pieces of equipment or unit operations in the treatment process. Figure 14.3 shows a possible water treatment system.

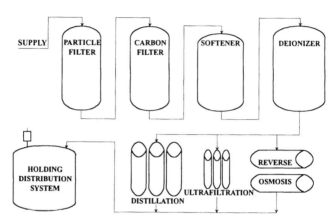

Figure 14.3 Representative Water Treatment System Schematic. The design of a water treatment system must take into consideration the contribution of each component and its effect on water quality.

14.4.2 System Sizing

Factors that need to be considered in the sizing of a water system include the requirements of the operation or facility for various grades and quantities of water, the average and instantaneous demands, the quality of the source water, and operating parameters of the treatment equipment.

Facilities may use water in operations such as medium preparation, equipment washing, component preparation, sterilization, compounding, and fermentation, and in laboratories. The operations have specific requirements for quantities of water, expressed as a quantity consumed, or a flow rate. The time of day when the water demand occurs indicates if the water consumption from the various operations is instantaneous. The grade of water used in the operations determines the type of system required.

The operating modes of the water treatment unit operations significantly influence the size of the system and of the individual units. The duration and frequency of processes such as backwashing of carbon or other media, deionizer regeneration, and equipment sanitization result in downtime and the consumption of water. These processes need to be taken into account for the sizing of the holding and distribution system.

Treatment unit operations should not be oversized to provide reserve capacity. Oversized equipment does not operate efficiently, has low flow, and is not sanitized or maintained adequately. This results in increased microbial levels within the treatment system and consequent contamination of the pharmaceutical water.

14.4.3 Holding Tanks

Water treatment systems utilize tanks for several purposes: to hold treatment chemicals, to provide atmospheric breaks for unit operations, to buffer flow or capacity, or to supply feed to process equipment. Tanks used for water that is being treated need to protect the contents from chemical and microbial contamination originating from the surroundings or being generated as a result of the design. Airborne contaminants are excluded by closed vessels with air vents protected by microretentive filters. Internal sources of contamination include instrumentation and fittings that create dead legs, gasket or seal materials, and leaks through the wall of a jacketed vessel.

14.4.4 Softeners

Water softening is an ion exchange unit operation used to reduce hardness in water by replacing calcium and magnesium ions with sodium ions. A softener is periodically regenerated with a brine solution. The associated piping and tanks provide a breeding ground for organisms in the stagnant brine and water. The organisms are introduced into the softener column with each regeneration and continuously inoculate the treatment water stream.

The brine solution, tank, and piping need to be cleaned and monitored routinely to minimize this source of contamination. Agitating the brine tank and maintaining a saturated solution help control microbial growth in the brine. The softener should be continuously recirculating, particularly when not producing water. Backwashing the softener resin during regeneration may help reduce the microbial level.

14.4.5 Multimedia filters

Multimedia filters are columns packed with sand or quartz, for the preliminary removal of silt and large particulates from feedwater. The media do not provide nutrients for bacterial growth, but the trapped material contains organic nutrients. Favorable conditions for elevated microbial levels exist if the unit operation does not have water flow. Circulation loops or operating modes need to be designed to ensure that water in the filtration equipment is not stagnant. Periodic backwashing of the media removes trapped silt and organics. The multimedia filter should be sanitized routinely, with hot water, steam, or chemicals. The sanitization process must ensure that all the interior surfaces, pipes, and valves are adequately treated.

14.4.6 Carbon Beds

Carbon beds are used to dechlorinate feedwater and to remove low molecular weight organic compounds. The removal of the chlorine from the water and the high nutrient level in the activated carbon column provide conditions for microbial proliferation. Similar precautions must be taken as with multimedia filters: recirculation, sanitization, and backwashing. The carbon bed should be sanitized using hot water or steam flowing in the same direction as the feedwater (typically from the top down), prior to backwashing. This provides maximum treatment to the area of highest microbial concentration, and minimizes distribution of bacteria throughout the bed during backwashing.

14.4.7 Deionizers

Deionizer columns or *beds* use synthetic resin beads to bind inorganic salts in the water and discharge a stream with low conductivity. Two types of resins are used, *anionic* and *cationic*. Anionic resin removes negatively charged ions such as bicarbonate, chloride, and sulfate, and replaces them with hydroxide ions. Cationic resin removes positively charged ions such as calcium, magnesium, and sodium, and replaces them with hydrogen ions. The resins, unless specifically designed, do not reduce the levels of organic chemicals, endotoxin, or microorganisms appreciably.

Deionizers are configured as dual bed units (with the cationic and anionic resin in separate columns), or as mixed beds (with the resins mixed). Which configuration to specify is based on the quality of source water and on economics. Generally, a mixed-bed deionizer provides a higher quality of low-conductivity water, but has a lower deionization capacity than a dual-bed system.

The deionization step is usually near the end of the treatment system. The feedwater does not contain any antimicrobial agents, and may have a considerable microbial burden. The deionizer columns provide a large surface area for microbial proliferation, and a source of nutrient in the resin beads and the ions.

Precautions to protect the deionizers and the water from microbial contamination must be taken in the sizing, design, and operation of the process. The capacity of the deionizer is based on the inorganic chemical concentration or hardness of the source water, the flow rate of the treated water, and the desired frequency of regeneration. The interval between regenerations is the length of time that the bed is in service and is therefore the time during which microbial growth occurs. This time should be minimized, usually to 2–7 days. Oversized deionizers should be short cycled, to regenerate based on time and not on performance.

Strong acid and caustic are used to regenerate the cationic and anionic resins, and are effective in sanitizing the beds. The regeneration process in a dual-bed system sanitizes more effectively because the individual beds are highly acidic or caustic; regeneration of a mixed bed results in a zone that has a neutral pH and is not sanitized. The regeneration cycle of a mixed bed can be modified to both fully exhaust the short cycles and sanitize the bed. The caustic kill cycle involves injecting the alkali from the top of the column through the entire bed. This starts the regeneration of the anionic resin and fully depletes the capacity of the cationic resin. The resulting alkaline condition in the entire bed effectively sanitizes the unit.

A resin column must be so configured that all surfaces of the unit come in contact with the regeneration chemicals. The freeboard or headspace used for expansion and mixing in the column may represent 50% or 100% additional height for a dual bed or mixed bed, respectively. This space is sanitized when the caustic solution completely floods the bed and some of it overflows from the top.

The rinse water used at the end of a resin regeneration cycle to remove the acid and alkali needs to be controlled both microbially, to minimize the reintroduction of microorganisms, and chemically, to not deplete the ion removal capacity of the resin prematurely with hard water. Compressed air used to mix the anionic and cationic beads in a mixed bed must be free of contaminants, such as oil and particulates, and be microbially controlled.

The design of a deionizer should include continuous recirculation of water through the unit. A nominal flow rate of 7 gallons per minute per cross sectional square foot of bed area (~5 m/s) should be used. This minimizes areas of low flow within the bed. The beds in a system containing multiple units should be configured to provide series flow. The more depleted unit is first in the series as the primary deionizer, and the freshly regenerated bed is last in the series as a polisher.

Canister deionization systems represent a potential source of microbial contamination, because they may not be adequately sanitized and they remain stagnant for prolonged periods prior to use. The procedures for handling and regenerating the resin, sanitizing the resin and canisters, using antimicrobial agents, storing them in the cold after regeneration, and protecting them from atmospheric contamination must be carefully reviewed. When a commercial vendor provides these services, there must be assurance that resins from other sources are not mixed with the resin from the pharmaceutical system, since this can be a source of cross contamination.

Instrumentation on deionizers includes conductivity or resistivity monitors, flowmeters and totalizers, and differential pressure gauges. A monitor for silica, located in the top one-third of a mixed-bed column, can be used as an indicator of decreasing quality for systems that require a low conductivity and silica level.

14.4.8   Continuous Deionization

Conventional resin deionizers require the use of strong chemicals, generate hazardous waste, and are labor intensive. A different approach to removal of ionic salts employs electricity to act as anode and cathode, instead of anionic and cationic resins. *Continuous deionization (CDI)*, illustrated in Figure 14.4, uses direct current to attract the ions from the feedwater across a selectively permeable membrane [19]. The ions are concentrated and continuously discharged. The resin within the system does not require regeneration, because the ions do not bind to the beads, but use them as a conductive path through the low-conductivity water to the transfer membrane. This system has an approximate ionic reduction efficiency of 95% to 99% that can be maintained continuously without regeneration. The feed rate and the voltage control the conductivity of the effluent. A schematic of the CDI process is in Figure 14.5.

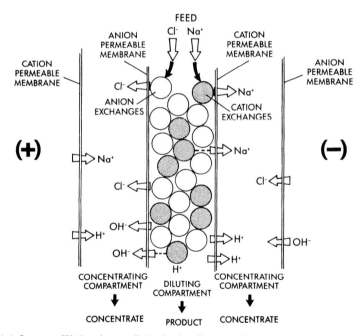

Figure 14.4 IONPURE™ Continuous Deionization Process. Ions are attracted by direct current to cross selectively permeable membranes, and are continuously discharged.

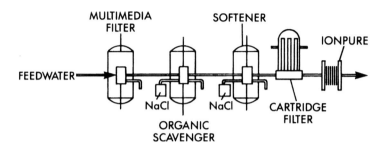

Figure 14.5 IONPURE™ Continuous Deionization Flow Schematic. Design of a CDI process includes maintaining a high quality of feedwater to prevent fouling of the resin and membranes. This configuration assumes feedwater Silt Density Index measured at 15 minutes (SDI_{15}) < 4.0 and free Cl_2 < 0.2 ppm.

14.4.9 Membrane Filters

Microbial retentive filters are used in water systems to prevent contamination of the system or to trap organisms in the water. These filters are used to protect vents on tanks and treatment equipment, to filter compressed gases for mixing or blanketing, and to control the microbial levels in the pharmaceutical water. The filters used in the water treatment system must be treated like those in any pharmaceutical manufacturing process. The filter must be properly selected for the application with respect to size, material, extractables, particulates, porosity, and stability during sanitization. The filter membrane and housing must also be maintained by proper installation, integrity testing, sanitizing, and inspection. Bacterial grow-through can occur in filters in wet areas such as tank vents or the water stream. These filters need to be sterilized daily so as not to contaminate the system.

Filters used to vent sanitary tanks need to be maintained above the dew point of the air being displaced from the tank, so that they remain dry. One way to maintain a dry tank vent filter and sterilize it is to use a jacketed sanitary housing (Figure 14.6) kept hot by steam or hot water. Methods for sterilizing in-line filters include heating the entire stream, steaming them in place, or doing it off-line and returning them with proper aseptic technique.

Membrane filters should never be placed at the point of use to control the microbial quality of the pharmaceutical water. This location provides a dead leg in the system that is typically not maintained properly. The use of any membrane filter in the pharmaceutical water system should be carefully evaluated, and any such filter must be properly maintained.

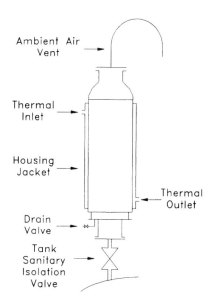

Ambient Air
Vent

Thermal
Inlet

Housing
Jacket

Thermal
Outlet

Drain
Valve

Figure 14.6 Jacketed Sanitary Filter Housing
Configuration. Sanitary vent filters use hot
water or steam jackets to keep the membrane
dry and sanitized.

Tank
Sanitary
Isolation
Valve

14.4.10 Ultrafilters

Ultrafiltration (UF) is a physical separation process across a semipermeable membrane. Since the pores are 1–20 nm, this process can separate high molecular weight species and particles, including colloids, particulates, microorganisms, endotoxin, and organic carbon, from the water stream (see Figure 14.7).

Feedwater flows through the ultrafilter tangentially to the membrane surface. A pressure differential forces product permeate water through the pores. The feed stream becomes concentrated and continuously flushes the surface of the membrane (Figure 14.8), cleaning the large material from the filter surface at a high flow rate. The flow is periodically reversed to backwash the filter pores (Figure 14.9).

The ultrafiltration process depends on maintaining both a balanced flow between the product and concentrate streams and a transmembrane differential pressure. If these factors are properly controlled, the filter has good retention properties and its surfaces do not become fouled. Design of an ultrafiltration system requires particular attention to the fluid dynamics of the treatment unit, to ensure that dead legs are not created as a result of piping, valves, and instruments.

Ultrafilters are periodically sanitized, typically with chemical agents compatible with the filter membrane. Hollow-fiber filter configurations can be integrity tested by a bubble-point test method.

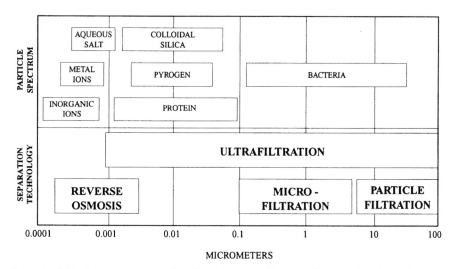

Figure 14.7 Particle Spectrum and Physical Separation Technologies. Particle sizes of different materials and the various technologies used for separation.

14.4.11 Reverse Osmosis

In *reverse osmosis* (*RO*), the application of external pressure changes the direction of the pressure gradient between low- and high-concentration solutions across a nonporous semipermeable membrane. The increase in pressure forces water across the membrane, but ions, microorganisms, endotoxin, and low molecular weight organic carbon are retained (Figure 14.7).

Feedwater flows through the unit tangentially to the membrane and continuously flushes the surface. The reject water stream is typically 30% to 50% of the feed flow. A portion of the rejected water may be recovered with a two-pass cartridge configuration (Figure 14.10). The separation process is at high pressure and requires large filter areas to achieve economical flow rates.

Reverse osmosis can be used to produce Water For Injection (WFI) [6]. The process is more economical than distillation, because it is less energy intensive. The maintenance of an integral reverse osmosis process is difficult to validate and keep in a state of control because of operational problems: the membranes may become fouled with organic or colloidal material, or they may leak or rupture because they are inherently fragile. Production of WFI with a validated reverse osmosis process often involves the use of a second unit, in series, to increase confidence in the operation. RO can be used to produce Purified Water, where the quality attributes are not as stringent.

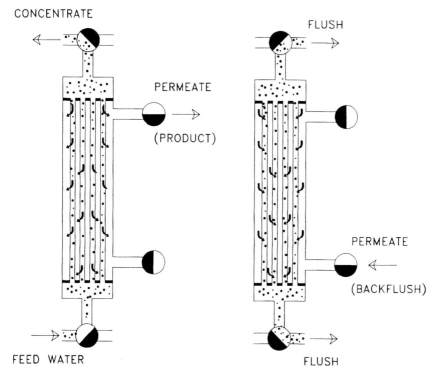

Figure 14.8 Ultrafiltration Cartridge in Forward-Flow Configuration.   The feed stream flows through the interior of the hollow-fiber ultrafiltration membrane, forces the permeate product through the pores, and flushes the contaminants from the surface.

Figure 14.9   Ultrafiltration Cartridge in Backwash Configuration.   The flow is reversed to force accumulated contaminants off the membrane.

14.4.12 Distillation

Distillation is used to produce Water For Injection.   Impurities are left behind when the water evaporates into steam.   Since particulates, microorganisms, endotoxin, and most organic and inorganic chemicals do not evaporate, they remain in the blowdown stream.   The water vapor is condensed, and volatile constituents are not.   It is then Water For Injection.

Distillation units are configured as single-effect, multiple-effect, and vapor-compression stills.   The major differences between the units are the method for separating entrained solids and liquids, and the energy requirements.

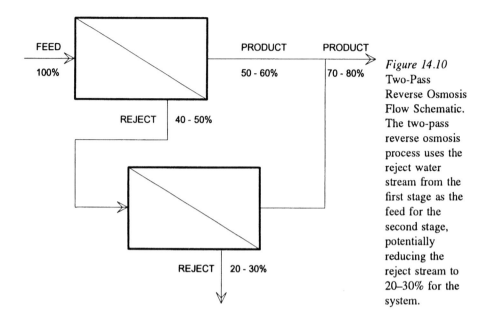

Figure 14.10
Two-Pass
Reverse Osmosis
Flow Schematic.
The two-pass
reverse osmosis
process uses the
reject water
stream from the
first stage as the
feed for the
second stage,
potentially
reducing the
reject stream to
20–30% for the
system.

Mist eliminators are used in low-velocity stills, to provide a surface for the liquid and particles to coalesce and fall back into the column. The velocity of the steam passing through the eliminator must not increase to the point that mist is reentrained upstream. Particulate material and water droplets entrained in the vapor stream may carry endotoxin and contaminate the WFI.

High-velocity distillation columns use gravity and centrifugal force to separate the heavier particulates and droplets from the water vapor. The steam enters the column tangentially and spirals down. The entrained matter is centrifuged and is separated as condensate. The steam travels up the column to be condensed (Figure 14.11).

Single- and multiple-effect distillation columns are high-velocity units that distill the water in one stage, convert the vapor into WFI, and use the bottom condensate to feed the next stage. The condensate from the last stage is the blowdown from the system (Figure 14.12). The use of a multiple-effect still is more energy efficient than a single-effect, because the energy recovered from the condensate and blowdown is used to preheat the feedwater to subsequent distillation columns.

Vapor-compression stills are single-effect units that operate at a low vapor velocity and low steam pressure. The feedwater evaporates on the exterior surfaces of the bank of tubes. The vapor passes through the demister and is mechanically compressed to a higher pressure. The steam passes through the inside of the tube bank, condensing in the tubes and transferring the latent heat to the feedwater (Figure 14.13). A vapor-compression still is more energy efficient than a multiple-effect still.

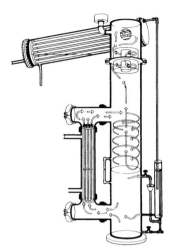

Figure 14.11 Spiral Removal of Particles in a High-Velocity Still. Gravity and centrifugal force are used to separate droplets and contaminants from the vapor stream. (Courtesy of Paul Mueller Co.)

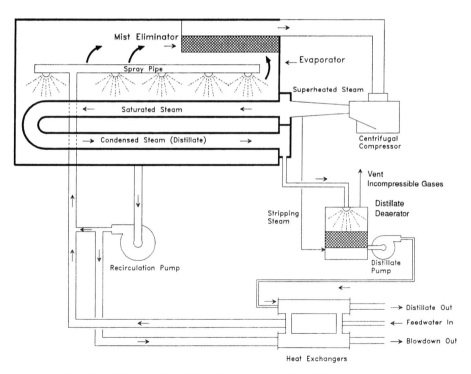

Figure 14.13 Vapor-Compression Still. Vapor-compression stills use mechanical energy to increase the operating pressure of the still. The latent heat recovered in the condenser is used to evaporate the feed stream. (Courtesy of Mechanical Equipment Co., MECO.)

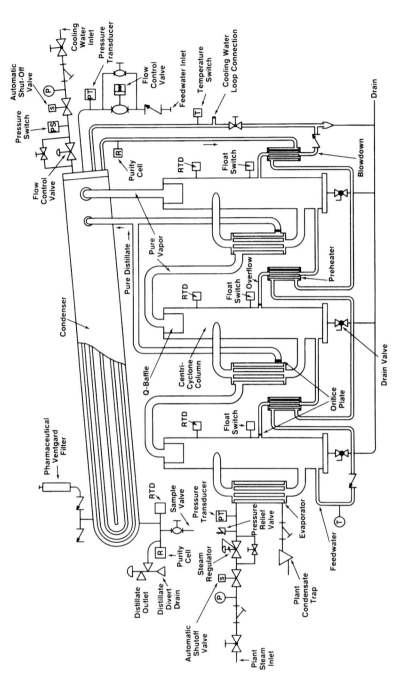

Figure 14.12 Multiple-Effect Still. The condensate from the first effect is the feed for the subsequent effect. Multiple-effect stills are more energy efficient than single-effect stills. (Courtesy of Paul Mueller Co.)

The quality of the feedwater to the still must be controlled to keep total dissolved solids (TDS) below 1 ppm. Such solids can form scale on the heat-transfer tubes, decreasing the efficiency of the unit and lowering the distillate quality. The scale can flake off and be entrained in the product water. Condensate can be blown down upon a signal from a high-level switch or continuously, depending on the design. A buildup of condensate with concentrated endotoxin and particulate matter can cause flooding of the column and contamination of the distillate.

The endotoxin level in the feedwater to the distillation unit should be monitored periodically to ensure that the reduction capacity of the process is not exceeded. Generally, the distillation process provides an approximate reduction in endotoxin level of five logs when operated in the midrange of its capacity, and three logs when operated at maximum capacity. The distillation unit can effectively produce condensate meeting the bacterial endotoxin attribute for WFI with a feedwater endotoxin level of 250 endotoxin units (EU)/mL. The efficiency of endotoxin reduction depends on the unit design, operation, and maintenance.

14.4.13 Vapor Generator

Clean-steam vapor generators produce steam free of boiler additives, particulates, and endotoxin. The quality of the condensed water vapor generally meets the quality attributes for Water For Injection. Vapor generators are designed similarly to a single-effect WFI still with the vapor condenser removed (Figure 14.14). Some systems use steam generators to produce small quantities of WFI, or WFI stills to produce clean steam.

The same design and operating concerns that exist for stills also exist for steam generators. The discharge steam pressure needs to be monitored to prevent a drop in the distribution system pressure, which can result in an upset in the operation of other plant equipment such as autoclaves and sterilization in place (SIP).

14.4.14 Condensers

Condensers are used in conjunction with distillation columns and clean steam generators to convert the steam into Water For Injection. The condenser should be of a double–tube-sheet type (Figure 14.15), to prevent cross contamination from the cooling medium in case of tube perforation. The condenser must be pitched toward the condensate discharge port, so that the shell side of the unit does not have a pocket of stagnant water. A sanitary, thermodisk-type steam trap should be installed on the clean steam supply pipe prior to the condenser inlet control valve, to prevent condensate stagnation when the unit is not in use.

A vent filter on the condenser prevents a vacuum from forming during the cooling process. The vent filter should have a jacketed, heated sanitary filter housing so

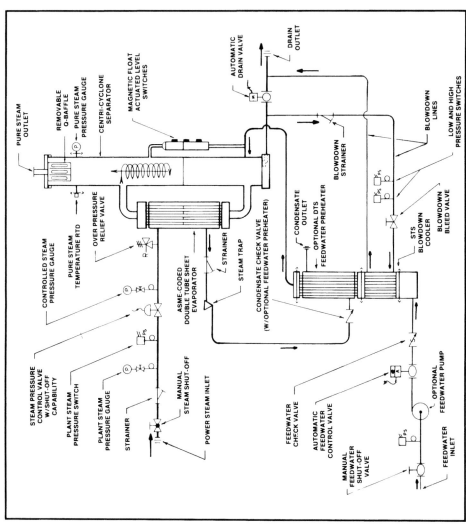

Figure 14.14 Clean-Steam Generator. Clean-steam generators are designed similarly to a single-effect WFI still with the vapor condenser removed. The quality of the condensed vapor meets the quality attributes of WFI. (Courtesy of Paul Mueller Co.)

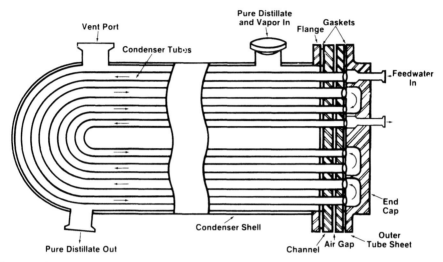

Figure 14.15   Sanitary Steam Condenser.   Double–tube-sheet condensers prevent cross contamination between the steam and cooling medium. (Courtesy of Paul Mueller Co.)

that the filter stays dry because the temperature of the unit is kept above the dew point of the air, and so that it can be sterilized in place or assembled aseptically.

14.5   Water Distribution System

14.5.1   Distribution Loop

The function of the distribution system is to provide the required quantity and protect the quality of pharmaceutical water to the point of use. The design of the system takes into consideration the sizing of the tanks, pumps, and tubing to accommodate the circulation of the water and the instantaneous demands.   The sanitary design considerations are incorporated to prevent microbial contamination and proliferation.

The configuration of the distribution system can be designed as a single-loop system with the use points in series (Figure 14.16), multiple sub-loops from a central distribution loop (Figure 14.17) with a parallel use point configuration, or dead-ended delivery pipe. The main consideration with any loop configuration is that the flow of water be balanced to ensure that fully developed turbulent flow exists, that no dead legs are created, and that the sanitization procedures incorporated into the operation of the system are effective at all locations.

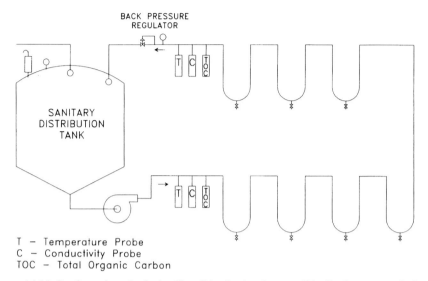

Figure 14.16 Configuration of a Series Flow Distribution System. Distribution system designed with all of the use points in series. The system must be sized to provide adequate flow with multiple use points in operation.

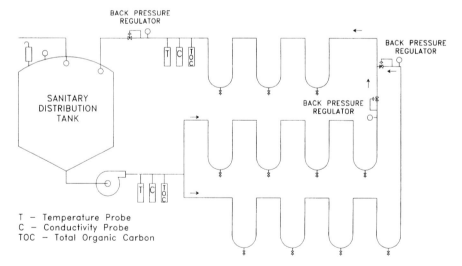

Figure 14.17 Configuration of a Parallel Sub-Loop Distribution System. Distribution system designed with loops in parallel with the main flow. The system must be designed to provide adequate flow in each loop with multiple loops in operation.

14.5.2   Holding Tanks

Sanitary tanks or vessels are used to provide a reserve capacity of water within the distribution system. They represent the buffer between the supply capabilities of the treatment system and the instantaneous demands of the user. This reserve also permits the treatment system to be off-line for regeneration, sanitization, or maintenance.

Distribution system tanks contain the Purified Water or Water for Injection to be provided at the point of use with minimum additional treatment. They must maintain the quality attributes of the water and prevent chemical and microbial contamination from the environment. The design of the tank conforms to sanitary design concerns, including smooth surfaces and welds; no dead legs from ports, valves or instruments; sanitary vent filter and pressure/vacuum safety device; isolation from the environment by sanitary gaskets and manhole covers; spray balls to flush the headspace; and capability of heating or sanitizing. Many of the design features of a sanitary tank are shown in Figure 14.18.

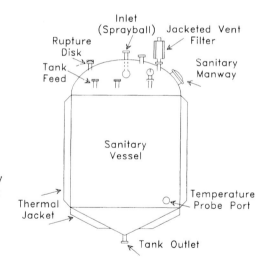

Figure 14.18 Features of a Typical Sanitary Vessel. Sanitary vessels contain the water, protect it from contamination, and conform to sanitary design practices. (Courtesy of Paul Mueller Co.)

14.5.3   Heat Exchangers

Heat exchangers are used to maintain the temperature of the water in the distribution system as a continuous microbial control method or for periodic sanitization. They are also used to satisfy production process requirements or safety concerns at the point of use. Heat exchangers are located in the distribution piping, with a continuous flow of water; or at the point of use, resulting in a potential dead leg.

Cross contamination between the heat transfer medium and the water is a major design concern with sanitary heat exchangers; it can occur if medium leaks past the mechanically formed seal between the heat exchanger tube and sheet. One way to prevent leakage of the medium is to ensure that the pressure on the water side of the heat exchanger is always greater than on the medium side, and to continuously monitor the differential pressure (with alarm point included). Concentric double-tube or double–tube-sheet heat exchangers are the preferred design to prevent cross contamination. These designs have an air break between the two fluids.

14.5.4   Pumps

Pumps used in the treatment and distribution system must meet sanitary design requirements. A particular concern is the type of pump shaft seal and material it is made of. The seal should not require external lubrication. This requirement is met by an external balanced or water-cooled balanced double seal. The cooling water must be of the same quality as that being pumped, or higher, and at a very low pressure so that it is not forced inside the pump housing. Pumps may also be equipped to permit steaming in place for sanitization. Such a pump has a low-point casing drain to discharge condensate, and the discharge connection rotated up 45° to minimize air entrainment during steaming. When multiple reserve pumps are connected in parallel, any pump not in operation at any given time must be considered a potential dead leg.

14.5.5   Valves

Valves are a potential source of microbial or chemical contamination anywhere in the treatment or distribution system. They are designed to stop flow and to prevent the mixing of fluid streams. The design of valves is critical to reduce potential areas of microbial growth and cross contamination.

The material of construction for valves must be compatible with the fluid stream to resist corrosion, prevent the leaching of materials such as plasticizers or lubricants, and meet the service requirements such as hot fluid or steam and repeated cycling. The design of parts that contact fluid must meet the requirements for sanitary design. These include smooth surfaces, no fluid contact with operator mechanisms, no trapped fluid when the valve is closed, and self-drainage of the downstream side. Diaphragm valves (Figure 14.19), radial-diaphragm valves (Figure 14.20), and plug valves (Figure 14.21) meet these sanitary design criteria.

Design of valve configurations and the modes of operation have to take into consideration the potential for creating dead legs whenever a valve is installed or operates. The sanitization must be designed for the valves and the associated upstream and downstream piping.

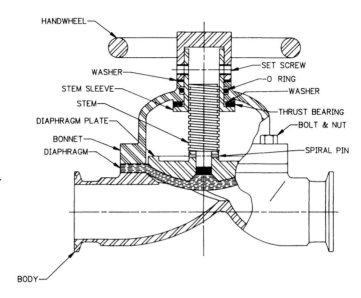

Figure 14.19
Sanitary
Diaphragm Valve.
The diaphragm
valve provides a
sanitary seal by
pinching the seal
against a weir.
(Courtesy of Tri-
Clover, Inc.)

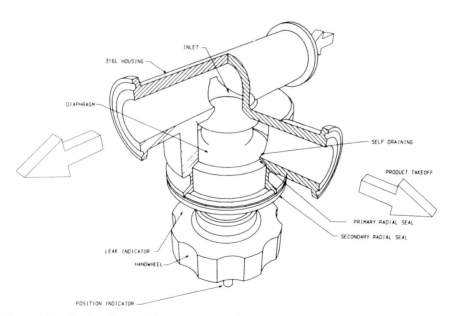

Figure 14.20 Sanitary Radial-Diaphragm Valve. The radial-diaphragm valve seals flush against the pipe or tank with no dead leg. (Courtesy of ASEPCO.)

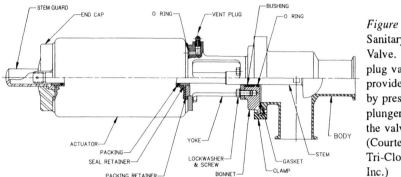

Figure 14.21 Sanitary Plug Valve. The plug valve provides a seal by pressing a plunger against the valve seat. (Courtesy of Tri-Clover, Inc.)

14.5.6 Passivation

Passivation is the treatment of stainless steel with strong oxidants to both remove free iron from the surface and protect the surface from further oxidation. During the fabrication, polishing, and welding of stainless steel, iron is imbedded or deposited on the surface of the material. This iron corrodes the surface and contaminates the water, and with time the iron in the alloy also corrodes because of the oxidizing property of high-purity water.

The process of passivation [20] includes the cleaning of the material to be passivated and the chemical treatment. The surface must be cleaned to remove oils and solvents used during the fabrication and installation of the tubing, valves, or tanks. Strongly alkaline agents, such as sodium hydroxide solution, are usually used. The solution is heated and recirculated in the system, or the parts are immersed in a bath. The cleaning solution is removed and the system or parts are rinsed with high-purity water. The passivating solution is a strong oxidizing agent, such as nitric acid or citric acid, or an organic acid chelating agent [21]. The system or parts are exposed to the solution, the solution is removed, and the metal is rinsed.

Passivation oxidizes and removes the free iron on the surface of the stainless steel and accelerates the formation of a protective layer of chromium oxide. This passive oxide layer is primarily responsible for the corrosion-resistant properties of stainless steel. Passivation may need to be repeated periodically to remove iron from the surface and to reform the passive layer.

14.5.7 Rouge

Rouge is the brownish rust film found on the surfaces of unprotected stainless steel in contact with hot high-purity water, in distillation columns, tanks, or distribution systcms. Thc rougc is powdery and is easily wiped from the surface. It is primarily ferric oxide, but may contain iron, chromium, and nickel [21].

Rouge can be removed from the system or part with a 1:10 solution of phosphoric or oxalic acid and water. The solution is heated to 71–82 °C and the system is treated for 3 to 4 hours [22]. The system is flushed with high-purity water. The surface passivity of the metal should be tested to determine the integrity of the passive layer. Passivation should be performed if indicated.

Rouge is best treated by prevention: specify material with low iron content, such as type 316L stainless steel; use proper welding techniques during installation; properly clean and passivate the items; and reduce the aggressive conditions for corrosion, such as heat.

14.5.8 Sanitization

The microbial quality of pharmaceutical water is maintained by the design of the system. The system minimizes contamination of the water and does not provide an environment for proliferation of organisms. Incorporated into the design are unit operations that sanitize or reduce microbial growth. The two approaches to sanitization are to control the microbial growth in the water with operations such as heat treatment, ultraviolet radiation, or ozonation; or to sanitize the equipment, tubing, and components with the same treatments as just listed or with various chemical agents.

14.5.9 Heat

Heating the system can effectively sanitize equipment, components, and water. A temperature of 80 °C controls the presence and growth of microorganisms in both the system and the pharmaceutical water. Heating the water does not introduce any "added substances" [6] that have to be flushed or decomposed, as would chemical treatment. The construction of the equipment and system needs to be able to withstand the elevated temperatures present during heat sanitization.

What temperature to use for heat sanitization depends on the type of microorganisms found in the particular system, and on the desired effect, whether sanitization or sterilization. Microorganisms are affected by heat in the following manner [23]:

- psychrophilic and mesophilic microorganisms do not generally proliferate above 50 °C;
- most pathogenic organisms do not grow above 60 °C;
- sustained temperatures of 55–60 °C are lethal to USP indicator organisms;
- most thermophilic microorganisms do not proliferate at temperatures above 73 °C;
- most vegetative organisms are killed at 60 °C after 30 minutes' exposure.

14.5.10 Ultraviolet Radiation

Ultraviolet (UV) light is used to reduce the microbial level in water that has a low level of particulates. UV radiation in the wavelength range of 200–290 nm (peak at 254 nm) disrupts the DNA of organisms and prevents reproduction. The effectiveness of UV is related to the dose applied, expressed in units of mW·s/cm². The dose is a function of the lamp intensity, velocity of the water, configuration of the lamps, and radius of the water line. A minimum dose of 30 mW·s/cm² is recommended to achieve a bacterial reduction of 99% or greater [24].

Ultraviolet units are configured as single- or multitube systems, and include light intensity meters, inspection ports, and running-hour meters to monitor the lamps. The intensity of the UV lamps is reduced by scaling of the quartz sleeves, solarization (hazing) of the sleeves, and time in service. The effectiveness of UV sanitization is also affected by the velocity of the fluid and by particulates scattering light or shielding the organisms. UV lights require maintenance to remain effective. This includes periodic cleaning of the lamps to remove scale, and monitoring of the UV light intensity.

14.5.11 Ozonation

Ozone, O_3, is an unstable, highly reactive form of oxygen used as a sanitizing agent for equipment, components, and pharmaceutical water. It is formed when air or oxygen is passed through an electrically charged field; the gas is dissolved in the water. The half-life of ozone in water is less than 30 minutes, and decreases with increasing turbidity, temperature, or pH.

Ozone is effective as a sanitizing agent for a broad spectrum of microorganisms. Ozone levels maintained between 0.2 and 0.5 mg/L in water, with a contact time of 20 minutes, have been shown to reduce the concentration of bacteria by 99.99% [25]. Ozone can be used to sanitize both the water and the system components.

Difficulties associated with the use of ozone are related to its generation, distribution, compatibility, and safety. The dissolution of ozone in the water must be carefully balanced to ensure that the desired concentration is maintained and that undissolved gas is not liberated during the process. The ozone concentration rapidly decreases as the water circulates through the system. Piping and components in an extensive system may not be exposed to an adequate dose and can permit microbial growth. The oxidizing properties of ozone can rapidly deteriorate components within the treatment system. Compatibility of gaskets, seals, instruments, piping, and filters must be evaluated. Human exposure to ozone must be limited because of its strong oxidizing properties.

The continuous use of ozone in a pharmaceutical water system as a sanitizing agent presents a unique situation. The monographs for Purified Water and Water For

Injection state that the water "contain no added substance" [6]. This, at times, has been interpreted to mean that any addition is subsequently removed from the water [23]. Ozone is easily removed because UV light decomposes it to oxygen. The process of ozone removal must be validated. Moreover, the stability of product manufactured using water saturated with oxygen must be evaluated.

Systems in which ozone use has been discontinued have shown significant increases in microbial numbers within 2 to 4 days. This rebound phenomenon has been attributed to the decomposition by ozone of complex organic molecules, not available as nutrients for the organisms, into assimilable organic carbon sources. A small number of microorganisms can use the relatively high nutrient level to grow rapidly. This may be prevented if a low total organic carbon level is maintained in systems being treated with ozone.

14.5.12 Chemical Sanitization

Chemical biocides are periodically used to sanitize equipment and components that cannot sustain prolonged exposure to elevated temperatures. Selection of the chemical treatment includes evaluation of the treatment on the microorganisms in the specific system, and its compatibility with the equipment, gaskets, and instrumentation, as well as the ability to remove the chemicals from the water system. Removal of all traces of the chemical agents may be difficult, and must be validated.

Compounds used are primarily oxidizing agents, such as chlorine, sodium hypochlorite, chlorine dioxide, iodine and iodophors, hydrogen peroxide, ozone, and peracetic acid. Other agents include quaternary ammonium compounds and formalin. Table 14.12 lists the concentrations and exposure times for some of these agents. These are hazardous substances that must be properly handled.

Table 14.12 Chemical Sanitizing Agents

Compound	Concentration	Exposure time
Chlorine compounds	50–200 mg/L	1–2 h
Formaldehyde	0.5–4%	2–3 h
Hydrogen peroxide	5–10%	2–5 h
Iodine	50–100 mg/L	1–2 h
Ozone	0.2–0.5 mg/L	1–2 h

14.6   Validation

Pharmaceutical water systems are considered to be a manufacturing process for a drug ingredient, and must be validated to conform to the requirements of the U.S. Food and

Drug Administration Current Good Manufacturing Practices [26]. The process of validation is the overall activity, composed of:
- defining the system and its functions,
- establishing operating parameters,
- qualifying the installation of the system,
- verifying the functions,
- establishing confidence in the operation and performance, and
- following up to verify that the system is operating in control.

Validation is defined as "establishing documented evidence which provides a high degree of assurance that a specific process will consistently produce a product meeting its predetermined specifications and quality attributes" [27].

14.6.1   Team

Validation of the water system should be implemented as a cooperative effort of the principals involved in its design, installation, start-up, operation, support, and use. The validation team brings together all the different perspectives and concerns of the disciplines represented. The team members involved in various aspects of the system and its validation are listed in Table 14.13.

Table 14.13 Roles of Members of a Validation Team

Engineering	System design, installation documentation, start-up
Laboratories	Testing services, troubleshooting
Operations	Start-up, operation, procedures, maintenance
Quality	Regulatory requirements, documentation review
Users	Quality attributes, consumption, operating modes
Vendors	Recommended practices, manuals, maintenance, troubleshooting

Responsibility for validation of the equipment and the system, as it has been installed and is being operated, may be given to the operators and users of the system, or to the engineers who designed and installed the system. Engineers, quality monitoring and assurance groups, and equipment manufacturers provide the supporting information and services that cumulatively result in a water system in a validated state of control.

14.6.2   Time Line

Validation is an overall process that encompasses the definition of the system, the installation and operational phases, and the performance qualification phases. Elements in the validation of a pharmaceutical water system parallel the activities necessary to design, install, start up, and produce acceptable product. The quality standards and functional requirements developed during the preliminary design represent the parameters used to develop the acceptance criteria for the validation protocol. Validation activities parallel the other water system implementation phases. A time line [28] for these elements is shown in Figure 14.22.

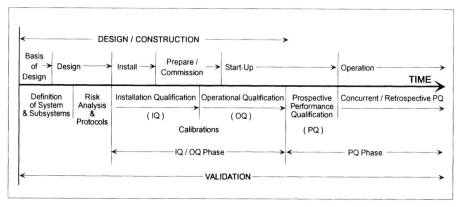

Figure 14.22   Validation Time Line.   Validation activities start at the time of the design of a system and proceed on a parallel track with the installation and start-up.

14.6.3   Criteria

Development of the validation program starts with a definition of the functional and performance requirements of the system. The analysis of the water treatment system results in a flow schematic that identifies all the variables, the procedures and operating conditions imposed on the system, and the relationships between the various operations and their effect on the quality of the water.

The first step in the analysis is to identify the components and subsystems within the overall system (Figure 14.23). The anticipated performance of the subsystem and the variables that can affect the operation are listed. The effect of each variable on the operation is evaluated, and those that significantly affect the operation or performance are identified for follow-up during a qualification phase. The subsystems are linked and their effect on each other is again evaluated.

14.6.4   Protocols

The validation program is formalized and documented in the form of *protocols*. These are the documents that identify the scope and objectives of the activity, describe the system, and list the test plans and acceptance criteria. The validation project plan or

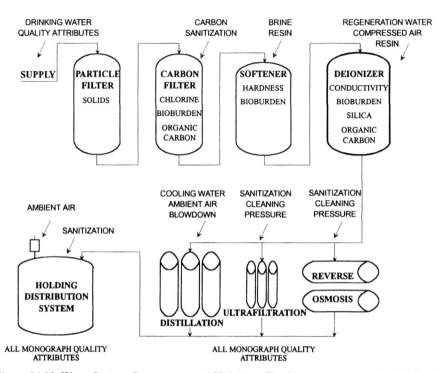

Figure 14.23 Water System Components and Variables. The inputs, outputs, and variables for each component of the system are identified. The effect of each item is evaluated against the overall performance of the system.

protocol may be a single document or it may consist of multiple protocols for the different qualification phases or subsystems. Protocols should be carefully developed, reviewed, and approved by the different principals of the validation team. Protocols should be concise and simple, referring to other documents when needed. The documentation and test plans identified in the protocols should be substantive and worthwhile activities for the operation and be able to withstand a regulatory review.

The major sections of the validation protocol are the installation qualification, the operation qualification, and the performance qualification. These identify the specific documents, tests, and acceptance criteria used to evaluate the system.

A summary report is prepared at the completion of the individual protocols. It presents the test data and documentation identified in the protocol, discusses the results, and provides a conclusion. Any follow-up activities resulting from implementation of the protocol or review of the data need to be identified and tracked until completed. The summary report must be formally reviewed and accepted.

14.6.5 Installation Qualification

The *installation qualification* (*IQ*) phase of the validation involves the documentation of the water system. The installed equipment, piping, and components are inspected and verified against the specifications and drawings developed during the design phase and the manufacturers' recommendations.

Implementation of the installation qualification begins with the fabrication of the equipment. Critical specifications need to be inspected prior to completion of fabrication of the components at the manufacturer's site. Items to be inspected may include tank jackets, interior components of deionizers or distillation columns, interior polish of tanks, and function of control panels.

Documentation for the equipment and components in the system is assembled and verified. This includes part and serial numbers, materials of construction, surface finish, operation and maintenance (*O&M*) manuals from the manufacturer, shop drawings and installation drawings (*as-built*), instrument manuals, control sequences, and control system documentation.

The drawings developed during the design and installation of a water system provide details required to construct the system, but may be of little use during the validation. A drawing of the water treatment and distribution system should be prepared that provides the information that will be useful during the validation and operation. This may be a system schematic or a process and instrumentation drawing (*P&ID*) that shows the equipment, process streams, and instrumentation.

The documents and system schematic are verified by an inspection of the actual system installation. This includes an inspection for conformance with sanitary design practices, including proper slopes, materials of construction, weld quality, and identification of dead legs.

Procedures that detail the operation, preventive maintenance, and calibration of instruments are implemented during the IQ. These procedures are used during the subsequent phases of the validation and the ongoing operation.

14.6.6 Operation Qualification

The *operation qualification* (*OQ*) phase of the validation involves verifying and documenting that the components and subsystems of the water systems operate properly throughout the normal operating range. The physical operation of the components and controls is verified. This includes the proper operation of valves, pumps, and controls.

The operation of the components is verified against the manufacturers' literature and the system specifications. The positions of flow control valves are adjusted to achieve the operating pressures and flows. Calibration of critical instruments used for monitoring, controlling, or logging the operation of the system is verified and adjusted. This includes the sensor, output signal or device, control loop, and data acquisition device. Control sequences for the various operations such as backwashing, regeneration, or sanitization are verified, including timers, set points, valve positioning, interlocks, and alarms.

After the operation of the components is verified, the system is ready to be loaded with materials and started. The procedures for initial loading, sanitizing, and regeneration of carbon and resin have to be carefully performed to minimize the potential for microbial contamination and to ensure that the full capacity of the unit operation is not compromised. The start-up includes the sanitization and installation of filters, sanitization of the distribution system, and balancing of system flows.

The proper operation of all the components is given an initial check to verify that water of acceptable quality is discharged from each component of the treatment system. This is not a performance verification, but an operational and control verification.

14.6.7   Performance Qualification

The *performance qualification* (*PQ*) phase establishes the data to be used to demonstrate confidence that the system can produce water of acceptable quality. The PQ establishes the parameters for evaluating the water system and for sampling.

Data to evaluate the performance of a system can be obtained prior to use of the system (*prospective*), generated in parallel with use of the system (*concurrent*), or from historical sources (*retrospective*).

The initial prospective test phase provides confidence that the system will operate adequately when put in service. Data are collected from the distribution system and selected locations in the treatment system. The use of the system and the modes of operation should simulate those to be used during normal operations. This phase lasts long enough to demonstrate that the system is stable, generally 2–4 weeks. Water produced during this time may be used for manufacturing or other purposes only if it is tested and released prior to use.

The prospective performance qualification phase should be used to evaluate and substantiate any performance or regulatory compliance claims made by equipment manufacturers or vendors, and to develop a profile of the system for process control and troubleshooting. This includes parameters such as particulate removal, deionization capacity, and endotoxin reduction.

The data generated during the concurrent/retrospective phase, with the system under normal operating conditions, demonstrate that the control strategies for the system are stable while the system is operated following routine procedures and maintenance, during seasonal changes, and after multiple regenerations and sanitizations. This concurrent phase may last from 6 to 12 months.

At the conclusion of the concurrent phase, sufficient data have been generated to characterize the treatment system and distribution system. These data should be reviewed to identify any procedures or practices that need to be modified. The sampling plan can generally be scaled down in response to trends that have been identified.

14.6.8 Sampling Plans

The sampling plan in the performance qualification is developed to characterize the quality attributes of the treatment system and the distribution system. One approach is to sample at all the identified locations, for all the parameters, every day. This would burden the company finances and resources, with no significant increase in knowledge about the system. A well-designed sampling plan, based on information about the design and operation of the water system, can yield a sample collection and test pattern that evaluates only affected parameters, at a frequency ranging from daily to weekly.

Samples taken from points on the treatment system should be evaluated for the attributes that are affected by the unit operation or that can adversely affect a subsequent step. A list of the equipment performing the unit operations and the attributes to be examined is in Table 14.14. The frequency of the sampling depends on any trends observed in the performance of the unit operation and the operating cycles. A deionizer column that requires a daily regeneration is sampled more frequently to capture data on its performance with regard to chemical attributes, while a carbon bed is sampled more frequently for microbial attributes.

Samples taken from points on the distribution system are tested to evaluate the quality of the pharmaceutical-grade water. The attributes tested are those identified in the compendia or procedures. The sampling plan is based on the configuration and operation of the system. A sampling plan for a series distribution loop with multiple use points can assume that all sampling locations are equivalent and that the test data are representative of the entire system. This can result in a plan that rotates the sampling points, with fewer samples collected. A loop with parallel use points or multiple loops results in a sampling plan that provides data representative of only individual use points or branch loops. The operation of the distribution loop affects the frequencies and the attributes in the sampling plan. A hot recirculating loop

Table 14.14 Sampling Attributes for Equipment Performing Unit Operations

Unit Operation	Attribute
Drinking water supply	drinking water attributes
Softener	microorganisms, hardness
Multimedia filter	particulates
Carbon bed	chlorine, TOC, microorganisms
Deionizer	conductivity, microorganisms, silica, Purified Water
Membrane filter	microorganisms
Ultrafilter	conductivity, microorganisms, endotoxin, Purified Water
Reverse osmosis unit	conductivity, microorganisms, endotoxin, Purified Water
Distillation unit	conductivity, microorganisms, endotoxin, Purified Water
Ultraviolet lamp	microorganisms
Ozonator	ozone, TOC, microorganisms
Use points	conductivity, microorganisms, endotoxin, Purified Water

TOC = total organic carbon

requires fewer microbial samples than a cold system. The minimum frequency for microbiological testing should be weekly for each point of use. Chemical attributes can be tested less frequently if on-line instrumentation is used to monitor the water.

14.6.9   Performance Parameters

The validation protocol identifies the parameters to be used to evaluate the water system and the product, a pharmaceutical grade of water. These should include the location in the treatment or distribution system to be used for monitoring, the test methods, and the acceptance criteria for the chemical and microbial parameters.

Performance criteria for the pharmaceutical water produced and supplied to the point of use are defined in the pharmacopeia monograph. The monographs may also require a certain quality for water used to supply the treatment system [6].

The acceptance criteria for the chemical attributes may be pass/fail, based on an acceptance limit. The microbial attributes are established by the user, based on the requirements of the process and the product, and are evaluated according to an alert/action level and follow-up responses (Figure 14.24) [29]. *Alert levels* are set on the basis of the microbial populations and characteristics observed during routine monitoring. Deviations are investigated to determine a cause, the situation is corrected, and the effectiveness of the corrective acts is evaluated. An *action level* is defined by the quality requirements of the product or process and indicates the need

Figure 14.24 Microbial Attribute Alert/Action Trend Chart. Microbial attributes for water are evaluated using trend charts developed from sampling data. Alert/action levels are established based on the performance of the system and the requirements of the particular product or process.

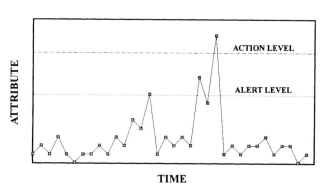

for an additional response. This could include identifying and quarantining product affected by equipment, components or compounding steps that were in contact with the water; performing additional tests to evaluate the water system and the product; determining the disposition of the product and the status of the system; implementing corrective measures; and documenting the alert level and investigation.

14.6.10 Change Control

A water system is validated on the basis of an established configuration and operational procedures. Significant changes to the equipment, operating parameters, or procedures can affect the validated state of control. An established mechanism for identifying proposed changes, evaluating their effect on the validation, and determining follow-up activities ensures that the system remains in a state of control. Change control should be implemented early in the validation process, to avoid undocumented operational or configuration changes, which could result in costly reassessments of the collected data or require repetition of the validation tests.

The change control process should operate in conjunction with the systems that effect change: engineering work orders, procedures, maintenance activities, operator instructions, and purchase requests. Those changes that affect the components, configuration, instrumentation, operation, or control of the water system need to be identified and evaluated. Changes that result from activities such as routine maintenance, calibration adjustments, or replacement of components by equipment with similar functional descriptions generally do not require any evaluation. A proposed change should be evaluated by qualified parties knowledgeable in the water system and validation principles. Follow-up activities for a proposed change may include updating drawings and procedures, performing additional testing, or requalifying the affected equipment or system.

A periodic review should be performed to evaluate the cumulative effect of all the changes on the operation and performance of the water system. This includes a review of the changes and successful completion of the follow-up activities, the test data, drawing and procedures, and a physical inspection of the system to determine whether any undocumented changes have been implemented.

Most changes are proposed to improve the operational or quality attributes of the water system. The change control system needs to be designed to facilitate these changes and not impede their implementation.

14.7   References

1. Title 40, *Code of Federal Regulations (CFR-40)*, Part 141; *National Primary Drinking Water Regulations*; Superintendent of Documents, Washington, DC, July 1, 1992.

2. Title 40, *Code of Federal Regulations (CFR-40)*, Part 143; *National Secondary Drinking Water Regulations*; Superintendent of Documents, Washington, DC, July 1, 1992.

3. SAYRE, I. M., International Standards for Drinking Water, *J. Am. Water Works Assoc. 80*, 56–58 (1988).

4. *Supplement to The Pharmacopoeia of Japan*; 11th ed., Japan Ministry of Health and Welfare, 1988; pp 1501–1506.

5. *Guidelines for Canadian Drinking Water Quality*; Minister of Health and Welfare, Ottawa, 1987.

6. *United States Pharmacopeia XXII and Supplements*; United States Pharmacopeial Convention, Rockville, MD, 1990.

7. MUNSON, T. E.; FDA Views on Water System Validation; *Pharm. Tech. Conference '85 Proceedings*; Aster Publishing Corp., Springfield, OR, 1985; pp 287–289.

8. PHARMACEUTICAL MANUFACTURERS ASSOCIATION WATER QUALITY COMMITTEE, Updating Requirements for Pharmaceutical Grades of Water: Microbial Concerns, *Pharmacop. Forum 18*, 4397–4399 (1992).

9. PHARMACEUTICAL MANUFACTURERS ASSOCIATION WATER QUALITY COMMITTEE, Updating Requirements for Pharmaceutical Grades of Water, *Pharmacop. Forum 16*, 1283–1285 (1990).

10. PHARMACEUTICAL MANUFACTURERS ASSOCIATION WATER QUALITY COMMITTEE, Updating Requirements for Pharmaceutical Grades of Water: Conductivity, *Pharmacop. Forum 17*, 2669–2675 (1991).

11. PHARMACEUTICAL MANUFACTURERS ASSOCIATION WATER QUALITY COMMITTEE, Updating Requirements for Pharmaceutical Grades of Water: Total Organic Carbon; *Pharmacop. Forum 19* (4), (July/Aug. 1993).

12. NATIONAL COMMITTEE FOR CLINICAL LABORATORY STANDARDS; *Preparation and Testing of Reagent Water in the Clinical Laboratory*; National Committee for Clinical Laboratory Standards, Villanova, PA, 1991.

13. INTERNATIONAL ASSOCIATION OF MILK, FOOD AND ENVIRONMENTAL SANITARIANS, INC.; *3-A Sanitary Standards*; International Association of Milk, Food and Environmental Sanitarians, Inc., Des Moines, IA; 1947–present. Alternatively, the Standards are

available from the 3-A Sanitary Standards Committees, 6245 Executive Blvd., Rockville, MD, 20852; (301) 984-1444.

14. MITTLEMAN, M. W., GEESEY, G. G.; *Biological Fouling of Industrial Water Systems: A Problem Solving Approach*; Water Micro Associates, San Diego, CA, 1987; pp 236–238.

15. SAKIADIS, B. C.; Fluid and Particle Mechanics; Chapter 5 in: *Perry's Chemical Engineers' Handbook*, PERRY, R. H., GREEN, D., Eds.; 6th ed., McGraw-Hill, New York, 1984; p 5-28.

16. ADAMS, D. G., AGARWAL, D., Clean-in-Place System Design, *BioPharm 2*, 56 (1989).

17. SLABICKY, R. O., unpublished cleaning validation data.

18. *Standard Specification for Seamless and Welded Austenitic Stainless Steel Sanitary Tubing*; ASTM A270-90; Annual Book of ASTM Standards, Vol. 1.01; American Society for Testing Materials, Philadelphia, PA, 1993.

19. GANZI, G. C., PARISE, P. L., The Production of Pharmaceutical Grades of Water Using Continuous Deionization Post Reverse Osmosis, *J. Parenter. Sci. Technol. 44*, 231–241 (1990).

20. BANES, P. H., Passivation: Understanding and Performing Procedures on Austenitic Stainless Steel Systems, *Pharm. Eng. 10*, 41–46 (1990).

21. COLEMAN, D. C., EVANS, R., Corrosion Investigation of 316L Stainless Steel Pharmaceutical WFI Systems, *Pharm. Eng. 11*, 9–13 (1991).

22. MENON, G. R., "Rouge" and Its Removal from High-Purity Water Systems, *BioPharm 3*, 40–43 (1990).

23. Pharmaceutical Manufacturers Association Deionized Water Committee, Protection of Water Treatment Systems, Part IIa: Potential Solutions, *Pharm. Technol. 7*, 86–92 (1983).

24. MITTLEMAN, M. W., GEESEY, G. G.; *Biological Fouling of Industrial Water Systems: A Problem Solving Approach*; Water Micro Associates, San Diego, CA, 112–118, 1987.

25. BLOCK, S. S.; *Disinfection, Sterilization, and Preservation*; 4th ed., Lea & Feibiger, Philadelphia, PA, 1991.

26. Title 21, *Code of Federal Regulations (CFR-21)*, Part 211; *Current Good Manufacturing Practice for Finished Pharmaceuticals*; Superintendent of Documents, Washington, DC, April 1, 1993.

27. *Guideline on General Principles of Process Validation*; Food and Drug Administration, Washington, DC, May, 1987.

28. CHAPMAN, K. G., A History of Validation in the United States, Part I, *Pharm. Technol. 15*, 94 (1991).

29. Pharmaceutical Manufacturers Association Deionized Water Committee, Protection of Water Treatment Systems, Part III: Validation and Control, *Pharm. Technol. 8*, 54–68 (1984).

Utilities for Biotechnology Production Plants

Dan G. Adams

Contents

15.1   Introduction

Utility services in a biotechnology facility are diverse. They include compressed air, various liquified or compressed gases, different types of water (some of which contact product and some of which do not), and two types of steam—clean steam (also known as pure steam), and plant steam or house steam. I describe these services, with the exception of purified water, in this chapter. Aspects of these systems that I cover include general system descriptions, selection criteria, and sizing methods.

15.2   Plant Steam Systems

Plant steam is used primarily as a heat source. It is seldom used for other purposes, such as power generation, and never in any product contact applications. This section of the chapter focuses on plant steam as a heating medium.

The major components of a plant steam system are the boilers, the feed water pretreatment system, the distribution system, and the condensate collection and feed system. I discuss each of these individually, as well as the major users of plant steam.

15.2.1 Boilers

Boilers in production facilities are normally industrial-type packaged units, which require minimal installation cost. Installation involves support and placement of the unit, connection of electrical power, and connection of piping and instrumentation. Only very large power-utility units, which can operate at supercritical temperatures and pressures, with capacities in the multimillion lb/h (or kg/h) range, are custom-designed and constructed from components. The plant steam requirements for a biotechnology facility do not warrant this size boiler.

The two principal types of industrial steam boilers are the *water tube boiler* (Figure 15.1) and the *fire tube boiler* (Figure 15.2). The difference between them, as the names imply, is the location of the combustion relative to the water being vaporized. A fire tube boiler passes combustion gases through the tubes of the boiler. The water is enclosed in a larger vessel, which surrounds both the water and the tubes. Because of the nature of the technology, fuel for this type of boiler is limited to gas and oil, high pressures are not practical, and superheating of the steam is not readily accomplished. The range of sizes for this type of boiler is 5–750 horsepower (hp), or about 150–25,000 lb/h (70–11,500 kg/h). A *boiler horsepower* is the amount of energy required to vaporize a pound of water (0.45 kg) at 100 °C (212 °F). Design pressures normally range up to approximately 250 psig (1700 kPa). This size range and pressure range are normally more than adequate to meet the needs of a biotechnology facility.

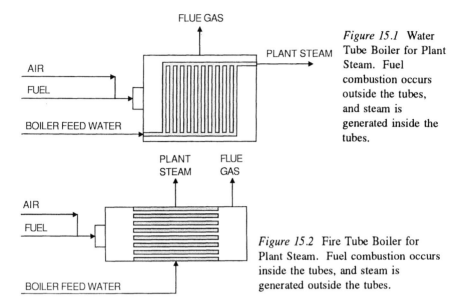

Figure 15.1 Water Tube Boiler for Plant Steam. Fuel combustion occurs outside the tubes, and steam is generated inside the tubes.

Figure 15.2 Fire Tube Boiler for Plant Steam. Fuel combustion occurs inside the tubes, and steam is generated outside the tubes.

The other principal type of industrial boiler, the water tube boiler, operates as the opposite of the fire tube boiler, in that the water is evaporated inside the tubes and the products of combustion pass through a firebox that has refractory materials on the outside of the tubes. These units are available in larger sizes than the fire tube type, and are also capable of higher operating pressures, with some boilers operating at up to 600 psig (4100 kPa). Capacities greater than 250,000 lb/h (115,000 kg/h) are possible, but rare. Shop-fabricated (packaged) units are available in the 10,000–250,000 lb/h range (300–7,500 hp).

The selection of the type of boiler depends on the capacity and pressure requirements of the facility, along with the cost of the units for the operating requirements under consideration. Most boilers in biotechnology facilities are of the fire tube type.

For smaller facilities, electrically heated boilers are sometimes used. The energy cost per pound of steam produced is higher for an electrically heated boiler; however, a facility with low plant steam requirements may be well served by this type of boiler. It is easily installed and requires little floor area, and no flue is needed for products of combustion. This type of boiler is available in sizes ranging from 10 lb/h to 5000 lb/h (5–2250 kg/h), with pressure ratings up to 250 psig (1700 kPa).

15.2.2 Pretreatment of Feed Water

Water for plant steam boilers is treated to protect the surfaces of the boiler, and other surfaces in contact with the steam, from scaling and corrosion. Which unit operations are involved in the pretreatment depends primarily on two factors: (a) the quality of the incoming feed water, and (b) the pressure of the steam being produced [1]. The impurities in the incoming feed water that are of primary concern are described by measurements of the total dissolved solids (TDS), alkalinity, hardness, silica, turbidity, oil, dissolved gases, and residual phosphate.

Dissolved gases in steam systems lead to corrosion of metallic surfaces in contact with the steam. Oxygen and carbon dioxide dissolve in water that is in contact with the atmosphere. These gases are less soluble at higher temperatures, as shown in Figure 15.3, which makes it possible to remove them by deaeration with steam [2]. A deaerator functions like a steam stripper: the water cascades over trays counter-current to steam that contacts the water as it rises through the deaerator. This is the most common method for removing dissolved gases. It is also common to remove oxygen with chemicals such as sodium sulfite, which is metered into the boiler feed water and combines with the oxygen to render it noncorrosive. If the boiler operates at a high pressure, the use of sodium sulfite is not recommended. Hydrazine is also used to control the oxygen content of steam, since it reacts with molecular oxygen to form water and molecular nitrogen. High concentrations of hydrazine lead to the formation of ammonia, and so must be avoided. Carbon dioxide forms corrosive carbonic acid in condensate collection systems if the boiler feed water is not treated properly. The feed water should be maintained in an alkaline condition (between pH values of 10.5 and 11.0) and, if necessary, amines can be added to further neutralize any carbonic acid that may form.

Constituents of the boiler feed water that cause scaling are primarily in the form of dissolved and suspended solids. As in the case of gases, some of these constituents are less soluble at higher temperatures. Chemical treatment provides a satisfactory means of preventing scale, as the chemicals reduce the scale-forming substances to a soft sludge, which is removed in blowdown. Frequently, softening or ion exchange is used to remove these impurities. Table 15.1 provides data for limits on the concentration of impurities in boiler feed water [4].

15.2.3 Distribution System

Plant steam distribution systems are well understood and developed. The velocity of the steam in the lines, insulation, materials of construction, and construction components are important considerations in the design of the distribution system.

The recommended velocity of steam in distribution piping is a function of the steam pressure in the line and of the equivalent length of piping through which it is flowing. Generally, the higher the plant steam pressure, the higher the allowable velocity, and the greater the equivalent length, the lower the allowable velocity. Table 15.2 shows the maximum pressure drop for various steam pressures and velocities per 100 equivalent feet (30 m) of schedule 40 carbon steel pipe, along with suggested velocity ranges. Velocities outside these ranges are certainly acceptable if a specific situation

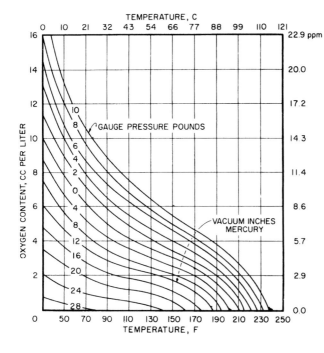

Figure 15.3 Dissolved Oxygen as a Function of Temperature and Pressure. The solubility of oxygen in water decreases with increasing temperature. (Courtesy of Betz Industrial, Trevose, PA.)

Table 15.1 Recommended Limits for Boiler Feed Water Constituents (parts per million)

Pressure, psi	Total dissolved solids	Alka-linity	Hard-ness	Silica	Tur-bidity	Oil	Phos-phate residual
0–300	3500	700	0	100–60	175	7	140
301–450	3000	600	0	60–45	150	7	120
451–600	2500	500	0	45–35	125	7	100
601–750	2000	400	0	35–25	100	7	80
751–900	1500	300	0	25–15	75	7	60
901–1000	1250	250	0	15–12	63	7	50
1001–1500	1000	200	0	12–2	50	7	40

warrants different operating conditions. If, for example, the equivalent length of a segment of pipe in a distribution network is very short, a higher velocity can be tolerated.

Table 15.2 Recommendations for Plant Steam Pressure Loss and Velocity

Steam pressure, psig	20	30	40	50	100	200
Steam temperature, °F	259	274	287	298	338	388
Steam viscosity, cP	0.013	0.014	0.014	0.014	0.015	0.016
Steam density, lb/ft^3	0.0834	0.106	0.128	0.15	0.257	0.468
Max. press. drop, psi/100 ft	0.2	0.3	0.3	0.3	0.4	0.5
Nominal pipe size, inches	Maximum lb/h x 1000					
(schedule 40 carbon steel)	(Velocity, ft/s)					
1	0.05	0.065	0.07	0.08	0.12	0.19
	(28)	(28)	(25)	(25)	(22)	(19)
2	0.3	0.4	0.45	0.48	0.75	1.1
	(43)	(45)	(42)	(38)	(35)	(28)
3	0.8	1.1	1.3	1.4	2.1	3.2
	(52)	(56)	(55)	(51)	(44)	(37)
4	1.7	2.3	2.6	2.8	4.3	6.5
	(64)	(68)	(64)	(59)	(53)	(44)
6	5	6.7	7.6	8.2	12.5	19
	(83)	(89)	(82)	(76)	(67)	(56)
8	10	14	15.5	17	26	39
	(96)	(106)	(97)	(91)	(81)	(67)

The values in Table 15.2 for pressure drop per 100 feet were calculated with the DARCY equation (here in English units); an arithmetic average for fluid properties was used:

$$\Delta P = 0.001294 \ Lv^2 f \rho/d \tag{15.1}$$

where ΔP is the pressure drop, psi;
 L is the equivalent length, ft;
 v is the velocity, ft/s;
 f is the DARCY friction factor;
 ρ is the density, lb/ft^3; and
 d is the pipe inside diameter, in.

Some physical and thermodynamic properties of steam are given in Table 15.3. The fluid properties at the initial conditions were averaged with the fluid properties at the final conditions, after a preliminary calculation of the pressure drop that would

Table 15.3 Properties of Saturated Steam and Saturated Water [3]

Absolute pressure, psia	Temperature, °F	Enthalpy of the liquid, Btu/lb	Enthalpy of the vapor, Btu/lb	Latent heat of vaporization, Btu/lb	Specific volume of the liquid, ft³/lb	Specific volume of the vapor, ft³/lb
0.8859	32.018	0.0003	1075.5	1075.5	0.016022	3302.4
5	162.24	130.20	1131.1	1000.9	0.016407	73.532
10	193.21	161.26	1143.3	982.1	0.016592	38.420
14.696	212.00	180.17	1150.5	970.3	0.016719	26.799
20	227.96	196.27	1156.3	960.1	0.016834	20.087
25	240.07	208.52	1160.6	952.1	0.016927	16.301
30	250.34	218.90	1164.1	945.2	0.017009	13.7436
35	259.29	228.00	1167.1	939.1	0.017083	11.8959
40	267.25	236.10	1169.8	933.6	0.017151	10.4965
45	274.44	243.50	1172.0	928.6	0.017214	9.3988
50	281.02	250.20	1174.1	923.9	0.017274	8.5140
55	287.08	256.40	1175.9	919.5	0.017329	7.7850
60	292.71	262.20	1177.6	915.4	0.017383	7.1736
65	297.98	267.60	1179.1	911.5	0.017433	6.6533
70	302.93	272.70	1180.6	907.8	0.017482	6.2050
75	307.61	277.60	1181.9	904.3	0.017529	5.8144
80	312.04	282.10	1183.1	900.9	0.017573	5.4711
85	316.26	286.50	1184.2	897.7	0.017617	5.1669
90	320.28	290.70	1185.3	894.6	0.017659	4.8953
95	324.13	294.70	1186.2	891.5	0.01770	4.6514
100	327.82	298.50	1187.2	888.6	0.01774	4.4310
110	334.79	305.8	1188.9	883.1	0.01782	4.0484
120	341.27	312.6	1190.4	877.8	0.01789	3.7275
130	347.33	319.0	1191.7	872.8	0.01796	3.4544
140	353.04	325.0	1193.0	868.0	0.01803	3.2190
150	358.43	330.6	1194.1	863.4	0.01809	3.0139

result if the steam behaved as an incompressible fluid. Only a single iteration was used. The friction factor, f, can be obtained graphically from Figure 15.4, or calculated from the COLEBROOK equation, which requires an iterative solution:

$$1/\sqrt{f} = -2\log(2.51/\text{Re}\sqrt{f} + \varepsilon/3.7\,d) \tag{15.2}$$

where Re is the Reynolds number;
   ε is the pipe absolute roughness, ft; and
   d is the pipe inside diameter, ft.
   Equivalent lengths for various valves and fittings are shown in Table 15.4. The selection of pipe size requires judgment as to the peak flow rate of steam in the line and the allowable pressure drop for that line so that an adequate delivery pressure at the point of use is maintained.

Table 15.4 Equivalent Lengths of Valves and Fittings [3]

Component	Equivalent length, pipe diameters
Gate valve	8
Globe valve	340
Angle valve	55
Swing check	100
Butterfly valve	45
90° bend	20 [a]
	30 [b]
Flow-through tee	20
Branch-flow tee	60
Pipe entrance	$K = 0.5$ [c]
Pipe exit	$K = 1.0$ [c]

Notes: (a) $r/d = 1$
   (b) standard
   (c) $K = fL/d$, so equivalent length is Kd/f (f is the friction factor, d the pipe diameter).

The pipe, valves, and fittings for plant steam systems are typically carbon steel (ANSI B36.10) with schedule 40 pipe. Valve selection depends on the service, with globe valves usually used in a throttling service, and gate and ball valves usually used in on/off service. Welded connections are preferred, but flanged connections are used

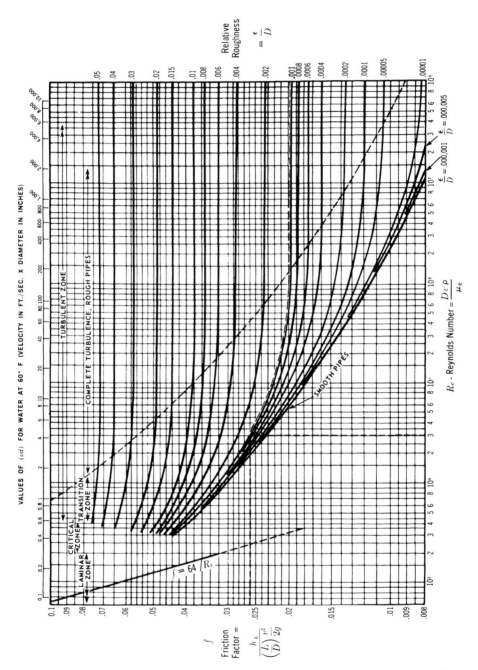

Figure 15.4 Darcy Friction Factor versus Reynolds Number. (Source: Crane Co., New York.)

in situations where moderate to frequent maintenance is required, such as at a control valve or steam trap. The routing of the distribution piping must take into account the allowance for insulation and removal of condensate at appropriate locations with steam traps. Steam lines should slope in the direction of flow, with the low points trapped.

The selection of the material and the thickness of insulation is primarily an economic decision. After an acceptable payback period for the investment in insulation is established, the insulation material and thickness can be selected based on the cost of energy for generating steam. Usually, glass-fiber insulation is used.

15.2.4 Condensate Systems

Wherever heat is rejected or lost in a plant steam system, the vapor condenses to the liquid phase (if the steam is saturated). Since there is an expense associated with treating the boiler feed water, and the condensate is at a much higher temperature than makeup water to the system, there is an economic incentive to collect condensate and return it to the boilers for reuse as boiler feed water. Condensate systems are similar to steam distribution systems in terms of insulation and piping materials, but there are additional considerations in an optimum design. Improved energy efficiency can be achieved if different pressures are used in the collection system when there are appreciable quantities of plant steam used at different pressures in the distribution system. For example, in a facility that uses 115 psig steam as the energy source for a clean-steam generator, and 15 psig steam as a heat source for recirculating hot water (for HVAC), a scheme can be devised wherein the 115-psig condensate is flashed in a flash drum that operates at 15 psig, to produce 15-psig steam and low-pressure condensate. The steam produced during the adiabatic flashing of the high-pressure condensate is not vented to the atmosphere in this case. Lower makeup water treatment costs and lower fuel costs are the net result, since the steam made in the flash drum supplements steam produced through a pressure-reducing station. Figure 15.5 illustrates this type of arrangement.

Condensate return piping must be sized properly to avoid back-pressure problems created by the flashing of the condensate in the line, with resultant two-phase flow. Correlations for calculating pressure drop for two-phase flow can be found in various references [4]. In cases where the condensate is condensed low-pressure steam, alternative means for returning the condensate to the plant steam boilers may be necessary. In Figure 15.6, a condensate return unit is shown. This device makes use of plant steam to provide a driving force for flow, thus enabling condensate, which does not have sufficient pressure of its own, to be returned to a central condensate collection drum. Biotechnology facilities do not always include a central condensate collection system. A deaerator, if present, can serve that function, provided it is sized properly. The condensate collection drum, if there is one, provides surge capacity for returned condensate and a means of disengaging the two phases commonly returned in the condensate return lines.

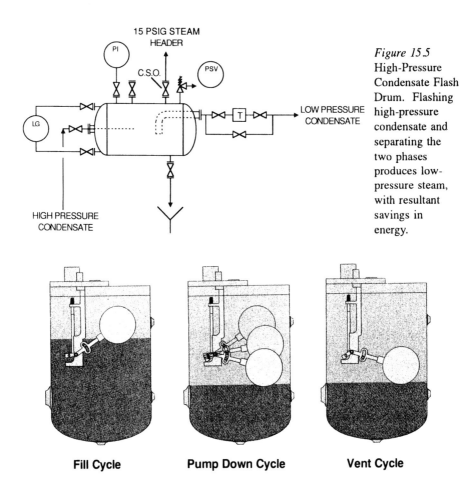

Figure 15.5
High-Pressure
Condensate Flash
Drum. Flashing
high-pressure
condensate and
separating the
two phases
produces low-
pressure steam,
with resultant
savings in
energy.

Figure 15.6 Condensate Return Unit. This unit is used to return condensate in lower flow rate systems. (Courtesy of Johnson Corp., Three Rivers, MI.)

Steam traps for plant steam come in various configurations; the most frequently used types are shown in Figure 15.7. The purpose of a steam trap is to maintain the appropriate pressure level in the plant steam line, while removing condensate produced by the condensing steam. Steam traps are often installed with strainers, to prevent scale or other debris from blocking the trap and causing a buildup of condensate in the line. When a trap is blocked, steam flow is reduced and steam velocity may become high enough to cause water hammer. Steam trap selection is based on the specific application [5].

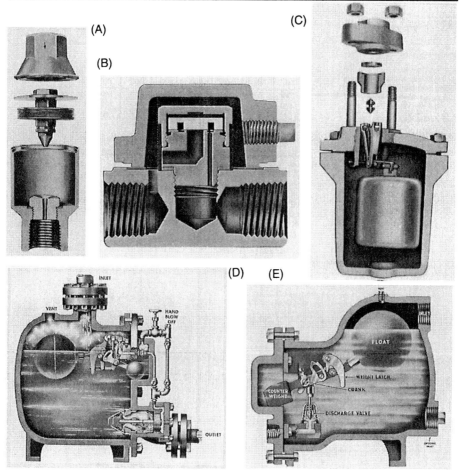

Figure 15.7 Steam Traps: (A) Thermostatic, (B) Thermoactive, (C) Inverted Bucket, (D) Piston-Operated, (E) Weight-Operated.   Steam trap selection depends on the specific application. (Courtesy of Nicholson Steam Trap, Inc.)

15.2.5 Major Users of Plant Steam

A biotechnology facility usually needs steam for the following services: WFI still, clean steam generator, HVAC heat, biowaste kill system, process heating, regeneration of desiccant or carbon beds, and jacketed autoclaves. The difficulty in sizing the plant steam boiler lies in the nature of the processes for which the steam is used, which tend to be intermittent. The quantity of steam required for each of the individual users is generally a straightforward calculation; however, an assessment of which requirements are simultaneous takes judgment. A conservative approach should be taken, as the incremental cost of additional capacity in the plant steam boiler is not excessive compared to the potential problems that could arise from an undersized boiler. This conservative approach should, however, be balanced against the firing efficiencies of the boilers under consideration. The efficiency of a boiler tends to be higher at loads closer to the maximum firing rate.

As an example of a sizing calculation, consider the steam required to heat hot water that will be distributed in an HVAC hot-water circulation system. Assume that the heat duty used as the design basis for the heat-transfer package is 1.5 million Btu/h (440,000 W). Further, assume that the steam pressure that will be used is 15 psig. The latent heat of vaporization for steam at that pressure is approximately 945.6 Btu/lb (2203 kJ/kg). Therefore, the steam requirement to meet that need is the duty (1,500,000 Btu/h) divided by the latent heat of vaporization (945.6 Btu/lb), or 1,586 lb/h (720 kg/h). Note that no credit was taken for sensible heat transfer. The condensate is assumed to leave the heat exchanger at essentially saturation temperature and pressure. This can be observed in practice, and is due to the much higher coefficient of heat transfer for vapor condensing on a surface than for liquid contacting the same surface.

15.3   Clean-Steam Systems

To comply with sterility requirements for FDA–approvable processes, steam sterilization is the current method of choice for the sterilization of equipment, glassware, and components. Clean steam is also used in a biotechnology facility for the decontamination of equipment, clothing, and components. Steam is used because of its high heat-transfer capacity and its inherent moisture, which render it highly lethal to microorganisms, while adding no contaminants to the surface being sterilized. Hot air is an alternative medium, but higher temperatures and longer exposure times are generally required to obtain the same lethality as with clean steam.

Clean steam, also known as *pure steam*, is not a necessity in all biotechnology facilities. However, if the facility is producing a substance that is incorporated in a parenteral drug product, there is usually a need for a clean-steam generator. If the steam contacts components, in-process materials, drug products, and drug product contact surfaces, the boiler feed water should not contain volatile additives such as amines or hydrazines. Additionally, the steam, if condensed, should meet the USP specifications for Water For Injection in terms of constituents such as pyrogens, bacteria, and dissolved solids [6].

To produce this high-quality steam, high-purity feed water and specialized steam generation equipment are required. In this section I describe the equipment used to produce clean steam, methods of pretreating the feed water, piping systems, and methods for sizing the generator.

15.3.1 Generators

There are three principal types of clean-steam generators: kettle, thermosyphon, and dry bottom. In all three designs, the fundamental components are the same (pressure vessel, entrainment separator, controls, and heat-transfer surface), but they are configured differently.

The *kettle* type of clean-steam generator is shown diagrammatically in Figure 15.8. This is a very simple design, in which a heat exchanger tube bundle is submerged in water inside the pressure vessel. Water droplets are removed from exiting steam at the top of the vessel by baffles or a mist eliminator pad; the removal of water droplets is essential, as pyrogens are carried by entrained water droplets. Since boiling occurs beneath the surface of the water in this type of clean-steam generator, droplets can be propelled into the kettle vapor space when bubbles break the surface. Therefore, the design of the mist elimination device is critical to ensure clean steam that meets specifications for pyrogens.

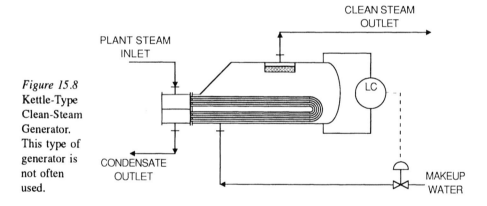

Figure 15.8 Kettle-Type Clean-Steam Generator. This type of generator is not often used.

Thermosyphon-type clean-steam generators have a vertical design; heat is transferred to the boiling water in a heat exchanger. An adjacent vessel provides feed water to the heat exchanger as well as a vapor space where deentrainment of water droplets takes place. Figure 15.9 illustrates this type of clean-steam generator. Water enters the heat exchanger through the bottom and rises in the tubes as steam is produced, because of the hydrostatic head of the water in the vessel on the right. Makeup water is introduced into the vessel upon a signal from the level controller. It is possible, in this design, for boiling to occur in the vessel as well as in the heat exchanger, if the water in the vessel is at its saturation temperature and a heavy demand for clean steam occurs. The pressure in the vessel falls below the saturation

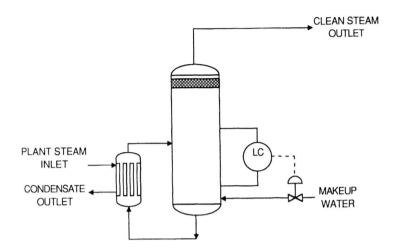

Figure 15.9
Thermo-
syphon-Type
Clean-Steam
Generator.
Liquid/vapor
separation
takes place in
the large
vessel;
heating takes
place in
the heat
exchanger on
the left.

pressure of the water, and the water boils. The design of the water droplet deentrain-ment device must be adequate to cope with this possibility.

The third type of clean-steam generator is shown diagrammatically in Figure 15.10. This type of clean steam generator, the *dry-bottom* type, operates on a principle similar to that of a falling-film evaporator. Feed water is introduced at the top of the unit and travels downward inside the tubes, vaporizing on its journey downward. At the bottom, the vapor reverses direction and travels upward, outside the heat-transfer tube bundle. As the vapor travels upward, baffles guide it in a spiral path; the centrifugal action forces droplets of water out of the vapor stream and into a collection zone, which is an annular space between the outside shell of the unit and a concentrically located cylinder inside the unit. This design relies on the dual action of the 180° reversal of the vapor at the bottom of the unit and the centrifugal action of the spiral baffles to eliminate entrained water droplets from the vapor stream, and has been proven effective.

Vendors of all three types of generators provide units with fairly uniform requirements for feed water quality, electricity, plant steam, and instrument air. An exception is smaller units, which use electricity for the heating medium instead of plant steam; these units are available in sizes ranging from 35 lb/h to 650 lb/h (16–300 kg/h) at 212 °F (100 °C).

For the common types of clean-steam generators, which use plant steam as the heating medium, the capacity is related to the pressure of the plant steam. Figure 15.11 shows how the efficiency of a clean-steam generator is related to the pressure of the primary steam (plant steam) and the pressure of the clean steam being produced [7]. This can be an important consideration in biotechnology facility design, as the vendor of a bioreactor or other packaged piece of equipment may impose a minimum

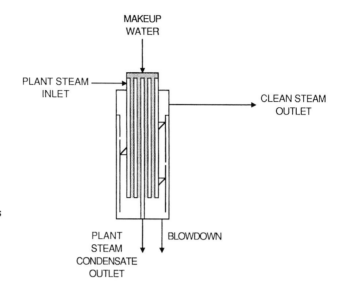

Figure 15.10 Dry-Bottom–Type Clean-Steam Generator. This unit operates on principles similar to those of a falling-film evaporator.

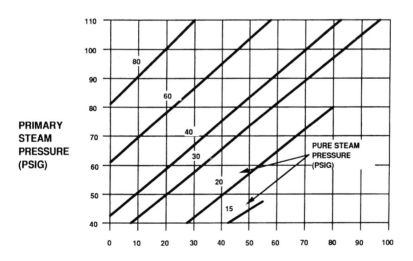

% OF MAXIMUM CAPACITY

Figure 15.11 Relationship Between Clean-Steam Generator Efficiency and Plant Steam Pressure. The efficiency depends on the temperature difference between the plant steam and the clean steam. (Reprinted from *BioPharm 1* (10), 36 {1988}, with permission of Aster Publishing.)

clean-steam pressure requirement on the user of their piece of equipment. The designer must then determine what pressure of plant steam is needed and apply an efficiency correction factor based on the clean-steam and plant-steam pressures. It may be most economical in some circumstances to maintain a relatively low plant-steam system pressure and incur the cost of a larger nominal capacity clean-steam generator. Of course, other users of the plant steam must be considered in this type of evaluation.

15.3.2 Pretreatment of Feed Water

Suppliers of clean-steam generating equipment are the best source of data for water quality requirements for feed water. In the absence of vendor data, Table 15.5 provides some typical values for constituents in clean-steam generator feed water. Values for individual vendors may differ from these typical data.

Table 15.5 Feed Water Quality Requirements for Clean-Steam Generators

Constituent	Units	Typical value	max or min
TDS	ppm	100	max
Pyrogens	EU/mL	100	max
Bacteria	CFU/mL	100	max
Chlorides	ppm	0	max
Resistivity	MΩ·cm	1	min
Silica	ppm	1	max
Amines	ppm	0	max

Other contaminants that may have to be dealt with are dissolved organics and particulates. If the feed water is not properly treated, scaling may occur on the heat-transfer surfaces; the result may be equipment downtime for cleaning, or off-spec clean steam may be produced, which can have the effect of shutting down the entire operation. Which unit operations are used to obtain properly treated water depends on the quality of the untreated water and the specifications of the clean-steam generator supplier. Generally, the minimum treatment steps are filtration, carbon adsorption (to remove chlorine), and softening. More typically, an ion exchange (anion and cation) or reverse osmosis step is also included. In some instances, mixed-bed ion exchange follows the anion/cation exchange, or WFI is used as feed water to a clean-steam generator. Very high purity feed water may have the effect of reducing the equipment life of a clean-steam generator, because the water is more corrosive.

15.3.3 Piping Systems

Both stainless steel piping and tubing can be used in clean-steam systems, with clamped, flanged, or welded fittings. Welded fittings are preferred over clamped or flanged connections, when possible, to avoid leaks at joints.

The choice of tubing or piping is frequently based on the designer's experience. Many engineers believe that a tubing system with an electropolished interior surface finish maximizes the usable life of the distribution system, but others believe that a lower cost and equally long-lasting system can be obtained using mechanically polished schedule 10S pipe. In either case the system must be cleaned and passivated. The material of choice is 316L stainless steel.

Sanitary ball valves are most frequently used in manual applications, but diaphragm valves are also used. Diaphragm valves suffer shortcomings with regard to membrane degradation, leakage, and frequent membrane replacement.

Steam traps must be constructed of 316 or 316L stainless steel, with condensate piping away from the trap to a collection system if the trap is located in a controlled environment area [9]. Otherwise traps should be followed by an air break.

Nonprocess-contact instrumentation should be used insofar as possible. Insulation typically is foam glass in areas where the piping is not exposed to a controlled environment, and polyisocyanurate covered with PVC sheathing in exposed areas.

General system requirements are that low points should be trapped and drainable, and the system should be generally cleanable. Branch tees should have run lengths as short as possible, but they do not have to be strictly limited to six pipe diameters or less. Sloping the distribution piping is recommended, because it makes drainage easier and prevents condensate buildup that can lead to hammering.

15.3.4 Sizing

The primary uses for clean steam are for humidification of controlled-environment areas, in autoclave operation, as condensate for WFI, for equipment sterilization, and as steam blocks. Establishing a capacity requirement for a clean-steam generator involves a knowledge of potential users of the clean steam, analysis to determine which of the users are likely to demand steam simultaneously, and calculation methods to estimate steam consumption for identified users. Generally, estimating the quantities required for the individual consumers is not difficult, but making a determination as to which of the users require clean steam simultaneously takes judgment and is not a simple matter.

The quantity of clean steam required for humidification is a function of the percentage of circulated air that is added as makeup air, the relative humidity of the makeup air, the relative humidity of the air inside the facility, the quantity of circulated air, the temperature of the circulated air, and losses from the facility. I provide an abbreviated estimating technique here. Generally, losses and the effects of room temperature are negligible as long as the rooms are held near typical conditions of between 62 °F and 72 °F (17–22 °C). Figure 15.12 illustrates the dependence of clean steam demand on relative humidity, percent makeup, and room classification. The conditions cited in Figure 15.12 as the design basis are intended to be conservative but still useful. For example, a facility that has a controlled-environment area (class 100,000) for component preparation with an area of 800 ft² (75 m²) and a 9-ft (2.75-m) ceiling requires roughly:

(800/100) x 2.5 = 20 lb/h of clean steam

to maintain a relative humidity of 50% at 68 °F (20 °C), with 20% makeup air. This is a simplified approach to estimating clean steam requirements for humidification, and care should be taken in its application.

Clean steam consumption in an autoclave depends on chamber size, whether or not clean steam is used in the jacket, and individual vendors' designs. The best source of data for peak clean-steam requirements for autoclaves is the vendor. However, if a vendor has not yet been chosen, then Figure 15.13 can be used to estimate the consumption of clean steam for autoclaves. The data shown are based on the use of clean steam in the chamber jacket as well as inside the chamber by Vendor A, and the use of plant steam in the jacket with clean steam inside the chamber by Vendor B.

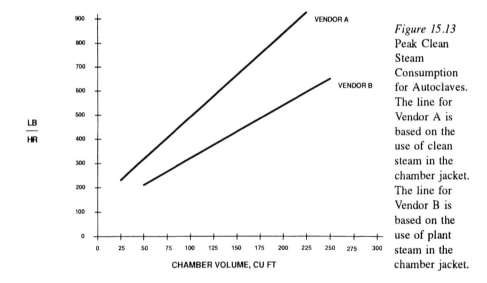

Figure 15.13 Peak Clean Steam Consumption for Autoclaves. The line for Vendor A is based on the use of clean steam in the chamber jacket. The line for Vendor B is based on the use of plant steam in the chamber jacket.

Clean steam can also be used to make WFI in a condensing unit. Estimation of clean steam consumption to make WFI is straightforward, as the quantity of clean steam used is set by the capacity of the condensing unit. Condensing units are

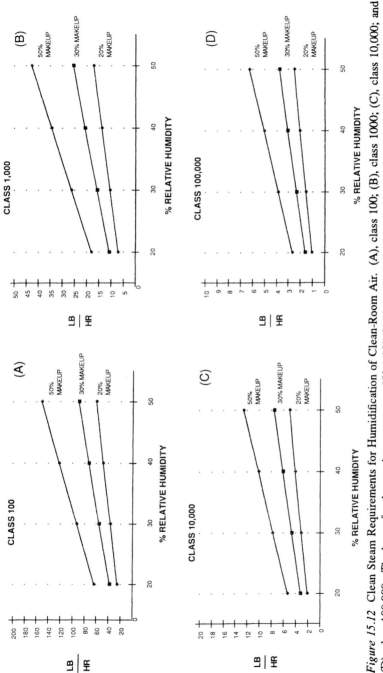

Figure 15.12 Clean Steam Requirements for Humidification of Clean-Room Air. (A), class 100; (B), class 1000; (C), class 10,000; and (D), class 100,000. The bases for the graphs are: a room 10' x 10', 9' high; dry-bulb temperature of 68 °F; outside air with 0% relative humidity; and room air changes per hour of (A) 600, (B) 170, (C) 50 and (D) 25. (Reprinted from *BioPharm 1* (10), 41 {1988}, with permission of Aster Publishing.)

available as stand-alone units that can be located remote from the clean steam generator, or as integral components of the generator. The rate of clean steam consumption is generally highest when the unit is diverting off-spec WFI to drain; at that time, the rate of consumption is limited by the cooling medium and the heat transfer area.

Equipment sterilization is an area where estimating the quantity of clean steam required can be difficult and possibly inaccurate. The sterilization of equipment in place (SIP) can be carried out in different ways. In fact, the procedure frequently varies from company to company, with different operating conditions, and variations in the way in which the process is validated. Even so, there are four fundamental steps to an SIP cycle. The equipment is purged of air, heated, maintained at sterilization temperature, and cooled. The first three steps require clean steam.

Air can be purged from equipment in two ways. A repeated cycle of pulling a vacuum on the system followed by introducing clean steam to the system is the most effective method, or, alternatively, the air may be vented, along with some clean steam, as clean steam is introduced. A technique for estimating clean steam consumption for the second method is presented here.

This is a theoretical approach, which should produce a value for clean steam consumption that is a fair estimate of losses that actually occur. The assumptions are that: (a) air and steam behave as ideal gases, (b) there is no pressure drop within the system being steamed, (c) there is no condensation of steam to account for (this is estimated separately), (d) the gases are perfectly mixed within the control volume, and (e) all air is purged from the system without any change in system pressure. With these assumptions, it is possible to draw an analogy between air purge during an SIP cycle and salt purge from an aqueous solution being fed with a pure water stream that is entering at the same rate that salt solution is exiting the control volume. The governing equation is [10]:

$$C_t / C_o = e^{-t/k} \tag{15.3}$$

where C_t is the concentration at time t;
  C_o is the concentration at time 0;
  t is the time; and
  k is a constant.

This equation is shown in Figure 15.14 as a family of curves. To use the graph, select a desired purge time, along with a level of air removal, and find the feed steam requirements in cubic feet per minute per cubic foot of system volume. If the total system volume and the system pressure during purging are known, a steam flow rate in pounds per hour can be calculated. Note that higher steam usage rates are required if the purge is carried out at higher pressures. This is due to the assumption that a volume of incoming steam displaces an equal volume of steam/air mixture.

The second step in an SIP cycle, heating the equipment, requires steam to heat the system metal to sterilization temperatures. Stainless steel has a heat capacity of

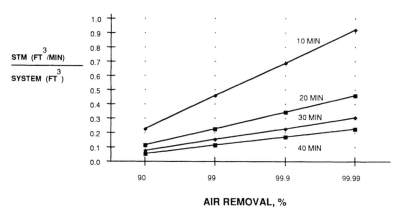

AIR REMOVAL, %

Figure 15.14 Clean Steam Venting Losses for SIP. The lines represent the requirements of a gravity-displacement operation. (Reprinted from *BioPharm 1* (10), 42 {1988}, with permission of Aster Publishing.)

approximately 0.12 Btu/lb·°F (0.5 kJ/kg·K), so to heat 1 lb of stainless steel from 70 °F to 250 °F requires 21.6 Btu. The clean-steam rate for this increase in temperature of the equipment can be calculated if the time to heat up and the weight of metal in the system have been determined. Piping, valves, fittings, vessels, and other equipment should be considered. Vendor data are the best source of information for the weights of the various components, although they may not be available in the early stages of a project. If the weight of vessels is unknown, the heat, Q, in Btu, required to heat vessels may be estimated by the following equations:

$$Q = 7600 + 69 \, V \tag{15.4}$$

when the volume, V (in gallons), of the vessel is 100 to 900 gallons, and

$$Q = 68,000 + 42.5 \, V \tag{15.5}$$

for 1,000- to 10,000-gallon vessels. These formulas apply to unjacketed stainless steel pressure vessels heated from 70 °F to 250 °F. The time needed for heating equipment is roughly the same as the time required to complete the purging of air. Therefore, to be conservative, it is best to assume that the two occur simultaneously.

Another, but usually minor, way that clean steam is consumed is through convective heat losses to the environment. Unless very long runs of piping are part of the process system, this form of consumption need not be calculated, because it is low compared to the other modes of clean steam consumption. Figure 15.15 can be used to estimate heat losses from insulated lines, if necessary. The graphs were generated with an iterative procedure in which the outside-film heat-transfer coefficient

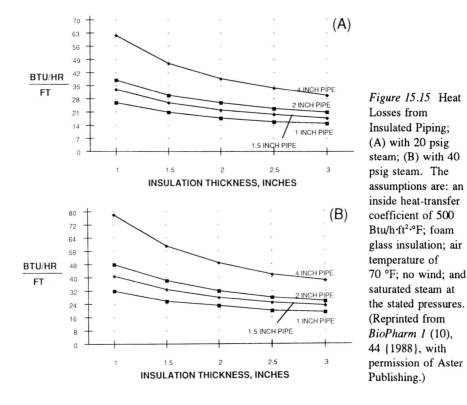

Figure 15.15 Heat Losses from Insulated Piping; (A) with 20 psig steam; (B) with 40 psig steam. The assumptions are: an inside heat-transfer coefficient of 500 Btu/h·ft²·°F; foam glass insulation; air temperature of 70 °F; no wind; and saturated steam at the stated pressures. (Reprinted from *BioPharm 1* (10), 44 {1988}, with permission of Aster Publishing.)

was calculated based on an assumed air-film temperature. The air-film temperature was back calculated until convergence was obtained.

To summarize, sizing a clean-steam generator requires analysis of simultaneous users coupled with calculation techniques for the known users. A conservative approach is warranted, as the incremental cost of a larger clean-steam generator is low compared to the potential costs that could arise from an undersized generator.

15.4   Chilled Water Systems

It is not unusual for a biotechnology facility to include capabilities for low-temperature processing of biotechnological products, particularly if the product is derived from mammalian cells. In addition to the process applications of chilled water, the heating, ventilation, and air conditioning (HVAC) system frequently makes use of chilled water as a means of cooling laboratory, office, and process areas. In this section, I discuss system configurations, hardware, and design guides for chilled water systems.

It may be useful to point out at the beginning that water is not always the circulating medium in these systems. There is a lower limit, in the neighborhood of 42 °F (6 °C), below which water cannot be used alone because of freezing in the heat exchanger where heat is rejected to the refrigerant. Below this temperature, a mixture of glycol and water is used. Propylene glycol is being used with increasing frequency because of its lower toxicity compared to ethylene glycol. In this section, I use the term *chilled water* to denote either water or water/glycol mixtures.

15.4.1 System Configuration

The fundamental components of a chilled water system are similar to those of a commercial or residential air conditioning system, in which a circulating air stream is cooled by rejecting heat to a boiling refrigerant in a heat exchanger. The major difference is, however, that water or water mixed with glycol, circulating in a loop, is the medium that is cooled, instead of air. The benefits of using chilled water instead of a conventional direct expansion of refrigerant in an air heat exchanger are that the facility geometry does not place constraints on a chilled water system as it can on a direct-expansion system, the cost per ton of cooling is lower, and the circulating chilled water can be used for other services, such as process cooling. The essential components of a chilled water system are the compressor, the condenser, the evaporator, an expansion valve, circulating pump(s), an air separator, and an expansion tank.

Two types of compressors are most frequently used: the reciprocating type and the centrifugal type; which is used is a function of system capacity. Smaller units, below about 150 tons (1.8 x 10^6 Btu/h or 527,000 W), generally are reciprocating or screw type, and units larger than 150 tons generally are centrifugal.

Two types of condensers are used. In some, the heat added to the compressed refrigerant gas is rejected to water, while other units reject the heat to air. Which type of condenser to select depends on the availability of cooling water, the size of the unit (operating cost), and, in some instances, aesthetics. Aesthetics can be a consideration because many facilities are located in areas where it is unacceptable to have a plume of water vapor emanating from a cooling tower. If cooling water can be used, a higher overall efficiency is possible because the temperature difference in the summer generally favors heat exchange with water instead of with air; also, the heat-transfer coefficient is higher for water, so the heat-transfer area is smaller.

The evaporator is merely a heat exchanger in which refrigerant vaporizes, cooling the circulating water. The expansion valve reduces the pressure of the refrigerant by isenthalpic expansion. The expansion tank is in the chilled water circulating loop to compensate for changes in the specific volume of the recirculating liquid when the unit is started up and shut down.

A schematic of a chilled water system is shown in Figure 15.16. This particular system rejects heat to cooling-tower water. The thermodynamic basis for the operation

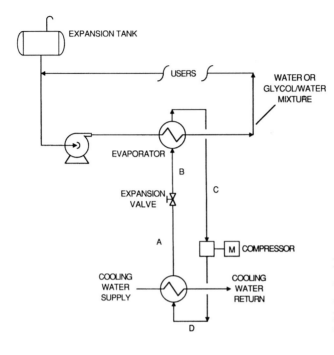

Figure 15.16 Chilled Water System. The thermodynamic condition of the refrigerant at points A, B, C, and D is shown in Figure 15.17.

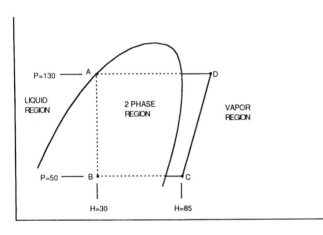

Figure 15.17 Typical Pressure/Enthalpy Diagram for a Refrigeration Cycle. This diagram depicts the thermodynamic condition of the refrigerant at points A, B, C, and D in the system shown in Figure 15.16.

of this system can be explained with a pressure–enthalpy diagram for a refrigerant, as shown in Figure 15.17. The points A, B, C, and D in Figure 15.16 correspond to the same points in Figure 15.17. The cycle may be described thus:

A to B—Condensed refrigerant expands across the expansion valve isenthalpically. The partial vaporization of the refrigerant lowers its temperature.

B to C—The refrigerant vaporizes in the heat exchanger at constant pressure. Energy is transferred from the circulating chilled water to the refrigerant in this exchanger: this is where the water is chilled.

C to D—The vaporized refrigerant is compressed, ideally at constant entropy. The energy input from compression raises the enthalpy and temperature of the refrigerant gas.

D to A—The hot compressed refrigerant condenses by rejecting its heat to the cooling-tower water.

A variation on this arrangement is to place a heat exchanger on the compressor suction line. The compressor suction gas exchanges heat with the warm liquid leaving the condenser. This can improve the overall efficiency of the process slightly, but, more importantly, it prevents slugs of liquid from reaching the compressor and damaging it.

15.4.2 Chiller Sizing

Determination of the required chiller size requires data for both the process and HVAC systems, assuming both are served by the chilled water system. For process loads, an energy balance around a piece of process equipment is all that is required to determine the load for the equipment. Attention must be paid to the composition of the chilled water fluid, if it is a mixture of glycol and water, because glycol lowers the heat capacity of the fluid in a mixture.

As an example of a process demand on the chiller capacity, consider the case where 176 °F WFI is cooled to 77 °F for medium makeup. Assume we wish to add water to the medium mixing vessel at a rate of 5 gal/min. To calculate the required rate of heat removal, multiply the flow rate of water by the density of water (8.34 lb/gal), the specific heat of water (1 Btu/lb·°F), and the temperature change in the water (99 °F). The result is 4,128 Btu/min. Expressed in the more conventional units for chiller capacities, this one process operation imposes an instantaneous load on the chiller of 20.6 tons (1 ton is equal to 12,000 Btu/h or 3517 W). Chiller sizing calculations must account for services that generally are considered continuous, such as the HVAC requirements, as well as those that are intermittent, such as the example just cited.

15.4.3 Piping Systems

Piping materials for chilled water systems are generally copper, carbon steel, or a combination of the two. The size of a given line usually determines what the piping material should be, with lines of diameter 2" (50 mm) or less being copper, and larger lines being carbon steel.

One special consideration for these systems is that there must be an adequate flow of chilled water through the evaporator (where refrigerant evaporates to chill the water) to prevent freezing of the water on the tube walls. This may be accomplished by including a recirculating bypass valve on the discharge of the chilled water from the chiller. Water is recirculated to the recirculating pump suction during conditions of low demand in the chilled water distribution system, thereby maintaining an adequate flow rate within the chiller.

15.4.4 Brines

The term *brine* refers to a fluid other than pure water used as a heat transfer medium. Brines are used only if the temperature of the chilled water leaving the chiller is low enough to pose a threat of freezing in the chiller evaporator, which translates into a chiller discharge temperature of about 42 °F (6 °C). Brines for chillers are usually a mixture of some type of glycol and water. Propylene glycol and ethylene glycol are most commonly used, at a concentration in the range of 20–40%. Recently, the trend in the industry has been to use propylene glycol because of its lower toxicity, which can be especially important in a facility in which portable equipment is frequently connected and disconnected from the coolant supply and return. Data for density, thermal conductivity, specific heat, and viscosity for propylene glycol brines of various compositions are presented in Figures 15.18–15.21.

15.5   Cooling Towers

Mechanical draft cooling towers are used in biotechnology facilities to provide recirculating cooling water for cooling applications where the lower temperatures available in a chilled water system are not required, or when cooling-tower water is used as the medium to which heat is rejected in the chilled water system. Typically, cooling water is cooled 10–20 °F (5–11 °C) in this type of cooling tower, with an outlet temperature of around 85 °F (29 °C) under design conditions. The outlet temperature varies according to the cooling tower design and the ambient conditions. Mechanical draft cooling towers belong to a class of systems broadly categorized as evaporative cooling systems, whose designs are founded on the principles of psychrometry. The other systems (atmospheric cooling towers, spray ponds, natural draft towers, and cooling ponds) are not often used for biotechnology facilities because

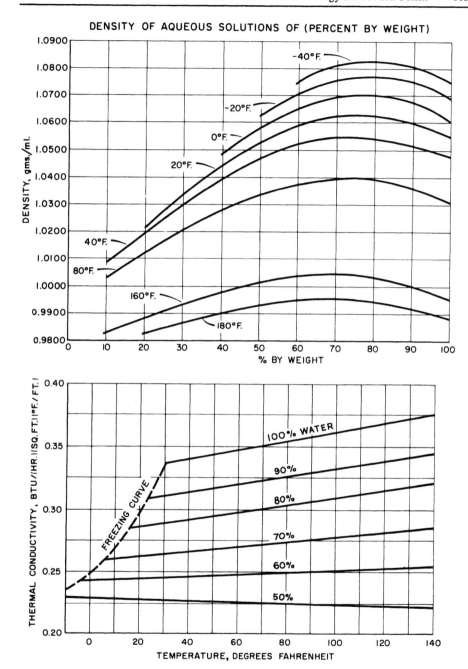

Figure 15.18 (above) Brine Density; and *Figure 15.19 (below)* Brine Thermal Conductivity. Both charts are for propylene glycol/water mixtures. (Both courtesy of Dow Chemical Corp.)

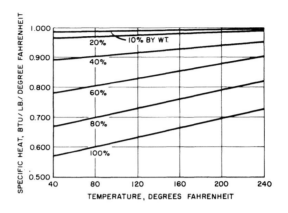

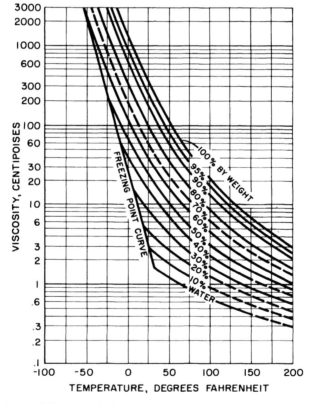

Figure 15.20 (above) Brine Specific Heat; and *Figure 15.21 (below)* Brine Viscosity. Both charts are for propylene glycol/water mixtures. (Both courtesy of Dow Chemical Corp.)

they are not as compact as the mechanical draft cooling towers, and the higher capacity is not needed (in the case of natural draft cooling towers).

15.5.1 Types

The configurations of mechanical draft cooling towers vary according to the location of the air-moving device, the path of air travel relative to the path taken by the water, and (in the case of induced draft towers) the number of banks of fill where water is contacted with air. The location of the air-moving device determines whether the cooling tower is an induced draft or forced draft tower. The *induced draft cooling tower* has the fan mounted at the top of the cooling tower, to pull air into the tower and expel it out the tower top; the *forced draft tower* has the air blower located near the bottom of the tower. The induced draft tower is the more common type, because the higher exit air velocity obtained if the fans are at the tower top decreases the possibility of moist warm air recirculating from the tower top to the tower bottom and decreasing tower performance. The second characterization of tower type, the path of air travel relative to that taken by the water, refers to whether the tower is a cross-flow tower or a counterflow tower.

15.5.2 Sizing

Cooling towers work on the principle that water, when intimately contacted with an air stream that is not saturated with water, is cooled as the water that stays liquid gives up heat to the water that vaporizes into the air stream (evaporates). A more complete discussion of psychrometry is given in the chapter on heating, ventilation, and air conditioning, but I review the basic concepts here. Air is capable of carrying water vapor in varying quantities, depending on the temperature of the air. The *dry-bulb temperature* is the measured temperature of an air/water vapor mixture. The definition of *wet-bulb temperature* is that temperature at which a dynamic equilibrium exists between air and a water surface, in which the rate of heat transfer to the surface equals the rate of mass transfer away from the surface. For practical engineering calculations, the wet-bulb temperature is approximately equal to the saturation temperature. The *saturation temperature* (*dew point*) is the temperature at which a mixture of air and water vapor is saturated.

A method for sizing cooling towers is based on the theory put forth by Merkel [4]. This analysis is based on the premise that heat transfer from a drop of water to the air that surrounds it is governed by the enthalpy difference between the water and the air.

The integrated form of the Merkel equation is:

$$KaV/L = \int_{T_1}^{T_2} (H' - H)^{-1} \, dT \qquad (15.6)$$

where K is the mass-transfer coefficient, lb water/h·ft²;
a is the contact area, ft²/ft³;
V is the volume, ft³;
L is the water rate, lb/h·ft²;
T_1 is the temperature of the water entering the tower, °F;
T_2 is the temperature of the water leaving the tower, °F;
H' is the enthalpy of saturated air at the water temperature, Btu/lb; and
H is the enthalpy of the air stream, Btu/lb.

The left side of the equation is referred to as the *tower characteristic*. The right side of the Merkel equation may be integrated graphically, or by the TCHEBYCHEFF method to obtain:

$$KaV/L = \left(\frac{T_1 - T_2}{4}\right)\left(\frac{1}{\Delta H_1} + \frac{1}{\Delta H_2} + \frac{1}{\Delta H_3} + \frac{1}{\Delta H_4}\right) \qquad (15.7)$$

where ΔH_1 is the value of $(H_w - H_a)$ at $T_2 + 0.1(T_1 - T_2)$;
ΔH_2 is the value of $(H_w - H_a)$ at $T_2 + 0.4(T_1 - T_2)$;
ΔH_3 is the value of $(H_w - H_a)$ at $T_1 - 0.4(T_1 - T_2)$;
ΔH_4 is the value of $(H_w - H_a)$ at $T_1 - 0.1(T_1 - T_2)$;
H_w is the enthalpy of the air/water vapor mixture at the bulk water temperature, Btu/lb dry air; and
H_a is the enthalpy of the air/water vapor mixture at the wet-bulb temperature, Btu/lb dry air.

Therefore, a value for the tower characteristic (KaV/L) can be calculated from enthalpy data for saturated air, enthalpy data for water, the entering and exiting water temperatures, and the ratio of the liquid and gas rates, L/G. Once a value for the tower characteristic has been calculated, determining a physical size for the cooling tower requires data specific to a particular cooling tower. A value for a, the contact area, and for K, the mass-transfer coefficient, can be determined only when the type of tower packing has been selected. At that point the engineer should work with a cooling-tower supplier to arrive at an optimum configuration for the cooling tower.

To determine the flow rate of circulating water, the needs of the facility for both intermittent and continuous users of cooling water must be assessed. The largest user of cooling water is usually the chiller. In fact, facilities have been designed where all cooling for process and HVAC uses is done with chilled water, which makes the chiller the only point of use for the cooling water. Determination of individual loads is calculated exactly the same way as was done in the example for chilled-water cooling of WFI in Section 15.4.2.

15.5.3 Types of Fill

Fill or *packing* for cooling towers is of two types, the *splash* and the *film* type of fill. The splash type of fill works by an impingement mechanism: falling water droplets encounter obstacles and splash as they fall through the tower. The obstacles are essentially baffles, of various possible geometries, made of wood, plastic, or metal. The cascading of the water from baffle to baffle provides an extended time for the falling water to contact the air flowing cross-currently or countercurrently to it; this action also increases the surface area available for air-water contact.

The film type of fill relies on a mechanism similar to that of a falling-film evaporator: the water falls over vertical sheets (usually corrugated), thereby exposing the water to the air in a way that creates a large air–water surface contact area within a relatively small volume of tower fill.

Most towers built today incorporate the film type of fill because it creates a larger water surface area, so the tower can be more compact. The splash type of fill cannot be ignored, however, as there are circumstances where it is the better choice. In particular, when the quality of the water being recirculated is not good, the splash type of fill has a lower tendency to plug and reduce tower efficiency [11]. Also, the film type of fill is not as forgiving as the splash type in terms of water distribution within the tower. If a problem with the distribution system arises, the performance of a tower with the splash type of fill probably does not suffer as much as a tower with the film type.

15.5.4 Water Treatment

The quality of water recirculated in a cooling-tower water loop is susceptible to degradation, because of biological growth and because the concentration of dissolved solids in the water rises as water is evaporated. To protect the system against scale buildup and corrosion, high-quality makeup water is added to the loop, and water is withdrawn as blowdown. In addition, chemicals can be added for specific purposes. Sulfuric acid is added to inhibit the deposition of calcium bicarbonate scale, since it converts the bicarbonate to the more soluble calcium sulfate. An excessive (above 1200 ppm) concentration of calcium sulfate can cause the deposition of sulfate scale, and should be avoided [12]. Corrosion control is effected primarily with blowdown, which reduces the electrolyte content of the water. Inhibitors that help protect the metallic surfaces can also be added. Biological growth can be retarded by using shock treatments of chlorine to kill algae. A continuous chlorine treatment is not recommended, as the chlorides tend to cause corrosion.

In summary, cooling towers are frequently needed as part of a biotechnology facility, to provide an efficient means of rejecting heat in the air conditioning system and to provide process cooling. However, not all facilities need cooling towers, especially if they are small, and the increased capital cost of installing and maintaining a cooling tower exceeds the benefits derived from lower energy consumption in the air conditioning and process cooling systems. The decision to include a cooling tower in the facility design is affected by the acceptability of having a cooling-tower plume, the rate of return on the capital cost realized from lower energy consumption in the chilled-water system (if there is one), the availability of space for the tower, and the cost and availability of water.

15.6   Air and Gas Systems

Air has many uses in biotechnology facilities besides those for instrumentation and control purposes; it can be a process stream providing oxygen to a bioreactor, a motive force for making transfers of liquid streams, a cooling medium for vessels after sterilization, or a source of energy for air motors on agitators.   Usually, the air supplied by the compressor in biotechnology facilities is dry and oil-free, because it comes into contact with the product directly or it contacts a surface that is exposed to the product.   For the special case of a facility producing a sterile product, or where the introduction of any microorganisms has a deleterious effect on the product or operations, the air is sterile-filtered, in addition to being dry and oil-free. A commonly used design basis is dry air with a dew point of −40 °F.

15.6.1 Compressed-Air Generation Systems

The configuration of a compressed-air system depends on how much air is needed, the pressure at the compressor outlet, and the intended use. A fairly typical system is a *two-stage compressor*, with one heat exchanger located between the two stages (an *intercooler*) and a second heat exchanger located after the second stage (an *after-cooler*) to remove the heat of compression. Following the compression equipment are a receiver and an air dryer. Desiccant-based air dryers are commonly used, with any of three methods for regenerating the desiccant: steam coils, electric heat, or depressurization.   Lower flow rate systems can usually be installed with the economical depressurization type of desiccant air dryer. The air receiver is usually a carbon steel vessel provided to equalize flow and pressure variations in the system. It is also a safety-related piece of equipment in the event of a power or mechanical failure: process systems can shut down in orderly fashion if compressed air surge capacity is available.

System sizing is based on facility requirements in the areas of instrumentation, vessel pressurization, air motors, and other miscellaneous users such as stopper washers (for some types) and autoclaves. The best source of data for air consumption for individual users of compressed air is the vendor of the piece of equipment. The requirement for compressed air for instrumentation is difficult to estimate, but is usually minor compared to that of the other users, so adding 20–50 scfm (0.6–1.4 standard m^3/min) is normally adequate.

After the pressure (usually 100–110 psig, 700–770 kPa) of air for distribution and the flow rate have been established, the type of compressor can be selected. Since compressors installed in biotechnology facilities are usually 100–300 scfm (2.8–8.4 standard m^3/min), the two types of compressors most frequently used are the *rotary-screw* and *reciprocating* types. Either type should be specified as oil-free, to prevent contamination of the compressed air with lubricating oil.

The piping distribution system can be specified with different materials of construction, depending on the service. Materials of construction range from carbon steel for instrument air, to polished sanitary stainless steel for applications where the air is filtered to achieve sterility or is exposed in a clean room. A common method for designing distribution systems is to use K-type copper tubing with 20% silver solder (brazed) for most of the distribution network, and change to polished stainless steel at the point where the air is sterile-filtered or where the tubing is exposed to a clean-room environment where it receives frequent heavy cleaning.

15.6.2 Compressed Gases

Compressed gases find various applications in biotechnology facilities: they are used to pressure-test piping and equipment and to manufacture fermentor air, they provide an inert atmosphere in a vessel head space, and they pressurize vessels to effect material transfers. Nitrogen and oxygen are the most commonly found gases, nitrogen for its chemical inertness, and oxygen as a feed stream to a fermentor, to enrich an air stream, or to make artificial air. These gases can be supplied as high-pressure compressed gas or as liquified gas, which is vaporized as needed. The choice of liquified gas versus compressed gas is made on the basis of the quantity of gas used. If relatively low quantities of gas are needed, compressed gas cylinders generally provide the most economical solution. Liquified gases are not as cost-effective when consumption rates are low or when demand for the gas is infrequent, because there are losses in the range of 1% to 3% per day to keep the cylinders from developing overpressure. When the consumption of a gas is fairly regular, however, liquified gas cylinders provide the benefit of less frequent changeover and resupply.

Another option for supplying gases uses membrane technology. The advantage of this type of unit is that there are negligible operating costs. The disadvantages are the low delivery pressure and the low (relative to bottled) gas purity.

The distribution piping for compressed gases is generally similar to that for compressed air. K-type copper is frequently used for general distribution, with a change to stainless steel made when conditions warrant.

15.7   Conclusion

The sizing and specification of utility support systems is a systematic exercise that involves three principal areas. First, sufficient knowledge of the production facilities and equipment is required so that the engineer or designer can itemize what the points of use are for the various utilities. For each of the points of use, an assessment must be made as to the quantity or duty requirement, and whether the service is used intermittently or in a more continuous manner. Second, operating pressures and temperatures for the utility service at each of the points of use must be determined. The effective design of the distribution system, particularly in the case of steam systems, depends on pressures and temperatures at the various points of use. Third, the potential alternatives for supplying the utility services must be evaluated in terms of economic, spatial, technical, maintenance, and availability considerations. For example, the suitability of different types of compressor for different services depends on the desired delivery pressure and the required flow rate. As long as a coherent and systematic approach is used in the sizing and selection of these systems, the responsible engineer can provide, to the personnel who must operate the facility, adequate utility systems for smooth, reliable operation.

15.8   References

1. PAREKH, B. S., Get Your Process Water To Come Clean, *Chem. Eng. 98* (1), 70–85 (1991).
2. Personal communication: Betz Water Treatment Technology Seminar, June 5–6, 1984, Philadelphia, PA.
3. *Flow of Fluids Through Valves, Fittings and Pipe*; Technical Paper No. 410, 21st ed.; Crane Co., New York, 1982.
4. COREY, R. C., *et al.*; Energy Utilization, Conversion, and Resource Conservation; Chapter 9 in: *Perry's Chemical Engineers' Handbook*; PERRY, R. H., GREEN, D., Eds.; 6th ed.; McGraw-Hill, New York, 1984; p 9-76.
5. RADLE, J., Select the Right Steam Trap, *Chem. Eng. Prog. 88* (1), 30–36 (1992).
6. *United States Pharmacopeia XXII/National Formulary XVII*; United States Pharmacopeial Convention, Rockville, MD, 1990.
7. ADAMS, D., Sizing Clean Steam Generators, *BioPharm 1* (10), 36 (1988).
8. COLEMAN, D., EVANS, R., Fundamentals of Passivation and Passivity in the Pharmaceutical Industry, *Pharm. Eng. 10* (2), 43–49 (1990).
9. COLEMAN, D., SMITH, P., Effective Steam Trapping for Clean Steam Systems in the Biotechnology and Pharmaceutical Industries, *Pharm. Eng. 12* (2), 8–14 (1992).
10. RUSSEL, T., DENN, M.; *Introduction to Chemical Engineering Analysis*; John Wiley & Sons, New York, 1972.
11. HENSLEY, J., Maximize Tower Power, *Chem. Eng. 99* (2), 74–82 (1992).
12. HENSLEY, J.; *Cooling Tower Fundamentals*; The Marley Cooling Tower Company, Mission, KS, 1983.

Biowaste Decontamination Systems

Kim L. Nelson

Contents

16.1   Introduction

Biotechnology and fermentation plant operations result in liquid, solid, and gaseous waste streams that require treatment prior to discharge. Treatment processes may include biological inactivation, wastewater pretreatment, or, in the case of process exhaust vent streams, inactivation, filtration, or odor abatement. Processes involving recombinant or pathogenic organisms in particular require inactivation of the waste stream to destroy viable cells that might otherwise escape into the environment. My primary aim in this chapter is to discuss the inactivation of liquid biowaste streams.

Both thermal and chemical methods may be used to inactivate liquid biological waste. As is illustrated by Table 16.1, most commercial production and pilot-plant facilities employ thermal inactivation systems.

Table 16.1 Methods in Use for Biowaste Inactivation

Company or Institution	Method	Reference
Cetus	thermal	1
Genentech	thermal	1, 2
Amgen	chemical/thermal	2
Serono	thermal	3
Monsanto	thermal	1
NCI—Frederick Cancer Research Inst.	thermal	4, 5
DuPont—Glasgow, DE	thermal	

Although sterility is absolute (something either is sterile or it is not), sterilization processes are stochastic, not absolute. The success of a sterilization process is a function of the probability that any contaminating organism may have survived the treatment method. For medical devices, this probability of sterility assurance is generally accepted as allowing one article in one million (a probability of 10^{-6}) to be nonsterile. For the food industry, a more rigorous standard has been adopted of one in one trillion (10^{-12}), based on the survival of spores of *Clostridium botulinum* in canned foods. For many large-scale industrial fermentations, the accepted contamination rate for sterilization processes is one batch in one thousand (10^{-3}).

The NIH guidelines for work involving recombinant DNA organisms [6] require that all culture fluids containing recombinant organisms be sterilized by a validated method prior to disposal. The guidelines specify that the sterilization validation be based on the host organism used. This is clearly organism-specific, requiring revalidation for each new organism processed in the equipment or system. It does not require that the liquid be sterile (free of all viable organisms), merely that the host, and, by assumption, the recombinant organism, be killed. This allows for milder conditions to be used where the host is not a spore-former or resistant to the treatment method. Strictly speaking, the NIH requirements are more appropriately termed *pasteurization* than *sterilization*.

In contrast, validation of autoclave sterilization processes, such as those used in the pharmaceutical or medical industry, identifies the types of contaminating organisms present in the autoclave load, and determines the highest level of resistance in that population. A thermophilic spore-forming organism such as *Bacillus stearothermophilus* is often chosen for the validation. The process is then operated under conditions designed to kill the resistant organism. This approach is both more rigorous and more flexible, in that it does not require constant revalidation. It does, however, result in more sophisticated and expensive systems than what may be needed to satisfy the organism-specific validation requirements of the NIH guidelines.

16.1.1   Regulatory Requirements and Permits

Recombinant biowaste inactivation and containment requirements in the U.S.A. are regulated by federal, state, and local governments. The United States Department of Health and Human Services, through the *National Institutes of Health (NIH)*, has issued containment guidelines for recombinant organisms [6] and pathogenic organisms [7]. Although these guidelines are not legally binding, most biotechnology companies comply with them voluntarily. The NIH guidelines, established in 1978, have been revised periodically to reflect the improving knowledge of the technology, most recently with the introduction of a new containment level classification, *Good Large Scale Practices (GLSP)* [8]. In this instance, the GLSP guidelines are similar to existing British and European guidelines for large industrial-scale practices [9]. GLSP classification is recommended for organisms such as those that have built-in environmental limitation, which permits optimum growth in the large-scale setting but allows limited survival (without adverse consequences) in the environment. The GLSP guidelines do not specify sterilization, as with BL1-LS, BL2-LS and BL3-LS, leaving this to applicable environmental regulations.

Release of recombinant organisms is also regulated by virtually every national government and many state and local governments [9–11]. As a result, it is always necessary to investigate such state and local code requirements, and also to consider the community relations aspects of a biowaste treatment system.

16.1.2   Pretreatment Prior to Discharge

For healthy waterways and lake ecosystems, dissolved oxygen is absolutely required, and its concentration must be kept above some minimal level. Ideally, the concentration of the dissolved oxygen in the water should be at least 90% of the saturation level, and should not fall below 4 mg/L [12]. This generally requires treatment of waste on the plant site, or by the municipal waste authority. Failure to pretreat sewage and industrial waste discharges can result in high concentrations of organic materials, which serve as nutrients for microbial growth. The growth of the microorganisms then depletes the dissolved oxygen concentration in the water, killing fish and beneficial organisms.

The availability of a municipal sewer system and the local requirements for pretreatment prior to discharge are important considerations in the siting of biotechnology plants. Wastewater pretreatment can add millions of dollars to the cost of a fermentation plant. While wastewaters from biotechnology manufacturing plants often contain high levels of organic materials, they do not generally contain toxic materials. For example, the classical fermentation industries, such as breweries and antibiotics fermentation plants, produce large quantities of waste organic materials, since only a small portion of the raw materials is converted into products. The high levels of organic materials result in serious depletion of the residual dissolved oxygen in the water as these organics are oxidized by the microorganisms of lakes or waterways.

The measurement of oxygen demand in waste discharges is characterized by the *biochemical oxygen demand* (*BOD*) or the *chemical oxygen demand* (*COD*). The chemical oxygen demand is based on a complete chemical oxidation method using a strong oxidizer, and is a very rapid test, typically 2–4 hours. The COD value is numerically larger than the BOD because the chemical oxidation is nearly complete. For example, the COD:BOD ratio for domestic sewage ranges from 2:1 to 5:1. Because the COD method oxidizes organics nearly to completion, it is often the only suitable test if inhibitory compounds are present, as well as when a rapid assay is required. In the COD test method, the sample is treated with a known amount of potassium dichromate solution and is boiled for several hours. The residual dichromate is then determined by titration with ferrous sulfate or ferrous ammonium sulfate.

Biochemical oxygen demand is based on an aerobic incubation of the sample for several days under optimal growth conditions. In this method the oxygen concentration is determined in the sample before and after 5 days' incubation at 20 °C in the dark. The concentration of the sample must be suitably diluted to be within valid testing levels, and mineral nutrients and bacterial inoculants must be added to ensure optimal growth. The oxygen demand is then calculated as milligrams of oxygen consumed per liter of sample.

Pretreatment of wastes uses three processes to minimize the environmental impact of discharge streams: (a) physical treatment, (b) biological treatment, and (c) chemical treatment. Physical methods typically remove excess solids, which may then be taken to a landfill or incinerated. Biological treatment to reduce the organic material in the waste stream includes both aerobic digestion and anaerobic digestion. Chemical treatments include coagulation of fine suspensions and pH adjustment prior to discharge. The pH of the waste stream must often be adjusted prior to discharge to minimize possible upset to municipal waste treatment systems.

I do not discuss the treatment of wastewater for BOD and COD reduction or the adjustment of pH in this chapter, but direct the reader to the extensive literature in these areas.

16.2   Kinetics of Microbial Death

Models and kinetic expressions that quantify the death of microorganisms are necessary to the reliable design of sterilization processes, including those of biowaste inactivation. Fortunately, such expressions and models have been developed as part of the hospital/medical device industry and the fermentation industry. In models of sterilization operations and biowaste inactivation, it is the total number of viable organisms rather than their concentration that is important, since even a single reproductively viable organism is capable of reestablishing the population. The kinetic expressions for cell death relate the remaining viable organisms in a sample to the temperature or chemical concentration, the exposure time, and the sensitivity of the microorganism.

16.2.1   Kinetics of Thermal Inactivation

The most common model for the death rate kinetics of microorganisms is a logarithmic equation. In logarithmic cell death kinetics, the number of viable cells remaining declines exponentially over the treatment period (Figure 16.1A), corresponding to a straight line on a semilog plot of the remaining viable cells, N/N_0, versus time (Figure 16.1B). The data can be fitted by a logarithmic equation characterized by the specific death rate constant, k_d, equal to the negative slope of the plot. The logarithmic cell death, expressed as a differential equation, is first-order with respect to cell number; the rate of decrease in viable cell number is directly proportional to the number of remaining viable cells and is given by:

$$dN/dt = k_d N \tag{16.1}$$

where N is the number of viable microorganisms,

   t is time, and

   k_d is the specific death rate constant.

Integration of eq 16.1 at constant temperature yields:

$$N = N_0 e^{-k_d t} \tag{16.2}$$

where N_0 is the initial number of viable microorganisms. The logarithm of eq 16.2 is linear with time and is shown in Figure 16.1B. Nonlinearities may arise either as a

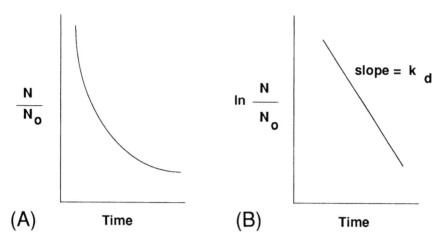

Figure 16.1 Logarithmic Cell Death Kinetics. (A) The fraction of surviving organisms declines exponentially over the treatment period, corresponding to a straight line on a semilogarithmic plot. (B) The semilogarithmic plot of the fraction of surviving organisms versus time; the slope is the specific death rate constant, k_d.

result of the activation of thermally resistant spores, which then germinate and begin to grow before the temperature overcomes and kills the vegetative (growing) cells (Figure 16.2A), or as a result of a mixed culture of organisms, with the more resistant organism in the minority (Figure 16.2B). The situation where the more resistant organism is in the majority does not generally display significant nonlinearities.

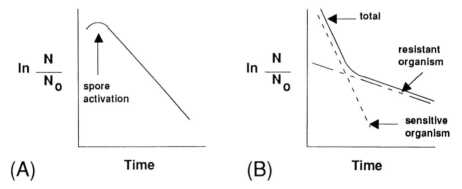

Figure 16.2 Nonlogarithmic Cell Death Kinetics. (A) When spores are activated, they germinate and grow, and initially offset the deaths of the vegetative cells. (B) A more resistant population initially in the minority overgrows the less resistant member of a mixed culture.

The specific death rate constant, k_d, of eq 16.1 and 16.2 is a function of temperature and is characteristic of each organism and its growth history. Figure 16.3 shows typical data for the influence of temperature on the kinetics of thermal death of vegetative bacteria. Each line corresponds to a different value of the specific death rate constant. The temperature dependence follows an Arrhenius relationship, given by:

$$k_d = k_{d0}\, e^{-E_d /RT} \tag{16.3}$$

where k_{d0} is a constant (for *B. stearothermophilus* spores, k_{d0} is $10^{36.2}$ s^{-1};
E_d is the thermal activation energy (for *B. stearothermophilus* spores, E_d is 67,700 cal/mol);
R is the universal gas constant, 1.98 cal/mol·K; and
T is the absolute temperature, K.

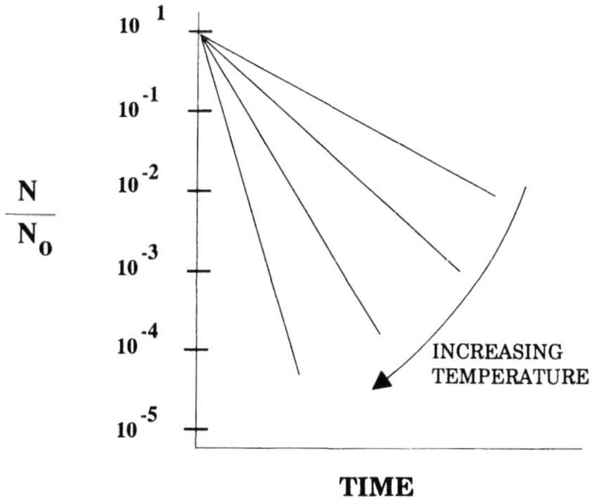

Figure 16.3 Temperature Dependence of Logarithmic Cell Death Rates. Increasing temperatures increase the specific death rate constant, k_d, corresponding to a steeper slope and shorter times to achieve a particular degree of population reduction.

Values of the thermal activation energy, E_d, range from 50 to 100 kcal/mol for many spores and vegetative cells. Table 16.2 lists values for several organisms.

Table 16.2 Thermal Activation Energies

Organism	k_{d0}, s^{-1}	E_d, cal/mol
Bacillus stearothermophilus	$10^{36.2}$	67,700
Bacillus subtilis	10^{42}	76,000
Escherichia coli	$10^{70.8}$	127,000

The temperature dependence of the specific death rate constant, k_d, is important in fermentation medium sterilization, where the medium must be exposed to elevated temperatures for the shortest possible time in order to minimize the destruction of the nutrients in the medium. In such cases, the time- and temperature-dependent kinetic expressions must be integrated in a rigorous fashion to account for the destruction of the microbes during the heating, high-temperature hold, and cooling phases [13]. Figure 16.4 shows a typical temperature profile for a batch sterilization process.

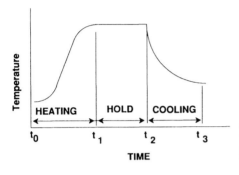

Figure 16.4 Batch Sterilization Process. The temperature profile for a batch sterilization process follows this typical course.

Biowaste treatment differs from medium sterilization in that the heating and cooling phases may be disregarded, since there is no concern for breakdown of nutritive value. In the design of biowaste inactivation systems, consideration of the time of the constant-temperature holding period, without consideration of the heating and cooling times, results in a slight overdesign of the biowaste sterilization system. This approach allows simplified calculations for batch systems, using equations rigorously appropriate for continuous sterilizers.

For continuous sterilizers eq 16.2 and eq 16.3 may be combined to yield expressions that solve for either the log reduction in viable cell number at a constant temperature for a set time (eq 16.4a), or, alternatively, the time required for a predetermined log reduction of the microbial bioburden at a given temperature (eq 16.4b):

$$\ln (N_0/N_1) = k_{d0}\, e^{(-E_d/RT)}\, t \tag{16.4a}$$

$$t = k_{d0} \ln (N_0/N_1)\, e^{(-E_d/RT)} \tag{16.4b}$$

Graphs of eq 16.4a and 16.4b can be plotted for specific organisms, as shown in Figure 16.5 for *B. stearothermophilus* spores (E_d = 67,700 cal/mol and k_{d0} = $10^{36.2}$ s^{-1}). Figure 16.5 can be used to conveniently estimate either:

(a)   the log reduction [ln (N_0/N)] at a set time and temperature, or
(b)   the time required to achieve a set log reduction at a given temperature, or
(c)   the temperature required for successful log reduction at a set time.

All the equations and figures discussed assume linear cell death kinetics, with no nonlinearities due to mixed populations or spore activation. The units of lethality

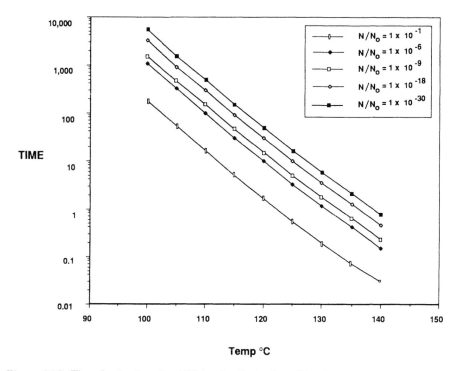

Figure 16.5 Time for Set Levels of Bioburden Reduction. The time, in seconds, to achieve various levels of total bioburden reduction depends logarithmically on temperature. The curves are based on heat-resistant *B. stearothermophilus* spores.

discussed above for the number of log reductions in microbial bioburden have been variously termed the *del factor* or *nabla factor* (since the name of the ∇ symbol is the del or nabla).   These terms are among many used in the literature on medical equipment and food industry sterilization practices.   To facilitate comparisons of the relative sterilization capacities of different thermal sterilization processes, various units for lethality (other than the del factor) have been defined. These include the *D value*, the *F value*, and their corresponding temperature dependence coefficients (*Z values*).

The *D value*, or *decimal reduction time*, is defined as the time required to reduce a microbial population by a factor of 10, thereby destroying 90% of the initial microbe or spore count. Figure 16.6 shows the semilogarithmic plot of survivor number versus heating time that yields the *D* value. The decimal reduction time, *D*, is related to the specific thermal death rate, k_d, by:

$$D = 2.303/k_d \qquad (16.5)$$

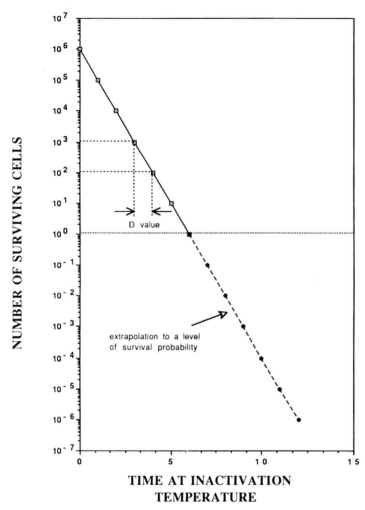

Figure 16.6 Decimal Reduction Time (D value), shown on a semi-logarithmic plot of surviving number of organisms versus heating time. The decimal reduction time (D value) is the time, in seconds, corresponding to one $\log_{10}$ cycle, and represents a 90% destruction of the initial number of organisms.

The *F* value is defined as *the period of time (minutes) to destroy **all** spores or organisms in suspension at 121 °C*, and is given by eq 16.6:

$$F = n D/60 \tag{16.6}$$

where D is the decimal reduction time,
  n is $\log_{10} N_0$, and
  N_0 is the initial number of cells.

D values and F values depend on the conditions during the treatment and sample recovery. Z *values*, which are temperature coefficients, are less dependent on these conditions. The Z value is equal to the number of degrees (°F or °C) required for the thermal destruction curve (D values or F values) to decrease by one $\log_{10}$ cycle on a semilogarithmic plot of D (or F) versus temperature, as shown in Figure 16.7. The Z value based on the D values is given by:

$$Z = (T_2 - T_1)/(\log_{10} D_2 - \log_{10} D_1) \tag{16.7}$$

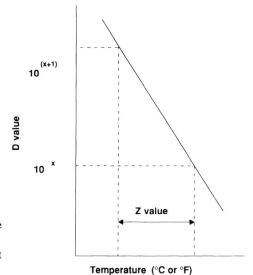

Figure 16.7 Z value. Z is defined as the temperature change required to shift the thermal destruction curve (D value or F value) through one $\log_{10}$ cycle on a semilogarithmic plot of D (or F) versus temperature.

Typical ranges for the Z *value* (*temperature dependence of bacterial death*) are 7–24 °C for moist heat [14], with a generally accepted average of 10 °C. By comparison, dry heat Z values have a range of 10–60 °C, with an average of 21 °C generally accepted. This difference in the Z values of moist heat and dry heat reinforces the importance of moist heat conditions (that is, saturated steam) for sterilization, or for decontamination of equipment or biowaste.

The Z value may be used to relate the decimal reduction time (D value) at one temperature to the expected value at another temperature:

$$D = D_{121}\ 10^{-\left(\frac{T-121}{Z}\right)} \tag{16.8}$$

where T is the temperature in °C.

Experimental techniques for measuring the thermal destruction rate of microorganisms have been developed for the food industry. These techniques are of two general types: (1) systems where successive samples are drawn from a suspension of the test microorganism and plated for enumeration, and (2) replicate systems where series of identically prepared, statistically significant numbers of samples are subjected to varying time and temperature programs, to determine growth/no growth. All techniques must take into consideration the sample size and numbers, and their statistical significance. Additionally, the presence of any inhibitory substances (such as antibiotics or decontamination chemicals) should be considered; if any are present, suitable inactivation or filtration separation techniques should be employed. As with all experimental designs, suitable controls should be included.

16.2.2   Variables Influencing the Effectiveness of Decontamination

Variables Influencing the effectiveness of biowaste treatment include the initial contamination level of the material (the initial bioburden), the accessibility of the organism (whether the organism is protected within any solids), and, lastly, the growth state of the organism (whether the organism is vegetative or spores). As was evident in the previous discussion of cell death kinetics, the effectiveness of a decontamination procedure depends on the initial number of microorganisms in the stream and the extent of bioburden destruction required.

Solids in the biowaste stream protect organisms trapped within by limiting diffusion (thermal or chemical) to the interior regions, as shown in Figure 16.8. Therefore, if the waste streams are known to contain solids, the sterilization procedures must be overdesigned. The engineering analysis to allow for the presence of solids in thermal sterilization systems is well documented in most engineering textbooks on transport phenomena and I do not discuss it here.

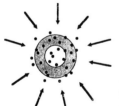

Figure 16.8 Protection of Organisms Within Solids. Solids in the biowaste stream protect organisms trapped within by limiting diffusion (thermal or chemical) to the interior regions.

16.3   Chemical Decontamination

Chemical inactivation of biowaste is troublesome, and quite sensitive to the nature of the waste stream. Variable amounts of proteins and organics make it difficult to estimate the inactivating chemical additions accurately. Solids may protect organisms trapped within them. Care must be taken to recognize organic compounds that may react chemically with the inactivation chemicals, resulting in potentially toxic compounds (for example, organics reacting with chlorine).

Chemical inactivation may be accomplished using a variety of chemicals, including, but not limited to: chlorine (as sodium hypochlorite), sodium hydroxide, glutaraldehyde, chlorine dioxide, and quaternary ammonium compounds. Of these, sodium hypochlorite is the most commonly used, and I discuss it in some detail. The effectiveness of chemical decontamination procedures depends on the susceptibility of the organism, the concentration of the treatment chemical, the initial bioburden level, the accessibility of the organisms to the chemical agents, and the treatment conditions (such as pH, temperature, and interfering ions). In the following sections, I discuss some of these factors as they relate to chlorine (hypochlorite) inactivation systems. The two most critical of these factors are the initial bioburden level and the pH during treatment. As with thermal inactivation systems, the initial bioburden level is a critical determinant of the ultimate effectiveness of chemical treatment methods. If the pH is not in the proper range, either the biocidal activity is dramatically reduced, or adverse side reactions may occur (such as the evolution of chlorine gas from sodium hypochlorite under acid conditions).

16.3.1   Chlorine Chemistry

Chlorine, when dissolved in water, hydrolyzes to form hypochlorous acid (HOCl), which may ionize to form hypochlorite. The extent of ionization depends on the pH of the solution, as shown below:

$$Cl_2 + H_2O \; \rightleftharpoons \; HOCl + H^+ + Cl^- \tag{16.9}$$

$$HOCl \; \rightleftharpoons \; H^+ + OCl^- \tag{16.10}$$

Chlorine is commonly added as sodium hypochlorite (NaOCl) or calcium hypochlorite (also known as chloride of lime). Sodium hypochlorite is generally available as a liquid solution; calcium hypochlorite is a granular solid.

$$NaOCl \rightleftharpoons Na^+ + OCl^- \tag{16.11}$$

$$Ca(OCl)_2 \rightleftharpoons Ca^{2+} + 2\ OCl^- \tag{16.12}$$

$$H^+ + OCl^- \rightleftharpoons HOCl \tag{16.13}$$

Chlorine reacts with ammonia and other nitrogen compounds such as amines and imines to form chloramines or N-chloro compounds.

$$HOCl + NH_3 \rightleftharpoons H_2O + NH_2Cl \qquad \text{(monochloramine)} \tag{16.14}$$

$$HOCl + NH_2Cl \rightleftharpoons H_2O + NHCl_2 \qquad \text{(dichloramine)} \tag{16.15}$$

$$HOCl + NHCl_2 \rightleftharpoons H_2O + NCl_3 \qquad \text{(trichloramine)} \tag{16.16}$$

The extent of the reaction of hypochlorous acid with ammonia depends on pH, temperature, time, and the initial concentrations. In the pH range of 4.5 to 8.5, monochloramine and dichloramine are generated, while below pH 4.4, trichloramine is produced.

Free available chlorine is defined as the total of elemental chlorine (Cl_2), hypochlorous acid (HOCl), and hypochlorite ion (OCl^-). *Combined available chlorine* is defined as the total of the chloramines or N-chloro compounds. The sum of the free and combined chlorine is defined as the *total residual chlorine*.

The stability of free available chlorine in solution is affected by such factors as chlorine concentration, temperature, pH, presence of organic materials, ultraviolet radiation, and presence and concentration of catalysts. Organic chloramines are more stable in solution than free chlorine compounds.

16.3.2   Susceptibility of Microorganisms to Chlorine

The relative sensitivities of various microorganisms to chlorine are listed in Table 16.3. Bacterial spores are more resistant to the biocidal action of chlorine disinfectants, and may require high concentrations and prolonged exposures for effective decontamination. Spores of *Bacillus subtilis* subsp. *niger* and other species can be killed rapidly by 100 ppm of available chlorine if the solution is buffered at pH 7.6–8.1 [15].

The mechanism of the destruction of microbes by HOCl has never been experimentally demonstrated. Speculation on the mechanism of action has included: (a) that HOCl liberates nascent oxygen, resulting in oxidation of critical macromolecules; (b) that chlorine reacts with cell membrane proteins, interfering with membrane function; and (c) that oxidative inhibition or destruction of certain critical intracellular enzyme systems occurs during chlorine treatment [16].

Table 16.3 Microbial Susceptibility to
Chemical Inactivation by Chlorine

Microorganisms	Susceptibility
Gram-positive bacteria	high
Gram-negative bacteria	high
acid-fast bacteria	moderate
bacterial spores	moderate
lipophilic viruses	moderate
hydrophilic viruses	moderate
amoebic cysts	high
algae	high
fungi	moderate

Increasing the temperature during treatment of biowaste with chlorine reduces the time necessary for effective decontamination. This can have practical implications, as the time required for sporicidal action of sodium hypochlorite may be reduced by as much as half for each increase of 10 °C [17].

The disinfecting efficiency of chlorine decreases with increasing pH, paralleling the concentration of dissociated hypochlorous acid. This indicates that hypochlorous acid (HOCl) has stronger biocidal activity than the hypochlorite ion (OCl$^-$). Alkaline solutions stabilize available chlorine at the expense of biocidal activity, as shown in Table 16.4 [12]. For example, 100 ppm of available chlorine at pH 7.6 is as effective in killing *Bacillus subtilis* spores as 1000 ppm at pH 9. Solutions at low pH (4.0 or below) decompose, evolving chlorine.

Table 16.4 Influence of pH on the Availability
of Undissociated Hypochlorous Acid

pH	% Cl$_2$	% HOCl	% OCl$^-$
4	0.5	99.5	0
5	0	99.5	0.5
6	0	96.5	3.5
7	0	72.5	27.5
8	0	21.5	78.5
9	0	1.0	99.0
10	0	0.3	99.7

Within the pH range of 6–8, combined available chlorine (chloramines, etc.) is much less effective than free available chlorine in biocidal activity, requiring up to 25 times more for the same temperatures and contact times. For example, the biocidal activity of monochloramine is only 1% to 2% that of free chlorine.

Effective treatment cannot be assured without a knowledge of the composition of the waste stream being treated, as both the ions present and the amount of organic material in the stream can have dramatic effects on the biocidal efficacy of chlorine (hypochlorous acid). Ferrous and manganous cations, and nitrite and sulfide anions reduce active hypochlorous acid to inactive chloride. Calcium and magnesium cations do not inactivate hypochlorous acid.

Chlorine compounds, both available and combined, react with all types of organic matter. Thus, it is important that sufficient chlorine be supplied to compensate for the organic load and still provide proper biocidal concentrations of available chlorine. Organics undergoing chlorination may sometimes form undesirable by-products. For example, chlorine can react with phenol, forming chlorophenol, a toxic chemical and suspected carcinogen. The importance of potential pollution problems such as this should not be overlooked. This is one of the many reasons for using thermal biowaste treatment systems.

16.4   Drain and Collection Systems

The biowaste drain collection system is a drainage utility distributed throughout the processing areas of the plant. It is similar to the common sanitary drain system, but it is a closed system, and as a result should not generally have floor drains or connections open to the atmosphere. It is used to collect liquid wash or flush, steam condensate, and liquid waste that could contain contaminant organisms. The drain collection system, shown schematically in Figure 16.9, is designed as a gravity-fed system, leading to a collection tank, which then feeds the inactivation system upon a signal from a level controller. The collection tank is generally located below the level of the drainage system. Alternatively, a small sump tank and pump system can be used to collect and transfer materials to a large collection tank, located at grade above the drain pipe system. The use of sump vessels should be avoided, if possible, due to the additional complexity of vessels, pumps, and controls, which require maintenance and may fail. The collection tank should be sized to accept approximately twice the maximum anticipated biowaste volume present in the plant at a given instant. This is generally assumed to be twice the total working volume of systems in simultaneous operation.

All collection and treatment tanks should be equipped with sterilizing vent filters or incinerators to prevent aerosol escape to the environment. The system should be designed for full decontamination of the drainage piping and collection tank system in the event that maintenance is required and the lines or tanks must be opened. The collection tank should be regularly sanitized or sterilized in order to control bioburden.

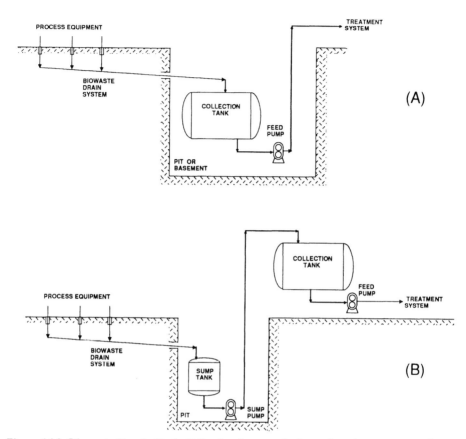

Figure 16.9 Biowaste Gravity Drain Collection System. In the preferred arrangement, shown in (A), the collection tank is below the level of the drain system. The system shown in (B) requires intermediate sump tanks, pumps and their controls.

Contained (or contaminated) drainage lines should be installed and used only for biowaste. Other process waste, such as purification buffers, should be directed to the neutralization system, where the pH is adjusted prior to discharge to the public sanitation system. Directing uncontaminated waste streams to the biowaste system results in an unnecessarily large system and higher operating cost. Additionally, the introduction of biowaste drain lines into clean purification rooms allows a possibility for back contamination into that room through the drain line. Under all circumstances, the biowaste drainage system should be a closed system, without open connections such as floor drains. If floor drains are necessary to provide for the possibility of a spill, these should be designed either as piping that is normally valved off and covered (to be opened to the biowaste system only if a spill occurs), or as a dedicated floor drain collection system leading to an emergency collection tank, which is then pumped to the biowaste collection tank. The piping and vessel for the latter case (a dedicated floor drain tank) should be designed for suitable decontamination, such as by a full flooding of the lines with disinfecting solutions.

Selection of piping materials for the drainage system depends on the chemical composition of the liquid stream, the normal operating conditions, the proposed method of decontamination (heat inactivation or chemical decontamination), the location of the drain line, and the expected service life of the system. Drain lines buried underground or below concrete slabs should be double-jacketed to provide a redundant containment system in case the drain line leaks. In a multistory building, where drain lines may be routed in the interstitial spaces, normal pipe may be used. Figure 16.10 shows a sectional view of a double pipe assembly. If the piping system is to be steam sterilized, the design must take into account the expansion of stainless steel piping, and provide suitable anchorage and expansion loops.

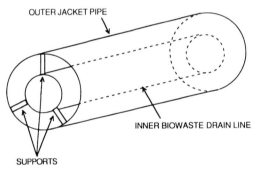

OUTER JACKET PIPE

INNER BIOWASTE DRAIN LINE

SUPPORTS

Figure 16.10 Double-Pipe (Jacketed) System for Containment of Buried Piping Systems. The outer annular space may be equipped with an electrical leak detector, which can identify the location of a leak, should one occur, and prevent leakage of waste into groundwater systems.

16.5   System Configurations

Biowaste treatment systems are designed as either thermal or chemical inactivation systems. Thermal inactivation systems may be configured for either batch or continuous treatment, while chemical treatment systems are generally designed strictly as batch systems. Relative advantages and disadvantages of batch and continuous treatment systems are listed in Table 16.5.

The primary advantage of batch systems is the ability to sample and test the discrete batches prior to discharge. Continuous treatment systems, on the other hand, are more energy-efficient, in terms of both steam and cooling water consumption. Both batch and continuous systems can be designed to provide some of the characteristics of the other; however, this is not normally done. For instance, batch

Table 16.5 Relative Advantages of Batch and Continuous Systems

Advantages	Disadvantages
Batch	
• discrete sampling of each batch prior to discharge without necessity of additional tanks • less complicated equipment	• higher energy costs • slow batch turnaround times require larger processing tanks
Continuous	
• lower energy costs (more energy-efficient heating and cooling) • more efficient treatment due to high temperature/short time (HTST) nature of process—requires smaller equipment and much lower peak energy demands	• more complicated equipment • no batchwise sampling prior to discharge without an additional holding tank • higher maintenance

systems can be designed with energy recovery heat exchangers on the discharge line, and continuous treatment systems can be equipped with a holding tank for sampling prior to discharge.

16.5.1   Batch Treatment

Batch systems have a collection tank, and a treatment tank that heats and inactivates the waste in batches. The collection tank may be a dedicated vessel for this purpose, or may be combined with the treatment system in what is termed a *swing-tank system*, as shown in Figure 16.11. Swing-tank systems use two tanks. Initially, waste goes into tank 1. When it is full, collection of the biowaste flow is switched to tank 2. Tank 1 is then batch-treated (heated, cooled, and sampled) while tank 2 collects biowaste. The process is repeated as required, using the tanks alternately to collect and treat the waste. This system requires two large tanks, and in the event that a tankful is incompletely inactivated and requires retreatment, the volume must be adequate so as not to disrupt manufacturing operations because waste cannot be dumped. The batch system can use a dedicated collection tank with a smaller treatment tank, if the collection tank is sized large enough.

The batch system may be heated directly by injecting steam, or indirectly by steam heating the vessel jacket. Direct steam injection has a higher heat transfer coefficient, and the batch therefore takes a shorter time to heat up, but the waste stream is diluted

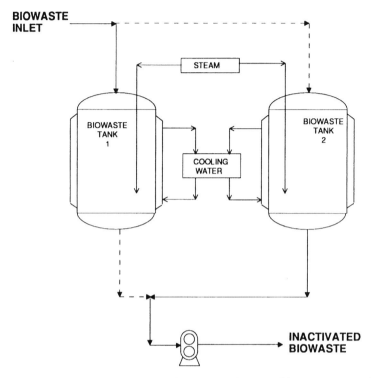

Figure 16.11 Batch Swing-Tank Biowaste Inactivation System. Biowaste collects in one of two tanks, while the contents of the other are pumped through a continuous sterilizer upon a signal from a high/low controller mounted on the collection tank.

with steam condensate. Indirect heating through the vessel jacket allows recycle of the condensate back to the steam system, but the jacket heat-transfer surfaces may be subject to fouling, and periodic cleaning of the interior surfaces of the vessel may be required.

The waste is cooled prior to discharge using the least expensive cooling medium available, generally cooling-tower water. The waste could be cooled by dilution with cold city or process water, but this is discouraged or banned by many municipal water authorities because of its impact on the sewage treatment plants and water conservation programs.

16.5.2 Continuous Inactivation

Continuous sterilizers for biowaste are designed according to the same principles as the continuous sterilizers for the fermentation and food industry. A typical continuous sterilizer system, shown in Figure 16.12, has a large collection tank that provides surge capacity; the contents of the collection tank go as feed to a continuous sterilizer upon a signal from the high/low level control. The continuous sterilizers use heat exchangers to recover energy and preheat the inlet biowaste.

16.5.2.1 Heat Exchangers

The waste stream is heated by steam. Heat exchangers and direct steam injection, alone or in combination, may be used. All continuous biowaste treatment system designs use heat exchangers for energy recovery, which can be in the range of 60–80%, depending on the amount of heat transfer area in the exchangers. Heat exchanger selection and size are based on engineering suitability and on an economic comparison of the energy costs versus the higher equipment costs for larger heat exchangers. Direct steam injection can supply most of the heat, although using steam results in higher boiler feed water requirements since no condensate is returned.

Heat exchangers for biowaste inactivation systems could conceivably be of any of the standard designs, including shell-and-tube, plate-and-frame, spiral, and concentric double-tube heat exchangers. Of these, shell-and-tube exchangers are the least practical for biowaste treatment systems because they tend to foul and plug, and are difficult to clean.

Plate heat exchangers have a large heat-transfer coefficient and therefore can be smaller, but have a number of disadvantages, one of the worst of which is that the gaskets separating the plates require a lot of maintenance. At high temperatures the gaskets have a limited service life, and are prone to leaks or ruptures. Furthermore, the crevice where the gasket meets the plate is subject to scale buildup and chloride corrosion, which may be high when the chloride content is above 150 ppm and the temperature exceeds 80 °C. Stress corrosion leaks do not show up as catastrophic failures, but as nonsterile output. Plate heat exchangers are constructed of corrugated (and therefore nonuniform) plates, which may cause plugging as solids settle out of the slower moving parts of the stream over the plate.

Spiral heat exchangers, although more expensive than plate-and-frame exchangers, are recommended because they require less maintenance and foul less readily (although they too can clog if the gap is small). They also have improved velocity profiles through the unit, and require less gasket material.

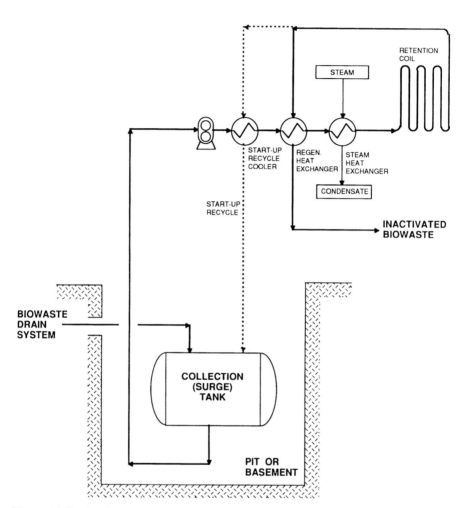

Figure 16.12 Continuous Biowaste Inactivation System. Biowaste, collected in a tank, is pumped through a continuous sterilizer upon a signal from a high/low controller.

The last type of heat exchanger is the concentric double-tube exchanger. This type offers the most advantages; corresponding to their preferred use in sanitary service, double-tube exchangers have the fewest crevices where corrosion or leaks can cause trouble, and they require the least cleaning. They are prohibitively expensive for the large surface areas required, however, so they are not normally used in biowaste treatment systems.

16.5.2.2 Retention Coil

Waste is inactivated effectively if it is at the design temperature for a minimum residence time. This is achieved in a long section of pipe called a *retention coil*. The length of the retention coil is calculated by dividing the cross-sectional area of the pipe by the average velocity (assuming plug flow) of the liquid in the pipe. The residence time distribution of the liquid through a pipe depends on both the flow rate and the velocity profile in the pipe. Flow rates should be chosen to ensure that a fully turbulent flow (plug flow) is achieved, as shown in Figure 16.13B. The flow regime, whether laminar or turbulent, is determined by the *Reynolds number* of the flow, N_{Re} or *Re*, defined by:

$$Re = \frac{d\,v\,\rho}{\mu} \tag{16.17}$$

where *d* is the inside diameter of the pipe;
    v is the fluid velocity;
    ρ is the fluid density; and
    μ is the fluid viscosity,
in any consistent units so that Re is dimensionless.

At Reynolds numbers below 2,000–4,000, flow is *laminar* and the velocity profile is parabolic, as shown in Figure 16.13A; material in the center of the pipe has a substantially shorter residence time in the pipe than the material near the wall. In *fully turbulent flow* (also called *plug flow*), at Reynolds numbers above 10,000–20,000, the velocity profile is nearly flat (Figure 16.13B); aliquots of liquid passing through the system have nearly the same residence time.

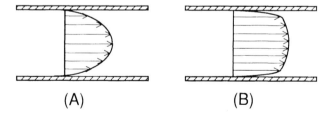

(A)   (B)

Figure 16.13 Laminar and Turbulent (Plug) Flow Through a Pipe. (A) Laminar flow has a parabolic velocity profile. (B) Turbulent flow has a nearly flat velocity profile. True plug flow would have a flat velocity profile across the full width of the pipe.

For economic reasons, the minimum velocity at which flow is turbulent should be used, so that the length of pipe required for the retention coil is as short as possible.

Flow should be turbulent, but the fluid velocity should not be so high as to cause erosion of the pipe walls. Generally, liquid velocities should be below 5–7 ft/s (~2 m/s).

Deviations from plug flow may be due to axial back mixing or turbulent diffusion, and can give rise to band spreading (the same phenomenon seen commonly in chromatography), as shown in Figure 16.14. These deviations are sometimes characterized by the dimensionless *Peclet number*, *Pe*, given by:

$$\text{Pe} = vL/D_z \tag{16.18}$$

where v is the average fluid velocity;
   L is the sterilizer length; and
   D_z is the axial dispersion coefficient.

As the Peclet number approaches infinity (as the axial dispersion coefficient approaches zero), the velocity profile approaches that of plug flow. Conversely, if the Peclet number approaches zero (as the axial dispersion coefficient approaches infinity), the flow is highly back mixed, and the retention coil must be longer if all the liquid is to be completely treated. LEVENSPIEL [18] has correlated dispersion coefficients to the Reynolds number for flow through pipes; some of the values are listed in Table 16.6.

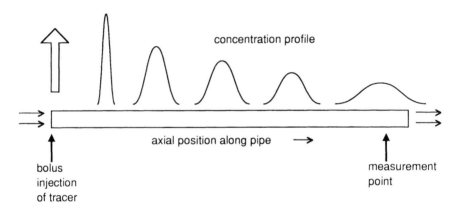

Figure 16.14 Axial Dispersion During Flow Through a Pipe.

Solids present in the waste stream affect the design in several ways. In particular, additional time is required to reach the desired treatment temperature at the center of the solids. Engineering approaches can be taken to minimize solids, or to break up the solid clumps to allow better penetration.

Table 16.6 Axial Dispersion Coefficients
for Turbulent Flow Through a Pipe [18]

Reynolds Number	$D_z / v\,d$
3,000	1.0–3.0
5,000	0.75–1.8
8,000	0.53–0.92
10,000	0.42–0.53
100,000	0.22

* Values listed are based on experimental data,
   and presented as a range. D_z is the axial
   dispersion coefficient; v is the average fluid
   velocity; d is the diameter of the tube or pipe.

16.5.3   Exhaust gas

Exhaust gases from biowaste systems must be suitably treated to inactivate contaminating organisms. This may be as simple as sterile filtration in cases of microbial fermentation facilities, or as complex as an incinerator system for installations where viruses are used or produced. Sterile filtration uses cartridge filters to remove particulates; exhaust line systems must be designed to minimize the chances of clogging or blinding the cartridges. This is usually achieved by using a steam jacket on the filter housing, and possibly on the exhaust line, so that the temperature is above the dew point.

16.5.4   Instrumentation and Controls

Biowaste systems should be designed for fully automated operation, with appropriate alarms in the event of equipment failure or out-of-specification operation. The instrumentation should measure and record critical processing parameters, such as the collection tank liquid levels, the tank being processed (if batch), the time at initiation of processing, a continuous temperature profile of the inactivation process, the valve position indication for critical valves, and the reading at the low-point thermocouple in the outlet line near the steam trap.

An automated system typically has a sequence controller, programmable logic controller (PLC), or distributed control system (DCS). The control system monitors the liquid level in the collection tank, and when a high-level mark is reached, initiates transfer to the batch processing tank or continuous kill skid. During the processing, the temperature of the inactivation processing tank or skid is monitored and controlled, and valves sequenced (such as for recycle, during start-up of a continuous treatment system). In addition, the control system ensures that the exit temperature for either the continuous or batch treatment systems is within discharge limits (this is often regulated by state or local plumbing codes as no greater than 125 °F or 52 °C). For chemical kill systems, typical measurement and control parameters might additionally include a continuous recording and control of pH during the inactivation process.

Process parameter and trouble alarms are critical to safe operation and should be installed, at a minimum, for the following: high level in the collection tank, feed pump failure (for continuous-kill systems), outlet discharge stream over the allowable temperature, utility system failure (steam, cooling water, compressed air), and pressure buildup in the collection tank or drain system.

16.5.5   Supercritical Water Oxidation

Supercritical water oxidation (SCWO) has recently emerged as a possible successor to the continuous sterilizer [19]. Although analogous in its continuous operation, it is radically different in how it achieves the inactivation. In SCWO, the liquid waste stream is compressed and heated to raise conditions above the critical point of water (218 atm or 22 MPa, 374 °C). All organic compounds in the liquid stream oxidize rapidly and almost completely to inorganic compounds, carbon dioxide, and other simple compounds. The system has been demonstrated on bench scale on biowaste streams, such as pharmaceutical aqueous wastes contaminated with organics, and also on fermentation waste streams containing spores.

One of the interesting features of SCWO is that the supercritical conditions are maintained by the energy produced by the oxidation of the organic materials (if the aqueous stream does not contain sufficient organics, supplementary hydrocarbon feed must be added). SCWO may in the future be the preferred method of treatment, particularly if the liquid waste stream contains toxic or hazardous organisms or their by-products (such as toxins), because it is nonpolluting and is able to completely destroy all organisms.

16.6   Engineering Design Suggestions

Several suggestions related to the design of waste systems follow:
  (a) Avoid pressure relief valves. Use rupture disks and have each one alarmed in case it ruptures.
  (b) Maximize to the absolutely greatest extent the use of welded pipe joints, and minimize the use of threaded or flanged pipe joints, as these are prone to leakage.

(c) Provide a sampling system.

(d) Consider the effect of hot steam condensate flashing to steam after being released by steam traps, or steam leaking through malfunctioning steam traps or orifices. Such steam may cause a back pressure in the collection tank, or potentially blind the exhaust vent filters. Providing a chilled heat exchanger on the inlet line, or cooling the collection tank may help avoid this, as will heating the exhaust filter housings using steam jackets or electric resistance heater mantles.

(e) Locate the collection tank below the lowest point of the collection drain system.

(f) Do not connect open drains (floor or sink) to the biowaste system; this prevents contamination from entering the rooms.

(g) Provide redundant exhaust filters on the collection tank, with differential pressure sensors to provide for switching between the filters in the event of blockage or blinding due to aerosols.

(h) Minimize cross contamination between systems or rooms by using separate drain lines and, if necessary, separate dedicated sump collection tanks and pumps.

(i) In chemical treatment systems, control pH for the most effective treatment, elevate temperature to shorten required treatment times, and use high-shear pump-around loops to help break up any clumps that might protect organisms.

(j) Design the automation system for a fail-safe condition in the event of a failure of steam, electricity, or instrument air. This requires that outlet valves close and the flow be diverted back to the holding tank.

(k) Provide an emergency backup of the control system and recorder for the biowaste treatment system, to ensure that, in the event of utility failure, the system continues to monitor valve positions, and goes through an orderly shutdown.

16.7 Validation

Validation is documented evidence that provides a high degree of assurance that a specific process will consistently produce a product of a certain predetermined quality. In the case of biowaste inactivation, validation assures the reliable, reproducible reduction in viable cell count of the target organism to a predetermined (safe) level.

Validation of biowaste inactivation systems includes the standard equipment validation tasks of Installation Qualification (IQ) and Operational Qualification (OQ), the process development studies that identify the required kill procedure and conditions, and finally the Process Qualification (PQ). PQ for recombinant processes uses the nonengineered host cell line, for example the nonrecombinant *E. coli* strain used for genetic engineering. PQ for nonrecombinant processes that require biowaste inactivation (for example, vaccine manufacturing using pathogenic organisms) uses a heat-resistant organism (spore) such as *B. stearothermophilus.*

16.8   Conclusions

The design of biowaste treatment systems must be based on the magnitude (volume) of the waste stream, the target organism(s) requiring inactivation, the impact on the facility and its utilities, and the corresponding economics.   Pilot plants, small manufacturing facilities, and R&D labs generally use an overkill approach, where the inactivation system is validated for the most resistant organism likely to be seen, perhaps using heat-resistant spores such as *B. stearothermophilus* as the indicator organism.   Such an overkill approach is appropriate for facilities where flexibility is required for rapid changeover between organisms, or where multiple organisms are in simultaneous use.   Manufacturing plants with large biowaste streams validate an inactivation protocol specific for the production organism, generally using a system designed around thermal inactivation with energy recovery systems.   For reasons of cost, the inactivation systems are often designed around the mildest conditions that can still reliably inactivate the target organism, conditions that may not ensure true sterility (that is, 121 °C).

Although chemical inactivation systems can be designed for inactivation of biological wastes, many problems must be addressed.   The chief difficulties are assuring adequate biocidal activity of the treatment chemical agent under the conditions of use, and assuring its adequate penetration into any solids in the stream. The most common chemical used in inactivation systems is sodium hypochlorite; if it is used, the pH of the waste must be adjusted for proper biocidal activity and safe operation.

16.9   References

1. Fox, S., Process and Equipment Requirements for Sterilization of Biological Wastes, *Genet. Eng. News 11* (1), 6–7 (1991).
2. ABREO, C., HOERNER, C., LUBINIECKI, A., WALES, W., WIEBE, M.; Recombinant DNA Containment Considerations for Large Scale Mammalian Cell Culture Expression Systems; In: *Advances in Animal Cell Biology and Technology for Bioprocesses*; SPIER, R. E., GRIFFITHS, J. B., STEPHENNE, J., CROOY, P. J., Eds.; Butterworths, Kent, UK, 1989.
3. NELSON, K. L., GEYER, S.; Bioreactor and Process Design for Large Scale Mammalian Cell Culture Manufacturing; In: *Drug Biotechnology Regulation: Scientific Basis and Practices*; CHUI, Y. H., GUERIGUIAN, J. L., Eds.; Marcel Dekker, New York, 1991.
4. RUNKELE, R. S., PHILLIPS, G. B.; *Microbial Contamination Control Facilities*; Van Nostrand Reinhold, New York, 1969; pp 136–148.

5. FLICKINGER, M. C., SANSONE, E. B., Pilot and Production Scale Containment of Cytotoxic and Oncogenic Fermentation Processes, *Biotechnol. Bioeng. 27*, 860–870 (1984).
6. NIH Guidelines for Research Involving Recombinant DNA Molecules, *Fed. Regist. 51*, 16958 (1984).
7. *Biosafety in Microbiological and Biomedical Laboratories*; Publication Number (CDC) 86-8395, Department of Health and Human Services, Washington, DC, 1984.
8. NIH Guidelines for Research Involving Recombinant DNA Molecules, *Fed. Regist. 55*, 53258 (1991).
9. Guidance Notes of the Advisory Committee on Genetic Manipulation, 1977–1988; HMSO, London, UK.
10. Proposal for a Council Directive on the Contained Use of Genetically Modified Microorganisms, *Off. J. Eur. Communities 199*, 9–27 (1988).
11. FROMMER, W., *et al.*, Safe Biotechnology. III. Safety Precautions for Handling Microorganisms of Different Risk Classes, *Appl. Microbiol. Biotechnol. 30*, 541–552 (1989).
12. ERICKSEN JONES, J. R.; *Fish and River Pollution*; Butterworths, London, 1964.
13. DEINDORFER, F. H., HUMPHREY, A. E., Analytical Method for Calculating Heat Sterilization Times, *Appl. Microbiol. 1*, 256 (1959).
14. MOLIN, G.; Destruction of Bacterial Spores by Thermal Methods; In: *Principles and Practices of Disinfection, Preservation and Sterilization*; RUSSEL, A. D., HUGO, W. G., AYLIFFE, G. A., Eds.; Blackwell, Oxford, 1982.
15. DEATH, J. E., COATES, D., Effect of pH on Sporicidal and Microbicidal Activity of Buffered Mixtures of Alcohol and Sodium Hypochlorite, *J. Clin. Pathol. 32*, 148–153 (1979).
16. KNOX, W. E., STUMPF, P. K., GREEN, D. E., AUERBACH, V. H., The Inhibition of Sulfhydryl Enzymes as the Basis of the Bactericidal Action of Chlorine, *J. Bacteriol. 55*, 451–458 (1948).
17. LEVINE, M., RUDOLPH, A. S.; Factors Affecting the Germicidal Efficiency of Hypochlorite Solutions, *Bull. Iowa State Univ. Sci. Technol. Eng. Exp. Stn. 150*, 1941.
18. LEVENSPIEL, O., Longitudinal Mixing of Fluids Flowing in Circular Pipes, *Ind. Eng. Chem. 50*, 343–346 (1958).
19. JOHNSTON, J. B., HANNAH, R. E., CUNNINGHAM, V. L., DAGGY, B. P., STURM, F. J., KELLY, R. M., Destruction of Pharmaceutical and Biopharmaceutical Wastes by the MODAR Supercritical Water Oxidation Process, *Bio/Technology 6*, 1423–1427 (1988).

Heating, Ventilating, and Air Conditioning (HVAC)

Dennis Dobie

Contents

17.1 Special Considerations for Biotechnology Projects

Biotechnology procedures are generally conducted in environments that are cleaner than standard buildings. The *heating, ventilating, and air conditioning (HVAC)* system shares a large part of the responsibility for maintaining these clean environments. Air used to condition a space is highly filtered, properly directed, and used to maintain pressure relationships within and between adjacent spaces. Humans are generally the largest contributors of airborne particulates in clean or aseptic spaces. Gowning and other sterile room procedures reduce this, but airborne contaminants still exist, and must be controlled and removed. Specially designed air-handling units, high-efficiency filters, high rates of air flow, unidirectional (laminar) air flow, and space pressurization are some of the HVAC techniques employed to provide the proper environment. Biotechnology facilities are usually subject to governmental agency regulation, which requires design approval in addition to ongoing frequent monitoring and reporting after commissioning. Some examples of specific HVAC considerations for areas in biotechnology facilities are as follows:

(a) Fermentation rooms have high heat gains to the space; capturing the heat at the point of generation requires high air flows, exhaust hoods, or both.

(b) Filling areas require very clean, monitored air (class 100) at the point of filling, and there must be positive air flow from the filling space to the adjacent area. Room temperatures should be 68 °F (20 °C) or less, since operators are heavily gowned. A lower temperature keeps them comfortable and minimizes the generation of airborne contaminants. Room humidity control is also employed to control microbial growth.

(c) Packaging areas are similar to filling areas, but if the material is a powder, the relative humidity may have to be very low to prevent the powder from absorbing moisture. Dust collection hoods, to contain airborne powders, may also be necessary.

(d) For containment, support labs (QC) must be at a negative pressure with respect to adjacent spaces. Hoods and clean rooms are usually installed in the QC lab as containment areas and work spaces. Hoods with multiple flow requirements or multiple hood installations may upset room pressurization, a challenge to the design that must be taken into account.

(e) Gowning areas are at a negative pressure with respect to the work space, and at a positive pressure with respect to unclassified areas, to keep airborne contaminants from entering the work space.

17.2 Basis of Design

A *basis of design* must be developed by the engineering staff in a joint effort with the client or facility user. This involves programming techniques in which the codes to be followed are established, design conditions for room temperature and humidity are set, room classifications are identified, and the operation of the facility is described. Any unusual or unique facility requirements, such as emergency backup or redundancy for HVAC systems, must also be designed into the HVAC system at this time. This is also the stage of the design process at which studies of alternatives are conducted to compare options for the HVAC system. The cost of a backup or redundant HVAC supply system may be compared with the cost of losing product or of having an experiment interrupted, should temperatures or air flow go out of control or specification. Heat recovery from exhaust systems or thermal storage are examples of other potential areas for study [1].

17.2.1 Regulations and Codes

Clean rooms provide clean air in aseptic spaces, and are defined or classified in the U.S.A. by Federal Standard 209D [2]. Other countries have similar classifications, generally adapted from 209D. Room classifications are established by measurement of the number of particles 0.5 μm and larger in one cubic foot of sampled air. Class 100 to class 100,000 rooms are generally used in the biotechnology industry. Rooms may be classified as clean as class 1 or 10 for other applications, particularly in the microchip industry. Table 17.1 (derived from Federal Standard 209D) describes the air cleanliness classes.

Table 17.1 Air Cleanliness Classes

Class		Maximum number of particles 0.5 μm or larger		Maximum number of particles 5.0 μm or larger	
English units	metric units	per cubic foot	per liter	per cubic foot	per liter
100	3.5	100	3.5	Less than 10	0.35*
1,000	35	1,000	35	Less than 10	0.35*
10,000	350	10,000	350	65	2.3
100,000	3,500	100,000	3,500	700	25

* Counts below 10 particles per cubic foot (0.35 per liter) are unreliable except when a large number of samples is taken.

Other regulations and guidelines may include those published by insurance underwriters, corporate or plant guidelines formulated from specific industry experience, and *Current Good Manufacturing Practices* (*CGMPs*) as published in the U.S. Government Code of Federal Regulations, Parts 210 and 211 [3]. The CGMPs apply to the manufacture, holding, or processing of drugs to assure safety, purity, and quality of the product. The CGMPs regulate space temperature and humidity, require a pressure differential between adjacent rooms of different cleanliness levels, and prescribe a minimum of 20 air changes per hour for classified rooms. The regulation also defines class 100 requirements for filling lines and microbiological testing sites, as well as parameters and testing for laminar flow control equipment at filling lines. Walls, floors, and ceilings for CGMP areas are to be constructed of smooth, cleanable surfaces, impervious to sanitizing solutions and resistant to chipping, flaking, and oxidizing. Horizontal ducts, pipes, or crevices that foster dust accumulation and cannot be easily cleaned are not permitted.

17.2.2 Temperature and Humidity

Generally, areas are designed to have room temperatures between 67 °F and 77 °F (19–25 °C), with a control point of 72 °F (22 °C). Humidity should be kept between 40% and 55% for several reasons: to keep people comfortable, to prevent corrosion, to control microbial growth, and to reduce the possibility of static electricity [4]. Internal heat gains from lights, people, and equipment must be calculated, along with transmission gains from adjacent spaces, to ensure sufficient air flow for a particular classification, so that the required temperature conditions are met. In calculating final system cooling load requirements, do not neglect fan and motor heat from recirculating fans, which can be a major source of heat in clean spaces.

Lower space temperatures may be required where people are very heavily gowned and would be uncomfortable at normal room conditions. People who are too warm are likely to perspire and generate significantly more airborne particulates. To achieve colder room conditions, the supply air to the rooms may have to be colder than the normal 50–55 °F (10–13 °C), and the air conditioning system may have to meet a greater dehumidification demand.

The amount of dehumidification required can be determined from a standard psychrometric chart if the conditions of room air, mixed air, outdoor air, and coil are plotted. Coil duty in Btu/h is determined from the enthalpy difference, ΔH, between the dry- and wet-bulb conditions of the air entering and leaving the cooling coil. In some cases where products sensitive to moisture are handled, the room relative humidity may have to be as low as 15% to 20%, and chemical dehumidifiers may be required.

It is necessary to establish these criteria early in the design process, to allow the engineer to select dehumidification equipment and to add allowances to the construction cost budget estimate. The lower the required temperatures and humidities, the more sophisticated the HVAC equipment becomes, and the higher are both first and operating costs. Range of temperature control also affects the cost of the control and delivery system. The closer the space temperature tolerances are, the more sensitive the controls must be. To achieve close temperature tolerances and prevent hot or cold spots, the air distribution system must be very extensive and elaborate. Table 17.2 is a sample of a preliminary design document with room temperature and

Table 17.2 HVAC Room Criteria

RM. NO.	NAME	AREA SQ.FT.	CLASS	NO. A.C.	CFM	PRESS.	ROOM TEMP F.	RANGE +/- F.	% R.H.	RANGE +/- %	REMARKS
100	AIR LOCK	140	100000	30	630	+	68	2	45	5	
101	CLEAN CORR.	240	100000	20	720	++	68	2	45	5	
102	AIR LOCK	140	100000	30	630	+++	68	2	45	5	
103	SERV. CORR.	240	100000	20	720	++	68	2	45	5	
104	PRODUCTION SUITE	600	10000	50	4500	+	64	2	55	5	LOCAL CLASS 100 BIOCONTAIN.
105	CELL CULTURE	150	10000	50	1125	+	64	2	55	5	LOCAL CLASS 100 BIOCONTAIN.
106	AIR LOCK	140	100000	30	630	+	68	2	45	5	
107	PUBLIC CORRIDOR	800	GENERAL	10	1200	0	72	—	50	—	

ALL AIR QUANTITIES BASED ON 9'-0" CEILING HEIGHT

humidity design conditions. This document is useful in communicating conditions to the design team and reviewing agencies. If the document is kept up to date it also becomes useful to installing contractors and the construction supervision team as a check on ultimate system performance and validation. The table can also be used as a guide for the generation of an installation qualification (IQ) protocol [5].

17.2.3 Room Classifications

During initial stages of design, each room must be assigned a cleanliness classification, according to its function. Clean rooms are classified according to cleanliness, and assigned the proper classification number as defined by 209D or other ruling agency. Laboratory spaces are classified by the activity for which the lab is used. These classifications range from *BL-1 (Biosafety Level-1)* through *BL-4*, with BL-1 the least hazardous. BL-1 compares to a school science lab, while BL-4 labs require glove boxes and redundant HEPA filtration [6]. These laboratory classifications were formulated in the U.S.A. by the Federal National Institutes of Health (NIH), and are published as guidelines in the *Federal Register*. The guidelines contain requirements for filtration of air, recirculation of air, containment, and biosafety cabinets. Table 17.2 is a sample of a form that, in addition to specifying temperature and humidity, can also define room classifications. BL requirements can be noted in the remarks column if applicable.

Table 17.3 indicates a typical range of air change rates generally used to achieve the desired room cleanliness classifications. These rates vary widely in practice, because the number and type of particulate generators in a room, such as people and equipment, as well as the room size and quality of air distribution, can differ over wide ranges. To calculate air flow requirements from room air change rates, the following formula is used:

$$CFM = VN/60 \qquad (17.1)$$

where CFM is the air flow, cubic feet per minute (cfm);
V is the room volume, ft^3; and
N is the number of air changes/h.

Table 17.3 Suggested Design Parameters for Classified Rooms

Class	Air changes per hour	Air flow*, cfm	Air flow*, m^3/h	Air return method	Unidirectional flow?
100	600	90	1646	air wall	yes
1,000	175	26.3	480	air wall	not usually
10,000	50	7.5	137	low register	no
100,000	20	3.0	55	low register	no

* Air flow rate per square foot or square meter of floor area, assuming 9-ft ceiling height.

17.2.4 Adjacency of Spaces

Figure 17.1 depicts a sample biotechnology laboratory arrangement in block diagram form. In the early stages of design the HVAC engineer must work with the architect, user, and layout planner, if one is involved, to prepare a layout of this type. The frequent manipulation that is required to arrive at an optimum arrangement can best be done using a computerized drafting system [7]. Spaces are classified as *clean*, *dirty*, or *contained*. A contained space is used to contain a product or material that could be spilled into an occupied space or find its way into the building air system.

From an HVAC standpoint it is desirable to keep similarly classified areas physically close together, so they can be connected to the same air-handling unit, and thereby minimize duct runs and air system complexity. Spaces must be so located that people can move between them without disrupting the cleanliness or containment of the spaces. Cleanliness and containment are maintained by air flow (or *leakage*) from one space to an adjacent one, caused by pressure differentials created by the air supply, return, and exhaust systems. Where major demarcations of pressure are required, *air locks* are used; these small rooms with controlled air flows act as barriers between spaces.

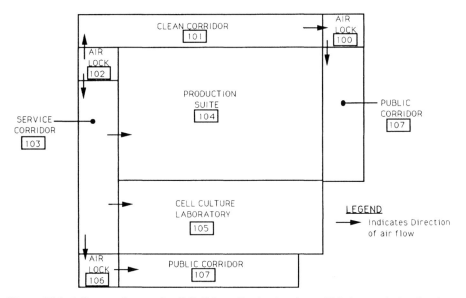

Figure 17.1 Adjacency Layout for Cell Culture Production Area. This is a typical cell culture suite. Arrows indicate the direction of air flow leakage to achieve pressure differences between adjacent spaces.

Ultimately, as the layout develops, suites or groups of spaces are established, and groupings of rooms can be made for assignment to separate air-handling systems. It is not desirable to mix dirty and clean systems, nor suites that may be a source of contamination to each other. Leaks can develop in a filter, or some source of contamination could find its way through the air system and potentially cross-contaminate different areas. Segregating contaminated exhaust eliminates the possibility of cross-contamination through the exhaust system, which otherwise is a likely source of contamination.

Figure 17.2 shows a typical block layout that can be made on architectural drawings or a letter-size sheet to indicate the coverage of rooms and spaces by various air-handling units.

17.3   Pressurization

Pressurization can be explained by the simile of air leaking from a balloon. Pressure inside the balloon causes air to escape from a small leak. If the hole stays the same

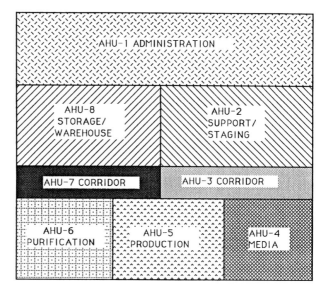

Figure 17.2 Typical Block Layout for Air-Handling Units. Coverage by each unit is denoted by the different cross-hatching. Labels identify the air-handling unit that serves each area.

size, a greater pressure difference between the inside and outside causes the air to escape at a higher velocity. As the air in the balloon is depleted, the pressure difference drops, the rate of air escaping diminishes, and the velocity decreases.

The same principle applies in biotechnology facilities: rooms are made as tight as possible, so the air-handling system can build pressure or reduce pressure in particular rooms. Standard 209D recommends maintaining pressures of 0.05" H₂O (12 Pa) between adjacent spaces with doors closed. When doors are opened the designed difference disappears, but air must continue to flow from the higher to the lower pressure space, even though at a greatly reduced flow rate. To maintain a difference of 0.05" H₂O (12 Pa), a velocity of approximately 540 ft/min (2.8 m/s) must be maintained through all openings or leaks in the room, such as cracks around door openings. If openings are larger, the required velocities are greater.

The conditions for air flow through cracks around a door resemble those for flow through a sharp-edged orifice; the pressure drop is the *velocity pressure* of the air at the stated velocity, multiplied by a suitable coefficient that adjusts for the physical size and configuration of the crack or opening between spaces. Figure 17.3 gives the velocity pressure, from 0" to 0.1" H₂O (0 to 24 Pa), at various air flows and can be used to estimate the air flow between rooms through large openings.

Limits on pressure drops (usually 0.05" H₂O) must be set during the design and observed. The pressure difference exerts a force, which can be calculated, on the door. If the force is too great, the door may not close fully or may be difficult to open. This is particularly important in large, complex facilities where many levels of pressurization may be required.

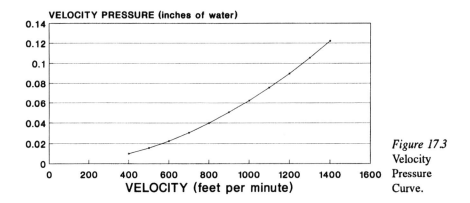

Figure 17.3
Velocity
Pressure
Curve.

17.3.1 Pressurization Diagrams

If the designer modifies Figure 17.1 by showing pressures in each of the spaces and air flows (leakage) in cfm (or m³/h) between the spaces, the result is a *pressurization diagram*. In practice the calculated air flows are approximate, since it is very difficult to build everything in a facility absolutely airtight. A very small, unexpected leak, or a seal that is tighter than expected in design, can affect the pressurization and air flow required to maintain the design value. Some flexibility must be built into the air-handling systems and controls to compensate for these variations.

One technique for establishing pressurization in separate rooms is to have a common reference for the control sensors. This common reference is usually a mechanical room or an interstitial space, where the pressure is not affected by changes in the controlled spaces. Referencing spaces to each other can lead to problems when doors are opened or frequent upsets in pressure occur; the controls never settle down, and tend to modulate constantly (called *hunting* or *hysteresis*).

17.3.2 Room Seals and Doors

In most facilities the openings around the doors between rooms are where leaks occur and the pressurization between rooms is created. Rooms are made tight by sealing any other openings with a proper sealant that does not promote growth of organisms and is easily cleanable. Areas to be sealed include ceiling tiles, lighting fixtures, pipe and telephone outlet penetrations, and any cracks or openings that appear in the structure.

A typical door has the following dimensions and crack area at the perimeter: door size: 3' 0" (0.915 m) wide by 7' 0" (2.41 m) high; cracks at top and sides: ⅛" (3.2 mm), with an undercut of ¾" (19 mm); calculated crack area around the door: 0.36 ft² (0.0357 m²). To achieve 0.05" H₂O (12 Pa) pressure difference across the door, approximately 200 cfm (350 m³/h) of air flow through the cracks is required. For a door of this type the appropriate air flow (cfm or m³/h) is shown on the pressurization flow diagram and included in the unit and room calculations.

Door seals around the top and sides are usually made of closed-cell neoprene, and should generally be used to reduce the crack area. To reduce the undercut a commercially available *drop-type seal* should be used; the drop type is preferred to a *wipe type*, since it does not mar or leave residue on the floor. With the seals added, the crack area can be reduced to about 0.119 ft² (0.011 m²) and the air flow requirement can be reduced to 65 cfm (110 m³/h). This reduction becomes significant when all facility doors are considered. The air used for pressurization generally must be made up from outside air, which is expensive because it requires greater energy for heating and air conditioning than system return air.

17.3.3 HVAC Components for Pressurization

All supply, exhaust, infiltration, exfiltration, and air that is returned to the system must be accounted for when the pressurization for a room is calculated. Table 17.4, prepared on a computer spreadsheet, is a sample format that can be useful in conjunction with air flow sheets to balance air flows in rooms and systems.

Table 17.4 HVAC Room Air Balance

RM. NO.	NAME	SUPPLY CFM	RETURN CFM	EXHAUST CFM	INFILTRATION CFM	EXFILTRATION CFM	RANGE +/- F.	% R.H.	RANGE +/- %	REMARKS
100	AIR LOCK	630	630	0	100	100	2	45	5	
101	CLEAN CORR	720	420	300	100	100	2	45	5	
102	AIR LOCK	630	430	0	0	200	2	45	5	
103	SERV. CORR.	720	120	400	100	300	2	45	5	
104	PRODUCT. SUITE	4,500	3,600	1,000	100	0	2	55	5	
105	CELL CULTURE	1,125	625	600	100	0	2	55	5	
106	AIR LOCK	630	630	0	100	100	2	45	5	
107	PUBLIC CORRIDOR	1,200	1,400	0	200	0	—	50	—	
	TOTALS=	10,155	7,855	2,300	800	800				

Balance is a term for the procedure used to ensure that all air flow quantities add up, which is no small task on a complex project that has values frequently changing during design. Of particular importance is exhaust air from equipment and hoods; they may be on or off at different times, or the air flow rate may vary during occupied periods. These variations must be dynamically compensated for, to maintain room pressurization. Numerous systems using manual and automatic dampers, constant- and variable-volume air control boxes, and elaborate air-flow sensing devices are combined with control systems and sensing devices to assure that room pressurization is maintained.

17.4   Air-Handling Systems

When block layouts, adjacency layouts, room design criteria, and pressurization levels are established, the number and type of air-handling systems can be selected. Air-handling systems commonly used include constant- and variable-volume, with and without reheat coils for control.

For controlled spaces the most reliable system is a *constant-volume system with terminal reheat (CVRH)*. This system is not used in less critical or commercial applications because the air temperature leaving the cooling coil is set at a fixed value, and the terminal reheat responds to a space thermostat by bringing on heating to satisfy the load. This can waste energy, since air is cooled and then reheated. Reheat systems require energy to simultaneously heat and cool the air, as can easily be demonstrated with a psychrometric chart. The advantages of reheat systems are that humidity is always controlled, because dehumidification always accompanies the cooling, and each space or zone that needs temperature control can easily be satisfied by adding a reheat coil and thermostat. Another advantage of the CVRH system is that air flow is constant, which makes balancing and pressurization easy to maintain in most cases. It is probably the simplest and easiest of all systems to understand and maintain.

A *variable air-volume system (VAV)* is generally used in administrative areas and some storage spaces, if pressure control is not critical, humidity control is not essential, and some variations in space temperature can be accepted. The VAV system works by delivering a constant-temperature air supply to spaces, with reductions in air flow to spaces as cooling loads diminish. This reduction in air flow results in a reduced cooling effect and allows the air flow or cooling to match the room load as established by the design space temperature condition. Perimeter heating must be supplied for spaces with exterior walls or large roof heat losses. The perimeter heating can be baseboard radiation or some form of air heating using heating coils. Finned radiation or convection heating devices should not be used in clean spaces, since they are not easily cleanable and have places where unwanted particulates can build up.

Combinations of systems can be used, especially if variable quantities of supply and exhaust air are required for fume hoods or intermittent exhausts. Biotechnology facilities usually require large quantities of air, to promote unidirectional flow and air cleanliness. In many cases the large quantities of air exceed the requirements for cooling, so it is possible to recirculate air in the system and pass through the air-handling unit only enough air to perform the heating or cooling.

Figure 17.4 is a typical depiction of an air-flow diagram for a CVRH system. The flow sheets should be developed as full-sized drawings using computer drafting. They become useful tools for conveying information to the owner/user, for agency reviews, and for transmitting information to HVAC designers and other engineers. The documents are also invaluable to construction contractors and for system checking by construction managers and balancing contractors.

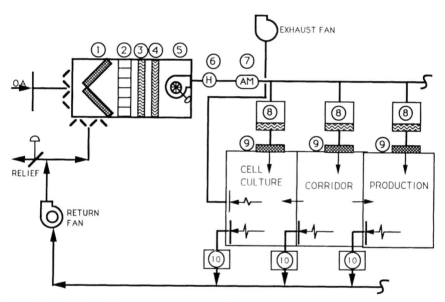

Figure 17.4 Air Flow Diagram for a Constant-Volume Reheat System. Mix box has (1) 25% filter, (2) 85% filter, (3) heating coil, (4) cooling coil, and (5) fan with inlet vanes or variable-speed motor; (6) humidifier; (7) air-flow monitor; (8) constant-volume control boxes with reheat coil; (9) room terminal HEPA filters; (10) constant-volume boxes.

17.4.1 Dehumidifiers

Low relative humidities are required in many cases, such as in drying or powder areas, where hygroscopic materials are handled and would absorb unwanted moisture from the air. To obtain sufficiently low humidity, a chemical dehumidifier may be required. Whether this is necessary can be determined by reference to the psychrometric chart, which establishes the dew point of the air leaving the coil. Normal chilled water for use in cooling coils is supplied at 42–44 °F (5–7 °C). Depending on the number of coil rows and fin spacing, a minimum dew point of about 50–52 °F (10–11 °C) can be obtained. This results in a minimum room relative humidity of approximately 50% at 70 °F (21 °C). If the humidity must be lower than that, a chemical dehumidifier is specified.

Chemical dehumidifiers are commercially available air-handling units that contain a sorbent material, either a solid or a liquid. *Wet dehumidifiers* use absorbents that change physically during the process. Lithium salt solutions are generally used to remove moisture from conditioned air, and are then regenerated by heat, usually a steam heat exchanger. *Dry dehumidifiers* use adsorbents that do not experience phase changes during the process, such as silica gel or activated alumina. A rotating wheel is commonly used to remove moisture from the conditioned air. The wheel is regenerated as heated outdoor air is passed over the wheel to dry it; steam or electric coils usually supply the heat for regeneration.

Depending on the amount of dehumidification required and the amount of outdoor air (usually with a high moisture content), it may be best to combine the dehumidifier with a conventional air-handling unit and dehumidify only a small portion of the air. The dehumidifier has a high initial cost compared to a conventional air-handling unit, and should be sized to do only the required duty. Knowledgeable vendors should be consulted, to select the best combination of equipment and specify a proper system for the application. These systems require considerable physical space, energy consumption, and utilities, so these criteria are also important in system selection.

17.4.2 Humidifiers

Air-handling systems that bring in quantities of outdoor air for makeup require the addition of moisture to the air stream to provide the desired humidity levels. There are many commercially available devices for providing humidity to air, but the most commonly used are *steam grid humidifiers*. It is important to use clean steam (discussed in Chapter 15), not plant steam, which may contain boiler chemicals and impurities from deteriorating piping and equipment. Humidifiers are controlled by modulation of a steam valve at the humidifier; the valve is controlled by a signal from a return or exhaust air stream, or a room humidistat. A high-limit stat is placed in the duct downstream of the humidifier to override the controlling stat and prevent condensate in the duct.

17.4.3 Selection and Location of the Air-Handling Unit

Conventional air-handling units consist of filters, coils, and fans in a metal casing, with an insulating liner inside the casing. For units in biotechnology applications, the casing must be a sandwich of metal, with insulation between metal sheets, and a smooth, cleanable interior surface that does not foster the growth of microorganisms.

Chilled water or glycol solutions are generally used for cooling and dehumidification at the normal relative humidity (RH) range of about 50%. Direct expansion refrigerant, where the refrigerant is in the air-unit coil, may be used, but these systems are less reliable than chilled water or glycol and are harder to control in narrow air temperature ranges. Figure 17.4 is a typical unit arrangement with prefilters, intermediate filters, heating coil, cooling coil, and fan drive system.

Units should have good access doors, viewing ports, interior lighting for maintenance, and fan isolation. The casings should be tightly sealed, and designed for pressures higher than those used in commercial applications, due to generally high system air pressures. All sealants and lubricants exposed to the air stream should be food-grade, to minimize the chance of air contamination.

Draw-through units have the coils on the suction side of the fan; *blow-through* units have the coil on the discharge of the fan, and are used when the unit filter is the last filter before the air is discharged to the space. On blow-through units an air distribution plate must be installed to distribute air evenly over the filter and coils. If the unit contains a *high-efficiency particulate air (HEPA)* filter and the air is ducted directly to the clean space, the duct should be constructed of a cleanable, smooth, nonflaking material, usually stainless steel.

Units located indoors are serviced more easily and tend to last longer. They should also be located as close as possible to the main rooms they are serving to keep duct runs short. Air-flow measuring devices and humidifiers must also be installed, and the location of a unit should allow enough straight run of duct for installation and proper operation of these devices. Location of inlet louvers for outdoor air must be carefully considered. Intakes are generally located high on the side of the building to keep dust intake low. Intakes should also be far from truck docks or parking lots where undesirable fumes and particulates are generated. The direction of the prevailing winds should be taken into account when an inlet is positioned; any nearby (particularly upwind) exhausts or fume concentrations should be avoided, to prevent recirculation of exhaust air back into the supply system. If suitable locations cannot be found for the outdoor air intakes, a makeup air handler should be used to introduce and precondition the outdoor air.

17.4.4 Return and Exhaust Fans

Return and exhaust fans are integral parts of the air-handling systems, and are also shown on the air-flow diagrams. Return fans are used on systems that have long duct returns or return system pressure drops greater 0.5" H_2O (120 Pa). Their use allows proper total system balance and minimizes the suction pressure required from the supply fan. If a return fan is not used, the capacity of the supply fan can be overextended and it may be difficult to limit and properly control the amount of outside air being admitted to the unit. Return fans are also needed to provide a negative pressure in some contained rooms. Return fans can be of the standard centrifugal type, or of an in-line type that works very nicely in crowded equipment rooms because it is installed directly into the duct. Return fans may also be required to handle varying quantities of air or provide a constant flow of air at varying pressure conditions, so some form of damper control, inlet vane, or variable-speed motor is generally used.

Building exhaust streams are generally collected and ducted to exhaust fans in groups or clusters. Exhaust fans should be located as near the building discharge as possible, since this keeps the duct under negative pressure, and any leaks are into the duct, and not from the duct into an occupied space or mechanical room. For this reason, roof locations for fans are preferred, even though it may make service difficult in severe weather conditions. Fans can also be located in mechanical rooms or interstitial spaces. Roof penetrations should be as few as possible, to prevent leaks. Fume exhausts and toxic exhausts should be extended through the roof and terminated well above the roof line in a suitable stackhead. Stackheads are used to direct exhaust vertically at high velocities, to enhance dispersion of fumes, and to prevent water from finding its way into the exhaust system; designs for stackheads are found in most duct manuals. Extremely toxic or dangerous active biological agents may require *bag-in/bag-out* HEPA filtration with containment, or other treatment such as incineration, before they can be exhausted to the atmosphere [8].

Another important consideration is the use of return and exhaust fans as part of the containment system. Should the fans be essential to maintaining containment, it may be desirable to have a backup fan or redundant system. An air-flow monitor or switch that gives a warning in case of a fan system failure is desirable on critical systems. A sensor in the air stream is better than an indicator on the electrical motor for proving that air is moving, because the motor could be running with a broken fan belt and the operator would not know that the fan is not moving air.

17.4.5 System Operating Procedure

When the air flow sheets are completed, it is appropriate to finalize the *system operating procedure*. This is usually a written document stating how systems are turned on and off, along with a description of their frequency of operation. For clean spaces, general practice is to operate the air system 15 to 30 minutes for purging before the start of production or use of the space. In some cases it is necessary to operate a system continuously to ensure that pressurization or containment is always maintained.

An analysis of system operation must also assess how a shutdown or flow reduction in the system affects an adjacent system or areas. Exhaust fans and fume hoods that may have intermittent or varying flows should be clearly noted in the operating procedures. It is essential that laboratory and production personnel understand and follow the established procedures, otherwise, if an exhaust or return system is inadvertently shut down, the air system might go out of balance. Alarms to indicate loss of pressurization are also valuable features—and sometimes essential—in the HVAC design.

At this stage of design, procedures should be developed so that the air system can accommodate a product spill or accident in a contained space. The ramifications of a spill on the air system, controlled space, and adjacent operations should be evaluated. Cleanup procedures could include a fumigation of the air system, which would require operation of a relief connection to the ductwork for venting fumigant. The fumigating agent should also be specified, so that the duct designer uses duct components and materials that are not deteriorated or destroyed by the fumigant. Fumigants must also be evaluated for environmental impact before discharge to the atmosphere. This entire step of establishing operating procedures should ask as many "what if?" questions as possible. System operation should assure that the HVAC systems can maintain cleanliness and containment requirements. These conditions must be maintained while operations are as close to normal as possible, in addition to those periods of unexpected scenarios. The reviewing agencies generally ask many of these questions, so it is essential that the design team be prepared with thoughtful responses.

17.4.6 Emergency Electrical Power

The final step in the HVAC flow sheet process is coordination with the electrical design team. Motor lists for HVAC equipment are prepared and reviewed, with motors that have provision for emergency power, variable speed, or other special characteristics designated. The flow sheet may also indicate which motors or systems are redundant or on emergency power. Any fans or equipment requiring interlocks should be identified on the motor list and flow sheet. This also is a good document to show to the reviewing agency and later on is a good device for training the personnel who will be operating the facility.

17.5   HVAC System Components

17.5.1 Terminal Air Control Devices

In most biotechnology applications the volume of air flow must be controlled to assure proper air changes and provide air for pressurization. The two options for controlling air flow are *variable flow*, which responds to some pressure- or temperature-actuated signal, and *constant volume*, which provides a uniform quantity of air under varying pressure conditions.

Numerous commercially available devices, usually called *boxes* or *terminal control units*, are available; they use a variety of air dampering designs, from a simple blade type to pneumatic bladder type and conical type. Each type has its own application and associated costs. The published damper design and test data for flow and repeatability should be studied against the projected operation and matched to the proper application. Dampers are controlled by pneumatically controlled operators or electric motors responding to electric/electronic signals. Pneumatic dampers generally react more quickly to changing air flow requirements. Dampers and controls should be selected for linearity of response, repeatability, and percentage of error (compared with desired flow).

The damper is generally housed in a device called a *volume control box*, which contains deflecting baffles and flow-measuring or pressure-sensing devices that are used for flow control. Accuracy of flow control boxes, combined with the controlling system, must be aggregated to assure that the flow is within acceptable limits for the control of pressurization. The accuracy of control devices can range between 5% and 10% of maximum flow. This built-in inaccuracy must be calculated when establishing flow differential rates. The design values of air flow rates may require adjustment to allow the system to operate and provide the pressure differential in the proper direction. For commercial applications, control boxes are lined with insulating acoustical material; these liners should be avoided on boxes in critical or controlled areas where dirt catchers or locations of unwanted microbial growth cannot be tolerated.

Some vendors provide a single box with multiple dampers, operators, and associated control boxes responding to different sensing points in the same room or area. An example of a good application for such a device is a room requiring a normal air flow, which is admitted through one supply damper in the box; another supply damper in the same box has its control tied into an exhaust hood. The first damper provides a constant volume of air, and the exhaust hood damper is opened only when the exhaust fan is operational, and closed when the fan is off. This combination can be adjusted to keep the same differential pressure between the lab and the corridor whether the exhaust fan is on or off.

17.5.2 HEPA and High-Efficiency Filters

HEPA is an acronym that stands for *high-efficiency particulate air* filter. HEPA filters are used in the final stages of air cleaning, to remove very fine particles. Rough and intermediate filters, not as efficient as HEPAs, are used to remove larger particulate matter; this is much more cost-effective than removing large particles with an expensive HEPA filter that then requires more frequent replacement. A HEPA filter, by definition, removes 99.97% of particles of $0.3\text{-}\mu m$ size. Even more efficient HEPA filters are commercially available at a higher cost.

The pressure drop across filters is rated for nominal flow at new, clean conditions and is usually about 1" H_2O (25 mm H_2O, 240 Pa). The systems are designed for filter changes at approximately 2" H₂O (50 mm H₂O, 480 Pa) or double the clean ratings, although filters can be operated at higher pressure drops (which some owners prefer). The higher operating pressures result in a greater use of electrical energy to power the fan against the increased resistance.

HEPA filters are usually located downstream of heating and cooling coils in air-handling systems, since coils are potential sources of contamination. The most popular HEPA filter location is in the room ceiling, with standard laminar flow outlets of nominal size 24" x 48" (0.6 m x 1.2 m). These outlets contain manual control dampers, testing sample ports, a diffusion panel, and a filter element. They are commercially available in permanent or throw-away units; in the latter, the entire assembly is disconnected from the flexible duct supply terminal and replaced when the filter is dirty. Figure 17.5 shows a typical terminal HEPA filter outlet.

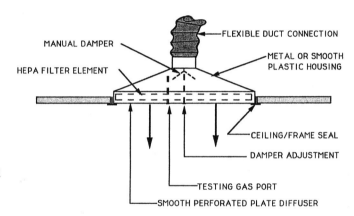

Figure 17.5 Terminal Ceiling HEPA Outlet, used to supply air to a classified or unidirectional flow space.

If the particles being filtered are potentially dangerous to maintenance personnel, filters are placed in special enclosures called *bag-in/bag-out housings*. These units are quite expensive and employ elaborate sealing methods, decontamination ports, and a double-bag arrangement that allows changing of the filter element(s) without ever exposing the changer to the filter or its collected contaminant.

HEPAs are also used in class 100 work stations, in single or double filtration of air, to provide sufficient cleaning before the unidirectional (laminar) air distribution to the work space. These work stations are commercially available with horizontal or vertical flow patterns, generally recirculating within the clean space. Filters are commercially available in many standard sizes and come with many frame arrangements, with different methods used to seal between the filter element and the frame. In actual installations, this seal is a major source of leaks or failures in the system, and should be carefully designed and installed to avoid problems and loss of system integrity.

17.5.3 Air Terminal Outlets

In clean spaces the desired distribution of air is unidirectional; the air acts as a piston, with little turbulence and few eddy currents in the air flow. This piston of air carries particulates to the floor or to the return, and helps to prevent airborne particulate matter from recirculating and contaminating the work space. In most cases it is desirable to recirculate air within a space through a filter, since the return air has a lower level of particulates than does outdoor air, and does not require extensive heating or cooling.

Air terminals should be selected of materials that are nonflaking, non-oxidizing, and easily cleaned. Since the cleanest air should be introduced above work areas, that is where outlets should be placed. If the product at the work station is hazardous to humans, the air should be supplied at the operator's back and pulled into the work area, to assure that flow is away from the operator and any particulates are captured in the air stream. Should the operator pose a threat to the product, the air flow pattern should be reversed, with the supply air entering behind the product and blowing over the product and toward the operator.

Containment velocity at the inlet of a biosafety hood or enclosure should be about 100 fpm (0.5 m/s). Air flow patterns should avoid any sudden changes of air direction, which could cause eddy currents or air turbulence and tend to spread particulates. Many cabinet manufacturers have done extensive testing of capture velocities and provide excellent data on operating conditions. Locations of terminals and enclosures should be selected after movement of personnel within the space has been considered, and frequently opened doors, which tend to upset air flow patterns, have been identified.

Return terminals are also an important consideration. They are generally located low in the walls for clean rooms. In class 10,000 to 100,000 rooms, low cleanable wall registers are generally used. In cleaner areas, low return wall systems termed *air walls* are used. The air wall is an almost continuous opening at the base of the wall, with the air ducted up in the wall and collected for return to the air-handling system. Air wall inlets are generally located not more than 15 ft (4.5 m) in plan view from a supply terminal, to reduce the possibility of turbulence. Figure 17.6 shows a typical air wall detail.

17.5.4 Duct Materials, Pressure, and Cleanability

Most systems have unlined galvanized steel ductwork in rectangular, round, elliptical, or flat oval configurations. Galvanized ducts can flake off or rust, so they should not be used downstream of the HEPA filters, since contamination from the duct system itself could result. Where the HEPA filter is located upstream of the room terminal, and a long run of duct is present, the material of choice for the duct is stainless steel,

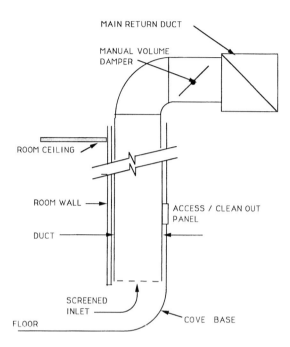

MAIN RETURN DUCT

MANUAL VOLUME
DAMPER

ROOM CEILING

ROOM WALL

DUCT

ACCESS / CLEAN OUT
PANEL

SCREENED
INLET

COVE   BASE

FLOOR

Figure 17.6 Low Return Wall Detail. Typical arrangement for low wall air return in a CGMP space.

but this is expensive and its use should be minimized.   Many systems are also fumigated or cleaned in place, so the duct material chosen should not be affected by the cleaning agent.

Ductwork systems in biotechnology facilities tend to have higher system pressure requirements than commercial systems, because they involve extensive use of filters, volume control devices, and physically complex arrangements.   The duct system pressures must be calculated and clearly stated on the contract documents, to allow the fabricator to provide the proper metal thickness and construction methods for the required system pressures.   System pressures also change as the system is operated with filters that get dirty or as space pressure conditions vary.   The duct system must allow for these pressure fluctuations, and the fans may require speed controls, inlet vanes, or variable-pitch blades to permit adjustment to varying flow and pressure conditions.

A duct system that might get dirty or contaminated must be cleanable.   Transverse joints with *Vanstone*® *connections* should be used for possible disassembly.   The design should include carefully located access doors in the duct for future cleaning. Access panels and doors must be in easily serviceable areas, preferably outside the clean space.   It is best to have facility system maintenance personnel entering the clean spaces as little as possible.

In very critical applications, the duct is *factory cleaned* and sealed after construction [9]. It is shipped to the site sealed; the end seals are broken and then quickly resealed during final installation. This strategy removes the oil and other contaminants present during duct construction, but is expensive. It may be difficult to find sheet metal fabricators willing to do this work, since their shops are not set up for such procedures, and shipping the completed product is also difficult.

17.5.5 Considerations of Insulation and Sound

Duct insulation in clean facilities should be externally applied, since the fibers used in the insulating material offer breeding grounds for microbial growth that are not cleanable and may break loose to become particulate contaminants. Any ducts in areas where the air around the duct might be humid or above the dew point temperature of the duct surface should be insulated to preclude the possibility of sweating. Insulation is usually applied to all air conditioner supply ducts and outdoor intake ducts. Ducts in concealed spaces (ceiling plenums) not subject to physical damage are usually insulated with duct wrap, which is a flexible glass-fiber blanket with an aluminum covering. Where ducts may be exposed to physical damage, such as in interstitial spaces or mechanical spaces, a rigid board material is usually used. Many excellent guidelines on insulation thickness are published by insulation manufacturers' associations, and should be consulted to select the type and thickness to be used. Energy costs, expected temperature differences, heat loss or gain, and installed cost of insulation are all figured into the calculations for determining the economic thickness. Smoke and fire ratings, which must comply with codes that apply to the facility, must also be considered during the selection process. Some insulating materials should be avoided where possible because they give off extreme quantities of smoke; they may be forbidden by code.

Sound becomes a consideration in all air-handling systems, particularly biotechnology facilities, since the systems may be complex and tend to generate more noise than commercial systems. The higher static pressures require larger motors, which produce more noise that must be dealt with. A sound analysis should be made; it should consider the air-handling unit noise, duct system attenuation, and acceptable noise criteria in the space. Many good forms of analysis are available (manual calculation and computerized) and should be used. Where noise reduction is required, the use of sound attenuators in the ductwork is not recommended. Since commercial attenuators use porous material exposed to the air stream, they are not acceptable. System attenuation may be obtained by locating air-handling equipment away from occupied space and using natural attenuation from duct runs and elbows.

17.6   Building Control and Automation Systems

17.6.1 Sequence of Operation

The control system that operates and monitors the HVAC system can be called the *automatic temperature control system* (*ATC*), the *energy management and control system* (*EMCS*), or the *building automation system* (*BAS*). The first element in the design of the system is the development of a *sequence of operation*, which is a written description of the HVAC and the intended system operation. A separate sequence is usually written for each air-handling system, describing the complete operation of that system from control of coils and humidifiers to control of the room temperature and humidity. Starting and stopping of the air-handling unit fans is outlined, along with interlocking of exhaust or return fans in relation to the main air system fan operation. Generally, all fans operate at the same time, which is necessary to maintain pressurization. The sequence also addresses abnormal occurrences, such as a smoke detection alarm or failure of an exhaust fan. The sequence describes what happens to other system components during an abnormal occurrence. It may be necessary to shut a supply fan down if a major exhaust fan should fail, to prevent or minimize the loss of pressurization. The sequence also describes any energy management strategies to be included in the system.

17.6.2 System Architecture

The control system of choice for major facilities and even some small systems is a *direct digital control* or *DDC* system. Most major control system vendors and many smaller vendors offer DDC systems that are similar but contain internal differences. The systems are computer-based, and communicate within and outside the system by coded digital signals.

System architecture refers to the major components of the DDC system and their interrelationship. The architecture is developed by determining what components are initially required, what may be required in the future, and how the system may grow or expand as additional requirements are added. Figure 17.7 shows a typical DDC system architecture.

When the sketch is developed, it is a starting point for designing the system and examining details and options with potential control vendors. Usually, when a vendor is selected, any addition or expansion to the system should be done by that vendor, since it is difficult to mix systems, and split responsibility is not recommended. The

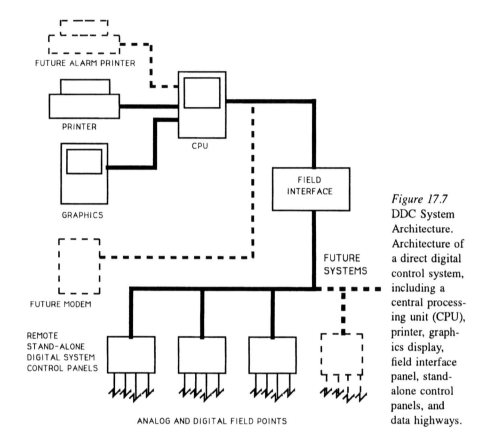

Figure 17.7
DDC System
Architecture.
Architecture of
a direct digital
control system,
including a
central process-
ing unit (CPU),
printer, graph-
ics display,
field interface
panel, stand-
alone control
panels, and
data highways.

original vendor should be selected on the basis that the owner has confidence that the vendor can expand the system in the future and has the capability to provide good service to the installed system at a reasonable cost. The owner should also define the programming knowledge required to make adjustments and ascertain whether it is possible to do it himself or whether vendor assistance is required.

17.6.3 List of Alarm, Control, and Monitoring Points

After the sequence is completed and the air flow diagrams are defined, the next step is to develop the list of alarm, control, and monitoring points, called the *points list*. This is an all-inclusive list of points that are to be connected to the DDC system. There are two major types of points: digital and analog. A digital point is simpler and generally less expensive, and works on a simple on/off or contact principle. Digital points are used to start and stop fans, indicate an on/off condition, or do anything that requires only a contact. An analog point is used to measure a variable such as temperature, pressure, or flow rate. These points generally use 4- to 20-milliampere signals, and give variable signals in response to the parameter measured. The points list should include analog control points, such as cooling coil valve and room temperatures. Monitoring points can be digital or analog; examples include fan run, room temperature indication, damper position, and room pressure indication. Alarm points can be either digital or analog, and can include smoke detection in an air-handling unit system, low cold-box temperature, high room humidity, or loss of room pressurization. Sensors in clean spaces should be unobtrusive and CGMP–acceptable.

17.6.4 Estimate of Control System Cost

The automatic control and monitoring system can become a major element in the overall HVAC system for a biotechnology facility. After the points list is developed, a reasonable guess for the system cost can be attempted. Several estimating numbers can be used to calculate the approximate cost, with a general range from a low of $500 per point to as high as $1200 per point. The larger systems have the lower costs, since the cost of the computers, monitors, and printers is spread over many points. Smaller systems have higher costs since this dilution effect does not come into play. Systems with a higher proportion of analog to digital points also tend to have higher per-point costs. A good first pass at a DDC system cost is $900 per point.

When a floor plan is completed, the system architecture is established, and the points list is completed, this information should be given to one or more DDC vendors for a budget estimate. Vendors make estimates based on information provided, take off wiring runs, and suggest panel locations in arriving at the budget cost. At the completion of this exercise, the cost of this major element can be considered settled. Should the cost be greater than desired, some points can be eliminated or the scope of the system reduced to bring the cost within the budget. When a system is purchased, it may also be desirable to obtain some unit pricing from the vendor, so that future costs are preestablished should additions or modifications be desired. It is essential that the vendor be required to give classes and instruction on his system before finally turning it over to the owner's operators. The cost of this varies, depending on the users' familiarity with the purchased DDC system.

17.6.5 Future System Flexibility

It is usually the case that, after a DDC system is installed, the number of points grows as the user becomes familiar with the system. Additional points may be added for maintenance, such as filter change-out pressure devices or equipment run-time logging. (Run-time logs are useful for lubrication schedules and belt checks.) This requires software to log in and integrate data, and to print out notices at preselected intervals of run time.

Another useful feature is the phone modem, which allows an operator to call in from home and perform a check on the system. This could save a trip to the facility in case of an alarm: someone at the plant, such as a security guard, who received the alarm could place a call to the operator at home; the operator could dial into the BAS by modem, and possibly analyze the problem by reading some key parameters; he could then determine the severity of the problem and either ignore it until the next shift, instruct the guard to perform some minor procedure, or make the trip to the facility if the urgency demanded.

Even though the original budget may not allow all desired features, it is good to look toward the future in anticipation of system expansion and be sure that what is purchased is capable of expanding.

17.7   Testing, Balancing, and Cleaning

Testing and balancing for biotechnology systems are more critical than for commercial systems, because of the pressure differences required between spaces that have different cleanliness requirements. Associations of testing and balancing agencies publish excellent standard forms and procedures for balancing air and water systems.

Air balancing consists of setting terminal devices and air-handling units to deliver the proper quantities of air. For biotechnology facilities the additional requirement of establishing pressure differences between adjacent spaces must also be set. These differences are obtained by adjustments of air flows and control set points. This effort can take some time, as each facility is different and each room has different leakage characteristics. As part of the balancing, it may be discovered that duct systems or rooms are not as tight as designed, and require sealing to obtain the required pressure differences. It should be remembered that air flows shown on drawings are design values, and may require field adjustment to give the required pressure differences.

A simple solution to many pressurization problems is to repeatedly increase outdoor air makeup to the system. This can lead to problems if design values are exceeded and coils cannot meet their duty, with the result that room temperatures are out of the specified range: the coils are using capacity to condition excessive quantities of outdoor air rather than to meet room loads. In these cases the rooms or systems should be tightened, to get closer to design values.

The best time to balance is when few people (whether workers or owner's personnel) are in the spaces. The balancing should be done with all doors closed, since openings and closings result in system pressure upsets and make balancing difficult.

Before the system is turned over to the owner, the rooms and ductwork system must be thoroughly cleaned, to allow certification of room cleanliness conditions. During the construction, the work areas should be broom-cleaned after each day's work, and open duct systems kept closed. These procedures, if followed during construction, result in an easier final cleanup before turnover.

17.8   Validation

An agency validating a facility peruses the HVAC documentation, and possibly communicates with the design engineers, to establish the validation protocol as it relates to the HVAC system.   If the design is proper, the contractor has properly installed the system, and the components perform as specified, the system should be easily validated.   The validator verifies the actual system operation against design values and intent.   The physical parameters reported by the BAS system are compared to measurements with calibrated instruments to verify accuracy.   Air-flow measurements are made at supply and exhaust outlets, as well as traverses across the face of hoods, to verify proper flows.   Filters are physically tested or challenged to find leaks in the filter media or at spaces between elements [10].   Any leaks are field-corrected. Samples of air are also taken by the validator to assure that the specified cleanliness requirement is maintained by the installed operating air system.

If problems occur they are usually in the form of pressurization or filter leaks. Even after the mechanical contractor had completed his work, a hole for a pipe could have been cut in a room and not been properly sealed, and the room pressurization would have been altered as a result.   Filter elements may leak because they are not properly sealed to the frame, or they may have small holes that are the result of handling or damage during installation but are not discovered until later.   Should any of these problems occur and not be easily resolved by the validator, it may be necessary for the contractor and engineer to help the system meet the specifications for validation.

Also important at this stage is the calibration of air flow monitors, which should be installed in systems that supply critical areas, to assure that air flow rates are as designed.   Temperature and humidity sensors at critical areas should also be checked for accuracy by comparisons of readings of space conditions against values reported by the BAS.

17.9   Summary

The HVAC system is an essential element in the successful operation of a biotechnology facility, and must be given a proper amount of attention during the planning and design stages.   The space requirements for air walls, control boxes, extensive duct runs, and large air-handling units must be established early in the design process, to prevent unsatisfactory compromises later on.   It is essential that the owner and operators work closely with the architect/engineer design team to establish the HVAC system philosophy and the proposed operation of the system.   The amount of energy management and control, as well as system redundancy, are economic decisions that should be agreed on during this design process.   The best projects from an HVAC standpoint result when team members invest time in considering and understanding each other's positions.   Compromises that are made during development of the design should make economic sense, be mutually beneficial, and be understood by affected members of the project team.   All parties involved must fully agree on the HVAC system, to avoid costly and time-consuming retrofits at commissioning or after the facility is operating.

17.10 References

1. STOCKER, A. C., EMBREE, J. A., Creating Cost Effective Biotech Facilities, *Pharm. Eng. 11* (3), 37–45 (1991).
2. *Clean Room and Work Station Requirements, Controlled Environment*; Federal Standard 209D, General Services Administration, Washington, DC, June, 1988.
3. Title 21, *Code of Federal Regulations (CFR-21)*, Parts 210 and 211; Food and Drug Administration; Superintendent of Documents, Washington, DC; 1991 ed.
4. Heating, Ventilating and Air Conditioning Systems and Applications; *ASHRAE Handbook*; American Society of Heating, Refrigerating and Air Conditioning Engineers, Atlanta, GA, 1987; p 32.6.
5. DESAIN, C., Equipment Installation and Identification, *BioPharm 5* (1), 23–25 (1992).
6. MESLAR, H. W., GEOGHEGAN, R. F., Jr., Biocontainment Facilities Theory and Practice, *Pharm. Eng. 11* (6), 27–33 (1991).
7. DEL VALLE, M. A., HVAC Systems for Biopharmaceutical Manufacturing Plants, *BioPharm 2* (4), 26–34 (1989).
8. NEWMAN, R. P., Air Pollution Control for Infectious Waste Incineration, *Pharm. Eng. 11* (3), 52–54 (1991).
9. ODUM, J. N., Construction Concerns for Biotech Manufacturing Facilities, *Pharm. Eng. 12* (1) 8–12 (1992).
10. ODUM, J. N., Project Schedule Evaluation for Biotech Pharmaceutical Manufacturing Facilities, *Pharm. Eng. 12* (3), 16–24 (1992).

PART IV

FACILITY DESIGN

Programming and Facility Design

Harry L. Johnson and David A. Stutzman

Contents

18.1   Introduction

Architectural facility projects typically begin with conceptual or schematic design phases, and progress through design development, working drawing, and construction phases [1–4]. Our intent in this chapter is to explore the use of *facility programming* as a predesign tool. In the predesign phase, an overview and definition of requirements for detailed design is the intended result. Predesign programming establishes the size and basic needs of the facility, and is frequently used by clients for strategic planning purposes prior to design, or as a part of a master planning effort. The predesign facility program permits basic design decisions to be made in an organized manner before major financial commitments have to be made. When completed, it provides the basis on which design can proceed in an efficient manner. This efficiency saves time in the design and construction periods.

18.1.1   Defining the Problem

An integrated biopharmaceutical facility is a very complex design problem, because of the multifaceted operations contained within. These facilities may typically have:

(a)   manufacturing and finishing areas with specialized process equipment, environments, and finish requirements; definitive material and personnel flows; and biological containment concerns;

(b)   wet and dry laboratories requiring specialized environments for cleanliness and containment;

(c)   animal facilities to support research and development activities, with unique concerns for the animals' welfare;

(d)   utility services such as water treatment, clean steam, high-purity gases, and waste treatment facilities;

(e)   warehousing for segregation, isolation, and handling of raw materials, components, and finished goods;

(f)   administrative areas to support management, finance, sales, technical services, and production needs; and

(g)   special spaces to support the entire facility, such as meeting and conference areas, cafeterias, and lunchrooms.

Typically, biopharmaceutical facilities are designed using intuitive methods that produce facility needs based on varying amounts of information from the owner and user, and trial-and-error development. Biopharmaceutical facilities can be described as being *process-driven*, because the production process heavily influences the design. In such cases, the production process is developed as a separate effort, usually before any consideration is given to the building that houses the process. This design process not only is slow, but also is one that most likely does not respond to all the needs of the owner and ultimate user.

As buildings become more complex, design by trial and error becomes less efficient. A better method is required, which responds more effectively to the needs of the owner. Available facts, perceptions, and opinions about the facility should be brought forth before design is started. This communication enables better decisions to be made during the design process.

Many experiments were done in an attempt to respond to this need. During the 1950s and 1960s, organized methods of architectural programming began to emerge. Many approaches were offered. Most continued the trial-and-error methods. Design decisions were made during the process of developing a design solution. The solution was reviewed against subjective criteria used to develop it. Modifications were made and a new solution was developed to satisfy the review. This development cycle continued until a study that began to classify various types of buildings was published by the National Academy of Sciences in 1964 [5].

In 1969 a method was presented that was based on a distinct separation of programming and design. This method, called *problem seeking* [6], used an organized methodology to search for information required to express the design problem clearly. Programming became a systematic, interactive process between the owner and the programmer. Programming became a focused activity, designed to discover and define problems and to seek solutions before design was begun.

In subsequent years the availability of personal computers permitted the combining of programming's data discovery with the ability to process large quantities of detailed information. This combination allowed a practical approach for programming the complex technical and spatial needs of biopharmaceutical facilities.

Programming is defined as an analytical expression of the design problem. It reduces masses of information to a comprehensive analytical statement that provides the basis for a responsive design. Programming permits discovery of and solutions to design problems, establishment of functional requirements, and an early forecast of facility costs.

Programming can be an effective tool for determining the needs of new grass-roots facilities or for expansion of existing ones. Any facility can benefit from the advanced systematic planning offered by a thorough programming effort. Programming provides a method to inventory and evaluate existing facilities and to make decisions for expanding or converting them, based on factual comparative information.

18.1.2   Interactive Programming

Programming is an interactive process between the owner/user and the programmer. It is a concentrated, focused activity, which requires a strong commitment from all parties involved in the process. The initial programming effort requires a relatively short period of time when compared to the overall project schedule. When design begins, the design must be constantly checked against the program to verify that the design solution satisfies all established parameters.

18.1.3   Programming for Biopharmaceutical Facilities

Programming has typically been used to define building needs for shelters and environments for people. As building technology advanced, programming began to define the increasingly more complex physical needs of the mechanical and electrical systems.

For the most part, programming has not been used for projects governed by process requirements (such as chemical and power facilities). Process-driven projects have typically been designed to enclose the process elements with little concern for the occupants. With the increasing regulatory requirements of CGMP (Current Good Manufacturing Practices), building codes, NIH, and biosafety guidelines, and the recognition of employee satisfaction in maintaining efficient operations, a method of satisfying these combined concerns was required.

Programming methods that combine to define both the people and the process concerns of the design problem were required. While the methods of architectural programming are generally known by many design professionals, the programming of biopharmaceutical facilities requires specialists who also understand the process and regulatory requirements.

Biopharmaceutical facility project schedules are frequently driven by the need to market a product as soon as possible, which usually requires shortening the design schedule. Schedule pressure tends to reduce the efficiencies that could be achieved by good design, because time for planning and assessing alternative solutions is curtailed. However, if both process and facility needs are defined in parallel, good design can be efficient yet not sacrifice the schedule.

Process production, finishing, automation, and support systems are all important concerns in a biopharmaceutical facility, but process concerns are not the only factors important to good design. If the facility designer waits for the process concerns to be defined, the ability to integrate the process with the facility concerns is lost.

To maximize the efficiency of the building, programming methods that integrate process and architectural requirements are necessary. Process engineering and architectural programming should progress in parallel to establish clear integrated objectives and direction for design. Ultimately, the integrated program statement provides the basis for the success of the overall facility.

Biopharmaceutical programming is an interactive problem-solving activity involving the process engineer, the architectural programmer, and the owner. All these participants are necessary for the preparation of a complete, integrated program statement for a biopharmaceutical facility.

Programming for biopharmaceutical facilities usually proceeds in five steps, shown in Table 18.1. Steps 1 and 2 are sometimes reversed, depending on the initial information available from the owner/user. Details of each step are given below.

Table 18.1 Programming Steps

Step	Method used	Data
Establish goals & collect facts	interview & questionnaire	quantitative
Define operation	interview & questionnaire	quantitative
Assemble	interview & questionnaire	quantitative
Uncover concepts & determine needs	interview, questionnaire, & work session	qualitative & quantitative
State the problem	work session	qualitative

(1) *Establish Goals and Collect Facts.* This step is very important. It establishes the basis for the use of the information received in subsequent steps. All available information is obtained, such as product descriptions, organization charts, annual reports, office standards, lab safety standards, and any other materials that may provide some insight into the owner's operations. Key individuals from the owner's operations are identified as major participants for later steps. This step is an on-site activity, conducted primarily by direct interview of the owner.

(2) *Define the Operation.* This step is intended to test some of the information received during Step 1 and to gather most of the qualitative data required to analyze the problem. This step involves the completion of questionnaires and on-site interviews of the key individuals in the owner's operation. Questionnaires are used to obtain details of operations, equipment, and processes. Interviews are conducted with individuals to review the questionnaires and to discuss their responses.

(3) *Assemble.* Sometime near the conclusion of Step 1, preliminary process block diagrams, process flow diagrams, and equipment lists are started. The diagrams and lists are used to help establish the scale for the process and support areas. The information received up to this point is assembled and organized into a program master list. The master list defines each required room or space, based on the information received to that point. The master list includes spatial requirements obtained from the process diagrams and equipment lists. Adjacency lists are also prepared, to establish required spatial relationships.

(4) *Uncover Concepts and Determine Needs.* During this step, the assembled materials are used to present and test concepts for each area. The information is also used to determine the needs of each space. This step is an exercise

requiring on-site decision making by the owner and the program team. The key to the efficient completion of this effort is getting the right people in the right place at the right time.

(5) *State the Problem.* The last step in programming is to present the report. Generally the report includes:

(a) a definition of the functional requirements of the project, including any unique and important performance requirements that shape building design or the quality of space;

(b) a definition of the spatial and physical requirements of each space;

(c) adjacency and isolation requirements for each space;

(d) process area block layouts, utility requirements, and equipment lists;

(e) preliminary process flow diagrams for each process;

(f) a statement of probable cost and the identification of budget constraints, quality of construction, and life cycle costs;

(g) implications of change or growth on long-range requirements for expansion and growth; and

(h) code analysis.

The systematic collection of programming data stimulates interaction within each group of participants and provides a framework of information covering the whole problem. The owner's aspirations, needs, conditions, and ideas are uncovered, to provide a statement that properly influences the design and results in an effective solution. The program provides specific spatial details, such as equipment requirements, environmental criteria (cleanliness, containment, visual, physical, and acoustical factors), and service requirements (mechanical, electrical, and process), to permit a clear, concise design process.

Programming is a team effort led by two responsible group leaders representing the owner/user and the architectural programmer. Each leader must be able to coordinate efforts of his or her group, make decisions or cause them to be made, and establish and maintain communications within and between the groups. Each team must be well managed. Participants must be willing to cooperate with one another. The owner/user must have a commitment to programming the facility. The effort requires substantial amounts of time from the participants, concentrated within a short period.

18.1.4 Management Control

While programming is a methodology that is used to define the design problem, it also provides a means for the owner/user to explore managerial changes or departmental reorganization. These changes may produce more efficient operational methods, with the direct benefit of reducing the long-term operational costs of the facility.

Programming, when properly used to seek responses from a broad range of the owner/user's team, results in a powerful management tool, which can result in improvements in employee satisfaction and functional performance. As a management tool, programming permits the owner/user to clearly see the collective expression of individual needs and their associated costs, before major commitments to design, engineering, and other resources are made. This permits adjustments to be made with minimal impact to construction schedules and professional fees.

Programming can provide management controls that help recognize and solve functional inefficiencies, plan space and services to meet future requirements, and establish proper relationships of space for maximum efficiency. As a separate exercise, programming can also be used to determine existing needs, to establish the basis for an appropriation request, or to plan for the future.

18.2   Regulatory Requirements

18.2.1   Federal Regulations

Biopharmaceutical manufacturing facilities in the U.S.A. are regulated by the Food and Drug Administration (FDA), which is empowered by the Food, Drug & Cosmetic Act of 1938 and its amendments of 1962 and 1976. The Act gives the FDA the authority to regulate manufacturing practices for finished pharmaceuticals. The Act addresses these requirements in the Code of Federal Regulations, Title 21, Part 210 and Part 211, referred to as Current Good Manufacturing Practices (CGMPs) [7]. These requirements are the most widely discussed portions of the Act, and relate specifically to the subject of biopharmaceutical facilities. Part 210 does not deal directly with facility considerations, but refers to Part 211. Supplemental CGMPs for the manufacture of medicated animal feeds (Part 255), medicated premixes (Part 226), blood and blood compounds (Part 606), and medical devices (Part 820) are similar for the products in question. Part 212 covers large-scale parenterals, small-volume parenterals, and certain sterile and aseptic processing applications. Part 212 is used to supplement the more general CGMPs defined in Part 211.

18.2.2   Current Good Manufacturing Practices (Part 211)

Part 211 of the Code has 11 Subparts, as follows:
    A—General Provisions,
    B—Organization and Personnel,
    C—Building and Facilities,
    D—Equipment,
    E—Control of Compounds and Drug Product Container Closures,
    F—Production and Process Controls,
    G—Packaging and Labeling Control,

H—Holding and Distribution,

I—Laboratory Controls,

J—Records and Reports, and

K—Returned and Salvaged Drug Products.

The Act also permits the FDA to issue guidelines, points to consider, and policy directives periodically on various drugs and drug-related products. These guidelines, which are revised and reissued from time to time, represent the FDA's official position on a particular topic. For this chapter the discussion of CGMPs is limited to issues that affect the facility design.

Subpart B (Organization and Personnel) defines the duties and authority of the quality control (QC) unit; it describes staff and personnel qualifications, manufacturing practices, and the use of outside consultants. The QC function must have access to adequate laboratory facilities, and must reside in an organization that is independent of the manufacturing group. The production work force must be regularly trained in the operation of the process equipment and in the CGMPs, particularly as they relate to the employees' jobs. All supervisors and outside consultants must also meet these requirements for regular training. The training requirements imply that space is available to conduct the training. By providing space for conference rooms and properly outfitted classrooms, owners can better conduct the CGMP training for all personnel.

Subpart C (Building and Facilities) provides guidance for designers and contractors in the design and construction of facilities housing pharmaceutical processing equipment. The most important guidelines contained in this subsection require:

(1) that the building be easily cleaned and maintained, and be capable of housing the required processing equipment;

(2) that space be provided to allow for the orderly arrangement of processing equipment to prevent product contamination and mix-up of products, components, in-process materials, and labels;

(3) that separate well-defined areas be provided for different stages of manufacturing processes to prevent contamination or mix-up, including: (a) areas for receiving and holding of raw materials and components before QC testing, (b) holding areas for rejected components, raw materials, drug products and labeling, (c) storage areas for QC-released raw materials, components, and labels, (d) storage areas for in-process materials, (e) areas for manufacturing and processing operations, (f) areas for packaging and labeling operations, (g) quarantined storage area for finished goods before release by QC, (h) areas for storage of released finished goods, (i) areas for control laboratories, and (j) areas for aseptic processing;

(4) that separate processing facilities be provided for the manufacture of penicillin or products containing the β-lactam ring;

(5) that adequate lighting and ventilation be provided;

(6) that equipment be provided to control humidity and to prevent the introduction of contaminants;

(7) that potable water be stored under positive pressure and away from process water;

(8) that sewage and refuse be disposed of safely;

(9) that toilet facilities be provided;

(10) that process buildings be clean and sanitary and have written maintenance procedures; and

(11) that process buildings be in a good state of repair.

Subpart D (Equipment) provides guidance for pharmaceutical equipment designers, fabricators, and vendors in the design and fabrication of pharmaceutical processing equipment. These guidelines are also used as the basis for detailed engineering design specifications for buildings, facilities, and equipment. The guidelines contained in this subpart about facility design are:

(a) that processing equipment perform its intended function and be designed to be cleaned and maintained easily. Therefore, the designer should provide adequate space around equipment for easy access and maintenance activities;

(b) that all processing equipment be calibrated, inspected, and checked periodically according to written procedures. The designer should try to locate noncritical items outside of the working space whenever possible; and

(c) that filters used in processing parenteral drugs not release fibers into products.

Subpart E (Control of Compounds and Drug Containers and Closures) serves as the basis for detailed procedures for operating plants and equipment. Controls are described for drug components and packaging materials. These controls include storage of materials to prevent contamination and to allow inspection, identification, and labeling by distinctive coding; storage before QC testing and release; sampling and testing; and handling if they are rejected. The controls require separate, well-defined spaces to segregate materials and product during the different stages of production.

Subpart H (Holding and Distribution) prescribes warehouse procedures and the conditions of storage for finished pharmaceutical products after their manufacture. A *first-in-first-out (FIFO)* system must be used. The final destination of each lot of finished goods must be recorded. Drug products must be held in quarantine until they are released by the QC unit. After release, they must be stored under conditions that preserve their identity, strength, quality, and purity. These requirements have a significant effect on the design of the warehouse. Consideration must be given to the type of storage and retrieval procedures to comply with the regulations.

Subpart I (Laboratory Controls) outlines the requirements for laboratory controls for components, drug product containers, closures, in-process materials labeling, and drug products. Samples must be retained for each lot in each shipment of active ingredients, for at least one year after the finished drug product's expiration date. Samples should contain twice the amount required to determine whether the materials meet predetermined specifications. The quantity of samples required to be maintained could be substantial. Sufficient space must be provided for these samples and must allow for periodic inspection of the samples.

Subpart J (Records and Reports) requires that complete records and reports be maintained for all production, including failed batches, for one year after the expiration date. Other records must also be maintained, including logs of equipment cleaning and use, records of components, drug product containers, closures, and labeling. Archive storage must be provided to accommodate the maximum anticipated number of records.

Subpart K (Returned and Salvaged Drug Products) provides for the destruction or salvaging of returned drug products. Returned drugs must be destroyed unless examination, testing, or other investigation proves that the drug meets pertinent standards of safety, identity, strength, quality, and purity.

18.2.3   Current Good Manufacturing Practices (Part 212) [8]

Part 212 covers large-scale parenterals, small-volume parenterals, and certain sterile and aseptic processing applications. These guidelines supplement Parts 210 and 211. Additional guidelines or "Points to Consider" are issued by the FDA under its authority (21 CFR 10.85). While compliance with the guidelines is not required by law, they do represent the latest approved system/process, and compliance is recommended. Part 212 carries many of the same subpart titles and contains requirements similar to those in Part 211.

Subpart C (Building and Facilities) provides guidance for designers and contractors in the design and construction of facilities housing equipment for processing parenterals and certain sterile and aseptic processing operations. The most important guidelines contained in this subsection require that:

(a)   the walls, floors, ceilings, fixtures, and partitions in controlled environments shall have smooth cleanable surfaces, and be constructed of materials that resist flaking or oxidizing. Floors in spaces handling liquid products or cleaned by flooding must be sloped to drains and have coved bases. Exposed fixed horizontal piping or conduit may not be located over process equipment. All piping for air or liquids must be identified as to its contents;

(b)   separate areas must be provided for sterile and nonsterile materials and products;

(c)   lighting must be cleanable, and lamps must be covered;

(d)   gowning areas must be provided for those entering controlled areas. Gowning areas shall be provided with negative air pressure, relative to the controlled area served, and positive to adjacent noncontrolled space; and

(e)   special requirements for water and other liquid handling systems must be provided.

Subpart B (Organization and Personnel) is similar to 211 in defining the duties and authority of the quality control (QC) unit. It also includes requirements for controlled environmental areas, for gowning areas, and for adjacent washing facilities.

Subpart G (Packaging and Labeling Control) requires packaging materials be adequate to protect the containers from damage during handling. This requires packaging, handling, and storage methods that are normal in pharmaceutical packaging lines.

Subpart J (Records and Reports) requires that complete records and reports be maintained for all production for a period of two years after the expiration date of the batch. Other records must also be maintained, including logs of equipment cleaning and use, records of components, drug product containers, closures, and labeling. Because the length of time records must be maintained is longer than that specified in Part 211, additional space for archive storage must accommodate this requirement.

Subpart L (Air and Water Quality) does not correspond to a subpart of 211. It defines the monitoring and testing for air in controlled spaces, compressed air, and water.

18.2.4   National Institutes of Health (NIH) Guidelines [9]

Four physical containment levels are established for large-scale recombinant DNA research or production involving more than 10 L of culture; these levels are *GLSP*, *BL1-LS*, *BL2-LS*, and *BL3-LS*. Four levels of containment are established for laboratories working with less than 10 L of culture; these levels are *BL1*, *BL2*, *BL3*, and *BL4*. Requirements for containment of vessels, validation of kill systems, and other considerations are included. Architectural containment requirements are:

(a)   GLSP, BL1-LS, and BL1 require a room for containment where the door must be kept closed while experiments are in progress. A sink is required, to permit personnel to wash their hands before they leave the room.

(b)   BL2-LS and BL2 require the same provisions as BL1-LS and BL1. BL2 also requires the posting of the Universal Biohazard sign on all access doors. Lab gowns, coats, or uniforms must be worn inside the lab. Lab clothing may not be worn to the lunchroom or outside of the laboratory building.

(c)   BL3-LS and BL3 require a closed system located in a controlled area, with air locks and changing rooms. Walls, ceilings, and floor finishes must be easily cleaned and decontaminated. Utilities, services, and wiring must be enclosed and protected against contamination. Hand washing and shower facilities must be near. The area must be designed to contain culture fluids inside the controlled area. Biofiltered ventilation must be provided for the controlled areas.

(d)   BL4 requirements are more stringent than BL3. The NIH guidelines should be consulted for the applicable requirements.

18.2.5   European Issues

Biopharmaceutical facilities manufacturing products for sale in the U.S.A. are regulated by the FDA. Validation of such facilities is required by the FDA.

Biopharmaceutical facilities manufacturing products for sale in the U.K. are regulated by the Medicines Acts 1968 (HMSO 1968) and 1971 (HMSO 1971) [10] and other Regulations and Orders. Biopharmaceutical facilities manufacturing products for sale in Europe are regulated by European Community Directives 65/65/EEC and 75/319/EEC [11]. Regulations for biopharmaceutical facilities manufacturing products for sale in other locations must be verified.

18.2.6   Conclusions

The CGMPs and biocontainment must be incorporated into the design to permit the validation and proper operation of biopharmaceutical facilities. Both Current Good Manufacturing Practices and biosafety guidelines are constantly evolving. Owners and designers of biopharmaceutical facilities must keep current on the changes to these regulations.

18.3   Biopharmaceutical Facility Design Considerations

For an inexperienced facility designer, a biopharmaceutical facility may be a mysterious subject. The design considerations include vague requirements imposed by the biochemists, indefinite spatial needs of the process engineers, limited knowledge of process hazards, and nebulous definitions of GMPs. To take some mystery out of the subject, we attempt in the following to describe typical operations and requirements that affect the facility design.

There are two classic fermentation processes: cell culture and microbial fermentation. From a facilities standpoint, the primary difference between the two is often the scale of production. Cell culture processes involve low growth rates and a high sensitivity to operating conditions; the result is rather low volumes of finished product. Microbial fermentation is usually on a larger scale, with higher volumes of finished product. While this is a very simplistic description, the primary difference is the scale or size of the equipment, and the resulting size of the facility. We demonstrate the needs of a typical facility by describing a mammalian cell culture facility.

18.3.1   Flow of Personnel, Equipment, and Raw Materials

To maintain cleanliness, the manufacturing staff is required to enter the controlled environment of the process areas through a plant locker and wash-up space. Personnel are required to change into appropriate garments. Then they pass through an air lock to access controlled clean areas such as the bioreactor, cell culture, and purification areas.

To assure proper flow of equipment and materials, the facility must be carefully designed. Soiled equipment is removed from each suite through an air lock, and is staged in a soiled equipment area before cleaning. Cleaning is done in an equipment washing or glassware washing area, outside the staging area. After being washed, the equipment is stored in a designated clean storage area. Before use, the equipment is sterilized in an autoclave or a depyrogenation oven. Clean, sterile equipment then leaves the clean equipment area and goes into the production area through air locks.

Raw materials from the warehouse, once released by QC, enter the production area through an air lock into the dry storage area. Nonchemical materials may enter through the personnel entrance. Materials are held in dry storage at ambient temperature, or in a cold box until needed in the weighing room. After the various components are weighed, the cell culture media and buffers are prepared in the media preparation area. The solutions are then filtered, sterilized, and transferred to the appropriate cell culture process. Solutions can be transferred by aseptic containers, but are usually pressure transferred through piping.

18.3.2   Final Formulation and Finishing

Final formulation or finishing operations can be accomplished in many ways, from hand filling operations to semiautomated equipment. Final finishing operations are always conducted under aseptic conditions.

Because the area is clean, particulate-generating or dirty items (such as packaging) should be removed before materials are brought into the formulation or finishing areas. Cardboard boxes containing material to be filled are moved from the warehouse into the material staging area to be unpacked. All exterior cardboard materials are removed and the plastic-wrapped material is moved next to the washing or sterilization equipment. Shrink wrap and internal cardboard separators are removed at the infeed tray of the washer or autoclave. Components such as fill heads or tubing, required in the filling suite, are cleaned and sterilized in a double-door autoclave. The second door opens into the clean room. Formulation tanks are outside the sterile filling room, in an adjacent room. A single fill line passes from the bulk tank into the filling area, where a sterile connection is made with the filling equipment. A local sterile class 100 environment is maintained continuously above the filling and capping operation.

If lyophilization is required, filled vials are loaded and a vacuum is pulled within the chamber to remove all moisture from the product. The amount of time required for this operation depends on the amount of moisture to be removed. The materials are then capped and collected into tote bins, placed in a cage, and transported to a cold storage box for holding until released by QC. Once released, they are moved into the packaging area for inspection, labeling, and final package assembly.

A major design concern for the manufacturing areas is to assure optimal personnel flow patterns and material movement. This is important for efficient operations and for GMP compliance. GMPs require that separate well-defined areas be provided for different stages of the manufacturing process. GMPs also require that the manufacturing area, and each space contained therein, be designed to allow for the orderly arrangement of processing equipment to prevent product contamination or mix-up.

18.3.3 Laboratories

QC laboratories are charged with maintaining the consistent quality of the product during production, and verifying the quality of the finished products. A relationship to the manufacturing area should be provided, due to the support function provided by QC. QC requires environmental and cold rooms for retention of samples. Areas are also required for records storage. Some QC labs require controlled environments with clean rooms, HEPA filtered air, or biological containment requirements.

The following considerations apply generally to all laboratory areas:
(a)   A central glassware washing facility is required, to support the lab operations. This should be separate from the facility used to support manufacturing.
(b)   Labs require office or work station space for scientists and technicians. The use of in-lab work stations is a subject that must be carefully considered.
(c)   Careful consideration must be given to location of safety and eye showers. Generally, areas with floor drains near doors are recommended.
(d)   Interconnecting doors between labs use valuable space and should be limited to areas where they are absolutely necessary.

Since future requirements for testing procedures and equipment are difficult to predict, laboratories should be designed for flexibility. This requires solutions that carefully consider the requirements of services and utilities.

18.3.4 Animal Facilities

Animal facilities may be required as support for R&D operations. Depending on the size of the facility, requirements can range from a separate building to a separate room where animals can be housed apart from areas of human occupation. To avoid contamination, animal facilities should be isolated by separate air lock and gowning facilities. Generally, an area equal to 25% of the animal housing space should be set aside for service functions such as cage washing and sterilization, storage, diagnostic laboratory, office facilities, receiving and quarantining of animals, and refuse disposal. Where facilities are 1,000 ft^2 or less, it is possible to combine some of the service functions. Cage washing and sterilization should remain separate, with a separate exterior access to this space.

18.3.5 Warehousing

Warehousing needs are based on anticipated pallet quantity, size, and stacking configuration. Inventory analysis, peak storage requirements, and bulk storage must also be considered in the design of the warehouse facility.

Rack storage alternatives affecting warehouse operations require consideration during design. Warehouses should be designed for high racking (20 ft), with large-drop sprinkler heads. This eliminates the need for in-rack sprinklers and provides more flexibility in rack arrangements.

A raw materials and finished goods warehouse, adjacent to the finishing areas, is required. The flow of raw materials to production staging areas should be as direct as possible. Finished products should go directly into packaging areas.

Bulk material is stored as a concentrate, often at freezing temperatures. The total quantity of containers to be stored in the warehouse represents some number of weeks' inventory. The acceptable stacking height also has to be determined.

Finished materials are packaged in cardboard boxes. Storage space for finished material is generally required to be refrigerated. Total storage capacity required for quarantined and released materials ready for shipment has to be determined. The acceptable stacking height also has to be determined.

Forklift trucks or other handling devices are used to move materials between the loading dock and storage racks; the type of device and its supporting requirements must be determined. Pallet jacks or other devices are used to move cardboard boxes and pallets from and to areas of use; these should also be defined.

Separate areas should be provided in the warehouse area for quarantined material, released raw material, finished product, chemical storage, and hazardous waste. Ideally, these areas should be directly accessible from the exterior.

Since the scale and proximity of the warehouse space are compatible with future conversion to manufacturing space, the warehouse should be designed in a manner that facilitates such conversion.

18.3.6   Loading Dock

A breakdown and staging area for raw materials is required adjacent to loading docks. If possible, provide a separate dock area for receiving general supplies, with access to central utilities, cafeteria kitchen, and maintenance areas.

Consider a trash compactor, located near the dock area, for kitchen garbage, paper wastes from office operations, and disposables from laboratory and manufacturing operations.

18.3.7   Central Utilities

A central utilities area should be located with access to the loading dock for convenient access for supplies and equipment. Pad-mounted equipment on the building exterior should be screened from public view. Roof-mounted equipment should be acoustically isolated and should be screened from view.

18.3.8   Administration

Office functions can be categorized into six groups: management, finance, sales, general services, technical services, and production.

The top management group is usually arranged together, often in a manner that reflects the chain of command. This area is somewhat isolated from general office traffic and casual interruptions. A conference room is accessible to this group.

The finance group includes accounting, purchasing, personnel, and data processing functions, and is often a considerable percentage of the administrative staff. Accounting should be located near sales. Purchasing has frequent contacts with vendors and should be accessible to the facility entrance. Likewise, personnel requires proximity to the entrance, so that applicants can be received and interviewed without interruption to the general office area. Data processing should be located out of the major traffic patterns.

The sales function requires a considerable amount of communication with other office functions. The sales group generally has frequent visitors. Sales should be accessible to a conference room for demonstrations and conferences.

The general office group provides services for all of the other functions, such as central files, general stenographic services, library, mail handling, duplication, telephone, and reception services.

The technical services group provides such services as facilities engineering, drafting, maintenance shops, utility services, and warehousing. Technical services are generally located near the services that they support.

The production support group for a biopharmaceutical facility is located in the production and R&D areas. The same space considerations and support requirements provided in the administrative area should be provided in production areas. Where this is done, a better relationship usually exists between the administrative and production staff.

Private offices should be assigned for functional reasons such as nature of work, visitor traffic, or security. Private offices are provided for all executives, senior managers, managers, and professionals. Open landscaped offices with low, half-height partitions should be considered for general office areas, with managers located in interior enclosed office spaces adjacent to their respective departments. Perimeter window areas should be carefully utilized to provide as much natural light to interior spaces as is practical.

18.3.9   Special Spaces

In addition to the six functional office groups, service facilities or special areas must also be accommodated. These special spaces must be determined by the owner in advance, and reflect the operation's policy. These spaces may include: lobby/reception area, conference rooms, security, lockers/changing rooms, medical/first aid, cafeteria/lunchrooms, break areas, copy centers, storage, and file areas.

A lobby or reception area is required, to control access to the facility. The nature of the space is determined by the image to be presented.

Meeting or conference rooms should be strategically located to serve all functions in an equal manner. Meeting rooms may be used for training required by GMPs. The space should contain a screen and projection area. The location should also be convenient to rest rooms.

A security area should be placed at a strategic location near the entrance. The security area should be capable of serving as a reception function after regular hours. Closed-circuit TV should be considered for monitoring of the loading dock for after-hours deliveries and security.

Locker and changing rooms for production and maintenance personnel should be located with convenient access to the main entrance and to the production areas.

Medical or first-aid requirements depend on the scale of the facility and the nature of the operation.

A cafeteria should be sized to accommodate the maximum number of people with one turnover for dining seating. This should be anticipated as approximately 60% of the total population. Most cafeterias are not designed as a full-service facility. Some grill and cold table facilities may be required, but generally food is prepared at a remote location and delivered daily. Consideration should also be given to arranging the cafeteria to permit its use as a large meeting room.

Coffee stations and copier stations should be dispersed throughout the facility and located for easy employee access. This minimizes the need for employees to leave their general area for coffee breaks and copies.

Files and storage areas are required in support areas. Fire-resistant storage should be provided for storage of QA/QC documentation and other files and records.

18.4   Project Design

The facility design process follows a definite prescribed pattern. The American Institute of Architects (AIA) promulgates the process through standard Owner-Architect Agreements [12]. Most architects and engineers follow this process in a logical progression during the design and construction of almost every building. The process is divided into five phases: *conceptual design*, *design development*, *construction documents*, *bidding*, and *construction administration*. Facility programming as described earlier is considered a predesign activity and occurs prior to conceptual design.

During the conceptual design phase, the design professional assimilates the program statements and develops concepts to solve the stated problem. Different approaches are explored in an attempt to find the most appropriate solution to the design problem. The process is then advanced to the design development stage. The interior arrangement and the building mass are studied, to accommodate finer details of individual spaces within the building, and arrangement of the building on the site. Once the arrangement has been defined, construction documents are started. These documents are used to describe the building in sufficient detail so competitive bids can be obtained and construction completed. The bidding phase is somewhat flexible. Not all projects are competitively bid. Some are negotiated; others may be turn-key, where the design and construction become the responsibility of a single entity. The construction phase concludes with the physical building being completed and turned over to the owner for production to begin.

Throughout the project phases, the design professional and the owner must work cooperatively, honestly exchanging ideas and reactions to the design being presented. Both the owner and the architect must be realistic in their expectations. Each must work to satisfy the most important considerations completely. Where compromise is necessary, the impact on the facility operation must be minimized.

18.4.1   Design Concepts

It is important to establish certain design concepts or approaches to solving the design problem as stated in the program. Concepts, along with the program statements, are what drive the design development. The arrangement, efficiency, and ultimate complexity of the building are drastically affected by concepts adopted in the design approach.

An integrated facility is a difficult design problem. Concepts should be developed and tested against the particular programming statements, before the design is advanced beyond the conceptual level. During this period, the building is treated as blocks of space rather than individual rooms or suites. Using block or bubble diagrams, as illustrated in Figure 18.1, the designers can establish required relationships between the major areas of the proposed facility. Generally, it should be looked at as several primary blocks of space such as Administration, Laboratory, Animal Facility, Manufacturing, Warehousing, and Central Utilities. Each of these areas has special design requirements that set it apart.

It is important not to concentrate on the details of the design at this early stage. The details are developed as the design progresses into design development. The diagrams (see example in Figure 18.1) are not meant to represent a building form, but only the relationships within the building. The final form of the building is governed by multiple considerations and compromises among the many factors affecting the building arrangement.

When concepts such as separation of product and personnel flow, visitor and employee flow, and clean and dirty environments are enforced, the complexity of the final building design is likely to be maximized. As complexity increases, cost and size also increase. Strict application of concepts should not be arbitrarily decided by design professionals alone, but rather with the owners and users of the facilities.

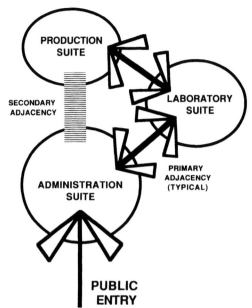

Figure 18.1 Bubble Diagram. Bubble diagrams graphically depict the relationships between and the relative size and position of the elements of the project. Bubble diagrams can be used at various levels of complexity to portray elementary or sophisticated relationships.

18.4.2 Adjacency

The program always considers adjacency of spaces or groups of spaces. Areas that are related functionally should also be physically adjacent to one another. Maintaining the required adjacencies improves the efficiency of the facility operation and reduces the amount of circulation space between related areas.

Problems of adjacency must be considered at all scales of the building design. Beginning with the block or bubble diagrams, the designer must arrange next to one another those blocks that require adjacency. Then, within each block, requirements of individual spaces must be considered. Certain spaces have a direct external relationship to areas within their own block and may have direct relationships with spaces outside of their block (Figure 18.2). A good example of this is a senior group manager's office, which is directly related to the junior manager's office within his or her own group and to the general administrative offices of the facility. For this reason the relative position of a space within its own block may be less important than the location relative to other blocks.

As the design progresses, more and more finite detail must be considered, and compromises must be reached based upon the practicality of satisfying the design constraints. Within each block, individual spaces that are related to one another must be arranged for efficient operation and interaction.

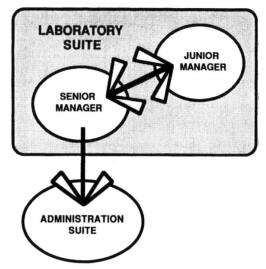

Figure 18.2 Internal and External Relationships. Relationships can be expressed as internal or external relative to the context. Project elements within the same functional group maintain internal relationships among themselves, while these same elements may require an external relationship with another functional group.

18.4.3   Separation of Public and Private Areas

With the tendency of the facility owners and users to show off their facilities to visitors, it becomes increasingly important to separate the public and private access and circulation. The security of the facility begins at the point of initial entry. This initial entry point could be at the building lobby or a site perimeter fence. It is important to provide separate entry controls for the public and the employees. Once this separation is accomplished, the facility security force can concentrate on the control of the public.

If separate entry points to the building are used, control can be immediately established because the available circulation paths are limited. Separate circulation throughout the facility probably is not required. The owners should designate which areas are to be used as showpieces for the public. If the areas available to the public are not limited, separate circulation paths may be required for the entire facility.

Circulation space is no small amount of space (Figure 18.3). Normal amounts of circulation can easily account for half of the unassigned space or 20–30% of the entire gross building area. Any consideration of providing completely separate circulation paths for employees and visitors should be carefully reviewed for its effect on the efficiency of the facility.

Visitor access should be limited to administrative areas and designated process viewing areas. Visitor circulation should be arranged to avoid passing through other areas that are off limits. It is also desirable for visitors to be within sight of a security point while they view the process areas.

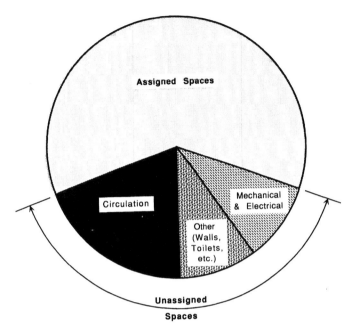

Figure 18.3
Assigned and Unassigned Building Spaces. Unassigned space encompasses a significant portion of the project. Building efficiency is measured by the ratio of net usable area to total building gross area. The more space that can be programmed directly, the more positive control can be maintained over building efficiency.

18.4.4 Isolation of Clean Areas

Some facilities use a *clean corridor* and *return corridor* concept. With this concept, bioreactor, cell culture, and purification areas are organized as separate suites, and may be accessed only through entry air locks from the clean corridor. Once having entered any suite from the clean corridor, all personnel and equipment are considered to be in contact with biologically active material, and they must leave through an exit air lock leading to a return corridor. This circulation pattern minimizes contamination of any other clean area.

Maintaining control of the flow is critical once the concept is adopted. Control may be established by a standard operating procedure to explain the rules governing access to and exit from biological suites, or control may be maintained by sophisticated interlocks integrated with door hardware and card readers or other devices, which permit access in only one direction. There are many variations on the theme, utilizing selected hardware and interlock mechanisms to provide different levels of control. The philosophy of control for this concept must be established early in the design process to allow enough room and the proper facilities to accommodate the requirements.

18.4.5   Product Flow

The primary concern of a biotechnology facility is the integrity of the product, from raw material delivery to final manufactured product shipment. Well-defined flow paths enhance the ability to maintain the product integrity. Minimizing crossing and circuitous paths also reduces the potential of contamination and mix-up of materials and products.

Raw materials classified in several distinct categories exist within the facility at the same time. At the time of receiving, the raw materials need to be kept separate from other materials and product. After QC samples and verifies the quality of the raw materials, the materials should be segregated according to whether they are released for manufacture or rejected.

During the processing operation, QC monitors the progress of the product. At any time, the product could be classified as suitable for or rejected from further processing. After the final processing step, the product is quarantined until QC verifies that it is suitable for release or that it is rejected. There is also the potential that manufactured product previously released and shipped may be returned to the facility as rejected material. This rejected material is also sampled to determine whether it meets the product standards or not. If not, the product is held separately to be destroyed.

From the description above, the importance of clear separations is evident (Figure 18.4). Once the raw material or product is determined to be unsuitable for further processing, it must be immediately removed from circulation. Rejected materials must be held separately from all other materials to prevent possible contamination of materials released by QC.

Separate storage areas for rejected materials must be provided. Rejected materials must be easily moved to these locations without the possibility of being mixed up with acceptable materials. Rejected materials must be able to be separately held until their final disposition is determined by QC.

The flow of raw materials and products through the facility should follow the production process. The flow should be arranged to eliminate backtracking and crossing paths, thus avoiding points of mix-up or contamination. As soon as a material is determined to be unsatisfactory, it should be removed and quarantined as a rejected material without crossing the production material flow pattern.

18.4.6   Personnel Flow

Personnel flow is unlikely to parallel product flow. Product flow is located in the production area, with a prescribed path. Personnel flow occurs throughout the facility, with differing intensities and paths. People cross the boundaries of the different areas,

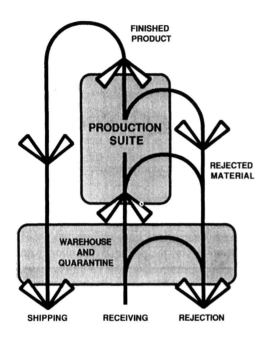

Figure 18.4 Material Flow in Production Suite. Clearly defined flow patterns for both materials and personnel are critical to prevent contamination of the finished product. Flow should be arranged linearly, without crossings between acceptable and rejected products.

going from production to administration, to warehouse, to research, and to other areas within the facility.

Just like product and material, people must be accommodated in order for the facility to function properly. The most common paths of travel must be identified based upon the strongest adjacency and daily communication. The critical paths must be given top priority, to minimize both travel time and interference with other operations. The number, frequency, and distance of personnel trips must be considered as the priorities are established. Optimizing rather than minimizing the travel time and distance is the better concept to follow.

Resolving circulation paths between the different areas defined in the block diagrams is important. Usually, the paths between areas are easily identified. As the circulation is examined at a smaller and smaller scale, the complexity increases. General circulation paths must incorporate access to and around furniture, equipment, work stations, and production lines. Sufficient space for all the separate circulation activities must be provided.

Where shared space serves multiple circulation functions, the optimum solution is a compromise. The goal is to provide the most practically functioning facility for the least amount of money. This usually equates to the smallest building in area. Circulation space can account for up to half of the unassigned space in a building program, and unassigned space in a biotechnology facility can be 40% or more of the total building area. The prudent design strives to minimize the circulation within the building. However, circulation areas cannot be reduced or constricted to the point of making the facility impractical to operate.

The efficiency of the building should be monitored throughout the design, but especially in preliminary design and design development. The efficiency should be measured by the ratio of net usable area to the total building gross area, expressed as a percentage. The efficiency should not vary greatly from the value assumed by the program. Although it may be desirable to maximize building efficiency, experience shows that this may have a detrimental effect on the building functionality (Figure 18.5).

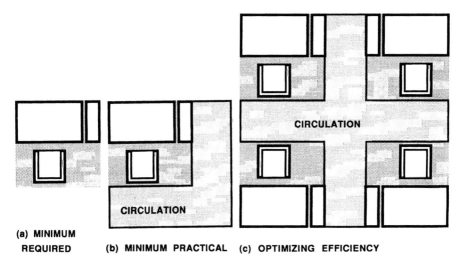

(a) MINIMUM
REQUIRED   (b) MINIMUM PRACTICAL   (c) OPTIMIZING EFFICIENCY

Figure 18.5   Work Station Efficiency. Optimum efficiency is a delicate balance between functionality and judicious use of unassigned space. Figure 18.5(a) indicates the minimum space required for an office work station with 100% efficiency, but in practice 100% efficiency is unrealistic. Figure 18.5(b) allows for circulation outside the work station to make the work station usable. Efficiency is reduced 44%. If circulation space is combined to take advantage of shared usage, as shown in Figure 18.5(c), efficiency is improved to 64% while functionality is essentially unchanged from 5(b).

To illustrate the point, think of a simple office situation. The typical work station includes a 3' × 5' desk, a 1' × 3' filing cabinet, and a chair. The minimum space required for this arrangement is 6' × 6', allowing a three-foot space behind the desk for the chair and the person occupying the work station. The minimum practical area required is 9' × 9'. This provides a three-foot aisle behind and beside the desk for circulation space. Although all the furniture and people could fit into the 6' × 6' space with 100% efficiency, the space is impractical. With a person occupying the chair, no one could walk by the work station. If someone was walking by, the file cabinet could not be opened.

To make the space comfortably usable, efficiency is sacrificed. However, providing more circulation space does not increase the space's usefulness. There is a balance that must be reached, based upon experience with similar kinds of spaces.

Personnel flow cannot be considered separately from product flow. The two coexist within the same spaces. It is important to keep the points of crossing to a minimum to avoid contamination of the product. The location of the personnel with respect to the product and the air supply is also critical.

The personnel are the greatest potential source of contamination within the process facility. People continuously release particulates to the surrounding atmosphere. The more contact personnel have with the product, the greater the chance of contamination. If the circulation paths are separated and properly placed, the risk of product contamination is reduced. If the personnel are downstream of the product with respect to the clean air flow, the chance of contamination is reduced. With a properly located circulation area and a properly designed air system, the contamination is swept away by the air flow before it has a chance to affect the product.

18.4.7 Raw Materials Flow

Raw materials are those whose reason for being at the facility is to participate in and be transformed by the process, to become product. As part of GMP requirements, raw materials must be kept separate from in-process materials and finished product. The potential for mix-up with other materials must be kept to a minimum. Raw materials should be received in an area separate from finished product shipping. This receiving area can also handle general maintenance materials and supplies. However, raw materials, once received, should be immediately separated from other materials in the receiving area and stored separately.

The raw materials must be held separately until QC samples and verifies them. The materials, once released for production, should be separated from rejected materials. The rejected materials must be kept separate from all other materials, to eliminate the possibility of contamination of materials released for production. Once raw materials are released for production, they are placed in the process circulation flow.

18.4.8 Process Flow

Process flow begins when raw materials are released for production. Because the primary objective of the GMP is to guarantee the quality and purity of the end product, the process flow is critical. As much as possible, the process flow must be separated from all other flows of people, raw materials, supplies, and utilities. The greater the separation, the more likely the product is to maintain its purity and quality.

The product may go through four separate classifications during production: in-process, released, rejected, and quarantined. *In-process product* is any product that is currently in a production stage. *Released product* is any product that has completed a production stage and has been checked and verified as acceptable for the next production stage. *Rejected product* is any product that at any time has been found unacceptable for further production. *Quarantined product* is product that has completed the production process and is awaiting final release for distribution.

To maintain the integrity of the product, rejected materials must be removed from the production flow, without interrupting the production suite. Rejected products should follow a prescribed circulation path to a separate holding area.

The process and personnel flows within the production suite should allow for easy removal of rejected product. The personnel within the suite should be able to remove the product without having to leave the suite. Product removal should not require other personnel to enter the suite simply to remove the rejected product. A separate passage, accessible from the suite, should be considered. The rejected product could be placed in the passage by the production suite personnel. Later, at a convenient time, other personnel could collect the rejected product from a utility corridor and remove it to a holding area (Figure 18.6).

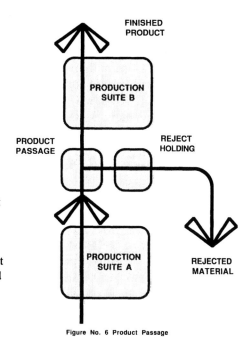

Figure 18.6 Product Passage.
The efficient handling of product without interruption to production operations is imperative. Suite personnel should be able to effectively segregate rejected materials from acceptable product without leaving the suite. Other personnel should be able to collect rejected materials at a convenient time without entering the production suite.

Figure No. 6 Product Passage

A similar concept can be used for product passing from one production suite to another. The use of a product passage allows production and cleanliness in both suites to be maintained. Personnel can pass completed product to the passage. Product samples can be evaluated while the product is in the passage. If rejected, the product can be removed. Acceptable product can go into the next production suite.

Once the production process is completed, the product is held in a quarantined area. The product is held until final QC evaluation and release.

The product flow should progress from the receiving of raw materials, through the production process, and finally to the shipping area. The path should be linear, continuous in one direction, without backtracking. The continuity helps to assure that contamination does not occur by mixing up released and rejected materials.

Product has one other classification outside of the production process: *returned product*. Returned product must be assumed to be rejected material until testing can prove otherwise. After its receipt, returned product should be immediately separated from any other received materials. The returned products can be held in the rejected materials storage areas until a final determination is made as to their disposition.

18.4.9 Sterile Environment

Product production is highly controlled to maintain purity and quality. The physical environment containing the process requires various degrees of control, depending on the stage in the production process. The critical area is the immediate area surrounding the product, especially when the product is exposed to the environment. The required sterility of the environment should be closely examined in relation to the process. Where the process is in a closed system, the facility environment should not need to be sterile to maintain the product integrity. Where the product is exposed to the environment, such as in filling and capping operations, the environment can be critical to maintaining product purity.

To help preserve the cleanliness of the environment, the process should be arranged so the level of cleanliness is progressive. The relative amount of control for the environment should become increasingly more stringent from preliminary to advanced process stages. This allows for a progression within the production suite for both product and personnel flow. Once within a clean environment, the product and personnel should remain in the clean environment until the production is complete. Facility arrangements that require flow through a clean area to a dirty area to another clean area should be avoided.

Clean areas are easily handled in a progression from an uncontrolled area to a very strictly controlled area. NIH has established specific requirements for the different biological containment levels. These are also progressive. They range from a simple hand sink for BL1 to air locks, changing rooms, and showers for BL3. It is imperative that the level of biological containment be clearly defined for each process activity, so that like areas can be grouped around common support facilities.

18.4.10 Support Facilities

Support facilities fall into two types, central and multiple. The concept of arranging support facilities in a central area takes advantage of economy of space by grouping like facilities into one large area. The kinds of support facilities that might be effectively centralized are shipping, receiving, warehousing, mechanical spaces, glassware washing, and break areas. Other support facilities, which might be better in multiple locations for operational convenience, include general supplies, equipment storage, cold rooms, and conference rooms.

Centrally located support facilities provide an increased efficiency for the overall building layout, but along with that efficiency comes an increased frequency of traffic to service the areas supplied from the central sources. If office supplies are maintained in a central stock room, there is frequent circulation to all parts of the building where offices are located. From an operations point of view, this probably is not the best solution for running the facility. An excessive amount of traffic is generated between the stock room and the office areas.

There is a middle ground between the central and multiple facilities concepts: utilize the central facilities to maintain large quantities of supplies, and provide multiple smaller facilities, strategically placed throughout the facility, for supplies required for daily operations. The smaller facilities can be stocked from the central facility by central facility personnel daily, weekly, or at whatever period makes sense for the facility operations.

This concept of satellite facilities can be used for many support operations. The office supplies example mentioned above is obvious. Laboratory and production services should also be considered. Central mechanical rooms can provide basic building services such as hot and chilled water; satellite mechanical rooms can provide local, specialized services such as clean steam and cleaning-in-place systems. Glassware washing can operate as a satellite facility. A central glassware washing facility can be used for the actual washing and the main storage area; satellite clean and dirty glassware storage areas can be provided for the daily operation of the laboratory and production areas, with central glassware washing personnel responsible for collecting dirty glassware and distributing clean glassware to the satellite storage areas.

18.4.11 Utility Service Space

Utility service space and layout are inseparable in the design of biotechnology facilities. Utilities and supporting services require careful consideration of the horizontal and vertical distribution of these services.

The production suites and laboratories require the support of many services to perform their function. The services include utilities, specialized equipment, portable vessels, gas cylinders, and maintenance equipment, among others. Locating these services within the lab or production suite causes unnecessary congestion and additional required circulation space, and service personnel are required to enter the lab and production suite to maintain the utilities and equipment.

To comply with the GMP requirements, it is necessary to limit the utilities within the production area. Overhead piping and conduits above the production area are prohibited. The cost of the specialized materials for the final utility connections to the process equipment is high, so the amount should be kept to a minimum.

18.4.12 Distribution of Services

Several methods have been used to distribute services. These are (a) utility corridor system, (b) multiple interior shafts, (c) multiple exterior shafts, (d) corridor ceilings with isolated vertical shafts, and (e) utility floor or interstitial spaces.

The utility corridor (Figure 18.7) concept provides a reasonable distribution of utilities and support services to a production suite or laboratory. Utilities are distributed in a space immediately adjacent to the lab or production suite. Utilities can be left exposed and accessible from the floor, because they are outside the controlled environment of the working area. Maintenance can be accomplished without interruption to activities within the lab or suite. With the process equipment arranged in the nearby utility corridor, service connections can be made where required and with minimal exposure within the suite.

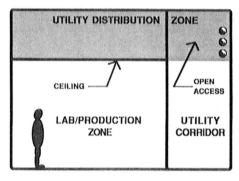

Figure 18.7 Utility Corridor. Utility corridors minimize interruption of laboratories and production suites caused by changeovers and routine maintenance of support services. Utilities may be run exposed in the corridor, which is not maintained as a controlled environment. Keeping piping, gas, electrical, and other services exposed within the corridor immediately adjacent to points of use minimizes lengths of runs.

The utility corridor can be used as a staging area for equipment needed in the suite or lab. The corridor serves to separate the circulation of personnel from the equipment. It also acts as a semicontrolled environment for the equipment before the latter is used in the lab or suite. The utility corridor also serves as a link between the central equipment and utility maintenance facilities, and the production suites and labs.

All service mains and ducts are brought to the various floors through a vertical central core. The horizontal distribution of utilities from the central core is at the ceiling of the utility corridor. This design provides for access of maintenance personnel to piping and ducts. Utility corridors also provide space for lab gases and miscellaneous pumps and coolers as needed. Utility corridors are usually double-loaded, with two labs backing up to the same corridor. Utility corridors are usually thought of as being most applicable to multistory buildings. However, the specialized needs of biotechnology facilities can often use this system to advantage.

Multiple interior shafts provide for utilities and duct work concealed in a series of regularly spaced shafts located along the circulation corridor. Services are brought vertically upward or downward from mechanical spaces, and distributed horizontally through walls and ceilings. Multiple interior shafts have a moderate net-to-gross-area efficiency and are usually a good selection for multiple-story buildings.

Multiple exterior shafts bring utility services and ventilation ducts to individual floor levels by exterior vertical wall shafts. Utility distribution and use of exterior shafts is similar to internal shafts. Usually, more shafts and longer distribution runs are required than for internal shafts.

Corridor ceilings with isolated vertical shafts have utilities located in the corridor ceilings, and in some cases above the ceilings of rooms on either side of the corridor. The utilities are supplied by one or two vertical shafts. Distribution is made horizontally, and may be made upward or downward in order to serve two floors from one distribution network. This system is usually used in office areas and simple lab areas. It is primarily suited to buildings of one or two floors. This system provides economy of construction and has a high net-to-gross-area efficiency, but does not offer good flexibility for lab or manufacturing areas.

Utility floor or interstitial spaces provide the greatest flexibility of all the systems described. This arrangement consists of a separate utility floor above or below each space to be served. Services can penetrate floors or ceilings where needed. The cost of this system is high and it has a very low net-to-gross-area efficiency.

18.4.13 Modular Approach

Facilities that are designed today must be designed to be flexible, in order to be capable of effective use tomorrow. The design cannot be so rigid and so specialized that only present needs are satisfied. A modular approach for the design contributes to maximizing the future flexibility of the facility. A standard module can be used for planning each class of space, such as laboratory and production suites. Building modules must also be consistent with the building structure and column layout and with building materials.

The module should be founded on the smallest unit that can accommodate a basic function. For instance, if a basic laboratory has two people, each requiring a fume hood and ten feet of bench top, then the lab module should be based on this minimum requirement. The module should allow for two work stations, two fume hoods, lab work tops for bench equipment, and circulation space within the lab. Once the area for these functions is established, the remaining labs should be developed as multiples of this basic unit. For practical purposes, the multiples could include one-half modules, which can be easily arranged within the building spaces. Division beyond half modules begins to defeat the modular approach.

Using a modular approach allows a convenient reorganization of space as physical requirements change over the life of the facility. The arrangement of the spaces in each of the office, lab, and production suites follows a prescribed pattern that fits within the building shell. Rather than having spaces designed to a specific need, adopting standard modules allows a better fit within the facility. There are fewer uniquely sized and shaped spaces to make reorganization of the building more difficult.

The other important advantage of the modular approach is that modular spaces can be adapted to other functions without reorganizing the space. Take, for example, a development laboratory that requires a double module to accomplish its function. Because of changes in product development, it may be decided that two new single-module research laboratories are required in place of the development lab. These two new labs can be set up in the one development lab with minor adjustments, and a wholesale reorganization of the space is avoided. Suites, labs, and offices can serve multiple functions with minimal disruption of services if arranged on a modular basis.

18.4.14 Special Conditions

Within a biotechnology facility, height requirements of individual spaces play an important role in the overall facility layout. Offices, R&D labs, and QC facilities generally require a one-story space. Production suites, pilot plants, and warehouses usually require a two-story space. The arrangement of the spaces can drastically affect the efficiency of the facility and the ability to expand individual spaces in the future.

The modular approach can be extended to the third dimension of the building—height. If the spaces requiring two stories can be accommodated in a multiple of the single-story spaces, then the expansion possibilities are not so limited. The structure can be easily designed to add the upper floor in a two-story space. Conversely, the upper floor can also be designed to be removed with no impairment of structural integrity. Then, all spaces within the facility are potential areas for expansion.

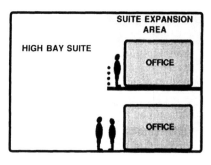

Figure 18.8 Two-Story Module. Suites that require two-story–high spaces can take advantage of built-in expansion potential. Offices and other accessory spaces associated with the high-bay areas can be located around the perimeter and arranged in a two-story configuration. The offices may be converted to high-bay space and the offices relocated to adjacent, less expensive, one-story spaces.

The offices associated with two-story spaces can be arranged two-high around the perimeter of the two-story space (Figure 18.8). The offices maintain direct contact with the associated production space, without requiring a large adjacent one-story space. The office spaces can then be an immediate area for expansion of the production area.

18.5 Interior Finishes

18.5.1 General Requirements

The GMP and NIH requirements for interior finishes in biotechnology production areas are stated rather broadly. The finishes must be smooth, be easily cleaned and maintained, and resist flaking and oxidation. Many finishes can meet all of these requirements. The designer, owner, and operator of the facility must determine what finishes are required for optimum performance without exorbitant cost. The real concern is how the finishes affect the finished product with respect to purity and quality.

Where the product is contained in a closed system, the concern for the finishes need not be so great. There is no potential of contamination of the product from the building environment. The general maintenance of the finishes should be the major concern. Good-quality architectural finishes withstand all normal maintenance such as washing. Rooms containing closed product systems should not require any special finishes.

The areas of critical concern must be identified. Usually, it is those areas where the product is exposed to the building environment. In these rooms, the extent of the product exposure is the controlling factor for finish selection. The potential for contamination is in direct proportion to the time the product is exposed and the area of exposure. For example, during vial filling the product is exposed for the time the vial is being filled, until it is capped. The length of time that the product is exposed is a factor of the production process. The area of exposure is the area of the mouth of the vial. The time and area may be adjustable, depending on the filling equipment selected. So, the finishes required are directly related to the process equipment performance.

The quality of finishes required is also related to the cleanliness of the environment. If the air quality in the immediate area of the exposed product directs the potential contamination away from the product, there should be less concern for the finishes in the area. The whole room containing the exposed product should be evaluated. It is not necessary to provide high-performance architectural finishes for the whole room, but rather in the areas of greatest potential contamination. If the room can be arranged so that the product area is a separately controlled environment, use of high-performance finishes required for exposed product can be minimized.

To maintain the product in a clean condition, clean air flow should be directed over and then away from the product area. Any contamination released in the rest of the room is immediately directed away from the product before having a chance to come in contact with the product. The result of this concept is a smaller area requiring the strictest environmental control, including architectural finishes (Figure 18.9).

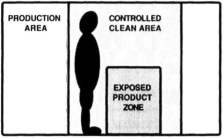

Figure 18.9 Clean Environment. Extensive controlled clean environments drastically influence the cost of the facility through the higher quality of space required. The architectural finishes and the HVAC systems are most strongly affected. Restricting the controlled environments to zones specifically required for product and personnel safety minimizes the additional expense.

Many considerations must be taken into account as the architectural finishes for the biotechnology production spaces are determined. The cleanability of the finish is very important. Moreover, the cleaning solutions, frequency and extent of cleaning, and production chemicals that may contact the finishes must be determined if finishes that can withstand such cleaning are to be properly selected. The permanence of the finish is equally important. An inexpensive finish can be selected that withstands the environment, but it may not last long and frequent refinishing is then necessary. Conversely, a high-performance finish, at a high initial cost, can be used; it requires much less refinishing and consequent process shut-down time.

Most architectural finishes have a definable useful life expectancy. Life cycle cost techniques should be used to evaluate the cost and the useful life of the candidate finishes before the one best suited to the application is determined. The most cost-beneficial finish could very well be the more expensive first-cost material.

The quality, performance, and cost of architectural finishes cover a wide range. The designer, in conjunction with the owner and users of the facility, should determine the level of finish required for the facility. Finishes may be classified as those that meet the minimum requirements of NIH and GMP, those that exceed all NIH and GMP requirements, and those that fall somewhere in between.

It is not always necessary to use the best finish, if a lesser one performs as well at less cost, but it may be prudent to provide finishes that exceed minimum requirements, to avoid having to replace them later if requirements are made more stringent. Knowing that the product and the process are subject to change over the life of the facility, those selecting the finish should be aware of the requirements of the next higher level of product protection. A higher overall level of finish may be provided, for a modest increase in the cost of the facility, to accommodate this potential future condition. This could avoid a costly replacement of finishes as the facility and product evolve.

18.5.2 Finishes

Each of the physical containment levels established by NIH and FDA requires a different level of treatment of the finishes. Production circumstances should be evaluated to determine the level of finish required. It is important to remember that NIH guidelines are concerned with protection of personnel; those of FDA are concerned with protection of the product.

The architectural containment and finish requirements shown in Table 18.2 are recommended minimums to meet NIH guidelines.

Table 18.2 NIH Requirements for Finishes

GLSP, BL1-LS, and BL2-LS	
Floors	concrete, sealed and hardened.
Base	none
Walls	concrete masonry units (CMU), painted
Ceiling	exposed construction, painted
BL3-LS	
Floors	concrete with troweled-on epoxy finish
Base	coved troweled epoxy, 4" high
Walls	CMU, epoxy paint
Ceiling	suspended drywall, painted
BL1 and BL2	
Floors	vinyl tile
Base	vinyl base
Walls	drywall, painted
Ceiling	suspended acoustical tile
BL3	
Floors	concrete with troweled-on epoxy finish
Base	coved troweled epoxy, 4" high
Walls	drywall, epoxy paint
Ceiling	suspended drywall, painted

Section 211 of the FDA regulations states that the surfaces of floor, wall, or ceiling in aseptic areas shall be smooth, hard surfaces that are easily cleanable. A proposed section of 212 states that the construction features of controlled environments should be regulated to preclude contamination. Walls, floors, ceilings, fixtures, and partitions must have surfaces that resist flaking and other deterioration and are suitable for cleaning. The architectural containment and finish requirements in Table 18.3 are recommended minimums to meet FDA guidelines.

Given the importance of the floor slabs, particularly in classified areas, proper subsoil conditions and compaction are vital to a good finished floor. Floor

Table 18.3 FDA Requirements for Finishes

FDA Part 211	Aseptic areas
Floors	epoxy terrazzo
Base	epoxy terrazzo, 4" high
Walls	concrete masonry units (CMU) with veneer plaster, epoxy paint
Ceiling	suspended drywall, painted
FDA Part 212	Aseptic, controlled areas and areas using recombinant DNA
Floors	epoxy terrazzo
Base	epoxy terrazzo, 4" high
Walls	CMU with veneer plaster, reinforced Descoglas finish*
Ceiling	suspended plaster, reinforced Descoglas finish*
Doors & frames	stainless steel, type 304, #4 finish
Sterile areas, class 10,000	Filling and stoppering room including air lock and gowning with class 100 directly over filling and stoppering operation
Floors	epoxy terrazzo
Base	epoxy terrazzo, coved
Walls	block with ¼" veneer plaster

* Provide coves at internal corners (wall to wall, wall to ceiling). Keep finishes flush (base with wall finish, door frame with wall).
Descoglas is manufactured by Master Builders, Inc., Cleveland, OH.

thicknesses, reinforcing, concrete mix designs, and curing methods are critical to ensure the compatibility of the finish and floor system selected.

18.6   Regulatory Issues

All buildings must be designed to comply with the applicable zoning ordinances, building codes, Americans with Disabilities Act requirements [13], and OSHA regulations [14]. Other regulations, such as EPA and individual state environmental, air, and emissions regulations, must also be considered. These regulations deal with the safety of the community and the welfare of the occupants and employees.

Facilities are expected to meet regulations of the National Institutes of Health intended to protect the health and welfare of the employees and the general public. Although these biosafety requirements are intended for use in a laboratory research setting, they are currently considered good laboratory practice standards for biotechnology production operations. Biocontainment issues are directly related to the scale of production and the type of organisms used. Design must identify work areas where physical containment of the manufacturing process is required to protect the product, environment, and workers. It must also define areas where containment is not required. The design must indicate the air flow direction of both controlled and uncontrolled areas, and the flow of materials, equipment and products, and biohazardous waste. FDA is charged with regulating the product, personnel, buildings and facilities, equipment, production, packaging, and QA/QC procedures. This is what governs the detailed activities of bioprocess manufacturing.

18.7 Conclusion

Dealing with these complex design and safety regulations can be difficult. Many requirements are conflicting, such as the principles of containment versus the principles of egress. To properly deal with these complex issues, one should start with as many facts defined as possible. Programming can provide the goals and facts required to achieve a good design response. These goals and facts permit the most balanced design response to these complex issues and offer the best solution for the public, the employees, efficiency in operations, efficiency in production, and good uniform quality of product.

18.8 References

1. *The Architect's Handbook of Professional Practice*; American Institute of Architects, Washington, DC, 1988.
2. LINN, C., Interactive Approach Produces Innovations, *Lab. Plan. Design 1* (1), 12–19 (1989).
3. NUNEMAKER, J., Planning Laboratories: A Step by Step Process, *Am. Lab.* (March), 104–112 (1987).
4. PALMER, M.; *The Architect's Guide to Facility Programming*; American Institute of Architects, Washington, DC, 1981.
5. *Classification of Building Areas*; Technical Report No. 50, National Academy of Sciences–National Research Council, U.S. Government Printing Office, Washington, DC, 1964.
6. PENA, W.; *Problem Seeking*; American Institute of Architects Press, Washington, DC, 1987.
7. Title 21, *Code of Federal Regulations (CFR-21)*, Part 210 and Part 211; Superintendent of Documents, Washington, DC, 1991 (revised April 1, 1991).
8. Title 21, *Code of Federal Regulations (CFR-21)*, Part 212; Superintendent of Documents, Washington, DC, 1991 (revised April 1, 1991).
9. *NIH Guidelines for Research Involving Recombinant DNA Molecules*; Department of Health and Human Services; Superintendent of Documents, Washington, DC, 1986.
10. *United Kingdom Medicines Act (1968)*; HMSO, 1968 and 1971.

11. *Council Directives 65/65/EEC and 75/319/EEC*; European Community.
12. *Owner and Architect Agreement*; AIA Document B141, American Institute of Architects, Washington, DC, 1987.
13. Code of Federal Regulations (CFR); *Americans with Disabilities Act (ADA)*, Superintendent of Documents, Washington, DC, 1991.
14. Code of Federal Regulations (CFR); *Occupational Safety and Health Standards (OSHA), Part 1910*; Superintendent of Documents, Washington, DC, 1991.

Project Planning

John P. Boland

Contents

19.1   Introduction

In almost all cases, the development of a bioprocessing facility is a complex project, involving the participation of personnel from all levels of the organization: executive management, marketing and financial analysts, scientists, laboratory technicians, manufacturing personnel, quality assurance and validation personnel, purchasing specialists, and engineers. In addition, the actual engineering, design, and construction of the facility is usually performed by consultants and contractors. Both singly and collectively, all these people must observe the principles of project management, planning, organization, execution, monitoring, and controlling.

In this chapter I provide an overview of the steps necessary to plan and execute a successful project, with particular emphasis on the planning process. I assume that the project under consideration is a manufacturing-scale facility, so that I may discuss all aspects. The principles I present are applicable in whole or in part to projects for other types of facilities, such as R&D laboratories or pilot plants.

19.1.1 Strategic Planning

The most significant issues in the commercialization of a new product are whether the product can be manufactured at a reasonable cost, available within a time frame permitting successful market penetration, and sold at a price that yields a satisfactory *return on investment* (*ROI*). Typically, the new product is identified through research and development operations on the laboratory scale. Marketing specialists assess the commercial potential of the product, and make a rough estimate of the market size, share, and pricing. For the preliminary economic analysis, rough estimates of capital and operating costs (usually based on historical data) are made, and, if the ROI is favorable, additional project work is authorized. A more detailed program and conceptual design is then performed, yielding more detailed and accurate cost and schedule data. The project is reassessed, and, if still favorable, design and construction are authorized.

19.1.2 Project Phases

The planning process organizes the project into discrete phases. The owner's organization, consultants, and contractors understand each other more easily and communicate more effectively if they adopt recognized conventions. For this discussion I have adopted the conventions of the *American Institute of Architects* (*AIA*) [1], with minor revisions. These conventions have the advantage of widespread use throughout the construction industry, and have traditionally been applied in pharmaceutical and bioprocessing applications. Using these conventions, we divide the project into (a) *Program Phase*, (b) *Conceptual Design Phase* (termed *Schematic Design* by the AIA), (c) *Design Development Phase*, (d) *Detailed Design Phase* (termed *Construction Documents* by the AIA), (e) *Construction Phase*, and (f) *Validation Phase* (not addressed by the AIA). Table 19.1 summarizes the phases of a bioprocessing facility project, and lists some of the more critical activities in each phase.

As part of the planning process, executive management must establish two essential project elements: economic feasibility and schedule performance. These are monitored during the course of project execution. The owner's project manager and each member of his project team must become familiar with the policies that were adopted in establishing these elements. These are discussed in the following sections.

19.1.3   Economic Feasibility

The initial decision to proceed with project development is based on very preliminary data. As the project proceeds, cost estimates must be reviewed to ensure satisfactory ROI. A cost estimate based on more project-specific details is developed at the conclusion of the conceptual design phase. This estimate should address costs for installation and operations, and is usually available upon expense of roughly 10% of the total engineering cost of the project. At this level of project development, the accuracy of the installed cost estimate is approximately ±25%, but is usually sufficiently detailed to allow analysis of alternatives. Thus, at the conclusion of conceptual design, the project cost is reassessed. If the ROI proves unacceptable, the project scope must be analyzed to determine whether downscoping can improve the return on investment. Alternative scopes are devised and estimated, and the most favorable is selected. At this point, the project scope must be fixed, and engineering may proceed to the next phase.

The importance of this procedure cannot be overemphasized. An example illustrates the potential financial consequences of changing scope at a later stage, say, at the conclusion of the design development phase. Let us assume a production facility with a total installed cost of $50 million. The cost for engineering such a facility ranges from 6% to 10%, or $3 to $5 million. Conceptual design of such a facility costs $300,000 to $500,000. If the project is properly executed, the next phase of design development requires an additional 20% to 30% of engineering, or $600,000 to $1,500,000. If the scope were not properly analyzed and fixed at the end of conceptual design, the engineering cost for design development could potentially be wasted, a potential loss on the order of $1 million! In addition, there would be significant delays in project schedule due to the need for re-engineering.

As the project moves through design development, an additional estimate of installed cost is performed, with a higher degree of detail and accuracy. This provides another checkpoint for verifying the potential return on investment.

Table 19.1 Phases of a Bioprocessing Facility Project

Program Phase
- Facility Program
- Site Analysis
- Design Criteria
- Utility Data
- Target Schedule

Conceptual Design Phase
- Process Flow Diagrams
- Equipment List
- Facility Layout
- Economic Analysis
- Fixing of Project Scope and Budget
- Project Schedule

Design Development Phase
- Process & Instrumentation Diagrams
- Control System Configuration
- Equipment Arrangement Drawings
- HVAC Air Flow Diagrams
- Electrical Load List and Single Line Diagrams
- Architectural Plans
- Equipment Specifications
- Instrument List
- Definitive Cost Estimate

Detailed Design Phase
- Construction Drawings (civil, structural, architectural, mechanical, instrumentation, electrical)
- Construction Specifications
- Construction Bid Documents

Construction Phase

Validation Phase
- Document Preparation
- Installation Qualification
- Operational Qualification

19.1.4 Schedule Performance

The planning process must establish a schedule for all project activities, and this schedule must be produced in summary form in the first (program) phase of the project. In most cases the schedule requirements dictate how the project is executed, and influence contracting strategies.

Time-to-market becomes a crucial factor in a competitive market, with manufacturers who first get their product into the hands of consumers having the distinct advantage in establishing market share. Due to the competitive situation, there has been a recent trend to execute projects on a *fast-track* basis. Figure 19.1 illustrates the differences between normal and fast-track approaches to project scheduling.

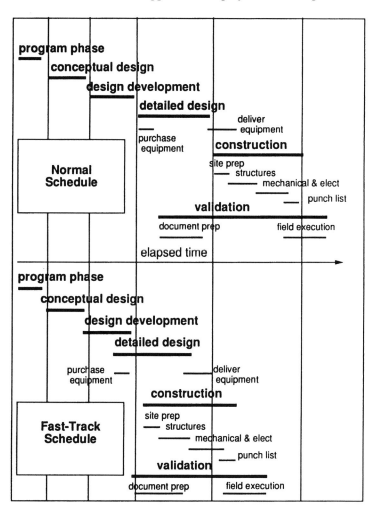

Figure 19.1
Normal versus
Fast-Track
Schedule

The normal schedule assumes sequential execution of the project phases, from the program phase through construction, with allowance for a review period after each phase to assess its results before proceeding to the next. In the normal approach, the detailed design is completed before construction begins. According to this schedule, a *General Contract* bid document is issued for construction of the entire facility, bids are received, and the contract is awarded, all before the start of construction.

The fast-track schedule begins in the same way, with design development proceeding sequentially after the program and conceptual phases are complete. However, in the fast-track approach, each of the subsequent phases is begun as soon as sufficient information is developed in the previous phase. Thus the phases of design development, detailed design, and construction overlap, saving considerable time in the schedule. Note that there is a considerable overlap in the execution of the detailed design and construction phases. The fast-track approach does not allow for the packaging of all construction work into a General Contract bid document. Rather, the bid documents are produced in a series of contract packages, for example, Site Preparation, Foundations, Structural Steel, Architectural, Mechanical, Electrical, etc. Consequently, a fast-track project is generally constructed using a *construction management approach*, with either the owner or a construction management firm directing the activities of the construction contractors on-site.

The advantage of the normal approach is that it is less expensive for the owner to revise the scope of the project as the design progresses. He may also get firm bids on the most expensive portion of the project, construction, before any construction work begins, thereby reducing financial risks.

The advantage of the fast-track approach is earlier project completion. The disadvantage is that it is more expensive for the owner to make scope revisions after the conceptual design phase. Also, it is more difficult to fix the total cost of construction, since it is bid in segments. It is possible, and desirable, to establish each construction package as a fixed-price contract. In addition, many design/construction contractors are willing to fix the cost of the remaining design and construction at the completion of the design development phase. If, at this point, the owner chooses to remain with his present contractor, he may thus proceed directly on either a fixed-price or cost-plus basis. If he wishes to obtain competitive bids, he has to accept the additional penalty of approximately 8 to 12 weeks for the bid/award cycle.

The project manager and his executive management must be aware of the above considerations in establishing a schedule and choosing a project approach. In all instances, it is essential that the project team members be cognizant of the importance of establishing and fixing the scope, and aware of the cost and schedule penalties involved in changing the scope of the project.

19.1.5 Special Considerations

Although the activities required in the design and construction of a bioprocessing facility are similar to those required for a chemical production plant, there are important differences to be taken into account in the planning process [2], including:

(a) The building and process systems must be integrated to satisfy *Current Good Manufacturing Practices (CGMPs)*. The design of the process is intertwined with the design and layout of the building. The flow of material and people is just as important as the mass and energy balance in the process. GMP requirements must be considered during all phases of engineering, design, and construction.

(b) Even in bioprocessing plants where GMP compliance is not mandatory (for example, fermentation processes for bulk chemicals), the requirement for the sanitary design and sterilizability of the fermentation system is crucial. The designer must possess comprehensive knowledge of sanitary design techniques for the process and piping systems.

(c) Where GMP compliance is required, the planning process should allow for periodic audits for GMP compliance during the engineering, design, and construction. Also, construction documents should be sufficiently detailed to allow review for GMP compliance and subsequent validation.

(d) The *Food and Drug Administration (FDA)* requires validation of pharmaceutical facilities in accordance with strict protocols, so validation requirements and activities must be anticipated in the planning process. Table 19.2 presents a list of some of the key technical considerations in the planning for a bioprocessing facility.

19.1.6 The Owner's Project Manager

When the owner's executive management has identified the prospective project and wishes to establish a project program, a project team is formed from personnel within the organization. The composition of this team varies through the course of the project life, and depends to a large extent on the nature and size of the project, the size of the owner's organization, and the plan for the execution of the project. The first member, however, that should be appointed is the owner's project manager. It is his or her mission to develop the project plan, form the balance of the task force, and proceed with the execution of the project. For projects of any significant size, this is a full-time assignment, and the selection of this individual is crucial. His or her qualifications should include experience with management of capital projects, intimate familiarity with the owner's organization, procedures, and operations, and a working knowledge of the process technology for the project.

19.1.7 Utilization of Consultants and Contractors

As the owner's project manager proceeds with his execution plan, he must examine the availability of resources within the company, and determine where outside

Table 19.2 Key Technical Considerations for Bioprocessing Facilities

Process Systems
- Cleanability
- Sterilizability
- Operability
- Maintainability

Utility Systems
- Clean Water
- Clean Steam
- High-Purity Gas
- Cleaning In Place (CIP)
- Steaming In Place (SIP)

Architectural System
- Compact Layout
- Containment
- GMP Compliance
- Efficient Material Flow Pattern
- Efficient Personnel Traffic Pattern
- Cleanable Interior Finishes

HVAC System
- Cleanliness
- Clean Rooms/Classifications
- Containment

Regulatory Compliance
- Quality Assurance
- FDA/GMPs
- NIH/Containment
- Building Codes

resources are required. Although the organization may have talented personnel, they may be burdened with other responsibilities and may be available only on a part-time basis. It is rare that a company is able to supply all the resources for the planning, engineering, and design of a major bioprocessing facility. Therefore, owners usually call on engineering and construction (E&C) companies to provide services. The services available from such contractors include process development (working with the owner's R&D personnel to provide scale-up design for production-scale facilities), programming and conceptual design services, definitive engineering and detailed design, construction or construction management, and validation.

Engineering contractors may be used in any or all of the above roles. When the owner has determined how they are to be utilized, he must also devise a plan to select them and contract for their services. The process of developing requests for proposals, interviewing, and selection may require anywhere from 4 to 12 weeks. An excellent in-depth presentation of this process is in references 3–5, and is recommended to any reader about to experience this process.

19.1.8   Project Team Formation

When the owner's plan for contracting for engineering services is established, he can proceed with the formation of the project team. Even if most of the engineering is contracted, extensive interaction between consultants and owner's personnel is required. Therefore, a team from the owner's organization must be formed, and procedures for interaction and review of the E&C contractor's work established.

An organization chart for an E&C contractor's team is presented in Figure 19.2. This represents a task force for performing the definitive engineering and detailed design, with the procurement and construction functions also reporting to the E&C contractor's project manager. Many variations of this scheme are possible, but this reflects the situation for a significant project.

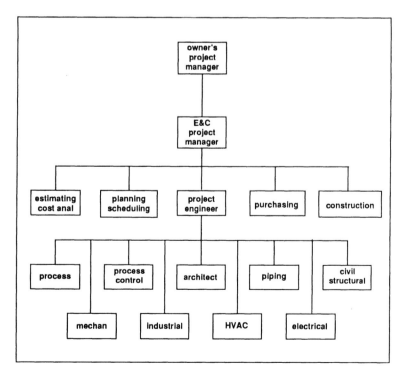

Figure 19.2 Typical Organiza- tion Chart for E&C Contractor's Team

The owner's team must be established to reflect the amount of work being performed in-house versus the amount being contracted. A minimal team from the owner's organization includes Project Manager, Process Engineer, Manufacturing Representative, QA/Validation Representative, Procurement Representative, Project Controls (that is, cost and scheduling), Instrumentation/Controls Engineer, Electrical Engineer, Architectural/Structural Engineer, and Mechanical Engineer. The owner's project manager may encounter a situation where even the minimal team is not available to act on the owner's behalf. For instance, the mechanical engineer(s) should have experience with clean HVAC systems, sterile piping systems, and the more traditional mechanical equipment such as vessels, pumps, and exchangers. A person with such qualifications might not exist within the owner's organization. In such cases the owner might consider hiring additional consultants to round out his project team.

The owner's team is charged with developing the preliminary scope and subsequent review of all work performed by the E&C contractor. A prime objective for the owner's team is to get all the pertinent design criteria defined at the earliest possible time; therefore, any group that can effect design or scope changes during the execution of the project should be represented on the owner's team.

The combination of the owner's team and the E&C contractor's team forms the *project team*. This concept may seem trivial, but is in fact crucial. It is the smooth interaction and communication among these individuals that determines the ultimate success or failure of the project. The owner's project manager must emphasize this point, and avoid the development of any adversarial relationships with his consultants and contractors.

19.2   Program Development

This first phase of the project may be executed solely by the owner's team, but is more commonly performed in cooperation with the E&C contractor. The owner collects and furnishes all the data for the program and provides the first definition of plant capacity (see Figure 19.1). Ideally, the program phase should provide at least a preliminary project definition by presenting a facility program, site analysis, design criteria, utility data, a very rough project budget, and a target date for project completion [4].

The facility program should contain the following information:

(a) *Operational Description*, which presents a list of the products to be produced, a brief description of the production process, a list of most of the major process equipment, expectations on performance of the process control system, and a discussion of the possibilities for future expansion of production capacity;

(b) *Organization Chart* of anticipated facility staff and work force, describing their functions;

(c) *Master List of Required Spaces*, including names and data sheets for each discrete space;

(d) *Functional Resolution Analysis*, which usually takes the form of an adjacency matrix, identifying the priorities of adjacency between related functions (for example, warehousing, ingredient preparation, bioprocessing);

(e) *Comparative Analysis of Spatial Sizes*, which compares relative net areas and volumes of spaces using modular increments;

(f) *Gross Area/Volume Predictions*, based on the professional's experience with past projects of a similar nature, taking the other elements of the program into consideration.

The facility program is best conducted by professional architects and process engineers from the E&C contractor team working in conjunction with the owner's personnel. It is important that facility requirements for all segments of the owner's organization be addressed in the master list of spaces.

Site analysis should be developed as a location and property survey, a topographic map of the site, and a preliminary subsurface exploration.

Design criteria for engineering and maintenance of the facility should be initially defined for each design discipline. This initial activity is best performed by the owner's team, as such criteria are definitely more costly and time-consuming for the contractor's team to develop. Much of this is dependent on owner preference, so that any criteria that a contractor develops independently are usually extensively revised by the owner.

The owner should also develop the initial utility data by tabulating the types of utilities required by the new facility, with a rough idea of peak demand for each. If utilities are available at or adjacent to the site, these should be identified, and the available capacity for each should also be defined (for example, currently 10,000 lb/h of 50 psig steam available from an existing boiler system).

The design professional can usually establish a very rough cost estimate, based on an estimate of gross area of the facility, using experience from past projects. This cost estimate, although the best that the owner may expect at this point in the project, has a high degree of uncertainty and is not sufficiently detailed to allow the economic analysis of alternatives.

The owner should identify a desired time frame for project execution and completion. This is more rigorously analyzed in the conceptual design phase, but it is important to understand the relative importance of early project completion versus avoidance of financial risks.

19.3 Conceptual Design

Conceptual design is the most important phase of the project. In the course of this phase, the process is scaled up and defined, utility systems are delineated, equipment requirements are detailed, the facility layout is developed, and the first rigorous economic analysis is performed. If appropriate, alternative schemes are devised and estimated, and the most favorable alternative is established as the project scope. The project budget is thereby established, and engineering proceeds to the next phase.

19.3.1 Process Development

The process is developed from either laboratory or pilot-plant process data furnished by the owner, and scale-up design is done by process engineers experienced in this discipline. The process is represented in conceptual form on *process flow diagrams* (*PFDs*), which show all major process equipment and material flows, with quantities.

A more detailed analysis of process utility requirements is then performed, taking into account requirements for heating, cooling, cleaning, and sterilization, and the cycle times required for each. Utility requirements are presented in a Utility Summary document showing average and peak demands. This analysis allows an accurate listing and sizing of utility equipment.

An equipment list is then compiled; each item is identified by name, the number required, size, materials of construction, and any other information pertinent to establishing an estimate of equipment cost.

19.3.2 Facility Layout

Once the process and basic utility equipment have been sized, a dimensioned facility layout is developed to scale. The layout must satisfy the technical considerations previously listed in Table 19.2, as well as provide for flexibility for future modification, expandability, ease of maintenance, constructibility, and economy of operation.

Several layout options are typically prepared and evaluated. The more favorable options should be marked (preferably color coded) to indicate personnel and material flow patterns, and the layout that best satisfies all the above considerations is selected.

The development of the layout is a critical procedure, since it directly affects the capital and operating costs, GMP compliance, operability, and maintainability of the facility. Perhaps the most complex facility layout is that required for a biopharmaceutical production, filling, and finishing facility. An organized approach to the layout of such a facility has been presented [6], and is recommended reading for those seeking additional guidelines.

19.3.3 Economic Analysis

With the completion of the process flow diagrams, equipment list, and facility layout, an economic analysis may be performed. This includes an estimate of capital, operating, and miscellaneous costs, and an analysis of these costs versus projected sales to estimate return on investment.

Capital investment usually includes the cost of the land and all improvements required to create the manufacturing facility. These costs are best estimated by the E&C contractor, who has experience in estimating and tracking costs of similar projects. The capital cost estimate developed at this stage usually has an accuracy of about ±25%, and is developed as follows:

(a) The cost of each piece of equipment is estimated. Quotes for more specialized or costly equipment, such as fermentors, sterilization systems, ultrafiltration systems, etc., may be solicited from vendors in order to assure accuracy, and to determine delivery schedules for those items requiring a long lead time. More routine items, such as pumps, vessels, and exchangers are usually estimated without quotes at this stage.

(b) Costs for piping, electrical, insulation, and painting are factored, that is, they are calculated as a percentage of the equipment cost. The factors used are a result of experience with previous facilities.

(c) The cost of buildings and room finishes are estimated from the projected area, with computed unit costs per square foot.

(d) If there are clean room applications and other specialized areas, the number, size, and classification of the air-handling units are identified to establish HVAC costs, with factors then applied for the ductwork and insulation.

Manufacturing costs include the recurring expense of producing the product, plus indirect costs and overhead. The owner's team usually prepares the estimate of manufacturing costs when plant capacity, process assumptions, and operating mode are established. Table 19.3 presents the various components of manufacturing cost.

Other costs might include such items as administration, marketing, research, and development. The handling of these costs is largely a matter of accounting practice. Sometimes the cost of validation is included in this category.

Cost analysis of the above cost elements is then performed, and a calculation of the return on investment is made. If favorable, the scope is accepted as is; if not, alternatives are considered.

19.3.4 Adjustments and Optimizations

The most obvious way to improve the return on investment is to reduce the capital and manufacturing costs. However, such steps also influence the plant's capacity and, therefore, potential sales receipts. Some of the more common alternatives considered are discussed in the following sections.

Plant location is a major determinant of the labor costs for both construction and manufacturing. Also affected are the availability and expense of utilities. Bioprocessing facilities often require large quantities of feedstocks for carbon and nitrogen sources, whereas the final products are usually relatively low in weight and volume.

Table 19.3 Components
of Manufacturing Cost

Direct Costs
• Raw Materials
• Utility Costs
• Supplies
• Production Labor
• Maintenance Labor
Indirect Costs
• Depreciation
• Taxes
• Rent
• Insurance
Overhead
• Administration
• Supervision
• Medical
• Regulatory Affairs
• Procurement

Therefore, locating closer to the source of raw materials might yield an economic advantage.

Phased expansion is a strategy that enables a company to establish some manufacturing capacity to allow initial market penetration and an earlier positive cash flow. Production and sales increase with plant expansion. There are many variations on this strategy. One commonly employed by biopharmaceutical companies involves the initial installation of only some of the manufacturing steps. Filling and finishing facilities, because of their rigorous requirements for cleanliness, are relatively the most expensive to construct per square foot. Therefore, they are usually the segment most often deferred, and the manufacturer must subcontract the finishing of his product. In recent years, many "toll manufacturers" have established themselves to satisfy these needs.

Process design modifications may yield cost savings, and usually involve compromises in operating efficiencies. Examples include the following:

(a) Bioreaction mode. Continuous or draw-and-fill techniques are generally more economical than batch processing, because they involve smaller and less expensive bioreactors. The drawback to continuous processing, however, is the higher risk of contamination.

(b) Use of portable equipment. This can reduce the amount of fixed piping required, but entails increased labor and less automation in manufacturing.

It can result in economies where sterile piping systems are required, and where operating labor is relatively less expensive.

(c)   "Campaigning" of equipment. Using a piece of equipment for more than one processing step or more than one product usually results in economies of capital cost.

(d)   Process automation. This results in lower costs for operating labor, but also necessitates a higher capital cost. The use of computer control also allows improvements in product quality and consistency, as well as automatic data acquisition for quality assurance purposes. The prospective benefits of automation must be weighed against the increased first cost.

The foregoing are only some of the many ways to improve a prospective project's scope and return on investment. Each project is unique and presents other opportunities. It is the job of the owner's project manager to identify the most favorable alternative through cost analysis, and finalize the project scope and budget accordingly.

19.3.5   Project Schedule

At the end of the conceptual phase, it is possible to define the project schedule more accurately. When the size and configuration of the facility are known, it is possible to determine optimal peak construction manpower and thus fix reasonable durations for construction activities. The engineering and design schedule may also be developed. The owner must establish an approach to the schedule, either normal, fast-track, or a variation thereof. The schedule should be prepared so that the project is divided into phases, as discussed here. The duration of each phase should be shown on the time line, and, if a review period is required between phases, that should be stated.

It is also important to devise a strategy for the procurement of equipment. This is important for two reasons. First, the construction schedule depends on the timely receipt of equipment at the plant site. Second, the detailed design schedule depends on the timely receipt of information contained on vendor drawings. Therefore, the schedule should identify the time period for purchase of equipment, receipt of vendor drawings, and delivery to the field.

Another critical schedule consideration is the process for obtaining environmental permits. Permit applications require plant descriptions and selected technical data, such as tabulations and analyses of waste streams and emissions. They therefore cannot be filed until the conceptual design has been completed, and sometimes not until the project is well into definitive engineering. The lead times on agency review and approval are considerable, and in some locales construction activities are not allowed to begin without permits. Requirements for all federal, state, and local permits must be ascertained early in conceptual design, and the project execution plan and schedule must include a strategy for obtaining permits.

19.4   Engineering, Design, and Construction

The definitive engineering and detailed design of the project may now be performed. These later phases, although comprising the bulk of the project expense, are not treated in complete detail here, since most of the planning activities have been performed in the preceding phases. The main functions of the project manager through the balance of the project are to execute the project according to plan, monitor the project's progress, and control project scope and costs.

19.4.1 Design Development

This phase is sometimes termed *Definitive Engineering*, which is probably a more descriptive name. In the course of this phase, which usually consumes approximately 20% to 30% of the engineering budget, additional engineering is performed in order to define all systems in detail (that is, by component). This in turn allows the project cost to be estimated at a higher degree of accuracy, a level that is sufficient for detailed analysis of costs versus budget.

Process and utility systems are fully defined by *piping (or process) and instrumentation diagrams (P&IDs)*, and the control system is defined by a detailed functional description. If a distributed control system is required, it should be fully developed and specified in this phase. A detailed instrument list is also prepared as the P&IDs are developed.

Heating, ventilating, and air conditioning (HVAC) systems are defined by air flow diagrams that show all air-handling units, the spaces that they serve, and the air flow quantities required throughout the system.

Electrical systems are defined by a motor list that shows the power requirements for all equipment; estimates of power for lighting requirements; identification of emergency power requirements; and single line diagrams defining the mode of receipt and distribution of power to each area in the plant, showing all substations, switches, transformers, and motor control centers.

Equipment requirements are refined by an update of the equipment list. During this stage all equipment is properly sized and dimensions and weights are established. Specifications for equipment are prepared and vendor quotations are obtained, with items that require a long lead time being purchased in accordance with schedule requirements.

Building system requirements are established by the parallel development of equipment arrangement drawings and architectural plan drawings. When these are sufficiently developed, preliminary structural designs may be prepared.

A definitive estimate of project costs is prepared at the conclusion of the design development phase. Based on more detailed information, this estimate has an accuracy of approximately ±15%. Takeoffs are made of piping, HVAC ductwork, steel, concrete, architectural finishes, and electrical cabling, conduit, and wiring. Vendor quotations are used to obtain accurate equipment pricing. Field labor man-hours are estimated as a function of material takeoffs. Detailed estimates of engineering and design costs and of construction staffing costs are also prepared. The definitive estimate provides a checkpoint for verifying the validity of the budget established at the conclusion of the conceptual design phase. It also serves as an essential element in breaking the project budget into cost accounts of sufficient detail to allow the analysis and control of expenditures.

19.4.2 Detailed Design

The detailed design phase normally requires 60% to 70% of the engineering budget. During this phase, the detailed construction drawings (see Table 19.1), specifications, and bid documents are prepared. When the sequence of activities for this phase is planned, it is essential to establish how the construction is to be performed (that is, as one general contract package, or as multiple construction contracts); the schedule for bidding and award of each package; and a definition of the *scope of supply* for each contractor, identifying which items will be supplied by the owner or contractor. Answers to the above influence the sequence of drawing development, as well as what the drawings contain. For example, some general contract packages specify that the contractor supply the air-handling units, but this limits the amount of dimensional detail shown on the drawings for these units. In such cases the construction contractor is obliged to finalize dimensional details for the mechanical systems. In fast-track projects, the owner usually procures the equipment, more design detail is shown on the detail design drawings, and the construction activities are subdivided into multiple contracts. This expedites the construction schedule and more clearly defines the responsibilities of each contractor.

With regard to the design of systems requiring sterility, such as Water For Injection (WFI) or pharmaceutical filling systems, a higher level of detail is recommended. Leaving the detailing of such systems to the construction contractor entails a higher risk of failure in the validation process. This could result in expensive rework and schedule delays.

19.4.3 Construction

Most of the project cost is expended during the construction phase. It is therefore essential that the construction activity be planned as thoroughly as possible. Table 19.4 presents a series of issues that must be addressed in preconstruction planning. Some of these issues, such as a general contract versus construction management

Table 19.4 Construction Considerations

Mode of Execution
- Self Perform/General Contract/Construction Management
- Direct Hire/Subcontract
- Construction Staffing
- Schedule
- Testing Plan
- Validation and Start-up Plan

Labor Relations
- Labor Availability
- Open/Closed/Merit Shop

Definition of Responsibilities
- General Contractor/Construction Manager
- Owner's Staff
- Engineer's Role

Site Considerations and Logistics
- Restrictions, Interface with Existing Operations
- Utilities
- Transport of Large Equipment
- Receiving
- Warehousing
- Trailer Space
- Lay Down and Fabrication Space
- Parking

Essential Programs
- Constructibility Review During Engineering
- Material Control
- Safety
- Inspection and Quality Assurance

approach, should be decided prior to detailed design. Others, such as construction staffing, the use of open-shop labor, utilities, or the need for a warehouse, might have an influence on project costs, and therefore should be addressed at the time of preparation of the definitive estimate.

A large project is usually constructed in a sequence that allows phased completion and start-up. For example, the steam system should be completed and started up prior to completion of the process areas. A detailed construction schedule is usually prepared toward the close of the detailed design phase, and this schedule should address the phased completion, validation, and start-up of systems and areas of the plant.

19.5 Validation

A biopharmaceutical production facility must comply with the current GMPs as required by the FDA. As previously mentioned, these requirements should be addressed during all project phases. The ultimate test of compliance is in the validation of the facility. Activities related to validation can amount to a significant expenditure of both time and money. Depending on the nature of the facility, validation expense may amount to as much as 10% of the total installed cost of the facility. These activities include:

(a) the preparation of documentation, such as checklists and procedures, to be used during the field validation;

(b) installation qualification of systems (IQ); and

(c) operational qualification of systems (OQ).

The project plan should integrate validation activities into the overall schedule (Figure 19.1). As system designs are finalized, documentation for validation use may be generated. As systems are completed in the field, installation qualifications may proceed. As the systems are commissioned, operational qualifications are performed. This integrated approach is much more efficient than sequential facility construction and validation.

It is also beneficial to notify the FDA in the early phases of the project, concerning plans for project completion and start-up. A review of the facility design with the FDA, perhaps after completion of design development, results in constructive commentary, which might facilitate the validation and acceptance process.

19.6 Summary

I have presented in this chapter an overview of the major steps required in planning and executing a capital project for a bioprocessing facility. Although similar to chemical projects in many respects, such facilities possess a number of unique requirements. A knowledge of the principles presented herein should help both the novice and the design professional to understand the planning process.

19.7 References

1. HAVILAND, D., Ed.; *The Architect's Handbook of Professional Practice*; 11th ed.; American Institute of Architects, Washington, DC, 1988; pp 6–9.

2. NELSON, K., BOLAND, J., personal communication; *Planning for the Design and Engineering of Biopharmaceutical Production Facilities*; Presented to the Project Management Institute Symposium, 1988.

3. SKIBO, A., Project Management and Contracting Issues, Part I, *BioPharm* 2 (4), 46–52 (1989).
4. SKIBO, A., Project Management and Contracting Issues, Part II, *BioPharm* 2 (5), 24–28 (1989).
5. SKIBO, A., Project Management and Contracting Issues, Part III, *BioPharm* 2 (7), 34–40 (1989).
6. BOLAND, J., LOOK, J.; *Considerations for the Layout of a New Biopharmaceutical Production Facility*; Presented to the PharmTech Conference; Aster Publishing Corp., Eugene, OR, 1988.

Containment Regulations Affecting the Design and Operation of Biopharmaceutical Facilities

Casimir A. Perkowski

Contents

20.1   Introduction

Biotechnology, as an industry, emerged with the production of fermented foods and beverages, and has evolved to include the production of diverse products such as antibiotics, vaccines, enzymes, specialty chemicals, and biopesticides. With the development of gene-splicing technology and the intense competition for capital by new biotechnology companies in the late 1970s and early 1980s, biotechnology and biosafety became more narrowly associated with those processes using recombinant DNA technology to produce a variety of products. Many have referred to these recombinant processes as the "new" biotechnology.

Antibiotics such as penicillin, neomycin, and bacitracin have been commercially produced in large fermentors (up to 50,000 gallons or 190 m^3) for over 40 years. These products were naturally produced by microorganisms recovered from environmental samples such as soil, compost, and water. Chemical and physical mutagens were used in the laboratory to develop highly productive strains. These were perceived as the result of natural processes, and, consequently, there was no explicit concern to contain these organisms except to protect a trade secret.

These antibiotic and similar facilities were designed to prevent wild contaminating microorganisms from entering the reactors. The primary focus of facility design was to protect the process and product. Facility designs and operating practices were not concerned with accidental release of organisms into the work space or environment, because the organisms and products were initially isolated from nature. As a result, workers occasionally developed allergic reactions to some of the products. When this occurred, they were reassigned to other areas within the company where they would not come in contact with the product.

The biosafety guidelines and requirements that have been developed for recombinants are based on a body of knowledge gained in the handling of medically significant organisms in research laboratories and in the production of products such as vaccines. The 1969 booklet *Classification of Etiologic Agents on the Basis of Hazard* [1] provided a classical reference for personnel working with infectious agents, and was the basis for classifying infectious organisms and different levels of laboratory activities based on potential risk.

In the booklet *Biosafety in Microbiological and Biomedical Laboratories* [2] and in the 1976 *NIH Recombinant DNA Research Guidelines* [3], various levels of biological safety are achieved by applying different kinds of controls either separately or in various combinations. These controls can be summarized as follows:

(1) **Biological Barriers:** This is particularly applicable with recombinant organisms where it is possible to select or develop enfeebled strains of microorganisms that cannot survive without specific and specialized media or environmental conditions. Recombinant mammalian, invertebrate, or plant cell lines are easily contained, because they cannot survive by themselves in

the environment. In this case, however, the potential for the presence of adventitious agents in the cell lines may warrant additional barriers.

(2)   **Operating Procedures and Practices:** The most basic and important element of biosafety and containment is strict adherence to standard principles of microbiological practice and techniques. Integral to this is the establishment of up-to-date written operating procedures that identify the actual and potential hazards. As part of initial and regularly scheduled training, employees must be required to read and understand these procedures and practices.

(3)   **Physical Containment Barriers:** These can be classified into two categories: (a) primary barriers, which include safety equipment such as biological safety cabinets, personal protection devices such as protective clothes and respirators, and closed operation systems; and (b) secondary barriers, which are integral to the design of the facility, and include special air handling systems, contained work areas, and kill tank systems, which are used to minimize or to prevent the escape of organisms into the environment.

Such barriers and practices form the basis for U.S. and various international biosafety guidelines and regulations that exist or are being developed.

Recombinant DNA (rDNA) technology permits selective removal of a specific gene (for example, human insulin) and insertion into a foreign microorganism, plant, or animal cell. The foreign host produces the alien protein in submerged culture. The uncertainties about the safety of this technology resulted in the development and publication in 1976 of the voluntary *NIH Guidelines for Recombinant DNA Research* [3]. These guidelines provided for the control of recombinant DNA organisms, both the deliberate release of these organisms into the environment and their containment during use in the manufacture of products.

The NIH Guidelines have subsequently been revised, and in some cases relaxed, due to the expanding knowledge and experience with recombinant DNA applications. The new revisions include guidelines for the large-scale cultivation of these organisms for the production of a variety of commercial products [4, 5].

Consequently, in addition to the traditional concern with protecting the process and the product from contamination, design considerations for new biotechnology facilities must focus on the containment of the recombinant organism within the process equipment.

In this chapter I provide an overview of U.S. government regulations on the use of recombinant organisms for the manufacture of products. Additionally, I include a brief summary of international regulatory trends in Japan and Western Europe.

Biotechnology itself is still evolving. Over the past five or more years, commercial biotechnology has advanced in product development, manufacturing technology, and the sophistication of equipment design and plant layout. The state of the art in biotechnology is continuing to change, and the regulations will continue to be affected by these changes for years to come.

The recent amendment to the NIH Guidelines to include the Good Large Scale Practice (GLSP) in 1991 [5], based on criteria originally proposed by the OECD, illustrates the trend toward relaxation of biosafety concerns based on the continuing growth of knowledge and experience in working with recombinant DNA processes. The Europe-based Organization for Economic Cooperation and Development (OECD) originally published a report in 1986 that described the criteria for more lenient containment practice [6]. Further relaxation of these guidelines may occur in years to come.

20.2   United States Guidelines and Regulations

The biotechnology industry in the U.S.A. is subject to a variety of regulations on a federal, state, and local level. There are five federal agencies that have a major role in regulating or providing guidelines for the biotechnology industry, especially in the area of biosafety and biological containment. These include the Food and Drug Administration (FDA), the Environmental Protection Agency (EPA), the United States Department of Agriculture (USDA), the National Institutes of Health (NIH), and the Occupational Safety and Health Administration (OSHA).

20.2.1 NIH Guidelines

The foundation of U.S. and many international biosafety and containment practices for laboratory and large-scale use of rDNA organisms is the NIH Guidelines for Recombinant DNA Research. These guidelines were voluntarily developed and adopted in response to uncertainties within the scientific community about the safety of the new rDNA technology. They address both containment measures and conditions for deliberate release of rDNA organisms into the environment.

These guidelines were based on principles and practices for medical research using pathogenic microorganisms. With increasing scientific knowledge and experience in handling recombinant organisms, the NIH Guidelines have been revised and relaxed in recent years to reflect the growing evidence that less stringent containment is adequate in many cases.

There are four levels of small-scale containment: Biosafety Level 1 (BL1), Biosafety Level 2 (BL2), Biosafety Level 3 (BL3), and Biosafety Level 4 (BL4). BL1 is the least stringent, while BL4 is the most intensive and reserved for organisms that pose a risk to life. Some organisms classified in BL2 and BL3 categories can, under certain circumstances, cause fatal diseases. The NIH Guidelines specify physical containment measures as well as operational practices for each biosafety level. These apply to the handling of recombinant organisms in volumes up to 10 liters.

Large-scale propagation of rDNA organisms (volumes greater than 10 liters) until recently included three biosafety containment levels: BL1-LS, BL2-LS and BL3-LS. The recent addition of the GLSP category to the NIH Guidelines [5] permits less stringent measures in the operations of certain processes.

To qualify for GLSP consideration, the organism must meet the following three criteria:

(a) The host organism must be nonpathogenic; contain no adventitious agents; and either have an established long-standing safety record in large-scale use, or possess intrinsic environmental limitations. These limitations should enable optimal process performance, but with limited survival upon release, without adverse environmental impact.

(b) The rDNA–engineered organism must meet the same criteria as for the host organism described above in (a).

(c) The vector/insert must: (1) be well characterized; (2) contain no known harmful sequences; (3) be limited wherever possible to the DNA required to perform the intended function; (4) not improve survivability beyond the intended function; (5) be poorly mobilizable; and (6) not transfer any resistance markers to microorganisms not known to acquire them naturally, especially if such acquisition might compromise the use of a drug to control human, animal or plant diseases.

Table 20.1 summarizes the revised physical containment requirements for large-scale fermentation using recombinant organisms, including GLSP. Many U.S. companies have avoided the use of organisms requiring BL3-LS containment, in part due to the perceived risks and the substantial incremental capital and operating costs associated with such facilities. Most companies have developed commercial processes using recombinant organisms that require BL1-LS containment measures. It is interesting to note that many have opted to design pilot and production facilities with most or all of the BL2-LS physical containment measures, to assure optimal flexibility for future processes.

20.2.2 Biotechnology Science Coordinating Committee

To minimize regulatory complexity and clarify jurisdictions, the President's Office of Science and Technology established the Biotechnology Science Coordinating Committee (BSCC). The BSCC includes senior policy representatives from the following agencies: the Department of Agriculture (USDA), the Department of Health and Human Services (FDA and NIH), the Environmental Protection Agency (EPA), and the National Science Foundation (NSF). It is alternately chaired by the Assistant Director for Biological, Behavioral and Social Sciences of the NSF and the Director of the NIH. The primary charter for the BSSC is to maintain interagency coordination in the following manner:

Table 20.1 Summary of NIH Requirements for Large-Scale Operations Using Recombinant DNA Organisms Including GLSP*

Containment Practice	GLSP	BL1-LS	BL2-LS	BL3-LS
Institutional codes implemented for personnel safety and enforcement of hygiene and safety	✓	✓	✓	✓
Adequate written procedures and training established for housekeeping and employee safety	✓	✓	✓	✓
Protective work clothing and appropriate clothes changing and hand washing facilities provided	✓	✓	✓	✓
No eating, drinking, smoking, mouth pipetting, or cosmetic application in workplace	✓	✓	✓	✓
Institutional accident reporting system	✓	✓	✓	✓
Control of aerosols during sampling, cell transfers, material addition or removal from closed system	MR/PC	MR/EC	PR	PR
All operations with viable organisms in closed vessel or other primary containment equipment	0	✓	✓	✓
Validated inactivation of organism before removal from closed process system	0	✓	✓	✓
Inactivation of process wastes and effluents from sinks and showers, appropriate to biohazard risks, in accordance with applicable governmental environmental regulations	0[1]	✓	✓	✓
Emergency procedures and system design for handling large accidental spills	0[2]	✓	✓	✓
Foam control systems to control aerosol releases	0	MR	PR	PR
Exhaust gases treated before leaving closed system	0	MR	PR	PR
Medical surveillance	0	0	✓	✓
Biosafety manual	0	0	✓	✓
Entry doors to work area posted with biohazard sign during all operations	0	0	✓	✓
Closed system to be permanently identified for record-keeping purposes	0	0	✓	✓

Table 20.1 Summary of NIH Requirements for Large-Scale Operations Using Recombinant DNA Organisms Including GLSP*, *continued*

Containment Practice	GLSP	BL1-LS	BL2-LS	BL3-LS
Monitors, sensors, and procedures used to assure integrity of containment in operations	0	0	✓	✓
System validated with host organism before use with rDNA organism	0	0	✓	✓
Agitator shaft and probe penetrations designed to control aerosol generation	0	0	PR	PR
Performance of mechanical systems and rotary seals	0	0[3]	PR	PR
Personnel decontamination facilities	0	0	0	✓
Airlocks on all entries	0	0	0	✓
Negative air pressure in work area	0	0	0	✓
Process systems operated at low pressures	0	0	0	✓
All surfaces in work area finished for ease of cleaning and decontamination	0[4]	0[4]	0[4]	✓
All piping, wiring and utilities protected against contamination with organism	0	0	0	✓
All operations in controlled work area	0	0	0[5]	✓
Air handling in work area designed to HEPA-filter air without recycling of air	0	0	0	0[6]

* Symbols
0 = not required
✓ = required
MR = minimize release
PR = prevent release
MR/PC = minimize release using procedural controls
MR/EC = minimize release using engineering controls

(1) In accordance with applicable government environmental regulations only.

(2) There must be a written emergency response plan which includes provisions for handling spills. In accord with prudent engineering practice, it is highly recommended to provide for containment of significant spills from process systems in the facility design.

(3) In practice, most companies specify double mechanical seals with hot condensate of live steam purge between the seals for BL1-LS fermentors. This, however, is driven by the need to prevent contamination of the batch and not to minimize release of aerosols that may contain the recombinant organism.

(4) GMP requirements for certain pharmaceutical production areas may make this a moot point.

(5) It has been industry practice to restrict access to BL2 areas.

(6) It has been industry practice to include a means of treating exhaust air from BL3-LS facilities to prevent release of viable organisms using validated means such as HEPA filtration or air incineration.

(a)   provide a coordinating role to identify and address technical and scientific problems, share information, and develop consensus among federal agencies;
(b)   facilitate consistency in the development of review procedures and assessments among federal agencies;
(c)   promote active cooperation among federal agencies on evolving technical and scientific issues;
(d)   develop generic recommendations on scientific issues applicable to recurring situations within the regulatory agencies; and
(e)   promote and coordinate scientific studies on issues of significance in biotechnology.

After its inception, the first major task of the BSCC was to assist the Domestic Policy Council Working Group on Biotechnology in the development of the 1986 Coordinated Framework for Regulation of Biotechnology [7]. Four basic points were established to provide this framework for regulation of biotechnology in the U.S.A.:

(a)   Wherever possible, existing laws for regulation of products developed with traditional genetic mutation techniques will be applied to the products produced by recombinant organisms.
(b)   All agencies will review and adopt consistent definitions applicable to the recombinant organisms.
(c)   All agencies will have scientists from each other's staff participate in scientific reviews to assure comparable rigor in those reviews.
(d)   The responsibility for the use of a product will lie with a single agency wherever possible.

These points are in concurrence with the broad goals proposed by the Ad Hoc Group of Government Experts convened by the European Organization for Economic Cooperation and Development (OECD). Their report was published in 1986 and was titled *Recombinant DNA Safety Considerations: Safety Considerations for Industrial, Agriculture, and Environmental Applications of Organisms Derived by Recombinant DNA Techniques* [6].

Also, the BSCC provided a common scientific basis for the definition of organisms that require regulatory oversight by the various federal agencies. This resulted in two categories of organisms requiring regulation: (1) genetically engineered organisms formed by deliberate insertion of genetic material from other species, and (2) microorganisms that are pathogenic or have been altered genetically through the insertion of genes from pathogens. A pathogen is defined as any microorganism or virus that can cause disease in any other living organism, be it animal, plant, or microorganism.

It is significant that the BSCC acknowledged that an organism in either category could be exempted from review by federal agencies. In this case the genetic information contained in this organism must not pose an increased risk to either human health or the environment.

The BSCC also began work on another significant issue that is crucial to the regulation by the EPA and USDA, of products produced by recombinant organisms that are to be used in agriculture or in other field applications, such as environmental bioremediation. A clear definition of what constitutes release into the environment is absolutely essential if we are to reap the full benefits of this technology. The BSCC established a working group charged with the development of technical recommendations for greenhouse containment and for small-scale field trials. Without such tests, widespread environmental releases of recombinant organisms could not be considered and approved.

While it can be argued that the BSCC should take a more proactive role, in accordance with its charter, the complexity of the scientific and political issues associated with recombinant organisms mandates a deliberate and thorough approach. This is a formidable task, especially with the number of federal agencies that are involved.

An illustration of federal agency involvement in the licensing of facilities manufacturing products using recombinant organisms is presented in Table 20.2. Each of these agencies has the authority to prepare and enforce regulations applicable to the industry and products within its jurisdiction. The NIH is not listed in this table because compliance with the NIH Guidelines is voluntary except for institutions that receive NIH support for rDNA research. Consequently, even though the NIH has no regulatory jurisdiction and enforcement over the biotechnology industry, the industry has universally accepted and applied these guidelines.

20.2.3 Food and Drug Administration

The Food, Drug and Cosmetic Act (21 USC 301-392) authorizes the FDA to regulate drugs, medical devices, foods, food and color additives, cosmetics, animal feed additives, and animal drugs. Human biologics such as blood products, vaccines, etc., are regulated under this act and the Public Health Service Act
(42 USC 262).

The FDA is primarily concerned with the potency, safety, purity, and efficacy of the products within their jurisdiction. The FDA requires premarket and marketing approval on a product-by-product basis. Therefore, when the FDA reviews a facility license application, they have a specific interest in the containment of the recombinant organisms, particularly from the standpoint of product purity and safety, as well as from an environmental standpoint, as required under the Coordinated Framework for Regulation of Biotechnology.

20.2.4 Environmental Protection Agency

The EPA has broad regulatory authority over water and air pollution, pesticides, herbicides, chemicals (including those produced by conventional biological processes), and hazardous substances.

Table 20.2 Federal Agency Involvement in the Licensing of a Facility That Uses Recombinant Organisms for Production

Product	Agency				
	FDA	USDA	EPA (TSCA)	EPA (misc)	OSHA
Human drug or biologic	✓			✓	✓
Processed food	✓	✓		✓	✓
Biopesticide			✓	✓	✓
Animal drug or biologic	✓	✓	✓	✓	✓
Plant		✓		✓	✓

The Federal Insecticide, Fungicide, and Rodenticide Act (FIFRA, 7 USC 136-136Y) empowers the EPA with the authority to review and register new pesticide products prior to manufacture. Biological pesticides produced by conventional or recombinant microorganisms are within this jurisdiction. Consequently, the EPA has specific requirements that govern small-scale and large-scale field testing, as well as the registration of all microbial pesticides before distribution and sale.

The Toxic Substances Control Act (TSCA, 5 USC 2601-2929) confers authority to the EPA for all chemical products, whether produced by chemical or microbial means, that are not regulated by the FDA or the USDA. TSCA was adopted to broadly empower the EPA to assess and control exposure to such substances through all phases of commercial development, including initial R&D, field tests, manufacture, distribution, and end use. Through this involvement, the EPA can require testing to assess risk before widespread damage can occur to the environment.

Intergeneric microorganisms (recombinant microorganisms) were defined by the General Counsel of the EPA to constitute a "new chemical" and subject to the provisions of TSCA. Consequently, TSCA applies to all recombinant microorganisms used for pollution control and remediation, bioleaching in the mining industry, improving nitrogen fixation in agriculture, chemical production, and pesticide production. Specifically excluded are drugs and biologics, foods and food additives, tobacco and associated products, and cosmetics. Jurisdictional disputes over the authority to regulate specific materials must be jointly resolved by the FDA, USDA, and EPA.

Under TSCA all manufacturers using microorganisms, conventional or recombinant, to manufacture a product for TSCA purposes must file a Pre-Manufacture Notification (PMN) with the EPA at least 90 days prior to initiation of manufacture. Included in the PMN is information on biosafety and containment of these organisms. This act gives broad discretionary powers to the EPA to obtain information from commercial users on items such as the type of organism used, product identity, safety and control measures, health and safety data, etc. Indirectly, TSCA affects overseas production as well as U.S. production because users of "chemical substances" must provide certification of compliance with TSCA.

On July 10, 1986, the EPA office of Pesticides and Toxic Substances released their *Points to Consider in the Preparation and Submittal of TSCA Pre-manufacture and Significant New Use Notifications for Microorganisms.* This 12-page informative document details the types of information required for filing of a Pre-Manufacture Notification (PMN).

TSCA also has a major impact on companies or individuals developing recombinant microorganisms for environmental use. The EPA must be notified prior to any testing involving environmental release of such microorganisms. Anyone proposing any environmental use of any recombinant microorganism not covered by other existing laws must notify the EPA prior to the use.

The EPA has not issued specific regulations for control of water pollution from manufacturing processes utilizing genetically engineered organisms. However, there are existing effluent guidelines for pharmaceutical manufacturers, hospitals, and pesticide producers (40 CFR 401-496).

There are no specific standards controlling air emissions from processes using genetically engineered microorganisms. However, manufacturing facilities must meet provisions of the Clean Air Act (42 USC 7401-7642).

In the U.S.A., manufacturers using recombinant microorganisms in their processes have diligently designed their plants to filter-sterilize or in some cases to incinerate fermentor exhaust gases in compliance with the NIH Guidelines for large-scale practice.

Consequently, it is unlikely that the EPA will have to enact separate regulations for facilities using recombinant processes unless an incident warranting such action occurs in the future.

20.2.5 U.S. Department of Agriculture

The USDA will use existing regulations for the biotechnology industry because it perceives that agricultural and forestry products developed from rDNA technology will be essentially the same as conventional products. The existing USDA regulations governing research, handling, shipping, and importation of plant and animal pathogens, and the NIH Guidelines for recombinants are adequate for regulating this industry at the present time. Like all regulatory agencies, it reserves the right to reassess policies and regulations in the future.

The USDA Animal and Plant Health Inspection Service (APHIS) is empowered under the Virus-Serum-Toxin Act (21 USC 151-158) to regulate all shipment and production of virus, sera, toxins, and related products developed to treat animals or to use in research and testing. None of these can be shipped interstate or imported into the U.S.A. without a license from the USDA. This licensing policy treats each biologics product, whether conventional or recombinant, on a case-by-case basis. All license applicants for products derived through the use of rDNA technology must comply with the USDA *Guidelines for Biotechnology Research*. USDA–APHIS licenses all biologics manufacturers. The USDA and FDA have established a mutual understanding for responsibility to regulate products that fall into the category between drugs and biologics.

20.2.6   Occupational Safety and Health Administration

OSHA is empowered by the Occupational Safety and Health Act (29 USC 651 *et seq.*) to establish compulsory standards to protect employee health in the workplace environment, including buildings, equipment, chemicals, toxic materials, and microbial agents. No extraordinary hazards to workers have been encountered in the biotechnology industry; consequently, OSHA has opted to use existing general standards for this industry. It is unlikely that this will change, unless specified hazards unique to biotechnology processes are identified in the future.

20.3   International Biotechnology Regulations

It is not my intent in this section to summarize the regulations on the worldwide biotechnology industry. The regulations vary from country to country and are evolving, especially those on the use of recombinant organisms for production.

In May 1986 the U.S. Department of Commerce, National Technical Information Service published a report prepared by J. D. Sakura of Arthur D. Little, Inc., titled *Review and Analysis of International Biotechnology Regulations* [8]. This report focused on biotechnology regulations in Japan, France, Germany, and the European Community. Sakura reached several conclusions from his analysis, including the following major trends:

(a)   There is considerable variation in the quantity and type of biotechnology regulations and guidelines, with Japan leading in the development and promulgation.

(b)   Government agencies were relying extensively on participation and technical input from academia and the affected industries. Consequently, he predicted a trend toward voluntary or consensus standards, especially in France and Germany.

(c)   The trend among many of the countries was to liberalize the rDNA research guidelines to provide incentive for the rapid commercialization of the biotechnology.

(d) Generally, all of the countries Sakura analyzed relied extensively on existing environmental and product approval regulations for regulating biotechnology products, except in the areas of deliberate release of recombinant plants and organisms.

(e) The major regulatory thrust within the countries studied was directed toward biotechnology products developed for the health care industry. Again, Japan was the apparent leader in this area, with particular regulatory focus toward developing guidelines and regulations and coordinating jurisdictions of government ministries for commercial-scale production of rDNA products.

In the 1990s we have seen refinements, clarifications, and amplification of policies and regulations, but Sakura's observations and conclusions remain substantively apt.

20.3.1 OECD Recombinant DNA Safety Considerations

The Organization for Economic Cooperation and Development (OECD) rDNA safety considerations are highlighted below. As Europe becomes a unified market force, these guidelines will in all probability become the standard for biosafety practices for the use of recombinant organisms in their biotechnology industry. Examination of these guidelines shows striking similarity to the NIH Guidelines.

In 1986 the OECD published a report, *Recombinant DNA Safety Considerations* [6]. It was prepared by an *ad hoc* group of experts that had been assembled in 1983 by the OECD Committee for Scientific and Technological Policy. This broad-ranging document reviewed the current and potential industrial applications of rDNA techniques; proposed technical considerations for assessing risks and selecting relevant safety measures for the use of recombinant microorganisms, plants, and animals; proposed containment principles for safe large-scale industrial use of rDNA organisms; and provided scientific considerations for the environmental and agricultural application of rDNA organisms and their products.

This report recognized that physical containment technology has been developed and successfully used for many years to contain pathogenic organisms. The overwhelming majority of industrial-scale applications of recombinant technology use organisms of "intrinsically low risk which warrant only minimal containment consistent with Good Industrial Large-Scale Practice (GILSP)". Criteria for classifying rDNA microorganisms suitable for use with GILSP were suggested and are summarized in Table 20.3.

In its safety considerations, this report recommends flexibility in the application of containment measures, as determined on a case-by-case basis. There is recognition that effective containment of rDNA organisms can be achieved through the judicious application of biological barriers ("enfeebled" strains) or physical and chemical barriers (facility design, equipment selection, and operating protocols and practices), or a combination of these.

Table 20.3 European (OECD) Criteria Suggested for rDNA GILSP Microorganisms

Host organism	• Nonpathogenic • No adventitious agents • Established record of safe industrial use • Intrinsic limitations against survival or harm to environment, but optimal growth under process conditions
Vector/insert	• Completely characterized; no harmful sequences • Minimal functional DNA segment; no enhancement of survivability unless crucial to function • Poorly mobilizable • Incapable of transferring drug resistance to microbes not known to acquire them naturally, if there is potential to compromise therapeutic use of drug
rDNA organism	• Nonpathogenic • Industrial characteristics same as parent, but "enfeebled" for survival in environment

In May 1988 the Commission of the European Communities presented a *Proposal for a Council Directive on the Contained Use of Genetically Modified Micro-Organisms* [9], which was passed by common position. This proposal included in virtual entirety the recommendations proposed in the 1986 report prepared by the OECD Committee for Scientific and Technological Policy.

These OECD containment recommendations have not prevented member states from reviewing and applying their existing regulations to the biotechnology industry. In some instances, member states have issued specific regulations or guidelines for use of recombinant technology. A summary of these actions is presented in the explanatory memorandum of the 1988 proposal. Despite discussions of a unified Europe in 1992, it appears that a unified European policy on the use of recombinant organisms will be a point of continuing discussion for years to come. However, with the 1991 adoption of the GLSP category into the NIH Guidelines for Large-Scale Recombinant Organisms, the United States and the OECD-recommended guidelines became quite similar.

20.4   Conclusions

Biosafety guidelines and requirements for recombinant organisms have evolved from knowledge and experience developed in the handling of pathogenic organisms and from the production of vaccines, toxoids, and similar products. The recombinant containment guidelines adopted by NIH in 1976 were quite conservative. An increasing number of products are produced using rDNA technology. This experience in large-scale applications, not only in the production of therapeutics but also in bioagricultural applications, has resulted in a gradual relaxation of containment guidelines within the U.S.A. and in other countries.

It is interesting to note the basic similarities of the recombinant containment guidelines between the U.S.A. and other countries. The major differences from country to country are based primarily on degree of enforcement, rather than on the basic practices and technologies applied to biosafety and containment of recombinant and other microorganisms.

With continuing scientific advances and the increasingly international focus on political and economic issues, it is reasonable to assume that in the foreseeable future we will reach an international accord on the issues associated with the regulation of the recombinant DNA technology and its products.

20.5   References

1. *Classification of Etiologic Agents on the Basis of Hazard*; 4th ed.; Centers for Disease Control, Office of Biosafety, U.S. Department of Health, Education and Welfare, Public Health Service, 1974.
2. *Biosafety in Microbiological and Biomedical Laboratories*; RICHARDSON, J. H., BARKLEY, W. E., Eds.; U.S. Department of Health and Human Services, U.S. Government Printing Office, Washington, DC, 1988.
3. NIH Recombinant DNA Research Guidelines, *Fed. Regist. 41*, 27902–27943 (July 7, 1976).
4. NIH Guidelines for Research Involving Recombinant DNA Molecules, Appendix K, Physical Containment for Large Scale Users of Organisms Containing Recombinant DNA Molecules, *Fed. Regist. 51*, 16959–16985 (May 7, 1986).
5. Revisions of Appendix K of the NIH Guidelines Regarding Establishment of Guidelines for Level of Containment Appropriate to Good Industrial Large Scale Practices (GILSP), *Fed. Regist. 56*, 33175–33183 (July 18, 1991).
6. *Recombinant DNA Safety Considerations*; OECD Publications, 2 rue André-Pascal, 75775 Paris Cedex 16, France, 1986; pp 1–70.
7. Coordinated Framework for Regulation of Biotechnology, *Fed. Regist. 51*, 23302–23393 (June 26, 1986).
8. SAKURA, J. D.; *Review and Analysis of International Biotechnology Regulations*, National Technical Information Service, U.S. Department of Commerce, 1986.
9. Proposal for a Council Directive on the Contained Use of Genetically Modified Micro-Organisms, *Off. J. Eur. Communities 199*, 9–27 (1988).

Validation of Biopharmaceutical Facilities

Robert Baird and Phil De Santis

Contents

21.1   Introduction

In the mid-1970s the United States Food and Drug Administration (FDA) began to require the validation of manufacturing processes in their Current Good Manufacturing Practices (CGMPs). The definition and applicability of the concept is still controversial. The following definition was offered by the FDA in May, 1987 in their *Guideline on General Principles of Process Validation* [1]: "Process Validation is a documented program which provides a high degree of assurance that a specific process will consistently produce a product meeting its predetermined specifications and quality attributes." This definition states what a validation program must accomplish, but it does not provide guidance on how to structure the program and format the documentation.

Since its introduction as a means to address the deficiencies of the USP sterility test, validation has been applied to numerous other areas of concern and has been used (and perhaps sometimes misused) as an umbrella term for a variety of formalized testing programs. The steadily expanding use of the term validation to include activities that were formerly referred to under different names has both advantages and disadvantages. The advantages are twofold. First, it systematizes the evaluation and documentation of the design and construction of facilities, the specification of equipment, and the development of products and processes. Second, it encourages the integration of the work done by everyone involved in the design, construction, and operation of biopharmaceutical facilities.

The disadvantage is that it has sometimes created an intimidating and somewhat mystical aura around activities that should be simple and straightforward. We hope this chapter will clarify the subject and provide the reader with a relatively simple framework by which one can produce an acceptable validation program.

21.2   Why Validate?

The obvious reason to validate is that it is required by the Center for Biologics Evaluation and Review (CBER) of the Food and Drug Administration (FDA) before an establishment or product license is granted. The cost of having a facility, worth perhaps tens of millions of dollars, sitting idle is enormous, and motivates us all not only to do validation, but also to do it right the first time. Seen in this light, validation does not become a roadblock to approval, but instead facilitates the FDA's review of the facilities and processes.

However, validation should not be viewed as merely a regulatory exercise or a necessary evil. There are a number of benefits to be gained from performing validation. In many companies, validation includes the responsibility for quality control during the engineering, procurement, and construction phases of a new facility. This provides assurance that good engineering and construction practices are utilized to produce a facility that meets the needs of its users and can be readily validated. These additional controls, and formalization of controls that previously existed, benefit everyone involved. In addition, the compilation of the engineering, construction, and vendor information into a validation file provides a data base that is of immeasurable use to those who have to run the facility in the future.

Operational qualification typically includes the *acceptance testing* of new equipment and systems. This formalized verification of conformance to specifications gives organizations a much higher level of confidence that vendors and contractors have provided what they were contracted to provide.

All the validation efforts provide the company with a much better understanding of the facility and equipment. However, it is during the process qualification that the most valuable benefits are reaped. The increased understanding and improved control that are obtained over the process provide the company with lower rejection rates, less waste, and a foundation for evaluating and implementing process optimization.

21.3   When Does Validation Occur?

21.3.1   Life-Cycle Approach

In the early days of validation (mid-1970s), the emphasis from a regulatory perspective was on sterilization processes. Some very good test methods, using multiple temperature sensors and microbiological challenges, were developed, which gave a high degree of assurance that the sterilization process was successful. As time went by and the necessity to validate nonsterile processes was recognized, it became evident that an approach that relied solely upon testing was not adequate.

In the early 1980s, the pharmaceutical industry became concerned about its ability to validate the newly emerging process control technologies being applied. After long consideration and discussion, led mainly by the Pharmaceutical Manufacturers Association (PMA), an approach toward computer systems validation, known as the *life-cycle concept*, was developed. This was derived from similar concepts in the software development and marketing fields. This concept is valuable for all validation, including sterilization processes. It recognizes the fact that quality cannot enter a process or system through testing; it must be designed and built in, as well as maintained. Figure 21.1 is a modification of PMA's life-cycle diagram, as presented in the concepts paper for computer system validation [2].

Life-cycle is an approach to validation that covers all aspects of the system or project, from start to finish. It can be roughly divided into three phases: (1) definition and design, (2) testing, and (3) operation and maintenance. The phases are not mutually exclusive. They depend upon each other and overlap. A computer programmer uses the life-cycle when he writes a program, debugs it, uses it, and continuously revises it.

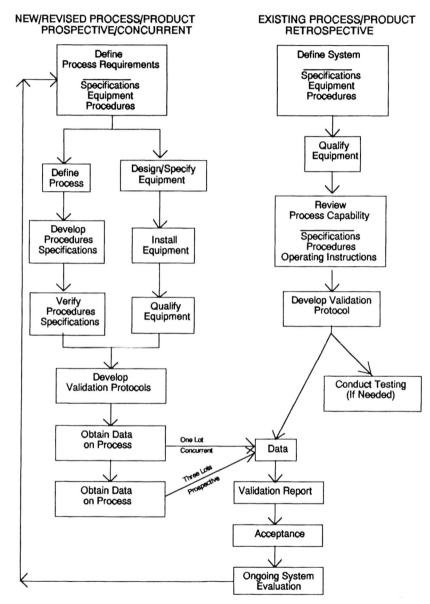

Figure 21.1 Validation Life-Cycle Approach. The life-cycle approach to process validation is one that has its roots in the software development field. The major emphasis is that validation is not merely a *testing* program. Validation is an important aspect of all phases of a project, from development and design, through installation and testing, to operation and maintenance.

The life-cycle approach is more than a consideration of the three phases of project life. It is a requirement to integrate these phases and to prevent their isolation from each other. It recognizes that each phase depends on the others. For example, design not only precedes testing and maintenance, it takes them into consideration and sometimes it even follows them. Notice that Figure 21.1 is a continuous cycle, not a one-time effort. By integrating the phases, we design systems that we can adequately test, our tests establish operating parameters we can maintain, and our ongoing evaluation establishes the need for updated design, retesting, or both.

Each phase of the life-cycle involves several activities leading to the successful control of the project. In some cases these activities may be carried out by a small group or an individual. More often a team is involved. In any case, there are three basic requirements for proper implementation of the life-cycle: planning, communication, and documentation. These requirements are imperative for the control and continuation of the project.

Looking at the three phases of the life-cycle in general terms, we may ask, "What do these phases entail and how do they interact?"

21.3.2   Design

The project or system is born as a problem that requires a solution. The design phase begins with a definition of the problem. Only then can the solution begin to form. In biotechnology, the production, isolation, and purification of a recombinant protein may be the simple problem. If this protein is to be used as a drug product, the problem becomes more complex. The drug must be produced according to CGMP, the process must be validated, and the plant must pass an FDA inspection. Clear definition of the problem or the scope of the project is an often overlooked critical point in validation.

Various solutions may be considered, until a basic or conceptual design emerges. The design team may decide upon a batch bioreactor, to be followed by a cell separation filter, and various concentration, diafiltration, and chromatography process steps. These are related to the simple solution. For the more complex solution, one needs to consider a process control methodology, plant layout, degree of documentation, utility and support systems, sampling points, and many other CGMP–related aspects.

The design phase continues as details are developed. These begin in the form of schematic representations of ideas, such as Process Flow Diagrams (PFDs) and Piping and Instrumentation Diagrams (P&IDs). They evolve into more complex forms, such as piping drawings, equipment specifications, and computer programs. Each of these becomes a part of the project documentation and should be structured so that it is of greatest value to the succeeding (testing, maintenance) phases of the project. For example, a piping designer prepares a water system isometric with welds identified, so that the validation team may better interpret weld inspection reports. Design documentation becomes the standard against which the installed system is measured during Installation and Operational Qualification.

The interface between engineering and validation is a critical one. Figure 21.2 provides a representation of how these two separate areas may work effectively together during the design phase of a project. Note that each step in the engineering process is paralleled in validation. Design documentation is reviewed, and validation and CGMP requirements are communicated to the engineers so that all information may be included in the design package. In turn, design documentation is used by the validation team to develop appropriate validation protocols.

21.3.3   Testing

The testing phase is required to determine that the system conforms to design and that it does what it is supposed to do. Testing is usually carried out according to a plan or protocol; results are compiled into a report. By itself, testing is not adequate to prove system acceptability. Coupled with good design, though, it provides the assurance required to put the system into use. In the past, the testing phase commanded the greatest interest in the field of validation. In fact, for a time the terms "testing" and "validation" were deemed synonymous, but today we recognize testing as only part of the equation.

The testing phase is generally divided into three major areas of concern:

(1) *Installation Qualification (IQ)*, in which the installed system is compared to design documentation;

(2) *Operational Qualification (OQ)*, in which the operational characteristics of the system are documented and, where appropriate, compared to design specifications; and

(3) *Process* or *Performance Qualification (PQ)*, in which the system is put to its intended use and challenged to perform as expected.

21.3.4   Maintenance and Operation

Validation does not end with testing. It seems ironic that the phase of the life-cycle that is the longest lived generally receives the least attention. Even if design and testing have resulted in a smoothly operating system, it is not permissible to consider the system validated. Continuing effort to evaluate and control the system is required.

Standard operating procedures, based on the validated operating conditions, are needed. Data should be reviewed periodically, with an eye out for trends, if applicable. A change control procedure evaluates the need for updating designs or additional testing. Changes should be documented and communicated. Hardware should be maintained and calibrated regularly. Routine retesting may be considered. The team involved in system design and testing should participate in maintaining validation.

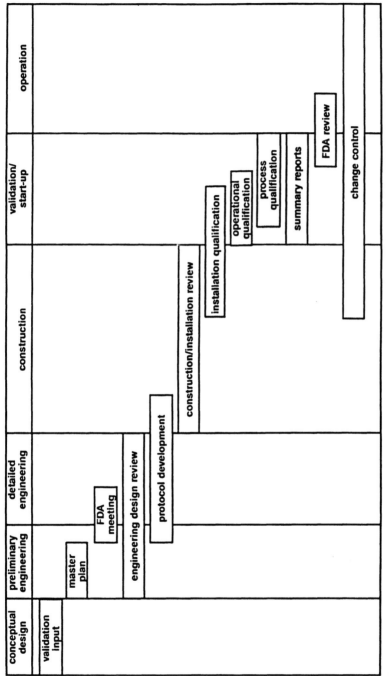

Figure 21.2 Validation Time Line. Validation input and activities are concurrent with the traditional engineering and construction activities. As design concepts evolve, validation documents based on them are prepared. A continuing dialogue between the design/construction team and the validation team must take place for the project to be successfully validated.

21.3.5   Validation Time Line

The schedule of events for validation following the life-cycle concept is closely linked to the overall project schedule. Figure 21.2 represents a typical validation schedule. Note that specific validation activities are occurring throughout the project design and construction. The validation team maintains a continuing interface with engineering and construction management. This involves review and comment on engineering drawings and specifications, field inspections, organization of field test reports, and setup of the validation files. Likewise, engineering and construction management continue their liaison with validation during the IQ, OQ, and PQ phases. The construction team may have the responsibility to carry out some of the preliminary testing required in the validation protocols, such as piping slope checks and pressure testing. They see to the correction of any deficiencies uncovered in the IQ process. All three groups (as well as contractors and vendors) may participate in the commissioning of systems. This phase is needed to get the plant up, running, and debugged prior to formal operational testing. Engineering may provide OQ assistance. In addition, engineering updates any drawings that are required in the validation protocols to be "as-built".

As the Process Qualification (PQ) phase begins, the operating and quality control groups responsible for production and continuing operation predominate. The product produced during this phase of testing should be suitable for clinical use or commercial distribution.

As the schedule shows, it is valuable for the validation team to establish liaison with the FDA early in the project. In this manner, comments regarding design, validation, and operating concepts may be received and acted upon before funds have been committed to detailed engineering, equipment purchases, and construction costs. Of course, this liaison must continue through the preapproval inspection.

21.4   Validation Structure

21.4.1   Master Plan

The *master plan* is a useful document that provides the framework for the validation project. It may be presented to the FDA to obtain their comments regarding the approach to be taken on the project, or it may be maintained as a strictly internal document. Once the approach has been defined in the master plan, the organization is able to identify and allocate the resources required to do the job.

One of the most important items that the master plan must define is who has responsibility for performing the different parts of the validation. Tasks must be clearly delineated: who prepares the protocols, performs the validation field work, performs laboratory analysis in support of the project, and, most importantly, who has responsibility for reviewing and approving the protocols and results obtained. All validation activities conducted and documented must be reviewed and approved by the appropriate levels of management. It is also the responsibility of these individuals to obtain additional reviews and approvals as required. These additional reviews may include Regulatory Compliance, Corporate Safety, Security, or Maintenance.

The master plan should provide a thorough description of the facility, processes, and equipment to be validated. This description is similar in depth of detail to that provided in the conceptual design phase of an engineering project. It may be extracted from the Establishment License Application and should focus on the regulatory issues and concerns that exist, and define how they are addressed in the design of the facility. This description provides a useful reference to project team members and those who become peripherally involved with the project.

Facility layout, personnel and material flow drawings, process flow charts, room classification and pressurization drawings, and simplified schematic drawings of critical utilities such as HVAC and water systems are useful. When P&IDs and other engineering drawings become available they should be incorporated into the master plan so that it serves as a project manual.

Identification of the utilities, equipment, process steps, and support systems to be validated is a critical aspect of the master plan. This may take the form of a protocol list, with the requirements for installation, operational, and process qualification identified.

It may also be appropriate to define the key acceptance criteria for each of the systems and subsystems to be validated. If these issues can be resolved at this time it streamlines the preparation of protocols. However, since the master plan is typically developed before the design is finalized, it is wise to identify the acceptance criteria on a preliminary basis.

A listing of the standard operating procedures (SOPs) and methods to be utilized in the performance of the validations is useful. This list may also identify the need to develop some procedures before the project gets under way. While the master plan is in preparation may also be a good time to achieve agreement regarding the format to be utilized for the documentation of the project.

The master plan should finally serve as a planning tool for the validation project. A tentative schedule for the validation should be drafted, key milestones identified, and a budget drawn up for the manpower resources required to implement the project. With appropriate controls in place at this early point in the project, any delays or cost overruns can be quickly identified and corrected before they harm the project.

21.4.2   Protocols

Specific protocols need to be developed for each of the systems to be validated. These should define the specific methodology to be utilized in the validation, and the criteria that the results of the validation study must satisfy for the system to be accepted as validated. These documents take a multitude of forms. In a small organization, where the people performing the testing are well qualified and familiar with the specifics of the validation study, there need not be an elaborately detailed procedure. However, if the level of experience of the validation group is not high, if a large number of groups or departments must be represented on the project team, or if a third party such as a validation contractor is performing the work, there is a real need for highly detailed and clearly written protocols. In the newly constructed plant, during startup is not the time to be learning or deciding how to perform the validation.

A part of protocol writing that is frequently underemphasized is the objective statement that begins each protocol. All too often the objective of the validation is stated as "to validate the system". If the only purpose in doing the validation is to get it done, then it is a useless exercise. Instead, the objective of the validation should be to demonstrate that the system, equipment, or process does what it is intended to do. What it is intended to do should be clearly and unambiguously stated. If the objective of doing the validation is poorly defined, the work lacks direction, and work performed without a clear direction has a high probability of going astray.

Another aspect of protocol writing that cannot be overemphasized is the KISS ("Keep it simple, Sam") principle. The protocols should be structured to minimize paperwork and streamline the project as much as possible. A common error made at many companies is to break the project into a large number of discrete unit operations and then prepare a protocol for each of these. An administrative nightmare is created when the decision is made to generate individual protocols for each aspect of the validation of each system (for example, 15 protocols to address the IQ, OQ, and PQ of the nitrogen, carbon dioxide, helium, oxygen, and compressed air systems). It is far simpler to define the work in one or two protocols and eliminate the paper chase that results when multiple approvals and signatures need to be obtained.

Finally, everybody involved must recognize that the protocol is a plan, and that all plans need to be revised as conditions change and additional knowledge and experience are gained. Therefore, it is important to develop a protocol addendum system that allows for changes to be made to the protocol and for approval of these changes by all parties.

21.4.3 Calibration

Accurate and thorough calibration of all critical instruments is a necessary first step in the validation process. Critical instruments are defined as those whose operation (or failure) affects the quality of the product being produced in the facility. It must be remembered that the requirements of the process should define the accuracy requirements of the instrument, not vice versa. For a cold room that always operates at 4–8 °C there is little need to verify that an instrument that has a span from −50 to

+100 °C is accurate to one one-hundredth of a degree at the upper limit of the range. It is critical, however, that a full-loop calibration be performed within the anticipated operational range. Merely inducing a voltage or resistance to the instrument instead of comparing the actual sensor to traceable standard is not adequate. All reference instruments should be traceable to a recognized standard such as National Institute of Standards and Testing (NIST) and, at a minimum, be of accuracy equivalent to, and preferably an order of magnitude greater than, that required by the process.

21.4.4   Installation Qualification

This phase of the validation determines that the system, as installed, meets the requirements of the design drawings, the specifications, and the vendors' recommendations. A system description in greater detail than the one in the master plan is included. Manufacturer, model numbers, capacities, materials of construction, and other pertinent information about each unit of equipment within the system, or about each process unit covered by the protocol, should be listed.

IQ checkout sheets are used to document that the installed equipment is, in fact, as described in the design documentation. Also included as part of the IQ protocol are installation checkout documents specific to installation requirements, such as pressure test logs, motor checkout sheets, and instrument loop checkout sheets. These are provided by the contractor responsible for the installation and document that the equipment has been installed properly and the installation verified. Each type of system is subject to a different set of checkout sheets. In addition, included in IQ are requirements and documentation for cleaning and chemical preparation of systems prior to use.

Installation qualification of the equipment and system components should be completed prior to the performance of the operational qualification, and certainly before the beginning of process qualification. However, in some instances some documentation may lag behind the completion of the installation qualification. Revised "as-built" drawings are a good example. There must be a mechanism for provisional approval of the package, to allow the work to progress while the overdue documentation is obtained.

21.4.5   Operational Qualification

OQ testing is intended to determine that the unit or system operates as specified and meets design requirements for control of operating parameters. This is carried out before actual product is introduced to the system. In the case of process steps, it may utilize a suitable placebo, or water. OQ testing may not necessarily follow the process. Instead, it is designed to establish a performance baseline, to provide assurance that the system can operate as intended, and for future troubleshooting. In the case of utilities, testing should be carried out to determine that the system can adequately service all users at normal peak load. Operational qualification of the equipment and system components should be completed prior to the process qualification.

21.4.6   Process Qualification

Process qualification is performed on critical utilities and key processing steps, to ensure the consistency and effectiveness of the operation. A minimum of three batches of product are evaluated to gain that assurance. Production parameters during these runs are varied, where feasible, to span the proposed acceptable operating range. A *challenge* may be performed, and the system or equipment operated at or below the proposed operating limit of a critical process parameter, to obtain additional assurance regarding the validity of established parameters.

21.4.7   Change Control

At an early point in the validation project, certainly before field work commences, a system must be established to control changes to the facility, processes, equipment, and the validation program itself. It is of crucial importance that this procedure operate smoothly and facilitate the completion of the project, rather than hinder it. A simple procedure, with a form that proposes the change to be made, reason for the change, the impact on the operation, and the validation, qualification, or documentation required to support the change, is all that is required. Once completed, the form is circulated for approvals at the appropriate level (usually by the persons who have the responsibility for approval of the validation protocol covering the area, or their alternates), and the change is then made and documented.

21.4.8   Summary Reports

One of the primary objectives of the validation effort is to provide a package for FDA (or other regulatory officials) to review. We stress that the packaging of the documentation should be such that the information is readily accessible and suitable for regulatory review. The documentation should facilitate the FDA's inspection. The goodwill obtained by making the inspector's job easier is of incalculable worth. Some companies have mistakenly used an approach wherein they inundate the inspector with documentation, to either impress him with the level of effort that was made, or intimidate him and discourage further investigation. This is counterproductive and invariably leads to confrontational investigations, even when the company is in compliance and has nothing to hide.

A layered approach to documentation is advisable. Concise summary reports should be prepared. These should either be brief enough to be reviewed in a few moments, or have an abstract at the beginning that provides the inspector a quick overview of your validation effort if that is all he desires. The inspector in many cases is interested only in obtaining an understanding of what was done to validate the system or process, and gaining confidence that the validation was conducted in a professional and technically correct manner. Forcing the inspector to wade through mountains of paper in order to attain that goal wastes everyone's time, annoys the inspector, and gives him opportunities to identify errors or discrepancies in the documentation.

Regardless of the documentation approach taken, all data must be fully traceable and accurately presented. If, for example, a tabulation is presented of minimum, maximum, and average temperatures during a bioreactor sterilization study, there must be no doubt that these values were accurately calculated and were obtained using an instrument that was calibrated against suitable standards, and that all documentation generated while the data were collected is complete and consistent with CGMPs with regard to signatures, dates, crossings out, etc.

21.4.9   Validation Project File

The summary reports presented to the inspector should not be accompanied by all the supporting documentation. The supporting documentation for any one system may rival the *Encyclopædia Britannica* in volume, and typically includes mountains of weld documentation, mill certificates, instrument calibrations, computer software listings, data logger printouts, laboratory test reports with supporting calculations, instrument printouts, and much more.

The volume of documentation generated during a validation requires that a document management system be in place from the very start of the project. A file should be set up based on the protocol list and with provisions for all types of documentation to be handled. A simple cross-referencing system is needed for documentation that is applicable to a number of systems.

A properly prepared validation filing system, in conjunction with a change control system that keeps it up to date, soon becomes essentially an owner's manual for the facility. This is a very valuable resource, and one that must be secured against the casual removal of documents. Critical documents (Protocols, Summary Reports, etc.) should be duplicated and maintained in a separate, secure, fireproof location. All the documentation should be maintained in a limited-access area. Anyone with access to the documentation should be made aware of the importance of maintaining the integrity of the files.

As computerized documentation systems become more and more used in the industry, thought should be given to utilizing this type of resource to make the validation information accessible to more users. Of course, the same need to protect the integrity of the data applies.

21.5   Resources for Validation

21.5.1   People

The successful implementation of any idea or concept requires that people work together, with a common goal. To put in place such a multifaceted approach as the life-cycle concept of validation requires the input and participation of a broad range of people and groups. All must be aware of and support the ultimate purposes of the project. These, too, are broad in scope and include such diverse objectives as ease of use, compliance to CGMP, and profitability. In the biopharmaceutical industry, so many different groups become necessary that we are in danger of being saddled with a committee approach to validation ("A camel is a horse that was designed by a committee." Anon.). Communication becomes difficult, documentation is haphazard, and planning is not followed. A poorly organized project results from too many, as well as from too few, members. We may avoid this by establishing a validation team. A core group of key members (numbers depending on project size and value) devotes day-to-day attention to the project. Other interested parties receive information, through memos or meetings, on an as-needed basis. Each team member may take responsibility to facilitate communications within his own organization.

Certain roles must be filled for a life-cycle approach to be implemented. The end user should participate in the project through the planning, execution, and review stages. The engineering function may be represented, to ensure that designs meet the needs of the user and that CGMP and validation requirements are incorporated. Quality Assurance must always be involved, to review and approve designs and validation tasks; this is a requirement of the CGMP. Lastly, someone must be appointed to see to it that proper communication and documentation take place throughout the life-cycle and that those tasks particular to validation are carried out. These include preparing the master plan and protocols, supervising and coordinating test execution, maintaining validation files, and preparing the validation report. This person (or persons) may be called a Validation Engineer or Validation Specialist. In some cases the role may be filled by a QA specialist or even an operating supervisor. The history of validation projects has shown, though, that the most successful ones have been led by someone whose sole responsibility is validation. Projects where the person responsible for validation retains responsibility for production, a lab, or some other area tend not to go as smoothly. It is possible to exist without a so-called partment, provided that validation projects are led by dedicated individuals. As firms grow, however, it is more common to define the responsibility for validation as a full-time function. Even after projects have been successfully started up and validated, the validation maintenance function still demands attention. How many people serve that function depends, of course, on the size of the firm or plant.

Validation has, in recent years, become a specialty or discipline. Interestingly, the people who have found a place in validation come from diverse backgrounds of education and experience. Microbiologists and others with a biology education become involved as a result of their interest in sterilization. Chemists and biochemists are involved based on their expertise in process concepts. Engineers add an interest in equipment, facilities, and control systems. Pharmacologists provide a background in formulation and dosage forms. A good validation group may draw upon all of these. Such people may learn from each other to ultimately handle any validation project.

Different firms manage validation in different ways. Some maintain large, full-service validation groups. These are equipped to prepare protocols, carry out testing, and write reports. Other firms have smaller groups, depending upon validation consultants or contractors to staff larger projects. Sometimes, for very large projects, the project management group may opt to contract for validation services as they might for engineering. This approach is valuable to smaller, start-up companies as well as to established firms in need of outside expertise.

21.5.2   Equipment

A validation group that intends to perform field execution should have certain basic equipment available. An abbreviated list follows:

(a)   a multichannel data logger capable of thermocouple input, used for logging temperatures during sterilization and depyrogenation trials; this item is even more valuable if it can perform calculations, such as for lethality;

(b)   a precision pressure gauge or transducer for pressure measurement;

(c)   an airborne-particle counter for qualifying clean rooms;

(d)   a slit-to-agar or centrifugal sampler for measuring airborne viable counts, for qualifying clean rooms;

(e)   precision RTD and RTD display for calibrating temperature-sensing equipment;

(f)   circulating baths with either high or low temperature, for calibrating temperature sensors;

(g)   ice point reference;

(h)   electronic amperage and voltage calibrator, for calibrating flow rate, level, and other instrumentation;

(i)   precision decade box for calibrating RTDs;

(j)   analysis (Draeger) tubes for gas, for measuring moisture, hydrocarbons, etc.;

(k)   hood-type velometer for measuring volumetric air flow (used in qualifying clean rooms—optional);

(l)   precision anemometer for measuring air velocity at a single point (for qualifying laminar flow devices);

(m)   precision manometer or Dwyer gauge for measuring pressure differentials in clean rooms;

(n)   humidity sensor/recorder for qualifying controlled environments; and

(o)   device or setup for filter integrity testing.

21.5.3 Other Resources

In addition to validation equipment, the validation department or group must have certain other services available. They may establish these within the department, obtain them from the plant, or contract for outside services.

All validation equipment must be calibrated regularly against standards traceable to the National Institute of Standards and Testing (NIST) or to an equivalent government standards bureau. Field transfer standards, needed by every validation group or calibration department, must be calibrated against laboratory standards. The laboratory standards may be purchased, but are expensive because of their high accuracy. Laboratory standards (or field transfer standards, if no lab standards are available) must be calibrated against even more accurate primary standards, directly traceable to the NIST. Primary standards are available at select contract calibration laboratories throughout the U.S.A.

Many validation activities require microbiological laboratory services. Analyses required include bioburden determinations, sterility testing, bioindicator preparation and evaluation, microbial counts in water samples, and pyrogen determinations (LAL testing). A qualified microbiology lab must be available to support validation testing. The closer this is to the operating site, the greater is the likelihood that microbial counts are accurate. If samples are to be shipped to a remote location, they often must be packed in dry ice.

Chemical analysis is also necessary to support validation; water testing, analysis of compressed gases, intermediate testing, and product testing are required.

In a biotechnology facility, highly specialized testing, such as cell morphology, protein analysis, and DNA characterization, may also be required. This is most often provided by the plant support laboratories.

21.6   Validation of Systems and Processes

21.6.1   Water and Steam Systems

These systems are variously defined by the usage and process requirements of the operations to be performed. The quality requirements range from city water and plant steam to Water For Injection and clean steam. Typically the systems must be demonstrated to provide water or steam of the required quality in terms of chemical purity, microbiological purity, and bacterial endotoxin levels. The specifications for the various quality levels are defined by the EPA, USP (United States Pharmacopeia), and in various FDA guidelines [3–6].

The quality of steam and water should be based upon the requirements of the processes to be performed. Although many older bacterial fermentation facilities for the production of antibiotics may use city water and plant steam (provided no additives inimical to the fermentation are present), a mammalian cell culture typically requires Water For Injection (USP) and clean steam (systems in which the condensed steam meets WFI specifications). The cost of constructing, operating, and monitoring these systems mounts rapidly as quality level rises. It is therefore important that the appropriate level of water quality be selected. However, there are good reasons to be conservative in specifying water systems. First, the FDA is very sensitive to matters of water quality. Second, it is extremely difficult and expensive to retrofit a system that does not meet requirements. Lastly, there are a number of companies who have attempted to utilize lower cost alternatives to the traditional distilled WFI system with a hot recirculating loop, and have found themselves unable to maintain the required level of quality.

High-purity water and steam systems generally require the most intensive IQ of any system in a biopharmaceutical plant. Typically, materials of construction, welding procedures, and construction practices are monitored closely as the systems are installed. It is critical that specifications for slopes, absence of dead legs, and trapping of steam systems be adhered to. Cleaning and passivation procedures must be rigorously followed and documented.

The OQ must demonstrate that the system meets design requirements for temperatures, flow rates, and pressures. The most critical and usually most problematic area on these sensing systems is the actual use points. Especially when hard-piped, the interfaces between process equipment and these utilities are often given insufficient attention during design and construction. A small holdup of water in piping or valves can result in large problems with microbial counts and endotoxin levels. Care must be taken that any segment of the system that is not kept recirculating or hot, be fully drained and dried.

In addition to the extensive IQ and OQ requirements, these systems usually have long Process or Performance Qualification (PQ) periods of sampling, where confidence is obtained that the equipment and the procedures used to operate the equipment can consistently provide water or steam of the required quality. Typically, sampling is performed for some set period (nominally six weeks), and then the results obtained are evaluated and the sampling plan redefined and sampling frequency reduced. This reduced sampling is more intensively focused on any areas that provided higher counts and on those sites that are more critical. Routine sampling may be followed after a period of time (nominally six months) by a second evaluation at which the sampling program is reevaluated and redefined.

21.6.2   Classified Environments

The importance of air quality in the processing environment steadily increases as the process progresses through fermentation, isolation, purification, and finally to the filling of the final sterile injectable dosage form. The prevention of contamination of product (primarily with microbial contaminants) depends first on the personnel working in the area, second on the design and construction of the process equipment, and third on the environmental quality of the facility. Traditionally, this environmental quality is provided by sophisticated HVAC (heating, ventilating, and air conditioning) systems, equipped with terminal HEPA (high-efficiency particulate air) filters, low-level returns, and local laminar-flow work stations. More recently, the industry has seen a trend toward achieving the same goals with nontraditional barrier systems and robotics that physically isolate the personnel from the product.

However, traditional clean-room technology is still the most popular method for protecting product from incidental contamination. Areas are typically classified according to Federal Standard 209D [7] as class 100,000 or class 100 (the number of particles per cubic foot of 0.5 μm in size or greater). For fermentation, purification, and the preparation of solutions and packaging components for injectable dosage form filling, class 100,000 is usual. Class 100 is used for operations such as purified final bulk product filling and injectable dosage form filling. Some companies utilize intermediate classifications of 10,000 or 1,000 for areas where the criticality of the operation does not readily fall into 100 or 100,000 classification, and in transition areas.

Guidelines for other air quality parameters in clean rooms are defined in the *Proposed CGMPs for Large Volume Parenterals* [5], as well as in other sources such as the FDA's 1987 *Guidelines on Sterile Drug Products Produced by Aseptic Processing* [8] and the old *Federal Standard 209B*. These include:

(a)   20 air changes per hour in class 100,000 areas;
(b)   air flow rate of 90 fpm (feet per minute) (75–105 fpm) in class 100 areas;
(c)   temperature limits for clean areas of 72 °F ± 5 °F;
(d)   relative humidity limits for clean areas of 30–50% RH; and
(e)   differential pressure between areas of 0.05" of water.

These guidelines are useful nominal parameters for design, even though the specific design and purpose of a facility may dictate that they not be used. The requirement for a differential pressure of 0.05" of water between areas is often not feasible if facilities are complex. This is because if multiple levels of pressurization are required, the sum of the differentials is often more than can be handled by conventional construction and room arrangement, and problems occur even with the opening and closing of doors. It is necessary during the validation to show that the intent of the guidelines has been complied with, especially where values other than those that have been proposed are used.

In biopharmaceutical facilities there are other considerations than that of product contamination. If spore-forming organisms are utilized, the *CGMPs for Biologics* [9] require dedicated facilities. The use of recombinant or pathogenic organisms requires provisions for their containment. These provisions are defined in the NIH *Guidelines for Research Involving Recombinant DNA Molecules* [10]. An interesting dilemma arises when the need for containment (requiring negative pressurization) must be balanced against the need to protect the product from contamination (requiring positive pressurization). Management must make a judgment call on where they feel the overriding concern changes from containment to product protection. This must then be applied consistently throughout the facility. Special designs using intermediate high-pressure anterooms between the controlled and ambient areas may be employed.

IQ of classified environments for a biopharmaceutical facility includes the documenting of the quality checks made during construction of ductwork and installation of HVAC units and HEPA filters. Controls should be in place during construction, to ensure that the ductwork is kept clean of dust and debris that may cause filter clogging and contamination problems during startup. Leak testing of ductwork, checking of HVAC units, DOP (dioctyl phthalate) testing of HEPA filters, and testing and balancing of the system should all be documented under the classified environments protocol. Because these activities are usually performed by contractors who have little familiarity with (or inclination toward) the documentation requirements of a highly regulated industry such as ours, it is critical that an extra effort be made to communicate the fact that these documents will become part of a regulatory package that will be given close scrutiny. If no communication takes place until the contractors have provided their reports, disastrous results are almost guaranteed.

If building automation systems control the HVAC system, the same validation as that required for any automatic system must be performed. Loop testing of dampers and sensors in the ductwork may be more easily accomplished before ceilings are in place, and often identifies interesting errors that can result in headaches during startup if not corrected.

The operational qualification of the classified environments must verify that the specified parameters of air quality can be maintained. Typically, the OQ evaluates room pressurization, air velocity, temperature, and relative humidity. The air flow patterns in class 100 laminar flow zones are evaluated using smoke sticks or CO_2 vapor to verify that laminar flow is maintained and that no movement of air from less clean areas into areas where product is exposed occurs. This testing may be conducted under both static (no activity) and dynamic (simulated normal production) conditions.

The process qualification of the classified area is conducted after cleaning and sanitizing of the area. The area must be under controlled conditions, with gowning procedures in place and limited access. Nonviable particulates are monitored with a suitable particle counter following the Institute for Environmental Sciences (IES) Recommended Practice, *Testing Clean Rooms* [11]. Viable counts are performed using an active air sampler such as a slit-to-agar (STA) or Reuters centrifugal sampler (RCS), and surfaces are monitored to validate the sanitizing procedures. The monitoring is conducted at an intensive level for some defined period and the results are evaluated. A reduced sampling plan is then implemented, with emphasis given to problematic and critical locations. An additional evaluation may be made once some seasonal data have been obtained, and the routine monitoring to be used in the facility during operations has been defined.

The training of personnel (especially in aseptic filling operations) is a critical aspect of maintaining the classified environment in a controlled condition, and may be addressed as part of the validation. Personnel monitoring and the validation of gowning procedures should also be incorporated into the protocol in aseptic filling facilities.

21.6.3   Compressed Gases

Compressed gases such as air, nitrogen, oxygen, carbon dioxide, and helium may be present in a biopharmaceutical facility and used in contact with product in a variety of applications. The IQ of such systems must verify that lines have been pressure tested and are correctly labeled, filters installed and integrity tested as required, and the correct storage tanks installed. The OQ should verify that pressure and flow requirements are met, and that interlocks, alarms, and automatic switching devices operate as required. The gas delivered must meet specifications for dew point, hydrocarbons, and chemical purity, where required. A PQ is performed where the gas is used in critical applications, and microbial and particulate level determinations may need to be performed. An STA can be used to determine microbial counts.

21.6.4   Environmental Chambers

Validation of cold or warm rooms that are used for purification or fermentation is required. Refrigerators, freezers, and incubators also require validation. The Installation Qualification of these units must verify that they are installed as required and that all calibrations have been completed. The Operational Qualification consists of a temperature mapping study, during which maintenance of the required temperature conditions is verified. This study should be run long enough to include defrost cycles. The operation of interlocks and alarms should be verified. In some applications it may be desirable to determine the time required for the room to recover after a door has been opened and closed. In critical applications, the length of time that the temperatures remain within limits during a power failure should be determined.

21.6.5   Drains and Kill System

Drainage and process sewer systems in traditional pharmaceutical facilities are of minor significance. The connection of vessels and processing equipment to drains is important, and having a drain sanitizing program is necessary, but, beyond that, little validation attention is given to the systems. However, in biopharmaceutical facilities where pathogenic or recombinant organisms are used, the drainage and kill systems are of critical importance.

The installation and operation of such systems must be validated to the same extent as any process system, and a process qualification must be performed, in which the ability to achieve an acceptable level of lethality for the organism in question is demonstrated. For virulent pathogens, an overkill approach may be used, wherein the destruction of an equivalent number of spores of *B. stearothermophilus* is demonstrated with a 10^6 safety factor. However, where desired, a less resistant challenge organism may be used, as long as data are developed to demonstrate that it has greater resistance to the sterilization process than does the organism to be used in the facility.

21.6.6   Steam and Dry Heat Sterilizers

The objective in sterilizer validation is to provide proof that adequate lethality has been delivered to a microbial population. For sterilization with saturated steam, it has become customary to define lethality over time in terms of the quantity F. F_{121} is defined as the sterilization time (in minutes) equivalent to exposure to saturated steam at 121 °C. The base sterilization temperature of 121 °C (250 °F) is chosen because it inactivates the most heat-resistant of organisms under the stated conditions.

The calculated value of F depends on time, temperature, and the heat resistance of the microbial population. It may be derived if two key properties describing heat resistance are known. These are designated D and Z. D, the *microbial death rate*, is defined as the time (in minutes) required to inactivate 90% of the microbial cells or spores at a given temperature. Since microorganisms die according to a logarithmic death-time or kill curve, each succeeding 90% kill takes as long as the preceding 90%. Figure 21.3 shows a typical kill curve at 121 °C. The D value of 1 minute shown represents the highest heat resistance likely to be found in most plant environments.

Note that D is calculated at a fixed temperature. As temperature changes, so does death rate. The relationship is defined by Z, the slope of the thermal death curve. The value of Z is the number of degrees required to change the D value by a factor of 10.

A typical thermal death curve is shown in Figure 21.4. The Z value of 10 °C is representative of several important heat-resistant strains (*B. stearothermophilus, C. botulinum*). This value has been chosen to be included in the calculation of lethality most often used, that of F_0. The general formula for F is:

$$F = \int 10^{\frac{(T-T_b)}{Z}} dt \tag{21.1}$$

where T_b is the base temperature and T is the process temperature. For F_0 we define T_b as 121 °C and Z as 10 °C.

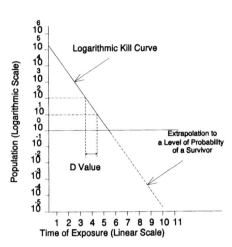

Figure 21.3 Microbial Kill Curve. The typical death curve for a microbial population at a constant temperature roughly follows a logarithmic curve. The time required to reduce the population by a factor of ten is termed the *D* value.

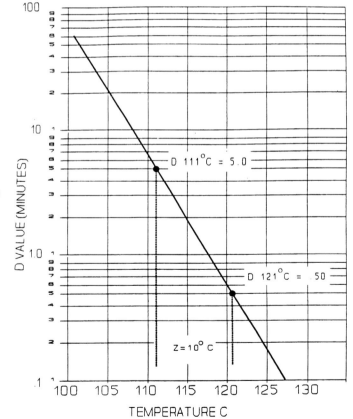

Figure 21.4 Effect of Temperature on *D* Value: *Z* Value. The logarithmic rate of microbial death, *D*, is temperature-dependent, increasing as temperature rises. The rise in temperature required to increase the *D* value by a factor of ten is termed the *Z* value.

$$F_0 = \int 10^{\frac{(T-121)}{10}} dt \tag{21.2}$$

The integral form indicates that we must integrate over time. Since this may not be practical, we may choose to simplify our task by taking an average temperature over a relatively short time period, say one minute, calculating the lethality for that time period, and adding up all the lethalities. Thus, the equation becomes:

$$F_0 = \sum 10^{\frac{(T-121)}{10}} \tag{21.3}$$

We may perform similar calculations for dry heat sterilization. Since dry heat is not as effective a sterilization method as saturated steam, a higher temperature is required. Dry heat lethality is defined as F_H, with a base temperature of 170 °C and a Z of 20 °C. Thus,

$$F_H = \sum 10^{\frac{(T-170)}{20}} \tag{21.4}$$

with temperature averaged over a one-minute interval.

Dry heat sterilization is suitable for only those items that can tolerate high temperature, but not moisture. This constitutes a very short list. Dry heat is more often used when depyrogenation is required. This needs to be accomplished at a much higher temperature, and is often required for glassware in which cultures will be processed. There is not a great deal known about the kinetics of depyrogenation in the form of microbial endotoxin destruction. It is known that a temperature of 250 °C for 45 minutes is adequate to destroy endotoxin. Studies have shown that the calculation of depyrogenation effectiveness may be estimated by

$$F_H = \int 10^{\frac{(T-250)}{54}} dt \tag{21.5}$$

In all applications requiring sterilization or depyrogenation, experience has shown that it is not a good idea to depend upon the equations very far below the base temperature. The main reason for this is that the assumed Z value may not be representative of the population present.

Understanding the time/temperature relationship involved in sterilization (and depyrogenation), we may proceed with our validation test program. The first aspect of this is the Installation Qualification of the sterilizer. This follows the same principles heretofore discussed. Since sterilization is such a critical operation, adherence to design criteria is particularly important. Also, since temperature is so critical to sterilization, the sterilizer's temperature sensor(s) must be calibrated against an NIST–traceable standard by the full-loop method. To do this, immerse the measuring element into a constant temperature medium and compare its reading to those of the sterilizer indicating and recording devices. Do this at a minimum of two, and usually three, temperatures.

For steam sterilizers, Operational Qualification begins with a check of the various cycles, timers, alarms, etc., that are included with the sterilizer. This may include vacuum and air overpressure. Leak testing, both under vacuum and at operating pressure, is important. Sometimes jacket mapping is done with surface thermocouples distributed over the jacket. OQ culminates in the empty-chamber runs, usually repeated in triplicate.

The temperature distribution within the chamber at operating or exposure temperature is determined and recorded on a multichannel data logger connected to thermocouples (type T are most often used). These thermocouples must be inserted through a special pressure-tight gland or fitting. There must be enough of them to map all parts of the chamber, including the drain (which is probably the cold spot) and anywhere a system temperature sensor is located. For most systems 10–20 thermocouples are enough. Most modern steam sterilizers should be expected to maintain a temperature distribution no more that ± 0.5 °C from the control temperature. In addition, the state of control should be within ± 0.5 °C of the set point once steady state has been reached. These runs should be at least as long as a normal sterilization run.

PQ, or process qualification, should proceed on representative proposed loads. Often, these loads are maximum in terms of mass; minimum loads should not be ignored. Also, studies have shown that slow heatup is related to slow steam penetration and air removal. Therefore, every type of item to be sterilized should be included in validation trials. Heat penetration runs are made with thermocouples distributed throughout the load in the most difficult-to-penetrate locations. The system cold spot can be determined in this way. Biological indicators may be included with each thermocouple in the initial penetration runs or in subsequent runs.

It is traditional to perform three consecutive successful trials with bioindicators and thermocouples for each load studied. The thermocouples are connected to a data logger. Temperature data are used to calculate F_0. The lethality required may be determined by a procedure in which the bioburden of the load is sampled and its heat resistance is determined. Then a cycle may be set to ensure that the probability of survival of any microorganism is one in one million. This is equivalent to an SAL (sterility assurance level) of 10^{-6}. Since most loads in a biotechnology plant's steam sterilizer are heat-stable, it is common to use the overkill approach. *Overkill* is defined as lethality adequate to sterilize a population of 10^6 spores per unit sterilized to an SAL of 10^{-6}. This would require a minimum F_0 of 12 minutes. In practice, most overkill cycles are run to a considerably higher F_0.

Bioindicators are always used to validate steam sterilization. This is because time/temperature data alone are inadequate to prove that the proper sterilizing conditions have reached all parts of the load. Saturated steam conditions require the presence of condensing steam at the surface to be sterilized. Bioindicators provide adequate proof of this. The most common bioindicators for steam sterilization are *B. stearothermophilus* spores. For loads of goods such as filling parts, bottles, stoppers, etc., the bioindicators are usually in the form of spore strips in glassine envelopes. Steam penetrates the envelopes but they provide a barrier against contamination when they are removed after the cycle is complete.

Dry heat sterilization may be validated in a similar manner, but *B. subtilis* as the indicator organism is applied to the spore strips. More often, dry heat ovens are validated for depyrogenation. Installation Qualification is usually simpler than for steam autoclaves, as dry heat ovens generally have very simple control systems. They have the added feature of HEPA filters, which must be certified, in the air supply ducts. The chamber should also be leak-tested under operating conditions; in the case of the door seal, this may be done with a smoke stick.

Empty-chamber runs are similar to those for autoclaves. Temperature distributions need not be nearly as tight for ovens; a good oven maintains a chamber distribution of $\pm$ 10 °C from the control point. Control may be as wide as $\pm$ 5 °C around the desired set point.

Loaded-chamber runs are also required. For dry heat depyrogenation, material heat capacity and mass are critical. Representative loads from each class of materials to be depyrogenated should be chosen. Glassware is the most common load. Loads of the largest and smallest glass containers, at a minimum, should be chosen for PQ runs. As depyrogenation temperatures provide an extreme overkill environment for microbial spores, no bioindicators are required. Thermocouples must be specially insulated to withstand the higher temperatures. Cycles are designed to deliver F_H of at least 45 minutes equivalent at 250 °C. The final run of three may include a pyroindicator. This should be placed in the slowest-to-heat zone (cold spot) of the load, and should contain enough bacterial endotoxin to indicate a three log (one-thousandfold) reduction. Some people require that endotoxin be deposited on the surface of articles in the load and be distributed throughout the load. Since dry heat depyrogenation depends only on temperature, and temperature can be accurately measured, this is probably not necessary.

21.6.7   Sterilization in Place (SIP)

In a biopharmaceutical plant SIP plays a very important role. Many fixed and large portable vessels, as well as other processing equipment and transfer lines, must be sterilized in this manner. Of particular importance are bioreactors and harvest vessels, as all equipment in which viable cells are to be processed should be sterilized. The method of choice for SIP is saturated steam. In this respect, SIP validation is similar in many ways to steam autoclave validation previously discussed. The objective again is to deliver adequate lethality, as measured by F_0, to sterilize all system contact parts to an SAL (sterility assurance level) of 10^{-6}. In SIP an overkill approach is almost always chosen.

The IQ of vessels to be sterilized must take account of several points. The vessels must be designed to operate at sterilization pressures. Saturated steam at 121 °C is at about 15 psig. They must also be rated to full vacuum or be equipped with a vacuum breaker. Failure of a vacuum breaker or pressure relief device can destroy an unprotected vessel. Vacuum breakers, pressure rupture disks, and other instrumentation must be of a sanitary design, presenting a smooth, flat surface to the interior of the vessel. A filtered gas source, air or nitrogen, should be installed to break vacuum as the system cools. This may also provide a purge at a slight overpressure while the vessel is awaiting use.

The vessel should be built of corrosion-resistant material, since highly corrosive clean steam is the sterilizing medium. The material of choice is usually 316L stainless steel. Interior polish of at least 150 grit, with welds ground smooth and polished, is necessary. More highly polished finishes and electropolishing are optional.

Vessel fittings should be of a sanitary design. Clamp-type fittings are best. Screw fittings are not suitable, nor are nonsanitary ASME flanges. Gaskets or O rings must be of resilient, steam-resistant material. Suitable elastomers are EPDM (ethylene-propylene-diene monomer), silicone rubber, Viton®, and Teflon®-envelope types.

A temperature control system should be in place, capable of measuring and controlling temperature to an accuracy of ± 0.5 °C. An RTD (resistance temperature device) is the temperature sensor of choice. A recorder of equivalent precision is also required.

The vessel vendor should provide certification of materials of construction (including elastomers), interior finish, ASME or equivalent pressure test report, and evidence of factory passivation and the method used. A thorough field inspection serves to provide additional assurance that the vessel has been delivered and installed as specified.

Product transfer lines to be sterilized are also subject to special IQ consideration. These lines are usually constructed of sanitary seamless tubing, not pipe. They are of the same material and interior finish as sterile vessels, and provide the highest degree of integrity, preventing loss of sterility, when they contain a minimum of fittings. Therefore, welded lines are usually specified. Inert gas arc-welding techniques are utilized to ensure that the steel maintains its corrosion-resistant properties at the welds, since areas of corrosion and surface imperfections provide sites for microbial growth. Welds must be uniquely identified by numbers indicated on a drawing. An isometric drawing is most suitable for this purpose. Documentation of weld conditions and visual weld inspection reports must be maintained.

Welds must meet ASME and 3-A [12] sanitary standards. They must be free of pits or inclusions, and show no evidence of crystallization or overheating. Where fittings are required in transfer lines (for example, at pumps, valves, and filters) they should be of the sanitary clamp design. Filters merit special attention. These must be fitted with bleeds, so that air may be removed during heatup and condensate removed during steam exposure. Filter media must be resistant to the temperature and pressures to which they will be subjected. Tubing arrangements must allow for adequate steam delivery to the filter housing. This may mean providing dedicated steam lines to filters.

Once the installation has been verified, the sterilization test runs may proceed. The same principles as with empty-chamber autoclave runs apply. Thermocouples should be distributed throughout the system interior, at all suspected cold spots. These are typically the bottom of the vessel, all filter housings (upstream and downstream of filter media), nozzles in the head of the vessel, all low points in transfer lines, and at the farthest point in the system from the steam source. It is common practice to place bioindicators (*B. stearothermophilus* spore strips) with each thermocouple. As the main body of the vessel is the easiest to reach with steam, only a few thermocouples are required therein, except where devices such as spargers, spray balls and dip tubes are included. A thermocouple should also be placed adjacent to the system temperature probe.

During the runs, thermocouples are connected to the data logger, from which time/temperature data may be used to calculate F_0. Time to reach final temperature is determined for all points in the system (see Section 21.6.6).

The specification of F_0 for filters requires particular care, especially if the filters are the hydrophobic type often used as vent or compressed gas filters on bioreactors. It is difficult for steam to penetrate hydrophobic media, and impossible to measure the temperature effects across the membrane. Studies have shown that, although an F_0 of 12–15 provides adequate overkill in vessels and lines, filters require a measured F_0 of 30–40.

Three consecutive successful runs are required at the minimum time of exposure and temperature to be used in normal operation. All points in the system must reach the specified lethality and all spore strips must be inactivated. To complete the study, three additional runs are carried out at the highest temperature and pressure and over the longest time cycle to be used. These are done without thermocouples or bioindicators. They are intended to show that all system filters retain their integrity under all allowable sterilization conditions. Therefore, filter integrity should be tested according to the manufacturer's recommended method before and after all SIP trials.

21.6.8   Cleaning in Place

Cleaning-in-place (CIP) processes are in widespread use throughout biotechnology facilities. Residues from the production of proteins and other biopharmaceutical processes are difficult to clean. Vessel interiors are the major objects of CIP, but product and medium transfer lines are also cleaned in this manner. CIP usually precedes the SIP operation.

The important factors in the validation of a CIP system are detergent, time, temperature, and turbulence. The proper combination of these four elements can clean just about anything.

The first step in CIP validation is to define a cleaning agent that can remove the process residue. In biotechnology plants this is often caustic soda (NaOH) solution. Laboratory studies must show that the detergent chosen is effective. These may be done in-house, or may be a service of CIP equipment vendors or detergent manufacturers. In addition to being effective, the detergent must not be corrosive to the systems being cleaned (usually stainless steel), and must be removable by water rinse. Written documentation of these preliminary tests, including methods and analytical tools used, is required.

The physical conditions for CIP may be provided in a number of ways. It is possible to clean a vessel by filling, heating, and mixing. In a bioreactor or harvest vessel this is not a totally effective nor an economical means. Residue on the head space may not be removed. In addition, because such vessels need to be rinsed with high-quality water (WFI), this method becomes prohibitively expensive. Instead, a spray-ball system is normally chosen.

Selection of a spray ball depends on vessel geometry. It is designed by engineers expert in CIP. After the normal IQ inspections have been performed, OQ testing determines the effectiveness of the spray ball in achieving surface coverage. The testing may be done in a number of ways, one of which is to place water-soluble indicator material (such as a dye) on interior surfaces throughout the vessel. After a rinse cycle, the surfaces are inspected for residual dye.

Times, temperatures, and pressures of the cleaning and rinse solutions are determined during OQ testing. Draining effectiveness must be demonstrated (whether gravity, a pump, or an eductor is used), and is required to show the absence of the "bathtub ring effect". Complete cycles are carried out, and the final rinse must be free of detergent residue.

Process transfer lines may also be cleaned in place. They require a high degree of turbulence to be cleaned effectively. OQ testing must determine the flow rate in these lines. An acceptable minimum is 5 ft/s in 1½–2" lines, as is a Reynolds number (N_{Re}) of 20,000. The Reynolds number is defined as the density, ρ, pipe diameter, d, and velocity, v, all divided by the viscosity, μ:

$$N_{Re} = \rho dv/\mu \tag{21.6}$$

Process Qualification (PQ) of CIP is more complex. First, a level of cleanliness is specified. For dedicated equipment, this is not a major problem. Product residue in the final rinse below an internal specification may be an adequate criterion. For multipurpose equipment, especially between campaigns of different products, the residue of original product must be reduced to a minimum level. For pharmacologically active or toxic materials, medical research may specify a level, called the PIEL or TIEL (Pharmacologically or Toxicologically Insignificant Exposure Limit) number. The objective of the cleaning process is to reduce total residues to below the level required to ensure that the succeeding product batch contains the impurity below the TIEL or PIEL number. Usually some fraction (0.01 or 0.001) of the number is specified, because of the difficulty in measuring residues.

In multipurpose equipment, rinse analysis for active products is considered inadequate. Surface analysis, by swabbing or other means, is required, to provide a high degree of assurance of product removal.

21.6.9   Process Equipment

The process equipment in a biopharmaceutical facility requires the same installation and operational qualification as do the facility and utilities. It becomes, in fact, more critical to implement these procedures as you get closer to the heart of the process and to the product. Unfortunately, the industry is, in many cases, in a state of transition, and research and development scientists are often running production facilities. If people yield to the urge to tinker with established processes and equipment, and have a relaxed approach toward the documentation, the establishment of a validated facility under change control can become very difficult.

The general principles of installation qualification are applied no differently to process systems than to utilities and facilities. Conformance to design documentation must be checked, calibrations completed, control system hardware and software qualified, and so on. One aspect of the qualification that acquires great importance is the documentation of materials of construction. Many of the smaller scale items of processing equipment are available with materials of construction that are not suitable for pharmaceutical operations. The cost of obtaining some of the equipment with "FDA–approved" materials of construction appears prohibitive, but may prove cheaper in the long run than the exercise of independently verifying that the materials that were used are safe in the intended application.

We have placed quotation marks around "FDA–approved" in the above paragraph because it is rather glibly used by manufacturers, with little or no justification. It is the end user's responsibility to satisfy himself that whatever materials are used are suitable for the intended application and will not leach undesirable extractables into the process stream.

Processes and systems that typically require validation are listed in Table 21.1. The operational qualification of these systems must show that they can perform whatever unit operation is intended. Medium and buffer preparation systems must be capable of charging metered amounts of water (within established limits). Controls for fermentation or cell culture temperature, pH, and dissolved oxygen must maintain the set parameters. Agitation and filtration rates of some solution (typically water) should be verified for future reference. The ability of vessels such as fermentors or

Table 21.1 Processes and Systems That Require Validation

Medium preparation
Bioreactors
Buffer preparation
Harvest/extraction
Purification
Ultrafiltration/diafiltration
Purified bulk filtration and packaging
Final dosage formulation

bioreactors to maintain sterility over an extended period of time, with ongoing sampling, nutrient addition and pH adjustment operations, must be verified with medium-holding tests that mimic actual production.

The process qualification (or validation) of these processes should be a straight-forward application of the FDA's definition of process validation quoted in the Introduction: "Process Validation is a documented program which provides a high degree of assurance that a specific process will consistently produce a product meeting its predetermined specifications and quality attributes."

Specific PQs such as sterilization, cleaning, or sanitizing must be completed as discussed previously.   Actual production processes are usually validated with demonstration batches that are extensively sampled and tested and placed into a stability program. Buffers and media prepared must meet limits for pH, conductivity, microbial counts, and endotoxins. Yields at various process steps must be verified to be within acceptable ranges (although if new processing equipment is being used for the first time this may be more of an exercise in defining process capability than in validating it). It is important that any objectionable materials used in the fermentation, isolation, or purification processes not remain in the final product, and that subsequent processing steps that reduce or eliminate these compounds be demonstrated to be effective and reproducible.

If limits have been established for a processing parameter that affects product quality, that limit must be validated by producing product while the parameter is at (or outside) that limit. Obviously, in even a relatively simple process, testing every possible permutation of parameters would soon develop into a massive project providing relatively little real benefit. It is important to determine those parameters that are most critical to the process, and those where flexibility is required. Both of these need to be validated at the extremes of their ranges. The validation should take advantage of data obtained during process development or clinical manufacturing tests on a smaller scale. During the process development those parameters were no doubt evaluated, and the new data should support whatever limits were established.

The validation should be an integral part of the development of the product or process, and should demonstrate that the production process remains within the limits established during process development. It is also important to have evidence supporting the fact that the product to be manufactured in the facility and marketed is the same as that used in the clinical trials.

The requirements for process qualification of the aseptic filling of sterile dosage forms have been very specifically defined by the FDA. The procedures are described in the Parenteral Drug Association's Technical Monograph No. 2, *"Validation of Aseptic Filling for Solution Drug Product"* [13]. These PQs consist of simulations of the aseptic filling process wherein vials or ampoules are filled with sterile medium (usually soybean casein digest medium). The filling process simulates the normal operation as closely as possible, with additional testing and controls to help identify any sources of contamination. The filled containers are incubated and examined for evidence of contamination. A minimum of 3,000 units are filled and a contamination rate not exceeding 0.1% is required. However, the FDA prefers that enough vials be filled to ensure that all normal operations are included in the simulation, and a contamination rate of 0.1% is considered the very worst level that is acceptable. This rate is based upon statistical confidence levels, and is not an acceptable level of contamination for filled product (rates under one in a million or 0.0001% are desired by the FDA).

Aseptic simulations are required on a biannual basis, and all filling lines and product/container configurations must be included in the program. Operations such as lyophilization, sterile powder filling, and preparation of suspensions must be evaluated, with simulations that include these steps. Irradiated sterile powders may be used as a placebo in some cases.

21.6.10 Automated Systems and Computers

Because of the complexity of computerized control systems, it is difficult to test them adequately to provide a high degree of assurance that they will always perform as intended. Perhaps in this type of system, even more than any other, the life-cycle concept becomes a necessity rather than a choice. Figure 21.5 represents the life-cycle as it applies to computerized systems.

It is important to remember that the control system is only a part of an overall system, which also includes the equipment and process being controlled. The ultimate goal of validation is to provide assurance that the process will be carried out properly and reliably. Thus, it is not possible to treat the control system separately from what it controls.

The first task in validation of an automated system is one of definition. This must be developed with information from many areas. In the CGMP environment, the ultimate users (operations, R&D, etc.) and Quality Assurance, as well as engineering and computer groups, all have something to contribute. The system definition may

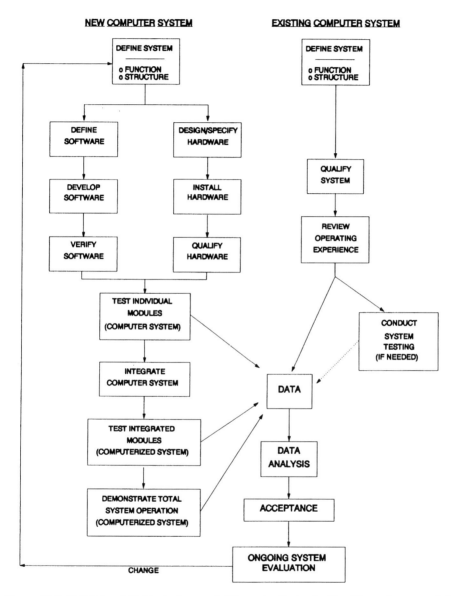

Figure 21.5 Validation Life-Cycle Approach (Computer System). The life-cycle approach to validation was originally applied to and is most necessary for computer systems. Testing alone is not adequate to ensure the reliability of any system; it is important as well that systems be properly designed, installed, and operated. Complete documentation of each phase of the life-cycle helps to ensure this.

take the form of a Functional Requirement document. This describes the entire process and the control philosophy to be employed. It should emphasize what needs to be done, not how it should be done. That is, consideration is given to types of data to be entered, process parameters to be controlled or monitored, operator interface, calculations, process documentation, communication to other devices or systems, etc. It is not appropriate that specific hardware or software solutions to the problem be proposed at this stage.

The definition phase continues through the conceptual or preliminary engineering phases. Process equipment design proceeds through the P&ID (Piping and Instrumentation Diagram) stage, where field instrumentation requirements are defined, and through instrumentation design, where field inputs and outputs are further defined. This information supplements the Functional Requirements and allows the addition of more detail. This detail may begin to address how the control task is to be accomplished. Still, specific vendors or control hardware selections are not yet required. A System Specification (or Engineering Specification or Purchase Specification) should be prepared, to allow vendors of control systems to propose specific solutions to the problem. It is the function of the validation team to review and compare the Functional Requirements and the System Specification. In particular, they should ensure that the two documents are specific enough to adequately describe what is required and what is desired, and to discern the difference. The documents must be consistent with the Specification, addressing all points in the Functional Requirements. Also, the Specification should inform the proposed vendors as to the type and level of documentation required of them and the extent to which they are expected to participate in the validation test program. It should be clear what standards are expected to be met in computers and computer-related hardware and in software development. Procedures used in design and development (especially of software) should be open to audit.

Once a vendor or system is selected, the developmental phase is entered. It is entirely possible that in certain situations very little development is necessary. Some control systems, especially single-function ones like an autoclave controller, may be off-the-shelf items. In this case, the vendor must still meet specified documentation requirements and be willing to demonstrate adherence to standards. History of the vendor and the proposed system may be taken into account.

For more complex, multifunction systems some development is usually required. Even though computer operating systems and basic, configurable software modules often exist, there is still work to be done to assemble these into an application program specific to the project. It is not usually necessary to pay a great deal of attention to operating systems or to the development of configurable software. This statement also applies to higher level shop floor and supervisory programs. The vendors must show, either by demonstrated experience or through audits, that they practice acceptable software quality assurance. The structural testing required in the development of a computer program may be left to the vendor in these cases.

For that level of system development specific to the process being controlled, more oversight is required. Programming and configuration may be taken over by the user firm or another contractor or consultant. The vendor/programmer must provide documentation as to how the system has been programmed or configured. In cases where a customized program is provided (such as for a programmable logic controller), the validation team may wish to examine the code for correct syntax, structure, adequate annotation, and the absence of dead code. The programmer must provide evidence of structural testing.

Each computer module is subjected to functional testing. That is, the module is run to determine that it performs as intended. This testing, usually done off-line from any process hardware, often uses a simulator, a device that provides inputs and outputs similar to that of field hardware. These tests should be performed and documented under observation from the validation team.

While software is developed and undergoes early testing, control hardware is received and installed. This then undergoes the typical Installation Qualification process. Among the important checks carried out as part of IQ are:

(a) component audit against the approved Bill of Materials—check models, serial numbers, and firmware designations;
(b) wiring checks from field devices to computer input terminals;
(c) checks for proper system grounding, shielding, and noise isolation;
(d) checks of communications network hookup and continuity;
(e) configuration check for proper setup of DIP switches and jumpers according to manufacturers' requirements;
(f) checks for proper power connections, including required fusing and disconnects;
(g) power loss and power backup tests;
(h) controller redundancy checks to see if bumpless transfer to backup units occurs when a control element fails; and
(i) similar checks when redundant communications networks are involved.

The next phase of the life-cycle integrates the software with the computer hardware in the operating environment. Integrated system qualification includes:

(a) verification that the software programs installed are identical to those verified in modular functional testing;
(b) verification that the application source code exists and is available for inspection if requested;
(c) online input/output testing (loop checks), where each I/O module is exercised to prove that each point in the system is addressed correctly and that the corresponding data are stored in the proper memory locations and appear properly on display screens;
(d) repeat of power loss testing to check system recovery, startup procedure, data status, and documentation; and
(e) alarm point checks.

The integrated system may then be subjected to functional testing online and in real time. A typical way to accomplish this in a process control application is through water or placebo operation. This testing verifies that the computer system executes the correct sequence of instructions, and that each control loop or function operates as required. The accuracy of calculations is rechecked, integrity of data is verified, documentation and reports are reviewed. It may be necessary in this phase to extend alarm checking to see that the system responds to alarm conditions with the proper priority and action. Throughout the functional testing of the modules and the integrated system, the validation team refers to the Functional Requirements and System Specification to ensure that the system meets design criteria.

The next necessary step is to use the computerized system to perform its intended function. For a process control system, this is the production of product. For an automated storage and retrieval system (AS/RS), this is the selection and delivery of materials from a warehouse to point of use. Without this proof of usability the validation is not complete.

Once the testing phase has been executed, as with any validated system, the computer controlled system is subject to all the important aspects of validation maintenance. These include utilization of validated SOPs, continuing calibration, and maintenance and change control. This last applies to changes in process, in process equipment, and to both control hardware and software.

21.7   Conclusion

The key to a successful validation is the perception of validation as an integrated effort, transcending design, construction, testing, and operation. Each member of the project team, no matter what his or her other responsibilities, is responsible for validation.

The concepts and techniques we presented in this chapter are meant to introduce and familiarize the reader with the necessity to validate and to chart an initial course. Along with the design and operational ideas presented elsewhere in this book, they should establish the foundation for a sound validation program.

21.8   References

1. *Guidelines on the General Principles of Process Validation*; Food and Drug Administration, May, 1987; Superintendent of Documents, Washington, DC.
2. Validation Concepts for Computer Systems Used in the Manufacture of Drug Products, *Pharm. Technol. 10* (5), 24–34 (1986).
3. Title 40, *Code of Federal Regulations (CFR-40)*, Parts 141.14 and 141.21; *Environmental Protection Agency Regulations for Drinking Water*; Superintendent of Documents, Washington, DC.

4. Water for Pharmaceutical Purposes; General Information; *United States Pharmacopeia XXII/National Formulary XVII*; United States Pharmacopeial Convention, Rockville, MD, 1990; pp 1712–1713.
5. Current Good Manufacturing Practices—Large Volume Parenterals; *Fed. Regist.* (June 1, 1976).
6. MUNSON, T. E.; *FDA Views on Water Treatment*; unpublished paper delivered to Proprietary Assoc., Philadelphia, October, 1984.
7. *Federal Standard Clean Room and Work Station Requirements*; Controlled Environment, Federal Standard 209D, June, 1988; Superintendent of Documents, Washington, DC.
8. *Guideline on Sterile Drug Products Produced by Aseptic Processing*; Food and Drug Administration, June, 1987; Superintendent of Documents, Washington, DC.
9. Title 21, *Code of Federal Regulations (CFR-21)*, Subchapter F, Parts 600 to 680; *Biologics*; Superintendent of Documents, Washington, DC, 1985.
10. NIH Guidelines for Research Involving Recombinant DNA Molecules, *Fed. Regist.* (November 23, 1984).
11. *Testing Clean Rooms*; IES Recommended Practice, IES-RP-CC-006-84-T; Institute for Environmental Sciences, Mt. Prospect, IL, November, 1984.
12. INTERNATIONAL ASSOCIATION OF MILK, FOOD AND ENVIRONMENTAL SANITARIANS, INC.; *3-A Sanitary Standards*; International Association of Milk, Food and Environmental Sanitarians, Inc., Des Moines, IA; 1947–present.
13. *Validation of Aseptic Filling for Solution Drug Products*; Technical Monograph No. 2, Parenteral Drug Association, Rockville, MD, 1980.

GLOSSARY OF ABBREVIATIONS

ABS	acrylonitrile-butadiene-styrene copolymer
acfm	actual cubic feet per minute
AIA	American Institute of Architects
AISI	American Iron and Steel Institute
ANSI	American National Standards Institute
APHIS	Animal and Plant Health Inspection Service (of the USDA)
API	American Petroleum Institute
ASME	American Society of Mechanical Engineers
AS/RS	automated storage and retrieval system
ASTM	American Society for Testing and Materials
ATCC	American Type Culture Collection
BLx	biosafety level x
BLx-LS	biosafety level x, large-scale
BPVC	Boiler and Pressure Vessel Code
BSCC	Biotechnology Science Coordinating Committee
Btu	British thermal unit (unit of energy, 1.06 kJ)
CADD	computer-aided design and drafting
CBER	Center for Biologics Evaluation & Research (of FDA)
CDI	continuous deionization
CFF	cross-flow filtration
cfm	cubic feet per minute
CFR	Code of Federal Regulations
CFU	colony-forming unit
CGMPs	current good manufacturing practices
CIP	clean-in-place
cP	centipoise
DCS	distributed control system
DI	deionized (water)
DIN	Deutsches Institut für Normung

DNA	deoxyribonucleic acid
DO	dissolved oxygen
DOP	dioctyl phthalate
dP	differential pressure
dw	dry weight
E&C	engineering and construction
EDTA	ethylenediaminetetraacetic acid
EEC	European Economic Community
EPA	Environmental Protection Agency
EPDM	ethylene-propylene-diene monomer
EPR	= ESR, electron spin resonance
EU	endotoxin unit
FDA	Food and Drug Administration
fpm	feet per minute
GILSP	good industrial large-scale practice
GLSP	good large-scale practice
GMP	good manufacturing practice
gpm	gallons per minute
h	hour(s)
HACCP	hazard analysis of critical control points
HEPA	high-efficiency particulate air (filter)
HEPES	name of a buffer
HMSO	Her Majesty's Stationery Office
hp	horsepower; unit of power, 0.75 kW
HVAC	heating, ventilating, and air conditioning
ID	inside diameter
IDF	International Dairy Federation
IQ	installation qualification
ISS	International Sanitary Standard
L	liter(s)
LAL	*Limulus* amoebocyte lysate
LVP	large-volume parenteral

MAb	monoclonal antibody	QA	quality assurance
MAWP	maximum allowable working pressure	QC	quality control
MF	microfiltration	R&D	research and development
MWCO	molecular weight cutoff	Ra	arithmetic average surface roughness
NAD, NADH, NADP, NADPH		rDNA	recombinant DNA
	nicotine adenine dinucleotides	RH	relative humidity
NBS	National Bureau of Standards	RO	reverse osmosis
NEMA	National Electrical Manufacturers Association	ROI	return on investment
		rpm, –s	revolutions per minute, second
NIH	National Institutes of Health	RTD	resistance temperature device
NIST	National Institute of Standards and Testing		
		s	second(s)
NPSH	net positive suction head	SAL	sterility assurance level
NPT	National Pipe Taper (pipe thread)	SAMA	Scientific Apparatus Makers Association
NSF	National Science Foundation	scfm	standard cubic feet per minute
NWP	normalized water permeability	SCWO	supercritical water oxidation
		SIP	sterilization in place
O&M	operation and maintenance	SLPM	standard liters per minute
OD	outside diameter	STA	slit-to-agar
OECD	Organization for Economic Cooperation and Development	SVP	small-volume parenteral
OQ	operational qualification	TDS	total dissolved solids
OSHA	Occupational Safety and Health Administration	TFE	= PTFE, polytetrafluoroethylene
		TFF	tangential flow filtration
P&ID	piping (or process) and instrumentation diagram	TIEL	toxicologically insignificant exposure level
PFD	process flow diagram	TIG	tungsten-inert-gas (welding)
PI	proportional/integral	TMP	transmembrane pressure
PID	proportional/integral/derivative	TOC	total organic carbon
PIEL	pharmacologically insignificant exposure level	TSCA	Toxic Substances Control Act
PLC	programmable logic controller	UF	ultrafiltration
PMN	pre-manufacture notification	UK	United Kingdom
ppb	parts per billion = μg/L	UL	Underwriters Laboratories
ppm	parts per million = mg/L	USDA	United States Department of Agriculture
PQ	performance (or process) qualification	USP/NF	United States Pharmacopeia/ National Formulary
psi	pounds per square inch (unit of pressure); also psig (gauge), psid (difference), psia (absolute)	VVM	volume per volume per minute
PTFE	= TFE, polytetrafluoroethylene	WFI	Water For Injection
PVDF	polyvinylidene fluoride		

Index